VOL.2 NUMERICAL METHODS FOR TRANSPORT AND HYDROLOGIC PROCESSES

ELSEVIER

COMPUTATIONAL
MECHANICS
PUBLICATIONS

OTHER TITLES IN THIS SERIES

4 J.J. FRIED
GROUNDWATER POLLUTION

5 N. RAJARATNAM
TURBULENT JETS

7 V. HÁLEK AND J. ŠVEC
GROUNDWATER HYDRAULICS

8 J. BALEK
HYDROLOGY AND WATER RESOURCES IN TROPICAL AFRICA

10 G. KOVÁCS
SEEPAGE HYDRAULICS

11 W.H. GRAF AND C.H. MORTIMER (EDITORS)
HYDRODYNAMICS OF LAKES: PROCEEDINGS OF A SYMPOSIUM 12-13 OCTOBER 1978, LAUSANNE, SWITZERLAND

13 M.A. MARIÑO AND J.N. LUTHIN
SEEPAGE AND GROUNDWATER

14 D. STEPHENSON
STORMWATER HYDROLOGY AND DRAINAGE

15 D. STEPHENSON
PIPELINE DESIGN FOR WATER ENGINEERS
(completely revised edition of Vol.6 in the series)

17 A.H. EL-SHAARAWI AND S.R. ESTERBY (EDITORS)
TIME SERIES METHODS IN HYDROSCIENCES

18 J. BALEK
HYDROLOGY AND WATER RESOURCES IN TROPICAL REGIONS

19 D. STEPHENSON
PIPEFLOW ANALYSIS

20 I. ZAVOIANU
MORPHOMETRY OF DRAINAGE BASINS

21 M.M.A. SHAHIN
HYDROLOGY OF THE NILE BASIN

22 H.C. RIGGS
STREAMFLOW CHARACTERISTICS

23 M. NEGULESCU
MUNICIPAL WASTEWATER TREATMENT

24 L.G. EVERETT
GROUNDWATER MONITORING HANDBOOK FOR COAL AND OIL SHALE DEVELOPMENT

25 W. KINZELBACH
GROUNDWATER MODELLING

26 D. STEPHENSON AND M.E. MEADOWS
KINEMATIC HYDROLOGY AND MODELLING

27 A.H. EL-SHAARAWI AND R.E. KWIATKOWSKI (EDITORS)
STATISTICAL ASPECTS OF WATER QUALITY MONITORING - PROCEEDINGS OF THE WORKSHOP HELD AT THE CANADIAN CENTRE FOR INLAND WATERS, OCTOBER 1985

28 M.K. JERMAR
WATER RESOURCES AND WATER MANAGEMENT

29 G.W. ANNANDALE
RESERVOIR SEDIMENTATION

30 D. CLARKE
MICROCOMPUTER PROGRAMS FOR GROUNDWATER STUDIES

31 R.H. FRENCH
HYDRAULIC PROCESSES ON ALLUVIAL FANS

32 L. VOTRUBA
ANALYSIS OF WATER RESOURCE SYSTEMS

33 L. VOTRUBA AND V. BROŽA
WATER MANAGEMENT IN RESERVOIRS

34 D. STEPHENSON
WATER AND WASTEWATER SYSTEMS ANALYSIS

36 M.A. CELIA, L.A. FERRAND, C.A. BREBBIA, W.G. GRAY AND G.F. PINDER (EDITORS)
VOL.1 MODELING SURFACE AND SUB-SURFACE FLOWS - PROCEEDINGS OF THE VIITH INTERNATIONAL CONFERENCE ON COMPUTATIONAL METHODS IN WATER RESOURCES, MIT, USA, JUNE 1988

COMPUTATIONAL METHODS IN WATER RESOURCES

VOL.2 NUMERICAL METHODS FOR TRANSPORT AND HYDROLOGIC PROCESSES

Proceedings of the VII International Conference, MIT, USA, June 1988

Edited by

M.A. Celia
Massachusetts Institute of Technology, Cambridge, MA, USA
L.A. Ferrand
Massachusetts Institute of Technology, Cambridge, MA, USA
C.A. Brebbia
Computational Mechanics Institute and University of Southampton, UK
W.G. Gray
University of Notre Dame, Notre Dame, IN, USA
G.F. Pinder
Princeton University, Princeton, NJ, USA

ELSEVIER
Amsterdam - Oxford - New York - Tokyo 1988
Co-published with
COMPUTATIONAL MECHANICS PUBLICATIONS
Southampton - Boston

Distribution of this book is being handled by:

ELSEVIER SCIENCE PUBLISHERS B.V.
Sara Burgerhartstraat 25, P.O. Box 211
1000 AE Amsterdam, The Netherlands

Distributors for the United States and Canada:

ELSEVIER SCIENCE PUBLISHING COMPANY INC.
52 Vanderbilt Avenue
New York, N.Y., 10017, U.S.A.

British Library Cataloguing in Publication Data

International Conference on Computational
Methods in Water Resources (7th : 1988 :
Cambridge, Mass.)
Computational methods in water resources.
Vol.2 : Numerical methods for transport
and hydrologic processes
1. Natural resources : Water. Analysis.
I. Title II. Celia, M.A. III. Series
628.1'61'01515353
ISBN 1-85312-007-3

Library of Congress Catalog Card number 88-70628

ISBN 0-444-98911-0(Vol.36) Elsevier Science Publishers B.V.
ISBN 0-444-41669-2(Series)
ISBN 1-85312-007-3 Computational Mechanics Publications UK
ISBN 0-931215-74-9 Computational Mechanics Publications USA

Published by:

COMPUTATIONAL MECHANICS PUBLICATIONS
Ashurst Lodge, Ashurst
Southampton, SO4 2AA, U.K.

Printed in Great Britain by The Eastern Press, Reading

PREFACE

This book forms part of the edited proceedings of the Seventh International Conference on Computational Methods in Water Resources (formerly Finite Elements in Water Resources), held at the Massachusetts Institute of Technology, USA in June 1988. The conference series originated at Princeton University, USA in 1976 as a forum for researchers in the emerging field of finite element methods for water resources problems. Subsequent meetings were held at Imperial College, UK (1978), University of Mississippi, USA (1980), University of Hannover, FRD (1982), University of Vermont, USA (1984) and the Laboratorio Nacional de Engenharia Civil, Portugal (1986). The name of the ongoing series was modified after the 1986 conference to reflect the increasing diversity of computational techniques presented by participants.

The 1988 proceedings include papers written by authors from more than twenty countries. As in previous years, advances in both computational theory and applications are reported. A wide variety of problems in surface and sub-surface hydrology have been addressed.

The organizers of the MIT meeting wish to express special appreciation to featured lecturers J.A. Cunge, A. Peters, J.F. Sykes and M.F. Wheeler. We also thank those researchers who accepted our invitation to present papers in technical sessions: R.E. Ewing, G. Gambolati, I. Herrera, D.R. Lynch, A.R. Mitchell, S.P. Neuman, H.O. Schiegg, and M. Tanaka. Important contributions to the conference were made by the organizers of the Tidal Flow Forum (W.G. Gray and G.K. Verboom) and the Convection-Diffusion Forum (E.E. Adams and A.M. Baptista) and by K. O'Neill who organized the Special Session on Remote Sensing. The conference series would not be possible without the continuing efforts of C.A. Brebbia, W.G. Gray and G.F. Pinder, who form the permanent organizing committee.

The committee gratefully acknowledges the sponsorship of the National Science Foundation and the U.S. Army Research Office and the endorsements of the American Geophysical Union (AGU) the International Association of Hydraulic Research (IAHR), the National Water Well Association

(NWNA), the American Institute of Chemical Engineers (AIChE), the International Society for Computational Methods in Engineering (ISCME), the Society for Computational Simulation (SCS) and the Water Information Center (WIC).

Papers in this volume have been reproduced directly from the material submitted by the authors, who are wholly responsible for them.

M.A. Celia
L.A. Ferrand
Cambridge (USA) 1988

CONTENTS

SECTION 1 - DEVELOPMENTS IN NUMERICAL METHODS

1A - Numerical Methods for Transport

SECTION 1 - DEVELOPMENTS IN NUMERICAL METHODS

SECTION 1A - NUMERICAL METHODS FOR TRANSPORT

Stability Analysis of Discrete Approximations of the Advection-Diffusion Equation Through the use of an Ordinary Differential Equation Analogy

A.A. Aldama

Mexican Institute of Water Technology, Insurgentes 4, Jiutepec, Mor. 62550, Mexico

Abstract

The existence of an analogy between the stability properties of numerical integration schemes of an ordinary differential equation and those of discrete approximations of the advection-diffusion equation is shown, and applied to simplify the stability analysis of the latter.

Introduction

The description of heat and mass transport processes is very relevant to geophysical and engineering applications (Fischer et al. [1], Fischer [2]). Thus, the study of the properties of numerical schemes used to solve the advection-diffusion equation has considerable importance (Roache [3]). On the other hand, it is well known that for linear problems the convergence of a finite difference scheme is guaranteed if it is consistent and stable (Richtmyer and Morton [4]). It is relatively simple to construct consistent schemes. In contrast, the stability analysis of some discrete schemes that approximate the advection-diffusion equation, based on the use of the classical von Neumann technique (Smith [5]), may involve the solution of algebraic equations of such complexity that the determination of practical stability criteria becomes very difficult. A method that allows the avoidance of those difficulties, through the use of an ordinary differential equation analogy, is presented in this paper.

The ordinary differential equation analogy

Let us consider the following ordinary differential equation:

$$\frac{dy}{dt} = \lambda y \tag{1}$$

where λ is a complex parameter. Equation (1) is commonly used as a model to study the stability properties of numerical integration schemes of ordinary differential equations (Gear [6]).

The most general N-level scheme that approximates (1) may be written as

$$\frac{1}{\Delta t} \sum_{k=0}^{N} \alpha_k y_{n+k} = \lambda \sum_{k=0}^{N-1} \beta_k y_{n+k} \tag{2}$$

where $y_n = y(n\Delta t)$, Δt is the discretization interval of the independent variable t and the values of α_k and $\beta_k (k = 0, 1, \ldots, N-1)$ are determined in such a way that (2) be consistent with (1) and to achieve a given order of accuracy (Gear [6]).

Now, in order to study the stability of the difference equation (2), a solution of the form $y_n = \xi^n$ (where ξ is complex in general) is assumed. Therefore, equation (2) becomes

$$\sum_{k=0}^{N-1} (\alpha_k - \mu\beta_k)\xi^k = 0 \tag{3}$$

where $\mu = \lambda\Delta t$. The stability condition is given by $|\xi| \leq 1$. Hence, the stability boundary of scheme (2) in the complex μ-plane may be determined in a simple way. In effect, that boundary is defined as the locus of the points satisfying the neutral stability condition $|\xi| = 1$. These points may be represented as

$$\xi = e^{i\theta} \tag{4}$$

where $\theta \in [0, 2\pi]$ is an angular parameter. Making use of (4) in (3), the values of μ in the stability boundary of scheme (2), $\mu = \mu_b$, are found to be

$$\mu_b = \frac{\sum_{k=0}^{N-1} \alpha_k e^{ik\theta}}{\sum_{k=0}^{N-1} \beta_k e^{ik\theta}} \tag{5}$$

These values define, in complete form, the stability properties of scheme (2) (Gear [6]).

Let us now consider the following periodic-Cauchy problem for the one-dimensional advection-diffusion equation:

$$\phi_t + V\phi_x = \kappa\phi_{xx} \quad , \quad 0 \leq x \leq L \tag{6}$$

$$\phi(x, 0) = f(x) \tag{7}$$

$$\phi(0, t) = \phi(L, t) \tag{8}$$

where V represents a velocity; κ, a diffusivity; x, the space variable; t, time; L, the size of the domain of interest in x; and ϕ, the dependent variable, which may represent temperature, concentration, etc. Thus, let us state the following:

Theorem. Any finite difference scheme that approximates the periodic-Cauchy problem, (6)-(8), for the one-dimensional advection-diffusion equation, through the use of an N-level integration scheme in t and an M-node discretization in x (where $N \geq 2$ and $M \geq 3$ for consistency), has a stability equation with structure identical to that of the stability equation corresponding to the same integration scheme in t, when it is used to approximate the equation $dy/dt = \lambda y$, where λ is a complex parameter.

Proof. The most general N-level, M-node approximation to equation (6), making use of the same scheme that was employed to obtain (2), is of the form

$$\sum_{k=0}^{N-1}\sum_{j=0}^{M-1}\left(\frac{\alpha_k}{\Delta t}\gamma_j+\frac{V}{\Delta x}\beta_k\delta_j\right)\phi_{m+j}^{n+k}=\frac{\kappa}{\Delta x^2}\sum_{k=0}^{N-1}\sum_{j=0}^{M-1}\beta_k\varepsilon_j\phi_{m+j}^{n+k} \tag{9}$$

where Δx and Δt represent space and time intervals, respectively; the notation $\phi_m^n = \phi(m\Delta x, n\Delta t)$ has been used; and α_k, $\beta_k (k = 0, 1, \ldots, N-1)$, γ_j, δ_j and $\varepsilon_j (j = 0, 1, \ldots, M-1)$ are determined in such a way that (9) be consistent with (6) and to attain a selected order of accuracy.

According to the boundary condition (8), rigorously speaking, the following discrete Fourier representation may be used for ϕ_m^n (Smith [5]):

$$\phi_m^n = \sum_{p=-P/2+1}^{\frac{P}{2}} \Phi_p \xi_p^n e^{i2\pi m\frac{p}{2}} \tag{10}$$

where Φ_p and ξ_p are the amplitude and the amplification factor, respectively, associated with the discrete wave number p; and P, the total number of independent values of ϕ_m^n in $[0, L]$, defined by $P = L/\Delta x$.

Substituting (10) in (9) and making use of the fact that Fourier modes are not coupled for linear problems, we get:

$$\sum_{k=0}^{N-1}\sum_{j=0}^{M-1}\left(\frac{\alpha_k}{\Delta t}\gamma_j+\frac{V}{\Delta x}\beta_k\delta_j\right)\xi_p^k e^{i2\pi j\frac{p}{P}}=\frac{\kappa}{\Delta x^2}\sum_{k=0}^{N-1}\sum_{j=0}^{M-1}\beta_k\varepsilon_j\xi_p^k e^{i2\pi\frac{jp}{P}} \tag{11}$$

Now let

$$\mu_p = C\frac{\sum_{j=0}^{M-1}(\gamma_j - \delta_j/P_g)e^{i2\pi j\frac{p}{P}}}{\sum_{j=0}^{M-1}\gamma_j e^{i2\pi j\frac{p}{P}}} \tag{12}$$

where $C = V\Delta t/\Delta x$ is the Courant number and $P_g = V\Delta x/\kappa$ is the grid Péclét number.

Employing (12) in (11) we arrive at:

$$\sum_{k=0}^{N-1}(\alpha_k - \mu_p\beta_k)\xi_p^k = 0 \tag{13}$$

that represents an algebraic equation in ξ_p of degree $N-1$, with structure identical to that of (3). ∎

Corollary. In view of the above stated theorem, the knowledge about the stability boundary of the scheme used to numerically integrate the ordinary differential equation $dy/dt = \lambda y$, may be used to determine the stability properties of finite difference schemes that approximate the advection-diffusion equation.

In effect, let us introduce the polar representation for μ_b (as defined by (5)):

$$\mu_b = |\mu_b| e^{i\psi} \tag{14}$$

where ψ represents the argument of μ_b in the complex μ-plane. In addition, let us represent μ_p in the following form

$$\mu_p = CF(p; P_g) \tag{15}$$

where, according to (12), F is a complex function of p and depends parameterically on P_g. Considering that the stability equations (3) and (13) have the same algebraic structure, the values of μ_p that make scheme (9) neutrally stable, must satisfy the condition $\mu_p = \mu_b$. Therefore, the determination of the critical Courant number $C_{cr} = V\Delta t_{cr}/\Delta x$, (where Δt_{cr} represents the critical time step) as a function of P_g, may be posed as the following optimization problem, in view of (14) and (15):

$$C_{cr} = C_{cr}(P_g) = \min_{p} \frac{|\mu_b|}{F(p; P_g)} \tag{16}$$

subject to

$$\frac{Im[F(p; P_g)]}{Re[F(p; P_g)]} = \tan \psi \tag{17}$$

the problem (16)-(17) may be easily solved numerically, allowing the computation of the stability curve $C_{cr} = C_{cr}(P_g)$.

Stability analysis of the Adams-Bashforth scheme

In order to illustrate the application of the ordinary differential equation analogy, the result of the stability analyses of the Adams-Bashforth scheme combined with various spatial discretizations, will be discussed here.

The Adams-Bashforth approximation is a second order-accurate, explicit scheme (Gear [6]), whose application to equation (1) yields:

$$\frac{y_{n+1} - y_n}{\Delta t} = \lambda\left(\frac{3}{2}y_n - \frac{1}{2}y_{n-1}\right) \tag{18}$$

Substituting a solution of the form $y_n = \xi^n$ in (18), the following stability equation is obtained:

$$\xi^2 - \left(1 + \frac{3}{2}\mu\right)\xi + \frac{1}{2}\mu = 0 \tag{19}$$

where $\mu = \lambda\Delta t$. The values of μ in the stability boundary of scheme (19), $\mu = \mu_b$, are determined by making $\xi = e^{i\theta}$ in (19). The result is

$$\mu_b = 2\frac{e^{2i\theta} - e^{i\theta}}{3e^{i\theta} - 1} \tag{20}$$

A graphical representation of μ_b in the complex μ-plane is shown in figure 1. Expression (20) contains the necessary information about the modulus of μ_b, $|\mu_b|$, and its argument, ψ, to apply the technique presented earlier for the determination of the stability properties of the Adams-Bashforth scheme, combined with any spatial discretization, as an approximation of the advection-diffusion equation (6).

Thus, the stability analyses of the following discrete approximation of equation (6) were performed through the use of expressions (16) and (17): Adams-Bashforth/Second Order Central Differences (AB2CD), Adams-Bashforth/Upwind-Central Differences (ABUCD), Adams-Bashforth/Fourth Order Central Differences (AB4CD) and Adams-Bashforth/Linear Finite Elements (ABLFE). In the case of the ABLFE scheme, the finite-difference analogue (that results when the finite element method with linear expansions and uniform element size is used, and the assembly procedure is applied) was employed.

The stability curves (in the form $C_{cr} = C_{cr}(P_g)$) for each of the above mentioned schemes, that resulted from the application of expressions (16) and (17) are shown in figure 2. To the knowledge of the author such curves have not been previously published for the advective-diffusive case. Stability analyses exist for the AB2CD scheme as applied to the pure advective (Lilly [7]) and pure diffusive (Aldama [8]) cases but not for the combined one, because of the algebraic complexity involved in imposing the stability condition to the solution of the stability equation.

Conclusions

The existence of an analogy between the stability properties of schemes for the numerical integration of the ordinary differential equation $dy/dt = \lambda y$ and those of finite difference aproximations of the advection-diffusion equation results in a powerful tool for the simplification of the stability analysis of the latter. This fact is illustrated through the determination of the stability curves for the Adams-Bashforth scheme combined with various spatial discretizations, which shows that the referred analogy allows the determination of practical stability criteria, even in cases when the complexity of the expressions resulting from the application of the classical stability analysis technique, makes those criteria difficult to obtain.

Aknowledgement

The computation of the stability curves was performed by Ana Wagner.

References

1. Fischer, H.B., E.J. List, R.C.Y. Koh, J. Imberger and N.H. Brooks (1979). Mixing in inland and coastal waters, Academic Press.
2. Fischer, H.B. (Ed.) (1981). Transport models for inland and coastal waters, Academic Press.
3. Roache, P.J. (1982). Computational fluid dynamics, Hermosa.
4. Richtmyer, R.D. and K.W. Morton (1967). Difference methods for initial value problems, Interscience.
5. Smith, G.D. (1978). Numerical solution of partial differential equations, Oxford University Press.
6. Gear, C.W. (1971), Numerical initial value problems in ordinary differential equation, Academic Press.
7. Lilly, D.K. (1965). On the computational stability of the numerical solution of time dependent nonlinear geophysical fluid dynamics problems, Monthly Weather Rev., 93, 11-26.
8. Aldama, A.A. (1985). Theory and applications of two- and three- scale filtering approaches for turbulent flow simulation, Ph. D. Thesis, Civil Eng. Dept., MIT.

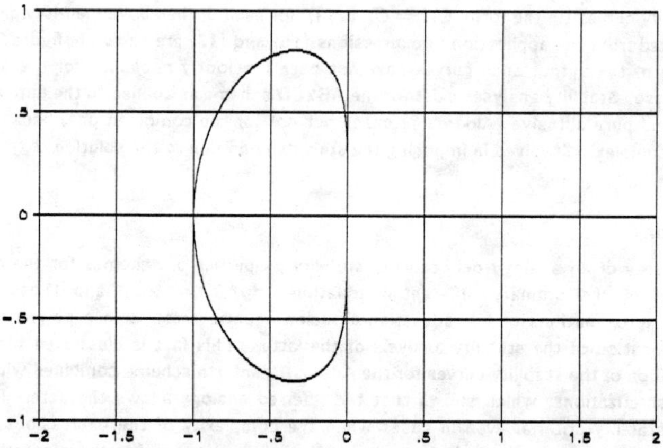

Figure 1. Stability boundary of the Adams-Bashforth scheme in the complex μ-plane

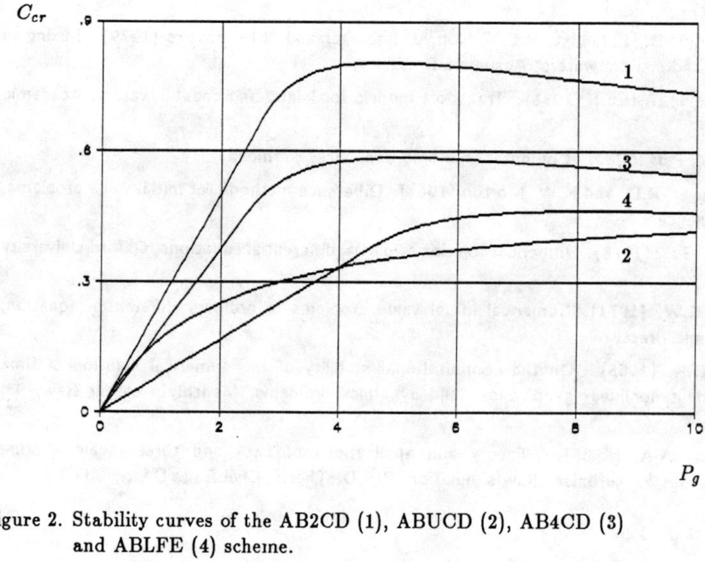

Figure 2. Stability curves of the AB2CD (1), ABUCD (2), AB4CD (3) and ABLFE (4) scheme.

Solution of the Advection-Diffusion Transport Equation using the Total Derivative and Least Squares Collocation

L.R. Bentley and G.F. Pinder
Department of Civil Engineering, Princeton University, USA
I. Herrera
Instituto de Geofisica, UNAM, Mexico

INTRODUCTION

The difficulties arising in the application of numerical approximations to advection-diffusion transport problems are well known. The difficulties arise because of the dual nature of the equation. When the transport is advection dominated, the equation behaves as a first order equation. When the transport is diffusion dominated, the equation behaves as a second order parabolic equation.

Recently many workers have turned to Eulerian-Lagrangian methods (ELM) in an attempt to satisfactorily capture both the second order parabolic and first order nature of the equation (Baptista (1987), Glass and Rodi (1982), Holly and Polatera (1984) and Neuman (1984)). The equation is solved in two steps. In the first step, past information is carried along characteristics, thereby decoupling the solution of the first order part of the equation from the second order parabolic part. In the second step, the second order parabolic problem is solved on a fixed grid.

The following method resembles an ELM in that information that is required in the difference equations will be brought from the last time step by tracking along characteristics. The advection-diffusion equation is written in Lagrangian coordinates. It is then approximated by a central difference in time and a least squares collocation (LESCO) (Joos, 1986) discretization in space. It is the collocation point locations which are backward projected along characteristics. A major difference between our approach and ELMs is that no intermediate solution is computed.

DEVELOPMENT

In a Lagrangian system, the one dimensional advection-diffusion transport equation is written:

$$\frac{DC(x,t)}{Dt} + LC(x,t) = 0 \tag{1}$$

where:

$$\frac{D}{Dt} = \frac{\partial}{\partial t} + v(t)\frac{\partial}{\partial x},$$

$$L = -D(x)\frac{\partial^2}{\partial x^2} - \frac{\partial D(x)}{\partial x}\frac{\partial}{\partial x},$$

$C(x,t) =$ concentration ,

$v(t) =$ velocity, and

$D(x) =$ diffusion coefficient .

We now approximate the above differential equation with a finite difference equation in time along flow lines:

$$\frac{C(x,t_n) - C(x^*,t_{n-1})}{\Delta t} + .5LC(x,t_n) + .5LC(x^*,t_{n-1}) = 0 \tag{2}$$

where:

$x^* = x - \int_{t_{n-1}}^{t_n} v(t)dt$, and

$\Delta t =$ time increment.

The total derivative is approximated by the difference between the concentration at a particle location at the present time minus the concentration at the location of the same particle at a previous time divided by the time increment. The spatial derivatives and the operator coefficients are approximated by an average of the values at the present time and location and the values at the location of the particle at the previous time.

The difference equation is solved using least squares collocation (Joos, 1986). Since the equation contains second order derivatives in x, approximating functions with at least C^1 continuity are required. Cubic Hermites are well known functions with this property. The concentration function is approximated by:

$$\hat{C}(x,t) = \sum_{i=1}^{nbas} \alpha_i(t)\Phi_i(x) \tag{3}$$

where:

$\Phi_i(x) =$ cubic Hermite base functions, two per node,

$\alpha_i(t) =$ coefficient of base function i at time t , and

$nbas$ = Number of base functions.

The approximate concentration function, Eq. (3), is substituted into the difference equation, Eq. (2), and this function is evaluated at a collocation point located at position x_k and time t_n :

$$R_k = \frac{1}{\Delta t}\sum_{i=1}^{nbas} \alpha_i(t_n)\Phi_i(x_k) - \frac{1}{\Delta t}\sum_{i=1}^{nbas} \alpha_i(t_{n-1})\Phi_i(x_k^*)$$

$$+.5\mathrm{L}\left\{\sum_{i=1}^{nbas}\alpha_i(t_n)\Phi_i(x_k)\right\}+.5\mathrm{L}\left\{\sum_{i=1}^{nbas}\alpha_i(t_{n-1})\Phi_i(x_k^*)\right\} \tag{4}$$

R_k is the residual associated with the kth collocation point. When the collocation point comes from a location within the domain of the last time step, the function and the operator of the last time step are simply evaluated from the cubic approximating function of the last time step at that location.

When the flow line intersects the domain boundary, the evaluation of the function and operator of the last time step is more complicated. If t_i is the time of intersection and x_B is the boundary coordinate, then Δt of Eq. (2) becomes $t_n - t_i$, x_k^* becomes x_B and $C(x_k^*, t_{n-1})$ becomes $C(x_B, t_i)$. The values of $C(x_B, t_i)$ and $\mathrm{L}C(x_B, t_i)$ must be approximated. If the boundary condition is of the first type, then $C(x_B, t_i)$ is simply the boundary value at time t_i . The spatial operator is approximated by:

$$\mathrm{L}\hat{C}(x_B, t_i) = \beta\mathrm{L}\hat{C}(x_k, t_n) + (1-\beta)\mathrm{L}\hat{C}(x_B, t_{n-1}) \tag{5}$$

where:

$$\beta = \frac{t_i - t_{n-1}}{\Delta t} .$$

Substitution of Eqs. (3) and (5) into Eq. (2) yields the error, R_k , associated with a collocation point that entered the domain during the last time step:

$$R_k = \frac{1}{t_n - t_i}\sum_{i=1}^{nbas}\alpha_i(t_n)\Phi_i(x_k) - \frac{\hat{C}(x_B, t_i)}{t_n - t_i} + .5(1+\beta)\mathrm{L}\left\{\sum_{i=1}^{nbas}\alpha_i(t_n)\Phi_i(x_k)\right\}$$
$$+ .5(1-\beta)\mathrm{L}\left\{\sum_{i=1}^{nbas}\alpha_i(t_{n-1})\Phi_i(x_B)\right\} \tag{6}$$

t_i never equals t_n, but as it approaches t_n, the formulation approaches fully implicit. As t_i approaches t_{n-1} the formulation approaches Crank Nicolson.

The sum of the squares of all of the errors, ε , is:

$$\varepsilon = \sum_{k=1}^{ncol} R_k^2 \tag{7}$$

where $ncol$ is the number of collocation points.

To minimize the sum of the squares of the errors, the derivatives with respect to the coefficients $\alpha_j(t)$ are set equal to zero:

$$\frac{\partial\varepsilon}{\partial\alpha_j(t_n)} = 2\sum_{k=1}^{ncol} R_k \frac{\partial R_k}{\partial\alpha_j(t_n)} = 0 \qquad j = 1, nbas \tag{8}$$

Combination of Eqs. (4), (6), and (8) yields the least squares collocation set of equations:

$$\left\{\sum_{k=1}^{ncol}\sum_{i=1}^{nbas}\mathrm{L}_1\Phi_i(x_k)\mathrm{L}_1\Phi_j(x_k)\right\}\alpha_i(t_n) = \sum_{k=1}^{ncol}\mathrm{L}_1\Phi_j(x_k)\sum_{i=1}^{nbas}\mathrm{L}_2\Phi_i(x_k^*)\,\alpha_i(t_{n-1}) \tag{9}$$

$$\textit{for} \qquad j = 1, nbas$$

The form of the operators L_1 and L_2 varies depending on the location of the backward projected collocation point. When the collocation point comes from within the domain of the last time step, the use of Eq. (4) yields:

$$L_1 = \frac{1}{\Delta t} + .5L \tag{9a}$$

$$L_2 = \frac{1}{\Delta t} - .5L \tag{9b}$$

When the collocation point enters the domain from the first type boundary at x_B and at time t_i, then Eq. (6) yields:

$$L_1 = \frac{1}{t_n - t_i} + .5(1+\beta)L \tag{9c}$$

$$\sum_{i=1}^{nbas} L_2\Phi_i(x_k^*)\,\alpha(t_{n-1}) = \frac{C(x_B, t_i)}{t_n - t_i} - .5(1-\beta)\sum_{i=1}^{nbas} L\Phi_i(x_B)\alpha_i(t_{n-1}) \tag{9d}$$

Cubic Hermites allow the specification of both the function and the first derivative at each node. Consequently, boundary conditions of both the first and second type are directly enforced in the matrix equations. The initial conditions are imposed by least squares fitting the cubic hermites to the initial values of concentration at the collocation points.

In summary, the computations required for each time step are:

1. Choose the collocation point locations (x_k) of the time step to be computed.
2. Back project the collocation point locations to the last time step (i.e., compute x_k^*).
3. Compute the coefficient matrix and the right hand side vector using Eq. (9).
4. Solve the matrix equation for $\alpha_i(t_n)$.
5. Compute the new set of C^1 continuous cubic polynomials that approximate the solution at the present time step by summing over all of the base functions in each element.

RESULTS

The results of two simulations are presented in Figures 1 and 2. In both cases velocity = .5, time increments = 192, total time steps = 50, element lengths = 200 and there are 8 collocation points per element. The Courant number is 0.48. Analytic solutions are solid lines and LESCO computed solutions are dash-double dot lines. The oscillatory Galerkin finite element solutions are shown as dash-dot lines. The Galerkin solution used Lagrange quadratic basis functions with element lengths of 200 and node spacing of 100.

In Figure 1, the diffusion coefficient is zero (the pure advection case) and a gaussian plume of standard deviation 264 was used as the initial condition. The right boundary has a zero concentration, and the left boundary a zero derivative. The analytic and LESCO computed solutions are coincident.

In Figure 2, a concentration front is propagated from the left boundary. The initial condition was zero concentration. The left boundary concentration is one, and the right boundary derivative is zero. The diffusion coefficient is one, and the grid Peclet number is 100. The analytic and LESCO computed solutions are essentially coincident.

CONCLUSION

Excellent results have been obtained using the total derivative and LESCO to solve the advection-diffusion transport equation. As can be seen from the two examples, the method works well in advection dominated transport. This is partially due to having eliminated the first order hyperbolic term that dominates when the Peclet number is large. In addition, numerical test results, not presented here, have demonstrated that the LESCO formulation reproduces the higher spatial frequencies in the concentration fronts in a superior way. As the velocity decreases, the equations reduce to the Eulerian equations for diffusion, so the procedure works well for diffusion dominated transport as well. Given the promising early results, the method deserves further investigation.

REFERENCES

Baptista, A.M. (1987) Solution of Advection-Dominated Transport by Eulerian-Lagrangian Methods Using the Backwards Method of Characteristics, Ph.D. Thesis, M.I.T.

Glass, J., and W. Rodi (1982) A Higher Order Numerical Scheme for Scalar Transport, *Comp. Math. in Appl. Mech. and Engr.*, 31: 337-358.

Holly, F.M., Jr., and J.M. Polatera (1984) Dispersion Simulation in 2-D Tidal Flow, *J. Hydr. Engr., ASCE*, 110: 905-926.

Joos, B. (1986) The Least Squares Collocation Method for Solving Partial Differential Equations, Ph.D. Thesis, Princeton U.

Neuman, S.P. (1984) Adaptive Eulerian-Lagrangian Finite Element Method for Advection-Dispersion, *Int. J. for Numerical Meth. in Engr.*, 20: 321-337.

FIGURE 1 - GAUSSIAN PLUME

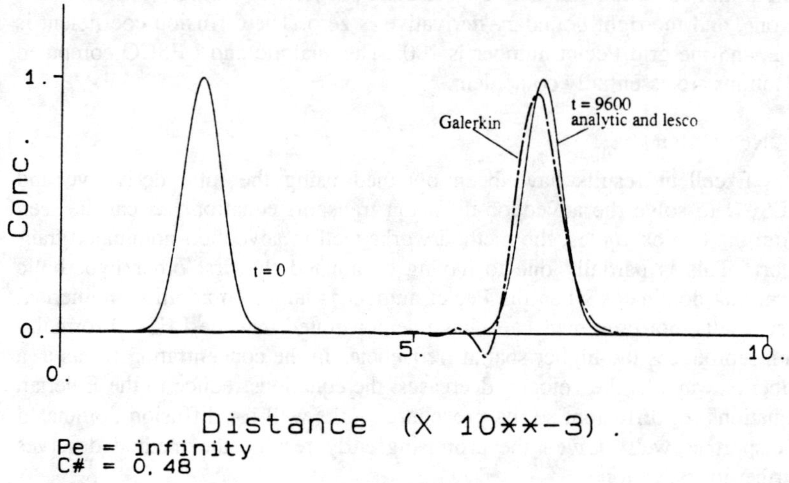

FIGURE 2 - ADVANCING FRONT

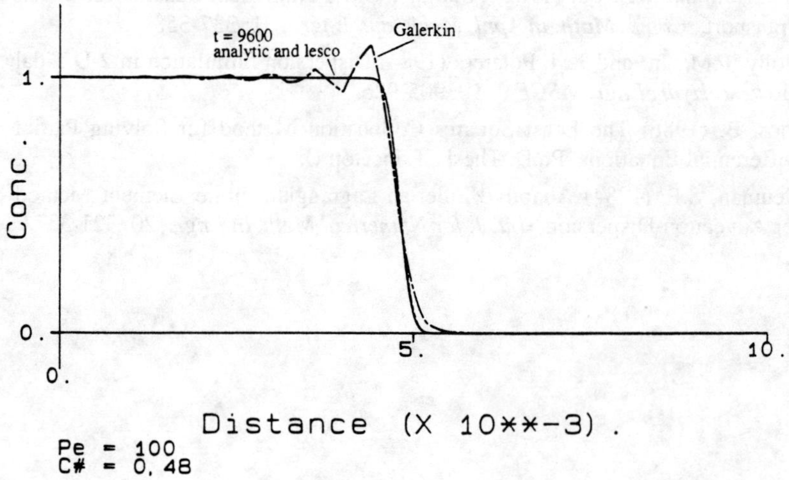

An Analysis of Some Classes of Petrov-Galerkin and Optimal Test Function Methods

E.T. Bouloutas and M.A. Celia
Ralph M. Parsons Laboratory, Department of Civil Engineering, Massachusetts Institute of Technology, Cambridge, MA 02139, USA

INTRODUCTION

Reliable numerical solutions to advection-dominated flow problems are of great importance to many engineering disciplines. Fluid flow at relatively high Reynolds number and convective transport in low diffusivity fields, are two of the important examples. The development of alternative weighted residual techniques, which give rise to upwind operators in a systematic framework, is one of the most important numerical contributions in this area. Petrov-Galerkin methods (Christie *et al*, 1976, Heinrich et al, 1977, Brooks and Hughes, 1981) and the newly developed Optimal Test Function methods (Herrera *et al*, 1985, Celia and Bouloutas, 1987), have proven to be very effective for the simulation of advection dominated flows. These methods retain the higher order accuracy in regions of smooth changes and have been shown to be quasi-optimal, even for cases of singularly perturbed problems (Griffiths and Lorentz, 1978).

The purpose of this paper is to systematically develop and analyze some of these schemes, and prove that, for model one dimensional steady state and transient advection diffusion problems, these diverse formulations produce similar or in some cases identical results. The methods considered are: Allen and Southwell difference scheme, quadratic Petrov-Galerkin, streamline upwind Petrov-Galerkin, exponential Petrov-Galerkin and optimal test function methods.

DEVELOPMENT OF THE NUMERICAL APPROXIMATIONS

A model problem

We consider the model stationary advection diffusion problem in 1-D

$$\boldsymbol{L}\,u = D\frac{d^2u}{dx^2} - V\frac{du}{dx} = f(x)\,, \qquad x \in [0,1] \qquad \text{with } u(0)=0 \text{ and } u(1)=1 \tag{1}$$

V and D represent the rates of the physical processes of advection and diffusion respectively. In what follows the case of zero forcing function will be considered unless otherwise stated. For the case of constant coefficients there exists a simple analytical solution to (1) given by

$$u(x) = (1 - e^{rx}) / (1 - e^{r}) \quad \text{with} \quad r = V / D \tag{2}$$

The solution exhibits a boundary layer of thickness O (1/r) near x=1. Let Π_h denote a subdivision of the interval [0,1] into N subintervals $e_j = [x_j, x_{j+1}]$, j=1,2,...N with $x_1 = 0$ and $x_{N+1} = 1$. For reasons of convenience only we choose a set of equally spaced points $x_j = (j-1)h$. On this mesh we seek numerical approximations of equation (1) . In particular we are interested in the approximate solutions for large values of r, i.e. r >> 1.

Allen and Southwell scheme

Allen and Southwell in 1954, as a part of an attempt to resolve the fluid motion around a cylinder, proposed the following scheme for equation (1),

$$\left(D + \frac{1}{2}\xi r h\right) \frac{u_{j+1} - 2u_j + u_{j-1}}{h^2} - V \frac{u_{j+1} - u_{j-1}}{2h} = 0 \tag{3}$$

It can be shown (Allen and Southwell, 1954, Christie et al, 1976) that equation (3) gives exact nodal values when ξ is chosen by the rule

$$\xi = \coth\left(\frac{rh}{2}\right) - \frac{2}{rh} \tag{4}$$

Equation (3) represents a second order approximation to a perturbed differential equation with added 'artificial' diffusion $D^* = (\xi r h) / 2$

Quadratic Petrov-Galerkin

In a Petrov-Galerkin method numerical solutions of the form

$$u_h(x) = \sum_{j=1}^{N+1} u_j \phi_j(x) \tag{5}$$

are sought with ϕ_j chosen from a set Φ_h of trial functions such that the residual is orthogonal to the set Ψ_h of test functions

$$\int_0^1 (Lu_h - f)\,\psi_i(x)\,dx = 0 \tag{6}$$

In general the set Ψ_h is different from the set Φ_h. After integration by parts equations (5) and (6) combine as

$$\left[\frac{du_h}{dx}\psi_i\right]_0^1 - \sum_{j=1}^{N+1} u_j\left\{ D\int_0^1 \frac{d\phi_j}{dx}\frac{d\psi_i}{dx}\,dx + V\int_0^1 \frac{d\phi_j}{dx}\psi_i\,dx\right\} = 0 \tag{7}$$

Christie *et al* (1976), introduced the quadratic test functions (figure 1A)

$$\psi_i(x) = \phi_i(x) + \xi\sigma_i(x) \tag{8}$$

with σ_i (x) a quadratic perturbation function and ϕ_i (x) the usual piecewise linear trial functions. Performance of the integrations in (7) results in the Allen and Southwell scheme of equation (3).

Streamline upwind Petrov-Galerkin

The streamline diffusion model of Hughes and Brooks (1979) was initally motivated by the form of the Allen-Southwell operator. Thus they introduced artificial diffusion in the direction of flow to damp the oscillations, while retaining the Galerkin formulation. However in a later paper Brooks and Hughes (1981) formulated the same procedure in terms of a Petrov-Galerkin method with discontinous test functions (figure 1B)

$$\psi_i(x) = \phi_i(x) + \frac{1}{2}\xi h \frac{d\phi_i}{dx} \tag{9}$$

Once again performance of the relevant integrations in (7) recovers the Allen-Southwell difference equation.

Optimal Test Function Methods

The underlying idea behind Optimal Test Function methods is to begin with the variational equation (6). Under the assumptions that u belongs to $C^1(\Omega)$ and ψ is in $C^{-1}(\Omega)$, with any discontinuities in ψ occurring at nodes x_j, the domain integration can be replaced by the sum of element integrations

$$\int_0^1 \{\mathcal{L}u - f(x)\}\psi_i(x)\,dx = \sum_{j=1}^{N}\int_{e_j}\{\mathcal{L}u - f(x)\}\psi_i(x)\,dx \tag{10}$$

Integration by parts is then repeatedly applied to each integral on the right side of equation (10). This leads to

$$\left[D\psi\frac{du}{dx} - \left(D\frac{d\psi}{dx}+V\psi\right)u\right]_0^1 + \sum_{j=2}^{N}\left\{u_j J\left[D\frac{d\psi}{dx}+V\psi\right]_j - \frac{du_j}{dx}J[D\psi]_j\right\}$$
$$+\sum_{j=1}^{N}\int_{e_j}u(\mathcal{L}^*\psi)\,dx - \sum_{j=1}^{N}\int_{e_j}f(x)\psi(x)\,dx = 0 \tag{11}$$

where **J** is a jump operator and subscript j implies evaluation at node j. Algebraic equations are generated by choosing test function $\psi(x)$ such that the homogeneous adjoint equation $\mathcal{L}^*\psi = 0$, is satisfied within each element e_j. For the model equation (1), this homogeneous equation has fundamental solutions $\{1, \exp(-r x)\}$. Test functions are constructed in each element as a linear combination of these foundamental solutions. A typical $C^0[0,1]$ test function is illustrated in figure 1C. This is chosen to exhibit the convenient properties $\psi_j(x_j)=1$, $\psi_j(x_i)=0$ $(i\neq j)$, and $\psi_j(x)=0$ outside $[x_{j-1}, x_{j+1}]$. One such function is chosen for each interior node x_j. Substitution of ψ_j into equation (11) gives the following algebraic equation

$$\sum_{j=2}^{N}u_j J\left[D\frac{d\psi}{dx}\right]_j = \frac{rD}{1-\alpha}[u_{j-1} - (1+\alpha)u_j + \alpha u_{j+1}] = 0 \tag{12}$$

with $\alpha = e^{(-r\,h)}$ It can be proven that (12) is another form of the Allen-Southwell operator.

Exponential Petrov-Galerkin

Hemker (1977) has made the observation that the pointwise error bound on a mesh Π_h is related to the capacity of the test space Ψ_h to represent solutions of the adjoint equation. Thus he proposes use of the exponential test functions of figure 1C in a Petrov-Galerkin format. Starting from the weak variational form (6) evaluation of the associated integrals gives the difference equation associated with node j as

$$\frac{V}{1-\alpha}\,[\,u_{j-1} - (1+\alpha)\,u_j\ + \alpha\,u_{j+1}\,] = 0 \qquad (13)$$

which is exactly the same equation as the optimal test function discretization.

Discussion / Comparison

In this section we try to outline some of the apparent similarities and point the differences of the various formulations. In particular it is evident that all Petrov-Galerkin methods are motivated by the Allen-Southwell scheme, and use different functional forms for test functions in order to acheive exact nodal results for the constant coefficient, zero forcing function case. It should be emphasized that when $f(x) \neq 0$ this superconvergence phenomenon breaks down. However the optimal test function and exponential Petrov-Galerkin formulations retain their optimality properties, resulting in exact nodal values, for any forcing function. The optimal test function method appears to be more flexible, since it naturally accommodates the nonconstant coefficient case. Also the quadratic and streamline upwind Petrov-Galerkin formulations yield the same difference equation for piecewise constant f(x) since the associated integrals of product of test functions with piecewise linear functions turn out to be the same. This result will explain the equivalence of the two methods for the time dependent case.

TIME DEPENDENT PROBLEMS

The methodologies presented in the previous sections can be easily generalized for the solution of time dependent problems. We consider , as a model problem, the transient advection-diffusion equation in one dimension

$$\frac{\partial u}{\partial t} = D\,\frac{\partial^2 u}{\partial x^2} - V\,\frac{\partial u}{\partial x} \qquad x \in [0,1],\ \ t \in [0,T] \qquad (14)$$

with appropriate initial and boundary conditions. A weighted residual formulation in space results to the following variational equation

$$\int_0^1 \frac{\partial u_h}{\partial t}\,\psi_i\,(x\,)dx = \int_0^1 \boldsymbol{L}\,u_h\ \psi_i\,(x)\,dx \qquad (15)$$

The right hand side of equation (15) represents the space dependent part, and is treated by all methods in exactly the same way as in the stationary case, with u_j in (5) now being a function of time. For u_h piecewise linear, evaluation of the integrations for the time term results in the same semidiscrete equation for the quadratic and streamline Petrov-Galerkin formulations given by

$$\left(\frac{1}{6}+\frac{\xi}{4}\right)\frac{du_{j-1}}{dt}+\frac{4}{6}\frac{du_j}{dt}+\left(\frac{1}{6}-\frac{\xi}{4}\right)\frac{du_{j+1}}{dt}=\text{R.H.S} \qquad (16)$$

For the evaluations of the associated integrals in the left side of (15) in the optimal test function method a spatial interpolation of the form (5) is introduced. When piecewise linear trial functions are used, a consistent Petrov-Galerkin method is formulated similar to Hemker's method (even though he did not consider transient cases). However this method tends to give a lumped mass matrix with the well known increased phase errors. If the trial functions are chosen to be piecewise quadratic Lagrange polynomials, that are designed to overlap the nonzero region of the test functions, it can be proven that the limiting values of the time related integrals in (15) when the grid Pechlet number goes to infinity take the values (5/12, 2/3, -1/12) which are the same as the limiting values in equation (16) (lim $\xi = 1$). Also all the formulations collapse to the traditional Galerkin method when the Pe number goes to zero. Furthermore the optimal test functions and the Petrov-Galerkin techniques presented in the previous sections give results that are very close in most practical situations. Even though these results represent a significant improvement over the Galerkin method (Figure 2 for a typical comparison in the case of a propagating steep front) they are characterized by some artificial diffusion. Improvements are currently being sought through alternative temporal formulations and multidimensional generalizations.

REFERENCES

A.N. Brooks and T.J.R. Hughes, 'Streamline upwind/Petrov-Galerkin formulations for convection dominated flows with particular emphasis on the incompressible Navier-Stokes Equations', *Comp.Meth.Appl. Mech. Eng.*, Vol 32 ,199-259, (1982)

M.A. Celia and E.T. Bouloutas, "An analysis of optimal test function methods for advection-dominated flows', to be presented in *International Conference on Computational Eng Science*, Atlanta, April 10-14 1988.

I. Christie, D.F. Griffiths, A.R. Mitchell ' Finite element methods for second order differential equations with significant first derivatives', *Int. J. Num. Meth. Eng.*, Vol 10, 1389-1396 (1976)

D.F. Griffiths and J. Lorenz, 'An analysis of the Petrov-Galerkin finite element method', *Comp. Meth. Appl. Mech. Eng.*, Vol 14, 39-64 (1978)

J.C. Heinrich, P.S. Huyakorn and O.C. Zienkiewicz, ' An upwind finite element scheme for two dimensional convective transport equation' *Int. J. Num. Meth. Eng.*, Vol 11, 131-143 (1977)

P.W. Hemker, 'A numerical study of stiff two-point boundary problems', Ph.D Thesis Mathematisch Centrum, Amsterdam (1977).

I. Herrera, M.A. Celia, E.T. Bouloutas, S. Kindred, "A new numerical approach for the advective diffusive transport equation, Submitted for review to *Int.J. Num.Meth. Fluids.* (1988)

T.J.R. Hughes and A. Brooks, ' A multidimensional upwind scheme with no crosswind diffusion', in *Finite element for convection dominated flows*, Ed (T.J.R. Hughes) , AMD Vol 34. ASME Pub (1979)

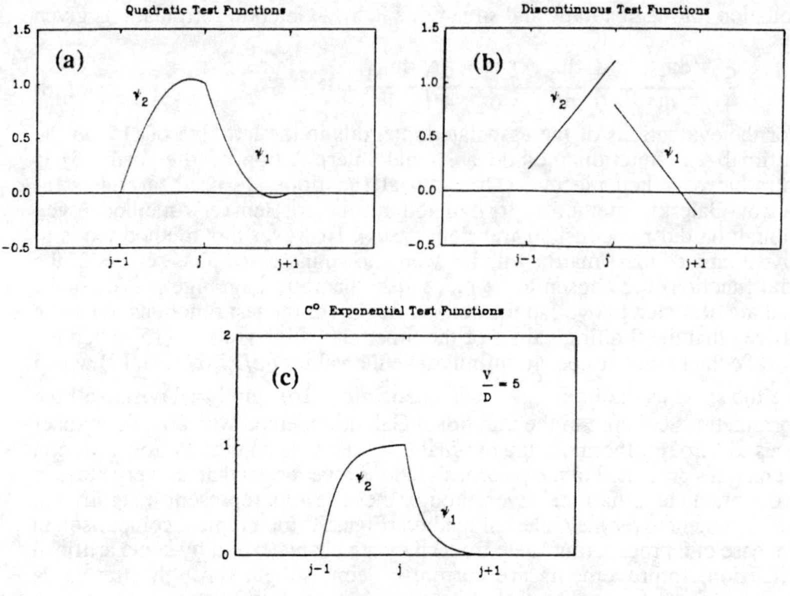

FIGURE 1 - Typical test functions used in various formulations

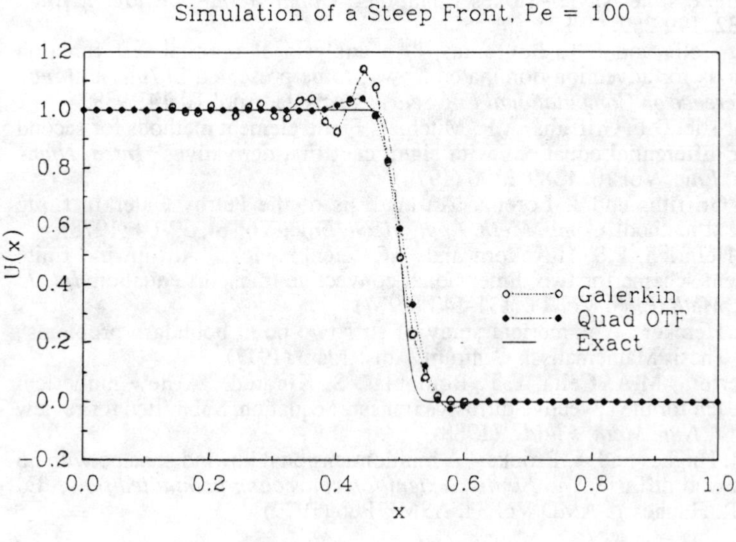

FIGURE 2 - Numerical comparison for a test problem

The Cell Analytic-Numerical Method for Solution of the Two-Dimensional Advection-Dispersion Equation

O.A. Elnawawy and A.J. Valocchi
Department of Civil Engineering, University of Illinois at Urbana-Champaign, Urbana, IL 61801, USA
A.M. Ougouag
Department of Nuclear Engineering, University of Illinois at Urbana-Champaign, Urbana, IL 61801, USA

ABSTRACT

A new numerical method, called the Cell Analytic-Numerical Method, is developed for solution of the Two-Dimensional Advection-Dispersion Equation by using a transverse integration technique followed by analytical solution of the transverse-integrated local equations. Continuity of the mass flux is then used to obtain a set of coupled tridiagonal equations which can be efficiently solved. This new method is demonstrated to have high accuracy, even when applied on coarse meshes, and to have minimal grid orientation error.

INTRODUCTION

In this paper a new numerical method, the Cell Analytic-Numerical (CAN) method, is developed and implemented for the efficient solution of the two-dimensional solute transport equation. The CAN method is akin to the modern, transverse integrated, nodal methods developed in the nuclear engineering area (Lawrence[1]).

The CAN method is based on decomposition of the entire solution domain into a number of rectangular cells (volume sub-domains). The essential idea is to transform the initial partial differential equation into a set of local coupled ordinary differential equations through the application of a transverse integration technique. This set of equations is solved analytically after approximation of the "source-like" expressions on the right-hand side. Mass flux continuity across cell boundaries is then used, along with the local analytical solution, to construct an algebraic relationship between the transverse-integrated concentration values at adjacent cell boundaries. Assembling all the cells together results in a set of coupled tridiagonal matrix equations. The CAN method shares with the alternating-direction finite difference and finite element techniques (Peaceman and Rachford[2], Daus and Frind[3]) the attractive feature of having tridiagonal matrix equations; but, the CAN method differs significantly from these traditional techniques because the spatial portion of the differential operator is represented by the local analytical solution, rather than a discrete approximation.

THEORY

To develop the CAN method for solving the two dimensional solute transport equation, we begin by decomposing the domain of the problem into M rectangular cells, $(-a_m, +a_m) \times (-b_m, +b_m)$, $m=1,\ldots, M$. Each cell contains a homogeneous medium of constant parameter values, the cell corners coincide with corners of adjacent ones, and the global boundaries of the domain coincide with edges of the adjacent cells. Then, we apply the transport equation locally over a cell to yield

$$D_{xx}\frac{\partial^2 c}{\partial x^2} - v_x\frac{\partial c}{\partial x} + D_{yy}\frac{\partial^2 c}{\partial y^2} - v_y\frac{\partial c}{\partial y} =$$

$$\frac{c - C}{\Delta t} - D_{yx}\frac{\partial}{\partial y}\left(\frac{\partial c}{\partial x}\right) - D_{xy}\frac{\partial}{\partial x}\left(\frac{\partial c}{\partial y}\right)$$

$$-a_m \le x \le +a_m\ , \quad -b_m \le y \le +b_m\ , \qquad m = 1,\ \ldots,\ M \qquad (1)$$

where the time derivative has been approximated by a finite difference in which C is the concentration at the old time level and Δt is the time increment. In eqn. (1), v_i is the average linear velocity component, and D_{ij} is the hydrodynamic dispersion coefficient. We assume that D_{ij} is related to v_i according to the classical model presented by Bear[4] and others. Note that the time discretization above is fully implicit since the spatial portion of the differential operator is evaluated at the advanced time level. Also, a constant porosity is assumed in eqn. (1). Then, eqn. (1) is converted to a set of coupled ordinary differential equations through the application of a transverse averaging procedure. Specifically, we separately multiply eqn. (1) by Legendre polynomials $P_\mu(x)$ and $P_\nu(y)$, then integrate over the cell with respect to x and y, respectively. The x-integration yields

$$D_{yy} \frac{d^2}{dy^2} \bar{c}^x_\mu(y) - v_y \frac{d}{dy} \bar{c}^x_\mu(y) = \bar{S}^x_\mu(y)$$

$$-b_m \leq y \leq +b_m \quad , \; m = 1,\ldots,M \qquad (2)$$

where the transverse-integrated concentration (i.e., the concentration moment) is defined by

$$\bar{c}^x_\mu(y) = \frac{1}{p_\mu(a_m)} \int_{-a_m}^{+a_m} P_\mu(x)\; c(x,y)\; dx \qquad (3)$$

where $p_\mu(a_m)$ is a normalization factor. The y-integration yields a similar eqn. for $\bar{c}^y_\nu(x)$. In our work, we have implemented the so-called higher-order method where $\mu,\nu = 0,1$. The moment of the source-like expressions on the right-hand side of eqn. (2) includes the approximate time derivative, the cross terms, and the resulting integrated-terms evaluated at the cell boundaries of the non-integrated direction.

In general, the solution of eqn. (2) can be written in each cell as the sum of the elementary solution and a particular integral. In order to evaluate the particular integral, we expand each moment of the source-like expressions in Legendre polynomials and truncate the expansion after the linear term. Therefore, the general solution of eqn. (2) takes the form:

$$\bar{c}^x_\mu(y) = A\, e^{\gamma y} + B + \sum_{i=0}^{1} \overline{\bar{S}^x_{\mu i}}^{\,y}\, Q_i(y) \qquad \mu = 0,1 \qquad (4)$$

where $\gamma = v_y/D_{yy}$, and $\overline{\bar{S}^x_{\mu i}}^{\,y}$ are the expansion coefficients for $\bar{S}^x_\mu$. The terms $Q_i(y)$ are the particular integrals for $P_i(y)$ and can be determined by the method of variation of parameters. The arbitrary constants A and B, can be expressed in terms of the concentration moments at the cell boundaries (i.e., $\bar{c}^x_\mu(\pm b_m)$)· Therefore, eqn. (4) is an expression for the moments of the concentration, $\bar{c}^x_\mu(y)$ in terms of their cell-boundary values and the expansion coefficients of the source-like terms. The general solution for $\bar{c}^y_\nu(x)$ takes a similar form.

In order to develop the discrete-variable equations, the continuity of the mass flux across cell boundaries is utilized. Upon applying the transverse-averaging technique to the continuity equation and substituting the expressions for the concentration moments and their derivatives into this equation, we develop four coupled systems of tridiagonal equations for the unknown concentration moments at the cell boundaries, two for each direction. For example, if cell m′ is located to the "north" of cell m, the tridiagonal systems in the y direction can be expressed as:

$$YL_m \bar{c}^x_\mu(-b_m) + YRL_{m'} \bar{c}^x_\mu(-b_{m'}) + YR_m \bar{c}^x_\mu(+b_{m'}) = F_{\mu m'} \quad (5)$$

The coefficients in the above equation are defined in terms of the velocity, dispersion coefficients, and the cell geometry. An expression similar to eqn. (5) applies to the x-direction. It is important to note that the expansion coefficients of both the source-like terms and the cross-terms, which comprise $F_{\mu m'}$, provide the coupling between the zero- and first-moment of the concentration from the adjacent cells in both the x and y direction.

The expansion coefficients of the source-like terms are evaluated by imposing two physical constraints on the solution: first, the conservation of mass in the cell, and second, the uniqueness of the cell-averaged concentration. These two constraints result in a set of eight coupled linear equations for the unknowns $\overline{\bar{S}^x_{\mu\nu}}^y$ and $\overline{\bar{S}^y_{\nu\mu}}^x$, $\mu,\nu = 0,1$; this system has been solved symbolically using the software "MACSYMA." The resulting expressions for the unknowns are functions of the concentration moments at the cell boundaries, $\bar{c}^x_\mu(\pm b_m)$ and $\bar{c}^y_\nu(\pm a_m)$, and the cell-averaged concentration $\overline{\bar{C}^x_{\mu\nu}}^y$ at the old time level.

SOLUTION PROCEDURE

Upon substituting the resulting expressions for the expansion coefficients of the source-like terms into eqn. (5) and implementing the global boundary conditions, we obtain four sets of coupled tridiagonal equations for the zero- and first-moments of the concentration at the cell interfaces in both x- and y-directions. These equations are used as the basis of an iterative method to solve for $\bar{c}^y_\nu(\pm a_m)$ and $\bar{c}^x_\mu(\pm b_m)$, $\mu,\nu = 0,1$, by sweeping back and forth in the x- and y-direction until convergence.

EXAMPLE PROBLEM

The formulation and solution procedure of the CAN method were presented in the previous sections. Here, we consider a test problem involving an instantaneous contaminant point source in an infinite homogeneous domain. The results are compared with an available analytical solution and also with another numerical solution obtained by the Alternating Direction Galerkin (ADG) Technique[3]. The steady flow field has a magnitude of 0.1414 m/d and is uniform at an angle 45° to the x-axis. The longitudinal dispersivity, α_L, equals 1.0 m, and the transverse dispersivity, α_T, equals 0.25 m. For the purpose of comparison, several different mesh sizes were used. For the sake of brevity, we present the plumes resulting from both models using the coarsest mesh ($\Delta x = 2.0$m) only. Also concentration profiles along the center of the plume in the longitudinal direction are compared with the corresponding analytical ones. All results are for 100 days of simulation time and for initial mass input of 40.0 units.

Figure 1 shows that CAN model yields an elliptical plume with minimal numerical error, whereas Figure 2 shows that the ADG plume exhibits pronounced grid-orientation error. On the other hand, both models result in accurate, elliptical plumes when either a fine mesh or a horizontal flow is utilized.

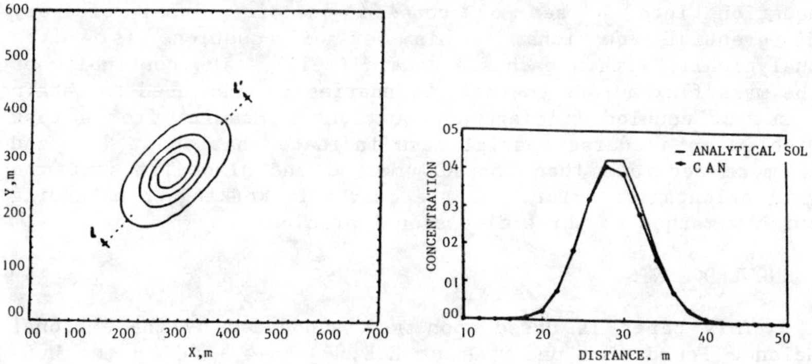

Figure 1. CAN Results at t=100 days. (a) concentration contour levels equal 0.20, 0.115, 0.21, 0.30, and 0.4. (b) profiles along L-L'.

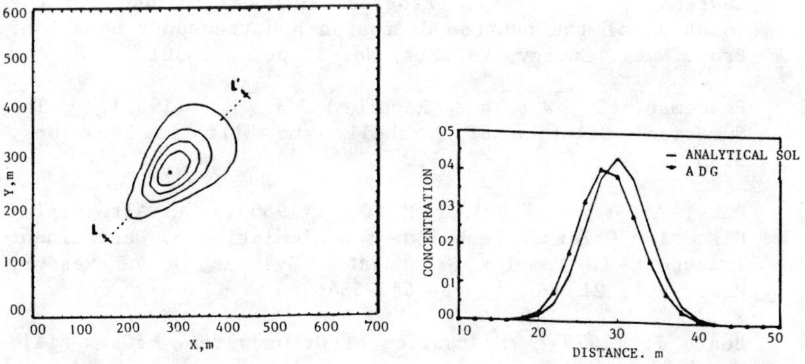

Figure 2. ADG Results at t=100 days. (a) concentration contour levels same as Figure 1. (b) profile along L-L'.

Additional tests have shown that the CAN method yields accurate results for large grid Peclet number, whereas the ADG method is known to suffer from oscillations and numerical dispersion errors under these conditions. However, due to the first order approximation of the time derivative, the grid Courant number must remain less than one in order to get an accurate CAN solution.

CONCLUSIONS

A new numerical method, called the Cell Analytic-Numerical (CAN) method, has been developed and implemented for the solution of the two-dimensional advection-dispersion equation. The method uses the so-called transverse-averaging procedure to transform the local two-dimensional partial differential equation into a set of coupled one-dimensional ordinary differential equations. This set of equations is solved analytically within each sub-domain (cell). The continuity of the mass flux across the cell boundaries is then used to obtain a set of coupled tridiagonal equations. Results from a test problem on a coarse spatial mesh indicate that the CAN method is more accurate than the ADG method and also free from the grid orientation error. We are currently working on extending our new method to three dimensional problems.

ACKNOWLEDGMENTS

This paper is based upon work supported by the National Science Foundation under Grant ECE 84-18644, and by the U.S. Geological Survey under Grant No. 14-08-001-G1299.

REFERENCES

1. Lawrence R. D. (1986). Progress in nodal methods for the solution of the neutron diffusion and transport equation, Prog. Nucl. Energy, Vol. 17, No. 3, pp. 271-301.

2. Peaceman, D. W., and Rachford, H. H. (1955). The Numerical Solution of Parabolic and Elliptic Equations, SIAM J., Vol. 3, pp. 28-41.

3. Daus, A. D., and Frind, E. O. (1985). An Alternating Direction Galerkin Technique for Simulation of Contaminant Transport in Complex Groundwater Systems, Water Resour. Res., Vol. 21, No. 5, pp. 653-664.

4. Bear, J. (1979). Hydraulics of Groundwater, McGraw-Hill, New York.

INVITED PAPER

Finite Element Techniques for Convective-Diffusive Transport in Porous Media

R.E. Ewing

Department of Mathematics, Petroleum Engineering and Chemical Engineering, University of Wyoming, Laramie, WY 82071, USA

ABSTRACT

Operator-splitting techniques are applied to convective-diffusive transport problems in porous media. The convection is treated by applying a modified method of characteristics to time-step along the characteristics of the convective-part of the flow. The non-symmetry in the spatial operator is addressed via a Petrov-Galerkin method which uses a test function to achieve stability through a balancing of the remaining convection, the diffusion, and any possible reaction terms. The use of time-stepping along characteristics allows the use of large time-steps in a stable but accurate fashion. If local phenomena are important, self-adaptive local grid refinement techniques can be coupled with the operator splitting.

INTRODUCTION

The modeling of many fluid flow problems involves very similar mathematical equations. Examples of mathematical and related physical properties of these models which must be addressed include: (a) the resolution of sharp moving fronts, like wetting fronts or contaminants, in convection-dominated convection-diffusion problems, (b) the stability and accuracy of discretizations of highly non-self-adjoint differential operators, (c) the need to have very accurate fluid velocities which largely control the flow, (d) the need to model dynamic local phenomena, such as mass transfer between phases, reactions, and diffusion, which may govern the physics, and (e) the emphasis on development of efficient numerical procedures for the enormous problems encountered.

Special techniques must be developed to treat convection-dominated parabolic problems with localized moving fronts. In a multiphase context, the location of a wetting front or a moving interface between non-soluble contaminants may be very important in understanding or controlling the fluid flow or contamination. In a miscible or multicomponent regime, the component mass transfer along the moving fluid interfaces often greatly affects the flow. Most standard numerical techniques for treating convection-dominated flows will tend to artificially diffuse or smear the sharp fluid interfaces, thus destroying the physics of the flow. Also, when advection dominates, there is a strong temptation to ignore the diffusion/dispersion mechanism which, although small in relative size, may be very important to the transport processes. In order to take advantage of the information from the dominant advection, but maintain the physical significance of even very small diffusion or dispersion, we will discuss an operator-splitting technique which addresses each phenomenon separately. In the first step of this procedure, the purely hyperbolic part of the flow is approximated by time-stepping along the associated characteristic of that part of the operator. Next, the effects of diffusion or dispersion are added to the solution to complete the approximation of the total flow. This operator-splitting technique has been shown to be very useful for both multicomponent [14–17] and multiphase flows [4,7]. Both methods will be described in the next section.

The convection-dominated flow processes, even when treated by operator-splitting techniques, often lead to non-self-adjoint operators. Special methods which take advantage of upstream-weighting concepts are routinely utilized to stabilize non-self-adjoint operators. Standard upstream weighted finite difference techniques which are frequently used in porous media simulations introduce an artificial, numerical diffusion which is of the order of the grid spacing and in the direction of the grid lines. These properties of the discretization can cause serious difficulties in the simulation if, due to large reservoir problems, large grid spacings are used. In the techniques presented here, we stabilize the non-symmetry in the operator by Petrov-Galerkin discretization techniques. In these methods, test functions, which standardly govern the stability of variational methods, are weighted in an upstream fashion in a way that reflects the relative size of the accumulation or reactions, the transport, and the diffusion/dispersion terms. These techniques approximate an optimal test function [2,5,7] via a combination of test functions for optimal use for various terms [4]. The ideas are also described in the next section.

In order to use a modified method of characteristics for the advection part of the flow, we must have accurate approximation of the characteristics. Since the characteristic direction depends heavily upon the fluid velocity, we must obtain very accurate approximations of the Darcy velocities during the flow. Mixed finite element methods have been very instrumental in obtaining accurate fluid velocities in complex flow regimes with rapidly changing reservoir and fluid properties [1,6,11,13,14].

If the physical mixing or dispersion between fluids or components is small, but important, it is often concentrated in a small internal layer which moves with the fluid flow. Usually the length scales of this layer are much smaller than the grid lengths which are used in the simulator. In this case, adaptive, local grid refinement methods are potentially useful. Efficient techniques for incorporating local grid refinement in simulators have been discussed in [3,9,10].

In order to be more specific about the methods discussed above, we consider prototype equations. The miscible displacement of one incompressible fluid by another, completely miscible with the first, in a horizontal porous reservoir $\Omega \subset \mathbb{R}^2$ over a time period $J = [T_0, T_1]$, is given by

$$-\nabla \cdot \left(\frac{k}{\mu} \nabla p \right) \equiv \nabla \cdot \mathbf{u} = q, \quad \mathbf{x} \in \Omega, \quad t \in J, \tag{1}$$

$$\phi \frac{\partial c}{\partial t} + \mathbf{u} \cdot \nabla c - \nabla \cdot \mathbf{D} \nabla c = q(\tilde{c} - c), \quad \mathbf{x} \in \Omega, \quad t \in J, \tag{2}$$

where p and $\mathbf{u}$ are the pressure and Darcy velocity of the fluid mixture, ϕ and k are the porosity and the permeability of the medium, μ is the concentration-dependent viscosity of the mixture, c is the concentration of the invading fluid, q is the external rate of flow, and $\tilde{c}$ is the inlet or outlet concentration. $\mathbf{D}$ is, in general, a diffusion-dispersion tensor which has two parts, molecular diffusion and a velocity-dependent dispersion term [8].

As is shown in [12], the equations which describe two phase, immiscible incompressible displacement in a horizontal porous media can be given by

$$-\nabla \cdot (\lambda \nabla p) \equiv \nabla \cdot \mathbf{u}_t = q_t, \quad \mathbf{x} \in \Omega, \quad t \in J, \tag{3}$$

$$\phi \frac{\partial S}{\partial t} + \nabla \cdot f(S)\mathbf{u}_t - \nabla \cdot D_1 \nabla S = q_1, \quad \mathbf{x} \in \Omega, \quad t \in J, \tag{4}$$

where p and $\mathbf{u}_t$ are the total pressure and fluid velocity, λ is the total fluid mobility, and $f(S)$ is a nonlinear, non-convex flux term which describes the fraction of the total flow attributed to the aqueous phase which has saturation S.

INTRODUCTION TO OPERATOR SPLITTING

In miscible or multicomponent flow models, the convective, hyperbolic part is a linear function of the fluid velocity. The operator-splitting technique applied to a variational method in this case leads to a symmetric bilinear form. In order to treat the time-stepping in this case, we will use a modified method of characteristics. In this method, the first and second terms in Equation (2) are combined to form a directional derivative along what would be the characteristics for the equation if the tensor $\mathbf{D}$ were zero. The resulting equation is

$$\nabla \cdot (\mathbf{D}\nabla c) - q(\tilde{c} - c) = \phi\frac{\partial c}{\partial t} + \mathbf{u} \cdot \nabla c \equiv \phi\frac{\partial c}{\partial \tau}. \tag{5}$$

The system obtained by modifying Equations (1)–(2) in this way is solved sequentially. An approximation for $\mathbf{u}$ is first obtained at time level $t = t^n$ from a solution of Equation (1) with the fluid viscosity μ evaluated via some mixing rule at time level t^{n-1}. Equation (1) can be solved as an elliptic equation for the pressure p or via a mixed finite element method for a more accurate fluid velocity. Let $\hat{C}^{\,n}(x)$ and $\mathbf{U}^n(x)$ denote the approximations of $c(x,t)$ and $\mathbf{u}(x,t)$, respectively, at time level $t = t^n$. The directional derivative is then discretized along the "characteristic" mentioned above as

$$\phi\frac{\partial c}{\partial \tau}(x, t^n) \approx \phi\frac{C^n(x) - C^{n-1}(\overline{x}^{n-1})}{\Delta t} \tag{6}$$

where $\overline{x}^{n-1}$ is defined for an x as

$$\overline{x}^{n-1} = x - \frac{U^n(x)\Delta t}{\phi}. \tag{7}$$

This technique, first described by Russell [17] for petroleum applications, is a discretization back along the "characteristic" generated by the first order derivatives from Equation (5). Although the advection-dominance in the original Equation (2) makes it non-self-adjoint, the form with directional derivatives is self-adjoint and discretization techniques for self-adjoint equations can be utilized. This modified method of characteristics can be combined with either finite difference or finite element Galerkin spatial discretizations. In applications where the contaminant forms an additional phase, the constitutive equations become more complex. For example, due to nonlinear relative permeability effects, the transport term in Equation (4) becomes a nonlinear, noncomplex function of the saturations of the separate phase. This nonlinearity complicates the formulation of the modified method of characteristics.

The operator splitting of Equation (4) gives the following set of equations:

$$\phi\frac{\partial\overline{S}}{\partial t}+\frac{d}{dS}f^{n}(\overline{S})\cdot\nabla\overline{S}\equiv\phi\frac{d}{d\tau}\overline{S}=0, \tag{8}$$

$$\phi\frac{\partial S}{\partial\tau}+\nabla\cdot(\mathbf{b}^{m}(S)S)-\epsilon\nabla\cdot(D(S)\nabla S)=\mathbf{q}(\mathbf{x},t), \tag{9}$$

$t_m \le t \le t_{m+1}$, together with proper initial and boundary conditions. The saturation S is coupled to the pressure/velocity equations, which will be solved by mixed finite element methods.

The splitting of the fractional flow functions into two parts: $\mathbf{f}^m(S)+\mathbf{b}(S)S$, is constructed [4,7] such that $\mathbf{f}^m(S)$ is linear in the shock region, $0 \le S \le S_1 < 1$ and $\mathbf{b}(S) \equiv 0$ for $S_1 \le S \le 1$. Further, Equation (1) produces the same unique physical solution as

$$\frac{\partial S}{\partial t}+\nabla\cdot(\mathbf{f}^{m}(S)+\mathbf{b}(S)S)=0 \tag{10}$$

with an entropy condition imposed. This means that, for a fully developed shock, the characteristic solution of Equation (8) always will produce a unique solution and, as in the miscible case, we may use long timesteps Δt without loss of accuracy.

The solution of Equation (9) via variational methods leads to the following Petrov-Galerkin equation:

$$\begin{aligned} B(S_h^m,\phi_i) \equiv\ & (S_h^{m+1},\phi_i)-\left(\frac{\Delta t}{\phi}\mathbf{b}(\mathbf{x},t^m)S_h^{m+1},\nabla\phi_i\right) \\ & +\left(\frac{\epsilon\Delta t}{\phi}D(\mathbf{x},t^m)\nabla S_h^{m+1},\nabla\phi_i\right) \\ =\ & (g_h^m(\mathbf{x},t^m),\phi_i), \quad i=1,2,\cdots,N, \\ & S_h^m\in M_h\subset H^1(\Omega), \quad \phi_i\in N_h\subset H^1(\Omega), \end{aligned} \tag{11}$$

where M_h and N_h are trial and test spaces spanned by $\{\theta_i\}$ and $\{\phi_i\}, i = 1,2,\cdots,N$, respectively. $B(\cdot,\cdot)$ given by Equation (11) is an unsymmetrical bilinear form with spatially-dependent coefficients.

In order to obtain Equation (11), we have used the characteristic solution from Equation (8) to approximate $\frac{\partial}{\partial\tau}S$ and the nonlinear coefficients in Equation (9). The nonsymmetry in the bilinear form $B(\cdot,\cdot)$ is caused by the nonlinearity of the convective part of the equation, represented by the term $\mathbf{b}(S)S$. This nonlinearity balances the diffusion forces in the shock region after a traveling front has been established.

We want to use numerical techniques which work well for the symmetric, coercive, bilinear forms to solve Equation (11). We consider a procedure, developed by Barrett and Morton [2], which symmetrizes the bilinear form $B(\cdot,\cdot)$ by defining a new set of test functions as follows:

$$B(S^m, \phi_i) = \left(a_{kl} \frac{\partial}{\partial x_k} S^m, \frac{\partial}{\partial x_l} \theta_i \right) \equiv B^*(S^m, \theta_i), \tag{12}$$

$$0 < a_{kl} < K.$$

The test functions ϕ_i defined by Equation (12) have nonlocal support and would thus cause serious computational problems for large-scale problems. However, a localization procedure was developed by Demkowitz and Oden [5] which allows efficient computational procedures. Since the bilinear form $B(\cdot,\cdot)$ is coercive, we obtain optimal approximation properties in the norms defined by the form. For computational reasons, it may be better to use an approximate form of the optimal test function ϕ_i [4]. An estimate for the error introduced by an approximate symmetrization of $B(\cdot,\cdot)$ is given by Barrett and Morton [2].

It seems natural to relate the size of the coarse domains to the solution of the pressure-velocity Equation [7], since the velocity varies slowly and defines a natural long space scale compared to the variation of the saturation S at a front. A local error estimate which determines if a coarse grid block must be refined, is given in reference [7]. Normally local refinement must be performed if a fluid interface is located within the coarse grid block in order to resolve the solution there. A slightly different strategy is to make the region of local refinement big enough such that we can use the same refinements for several of the large timesteps allowed by the method.

The difficult problem with these techniques is the communication of the solution between the fine and coarse grids. Because the pure convective part in Equation (8) gives a fairly accurate solution, this problem may be solved more easily in the dynamic case than in a pure static model.

We can use the characteristic solution at the coarse grid vertices [4,7]. This also means that we get information about the localization of a part in the first step in the procedure. The local description on the refined coarse grid blocks is obtained via patch approximation techniques described in the next section.

We obtain the following algorithm to build the preconditioner [7]:

1. Solve the characteristic Equation (8) at the coarse grid vertices.

2. Determine which of the coarse grid blocks, Ω_K, requires local refinement.

3. Solve local problems on each of the refined coarse blocks Ω_K.

4. Solve local zero-Dirichlet problems on each of the sides of the refined patches Ω_K.

5. Extend this solution on the boundary of Ω_K to the interior.

The solution at each of the coarse grid vertices and the calculations 4–5, may be sent to separate processors to achieve a high level of parallelism in the solution procedure.

REFERENCES

1. Allen, M.B., Ewing, R.E., and Koebbe, J.V. (1985), Mixed Finite-Element Methods for Computing Groundwater Velocities, *Num. Meth. for PDE's, 3*, pp. 195–207.

2. Barrett, J.W. and Morton, K.W. (1984), Approximate symmetrization and Petrov-Galerkin Methods for Diffusion-Convection Problems, *Comp. Meth. in Appl. Mech. and Eng., 45*, pp. 97–122.

3. Bramble, J.H., Pasciak, J.E., Schatz, A.H., and Ewing, R.E. (in press), A Preconditioning Technique for the Efficient Solution of Problems with Local Grid Refinement, *Comp. Meth. in Appl. Mech. and Eng.*

4. Dahle, H., Espedal, M.S., and Ewing, R.E. (1988), Characteristic Petrov-Galerkin Subdomain Methods for Convection Diffusion Problems, *IMA Volume 11, Numerical Simulation in Oil Recovery*, (Ed. Wheeler, M.F.), Springer-Verlag, Berlin.

5. Demkowitz, L. and Oden, J.T. (1986), An Adaptive Characteristic Petrov-Galerkin Finite Element Method for Convection-Dominated Linear and Nonlinear Parabolic Problems in Two Space Variables, *Comp. Meth. in Appl. Mech. and Eng., 55*, pp. 63–87.

6. Douglas, J., Jr., Ewing, R.E., and Wheeler, M.F. (1983), A Time-Discretization Procedure for a Mixed Finite Element Approximation of Miscible Displacement in Porous Media, *RAIRO Anal. Numér., 17*, pp. 249–265.

7. Espedal, M. and Ewing, R.E. (1987), Characteristic Petrov-Galerkin Subdomain Methods for Two-Phase Immiscible Flow, *Comp. Meth. Appl. Mech. and Eng., 64*, pp. 113–135.

8. Ewing, R.E. (1983), Problems Arising in the Modeling of Processes for Hydrocarbon Recovery, *Research Frontiers in Applied Mathematics, Vol 1*, (Ed. Ewing, R.E.), SIAM, Philadelphia, pp. 3–34.

9. Ewing, R.E. (1986), Adaptive Mesh Refinements in Petroleum Reservoir Simulation, *Accuracy Estimates and Adaptivity for Finite Elements*, (Eds. Babuska, I., Zienkiewicz, O.C., and Arantes e Oliveira, E.), Wiley, New York, pp. 299–314.

10. Ewing, R.E. (1986), Efficient Adaptive Procedures for Fluid Flow Applications, *Comp. Meth. Appl. Mech. Eng., 55*, pp. 89–103.

11. Ewing, R.E. and Heinemann, R.F. (1984), Mixed Finite Element Approximation of Phase Velocities in Compositional Reservoir Simulation, *Comp. Meth. Appl. Mech. Eng., 47*, R.E. Ewing, ed., pp. 161–176.

12. Ewing, R.E., Heinemann, R.F, Koebbe, J.V., and Prasad, U.S. (1987), Velocity Weighting Techniques for Fluid Displacement Problems, *Comp. Meth. Appl. Mech. Eng., 64*, pp. 137–151.

13. Ewing, R.E., Koebbe, J.V., Gonzalez, R., and Wheeler, M.F. (1985), Mixed Finite Element Methods for Accurate Fluid Velocities, *Finite Elements in Fluids, Vol. VI*, John Wiley and Sons, Ltd., pp. 233–249.

14. Ewing, R.E., Russell, T.F., and Wheeler, M.F. (1984), Convergence Analysis of an Approximation of Miscible Displacement in Porous Media by Mixed Finite Elements and a Modified Method of Characteristics, *Comp. Meth. Appl. Mech. Eng., 47*, (Ed. Ewing, R.E.), pp. 73–92.

15. Ewing, R.E., Yuan, Y., and Li, G. (1987), Finite Element Methods for Contamination by Nuclear Waste-Disposal in Porous Media, *Proceedings of Dundee Numerical Analysis Conference*, Dundee, Scotland, June 23–26, 1987.

16. Ewing, R.E., Yuan, Y., and Li, G. (to appear), Time Stepping Along Characteristics for a Mixed Finite Element Approximation for Compressible Flow of Contamination by Nuclear Waste in Porous Media, *SIAM J. Numer. Anal.*

17. Russell, T.F. (1985), The Timestepping Along Characteristics with Incomplete Iteration for Galerkin Approximation of Miscible Displacement in Porous Media, *SIAM J. Numer. Anal., 22*, pp. 970–1013.

INVITED PAPER

3-D Finite Element Transport Models by Upwind Preconditioned Conjugate Gradients

G. Pini and G. Gambolati
Istituto di Matematica Applicata, Università degli Studi di Padova, Italy
G. Galeati
ENEL - CRIS, Venezia-Mestre, Italy

INTRODUCTION

Numerical solutions to the dispersion-convection equation in subsurface systems have relied extensively on the Galerkin approach mainly because of its high versatility in handling irregular geometries. However in convection-dominated problems this approach leads to:

1) unstable solution with oscillating behavior
2) artificial dispersion which yields the smearing of the simulated contaminant plume.

The oscillations can be dampened out by weighting unsymmetrically the trial functions (i.e. with upwind or upstream test functions, see Heinrich et al.[13] and Huyakorn[16]) while the numerical dispersion can be controlled by an appropriate selection of the grid Peclet and Courant numbers (Daus and Frind[5]). This usually requires a refined mesh with a large number of nodes and some difficulty is easily met for a cost-effective solution in 3-D finite element simulations. The purpose of this note is to extend to 3-D problems a class of iterative solvers, based on the preconditioned conjugate gradients, which proved quite robust and efficient in 2-D dispersion-convection models (Galeati et al.[7]). In particular we will implement and analyse the performance of ORTHOMIN(K), the Generalized Conjugate Residual GCR(K) and the Minimum

Residual (MR) whose convergence is conveniently accelerated by appropriate preconditioning matrices. New unsymmetric test functions especially designed for tetrahedral elements are developed for 3-D meshes generated automatically starting from a 2-D triangular mesh. It will be shown that the convergence of the upwind preconditioned solvers is quite good and the solution is obtained also when a failure to converge occurs with the standard Galerkin approach.

UPWIND 3-D SCHEMES FOR THE DIFFUSION-CONVECTION EQUATION

Transport and dispersion of a conservative pollutant in a 3-D steady groundwater flow are governed by the equation (Huyakorn et al.[17]):

$$\frac{\partial}{\partial x_i}\left(D_{ij}\frac{\partial c}{\partial x_i}\right) - v_i \frac{\partial c}{\partial x_i} = \phi \frac{\partial c}{\partial t} + Q\,(c - c_o) \qquad (1)$$

where c is the solute concentration, v_i is the Darcy velocity, ϕ is the porosity and Q represents distributed sources (sinks) of flow with concentration c_o. In eq. (1) the indicial notation i,j = 1,2,3 stands for summation. The components of the dispersion tensor D_{ij} are defined as (Bear[3]):

$$D_{ij} = \phi \tilde{D}_{ij} = \alpha_T |v| \delta_{ij} + (\alpha_L - \alpha_T)\frac{v_i v_j}{|v|} + D^* \delta_{ij} \qquad (2)$$

where $|v| = (v_1 + v_2 + v_3)^{1/2}$, α_L and α_T are the longitudinal and transverse dispersivities, D^* is the coefficient of molecular diffusion and δ_{ij} is the Kronecker delta.
If we denote by $N_j(x_i)$ and $W_j(x_i)$ a complete set of linearly independent basis and test functions, respectively, and specify the appropriate boundary conditions (Gambolati et al.[9]), the finite element integration of (1) leads to an unsymmetric linear differential system as given in Huyakorn et al.[17]. The elements are tetrahedrons automatically generated from a set of triangles upon projection over several horizontal planes specified "a priori" and representing the vertical layering of the subsurface system (Gambolati et al.[10]).
The unsymmetric test functions $W_j(x_i)$ are the natural extension to tethraedral elements of the functions defined in Gureghian[12] and Huyakorn[16] for triangular elements. With reference to Figure 1, the upwind weights are given by:

$$\begin{Bmatrix} W_I \\ W_J \\ W_K \\ W_M \end{Bmatrix} = \begin{Bmatrix} N_I \\ N_J \\ N_K \\ N_M \end{Bmatrix} + 3 \begin{bmatrix} N_I & 0 & 0 & 0 \\ 0 & N_J & 0 & 0 \\ 0 & 0 & N_K & 0 \\ 0 & 0 & 0 & N_M \end{bmatrix} \begin{bmatrix} 0 & -\beta_{IJ} & \beta_{KI} & -\beta_{IM} \\ \beta_{IJ} & 0 & -\beta_{JK} & \beta_{MJ} \\ -\beta_{KI} & \beta_{JK} & 0 & -\beta_{KM} \\ \beta_{IM} & -\beta_{MJ} & \beta_{KM} & 0 \end{bmatrix} \begin{Bmatrix} N_I \\ N_J \\ N_K \\ N_M \end{Bmatrix}$$

The β coefficients are defined according to the formulation provided in Gureghian[12] and are related to the Courant number C_r assessed over each tetrahedron side (on condition that the Peclet number $P_e > 2$, otherwise they are equal to zero):

$$\beta = 0 \qquad C_r < 0.2$$

$$\beta = 0.0545 \exp 2.797\,(1 - \exp(-1.056\, C_r)) \qquad 0.2 \leqslant C_r \leqslant 1.6$$

$$\beta = 1 \qquad C_r > 1.6$$

With regard to the upwinding technique, it must be noted that this approach is effective in reducing the oscillations of the solution but it tends to increase the numerical dispersion and therefore it has to be used with some caution.

ITERATIVE SOLVERS

After integrating in time with the Crank-Nicholson scheme, the f.e. system arising from eq.(1) can be written in compact form as:

$$Ax = b \tag{3}$$

where A is a sparse unsymmetric matrix with the symmetric part being positive definite, b is the known vector and x is the wanted solution. Several iterative methods based on the generalized conjugate gradients have recently been proposed for the solution of unsymmetric systems (see Axelsson[1] and Gambolati[8]). These methods minimize the residual norm in a proper Krylov subspace and would require keeping in the core memory all the previous search directions. For the sake of simplicity however, only the latest vectors are usually saved. With ORTHOMIN(k) the most recent k directions are preserved and with GCR(k) a restart is performed every k iterations. MR is the particular case of ORTHOMIN and GCR with k = 0 (see Eisenstat[6], Gresho[11]). Since preconditioning is essential to obtain practical convergence, in the sequel only the preconditioned schemes will be discussed. Similarly to the

symmetric case (Gambolati[10]), the preconditioner will be provided by the incomplete Crout decomposition according to the strategy proposed in Kershaw[18].
Setting $B = L^{-1} A U^{-1}$, $y = Ux$, $c = L^{-1} b$ system (3) can be rewritten as:

$$By = c \tag{4}$$

ORTHOMIN(k) for the solution of (4) reads:

$$y_{i+1} = y_i + \alpha_i p_i \qquad r_{i+1} = r_i - \alpha_i Bp_i$$

$$p_{i+1} = r_{i+1} + \sum_{j=i-k+1}^{i} \beta_j p_j \qquad \alpha_i = r_i^T Bp_i/(Bp_i)^T(Bp_i)$$

$$\beta_j^{(i)} = -(Br_{i+1})^T(Bp_j)/(Bp_j)^T(Bp_j)$$

$$p_o = r_o = c - By_o \qquad y_o = L^{-1}b$$

GCR(k) uses the same equations as ORTHOMIN(k) except for a complete restart every k+1 iterations. MR is readily obtained from the above scheme upon replacing p_i with r_i and dropping the current relationships for both p_{i+1} and $\beta_j^{(i)}$. In literature values proposed for k do not exceed 10. In the present paper we have implemented and compared ORTHOMIN(1), GCR(5) and MR.

NUMERICAL RESULTS

Three test problems have been selected to analyse the performance of the CG solvers incorporated with 3-D diffusion-convection models. In the following examples consistent arbitrary units are adopted.
Problem A (Figure 2) is concerned with an aquifer initially clean and progressively polluted at the upper left corner. Two layers of tetrahedral elements are considered for a total of 1368 nodes. Problem B (Figure 3) simulates the radial propagation of a pollutant injected through a well fully penetrating an aquifer initially at zero concentration. The formation consists of 2 layers and the overall number of nodes is 2340. Problem C (Figure 4) analyses a two well system injecting contaminant and pumping out contaminated water from a 5-layer aquifer. The wells are partially penetrating and the concentration is set to zero on the boundary of the model. The nodes are 3168.
The models have been first run assuming zero velocity along the vertical axis so as to compare the 3-D solutions with the 2-D analytical response which can be found in Cleary[4], Hoopes and

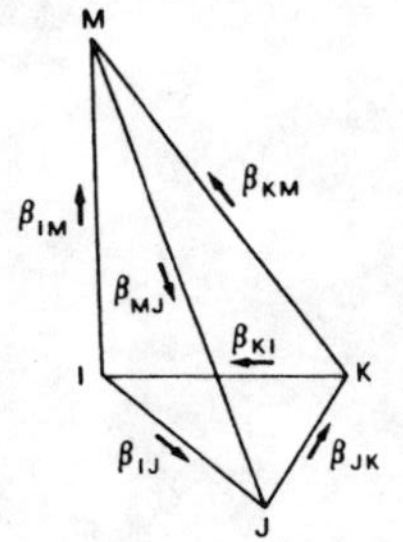

Fig. 1 - Upwinding weights. The arrows indicate the flow direction along the tetrahedron sides

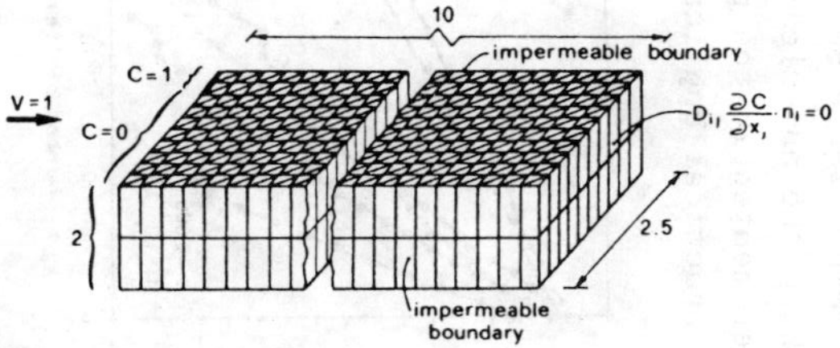

Fig. 2 - Problem A: Detail of the 3-D finite element mesh

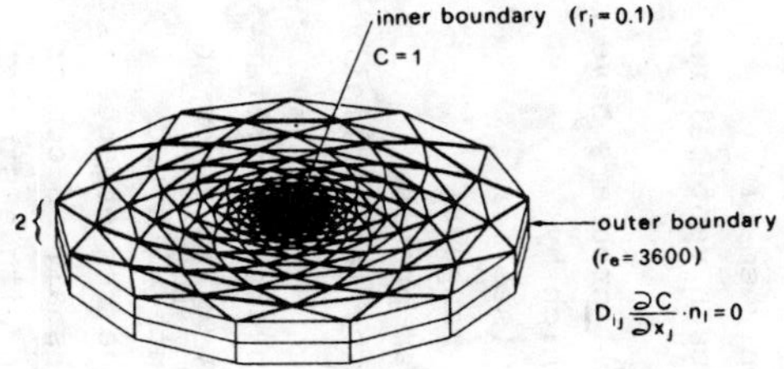

Fig. 3 - Problem B: 3-D finite element mesh

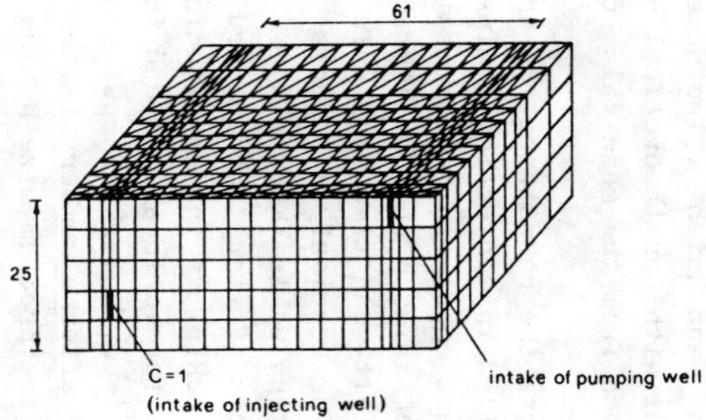

Fig. 4 - Problem C: Detail of the 3-D finite element mesh

Harleman[14] and Hsieh[15]. The numerical results are in excellent agreement with the theoretical solution with only small deviations from the mean value computed along each vertical row of nodes.

To explore the robustness of the CG schemes all problems have been solved in convection-dominated regimes, which by experience appear to be the most critical ones (see for instance Axelsson and Gustafsson[2]), for a wide range of dispersivity and time step values. In particular problem C allows for vertical components of flow induced by the partially penetrating wells (Figure 4).

The number of iterations needed to meet the convergence criteria with GCR(5) vs the longitudinal dispersivity α_L is given in Figure 5 for various Δt's and for finite elements with and without (e.g. Galerkin) upwinding. Note that the convergence is achieved after a number of iterations which is small as compared with the problem size N also in the examples with a strong convective component, i.e. for Peclet and Courant numbers far beyond the theoretical stability bounds. The upwinding improves the convergence of the CG solvers (Figure 6) and allows for the solution even when the standard Galerkin approach would fail to converge (Figure 5). These results are in keeping with those obtained from 2-D models (Galeati et al.[7]).

ORTHOMIN(1) and MR behave substantially in a similar way, although they appear to be slightly less robust in convection-dominated simulations and particularly so if upwinding is not implemented. Since large Peclet and Courant numbers are to be avoided to keep the numerical dispersion under control, the 3 schemes experimented in the present paper are all equally reliable in practical applications.

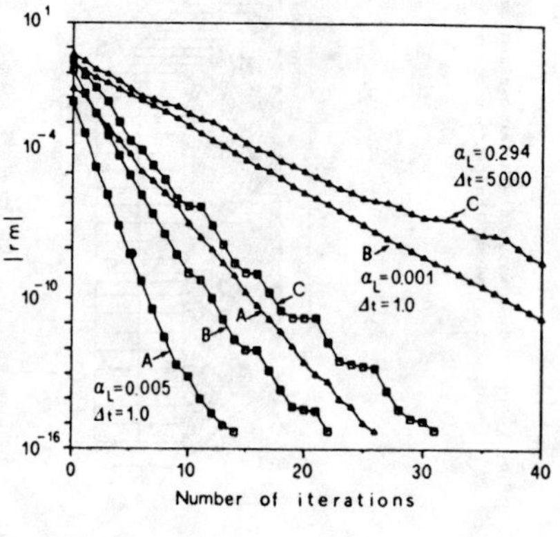

Fig. 6 - Acceleration of convergence with upwinding:
▲ without upwinding
▫ with upwinding
|rm| is the average residual defined as $[\Sigma r_i^2/N]^{1/2}$

It is worth noting that MR is the least expensive one in terms of CPU times with a saving with respect to ORTHOMIN(1) up to 30% (Figure 7). Therefore it is suggested to use the MR solver in general and to switch to GCR(k) in the exceptional case a convergence problem should arise (this may occur for very large Peclet/Courant numbers).

CONCLUSION

The CG schemes ORTHOMIN(k), GCR(k) and MR properly preconditioned appear to be robust, reliable and efficient solvers in the finite element integration of the diffusion-convection equation over 3-D subsurface systems. They allow for the easy treatment of large number of nodes, i.e. for an effective limitation of the numerical dispersion through the control of the magnitude of the Peclet and Courant numbers. Upwinding helps improve considerably the performance of the solvers. GCR(k) turns out to be the most robust one while MR is the most economical in the vast majority of applications. If the Peclet and Courant numbers are not too far from the stability limits, MR is to be preferred. Upwinding allows for convergence also in critical strong convection-dominated cases when the Galerkin approach fails to converge. However it tends to increase the artificial dispersion and hence it must be managed with some care.

REFERENCES

1. Axelsson O. (1985), Conjugate gradient type methods for unsymmetric and inconsistent systems of linear equations, Lin. Alg. Applications, 29, 1-16.
2. Axelsson O. and Gustafsson I. (1979), A modified upwind scheme for convective transport equations and the use of a conjugate gradient method for the solution of non-symmetric systems of equations, J. Inst. Maths. Applics., 23, 321-337.
3. Bear J. (1979). Hydraulic of Groundwater, Mc GrawHill, N.Y.
4. Cleary R.W. (1978), Groundwater pollution and hydrology: mathematical models and computer programs, Rep. 78-WR-15, Mat. Res. Program, Princeton Univ., Princeton, N.J.
5. Daus A.D. and Frind E.O. (1985), An alternating direction Galerkin technique for simulation of contaminant transport in complex groundwater systems, Water Resour. Res., 21, 653-664.
6. Eisenstat S.C. Elman H.C. and Schultz M.H. (1983),

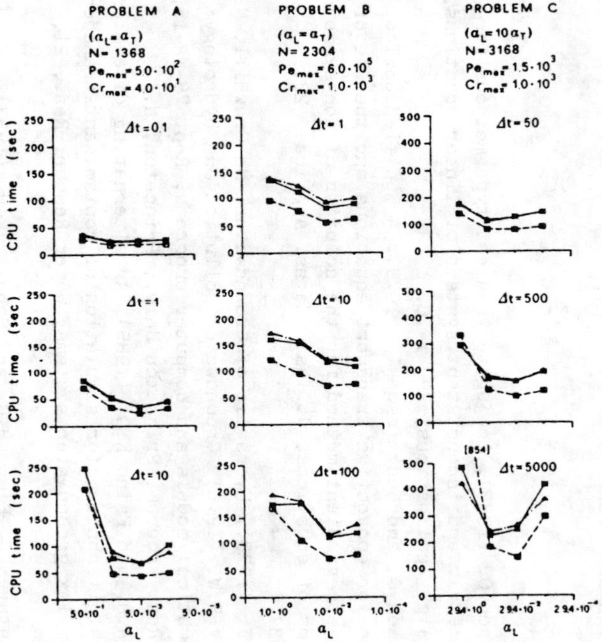

Fig. 7 - CPU time needed to solve the f. e. system at the initial time step with upwinding:
* ORTHOMIN (1) ▫ MR ▲ GCR(5)

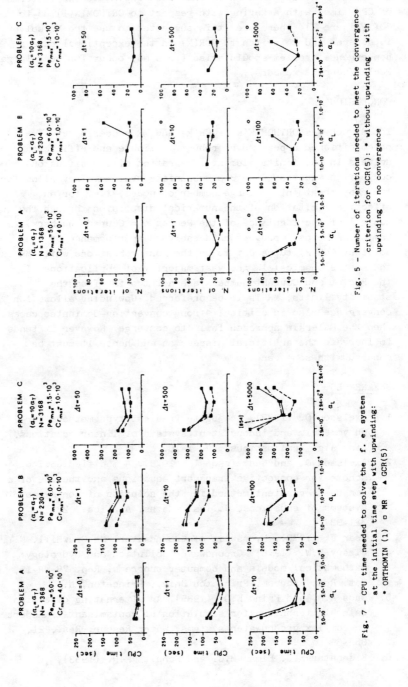

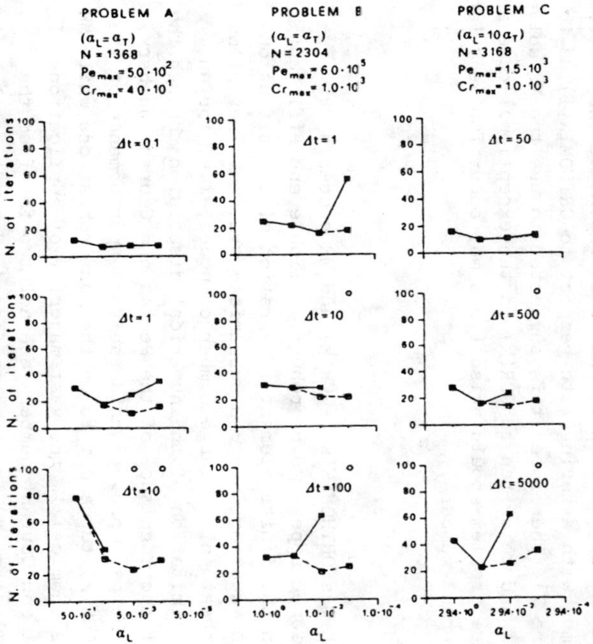

Fig. 5 - Number of iterations needed to meet the convergence criterion for GCR(5): * without upwinding ▫ with upwinding ○ no convergence

Variational iterative methods for nonsymmetric systems of linear equations, SIAM J. Numer. Anal., 20, 345-357.

7. Galeati G. Pini G. and Gambolati G. (1987), Upwind preconditioned conjugate gradients for finite element transport models, International Conference on Groundwater Contamination: Use of Models in Decision Making, Amsterdam, October 1987.
8. Gambolati G. (1979), Solution to unsymmetric finite element diffusive-convective equations by a modified conjugate gradient method, Adv. Water Resour., 2, 123-130.
9. Gambolati G. and Galeati G. (1987), On the finite element integration of the dispersion convection equation, International Conference on Groundwater contamination, Use of Models in Decision Making, Amsterdam, October 1987.
10. Gambolati G. Pini G. and Tucciarelli T. (1986), A 3-D finite element conjugate gradient model of subsurface flow with automatic mesh generation, Adv. Water Resour., 9, 34-41.
11. Gresho P.M. (1986), Time integration and conjugate gradient methods for the incompressible Navier-Stokes equations, VI Int. Conf. Finite Elements Water Resources, Lisboa, 3-27.
12. Gureghian A.B. (1983), TRIPM, a two dimensional finite element model for the simultaneous transport of water and reacting solutes through saturated and unsaturated porous media, ONWI-465, Office of Nuclear Waste Isolation, Columbus, OH.
13. Heinrich J.C. Huyakorn P.S. Zienkiewicz O.C. and Mitchell A.R. (1977), An upwind finite element scheme for two dimensional convective transport equation, Int. J. Num. Methods Engng., vol. 11, 131-143.
14. Hoopes J.A. and Harleman D.R.F. (1967), Waste water recharge and dispersion in porous media, J. Hydraulics Div. Am. Soc. Civ. Eng., 93, 51-71.
15. Hsieh P.A. (1986), A new formula for the analytic solution of the radial dispersion problem, Water Resour. Res., vol. 22, 1597-1605.
16. Huyakorn P.S. (1977), Solution of steady state convective transport equation using an upwind finite element scheme, App. Math. Modelling, 187-195.
17. Huyakorn P.S. Mercer J.W. and Ward D.S. (1985), Finite element matrix and mass balance computational schemes for transport in variably saturated porous media, Water Resour. Res., 21, 346-358.
18. Kershaw D.S. (1978), The incomplete Cholesky-Conjugate Gradient method for the iterative solution of systems of linear equations, J. Comp. Phys., 26, 43-65.

This work has been financially supported by CNR.

The Structure of Mass-Response Functions of Dissolved Species in Hydrologic Transport Volumes

A. Rinaldo and A. Bellin

Department of Engineering, University of Trento, Mesiano di Povo, 38050 Trento, Italy

A. Marani

Department of Environmental Sciences, University of Venice, Calle Larga S. Marta, Venice, Italy

ABSTRACT

Field solute transport of reactive solute species is investigated through a class of stochastic models termed as mass-response functions, which constitute a new and interesting way of solving large scale transport problems. Upon brief description of the major theoretical constraints and strengths implied by the computational schemes, a complete modelling example is given with reference to basinwide circulation of solutes. This is justified by the possibility of comparing the theoretical results with data collected in an experimental watershed in Japan yielding an unique reference for this approach for the accuracy of the experimental procedures and the refinement of the sampling procedures.

INTRODUCTION

In the wake of recent rationalizations (Sposito et al. [5]) it became apparent that transfer function models (TFM) of solute migration through soil at different scales may represent a new and interesting way of expressing results of transport experiments or calculations.

The TFM of field solute transport through unsaturated soils proposed by Jury et al. ([2]), and mass-response functions for basin-scale or aquifer-scale analysis developed for basin contaminant responses to rainfall pulses (Rinaldo and Marani [3], Rinaldo and Gambolati [4]) are used in this paper as predictive tools for solute migration in various phases of the hydrological cycle. The class of models investigated is related to mass balances interpreted in the context of probability theory, and is consistent with any mechanistic transport model with convection, dispersion and sorption (Sposito et al. [5]). Under a general assumption of ergodicity (Dagan [1]) of the processes involved (whose pathologies are still a subject of study), the approach makes use of probability density functions (pdf's) associated with the random holding time of a solute particle within the transport volume and conditional on the time of occurrence of particle injection.

The model (Rinaldo and Marani [3], Rinaldo and Gambolati [4]) of basin-scale solute yield in the hydrologic response or in a groundwater transport volume is:

$$Q(t) = \int_0^t f(t-t' \mid t')\, i(t')\, dt' \qquad (1)$$

$$Q_{out}(t) = \int_0^t G(t,t')\, i(t')\, dt' \qquad (2)$$

where: $Q_{out}(t)$, (M/T), is the outflowing solute load at the basin control section which represents all models of loss from the soil transport volume through the control boundaries or through physical, chemical or biological transformations; G(t,t'), (M/TL^3), is the (instantaneous unit) mass-response function, or the pdf of solute lifetime in the transport volume, conditional on the occurrence of the storm event i(t'); i(t'), (L^3/t), is the rainfall intensity time distribution, i.e. the solution flow rate through the boundary of the transport volume; f(t | t') is the pdf of residence time in the transport volume. Solutes enter or leave the basin mass balance in (2) because of sorption phenomena between fixed and mobile phases idealized in a Lagrangian framework in which water particles retain their individuality (quasi-particle behavior).

An important feature of the proposed approaches is related to consistency of statistical schemes with the basic mass conservation equation for a species in solution. Sposito et al. ([5]) have already addressed this issue for the general 1-D case and have concluded that the TFM is consistent with "any mechanistic model of solute movement". Nevertheless, Sposito's 1-D deterministic model, termed two-component convection-dispersion equation (CDE), is restricted to one dimensional solute movement, steady water flow and linear sorption processes. In some earlier work (Rinaldo and Marani [3]) a method has been explored to build such travel time pdf's for cases of engineering relevance.

This paper deals with the computational problems posed by the conditionality of the MRF's. The computation is explored of the instantaneous fraction of mass sorbed by the solid matrix of the transport volume, which governs the instantaneous equilibrium concentration with the mobile phase. The procedure leads to a first-order integral equation which is solved via an $O(\Delta t^2)$ scheme. A set of examples is presented in the end, based on comparisons with the results of field experiments for surface water quality.

THE COMPUTATIONAL STRUCTURE OF MRF'S

The reader is referred to the work of Rinaldo and Marani ([3]) for the description of the foundations of the theory on which MRF's are drawn. It is inferred that the link between the pdf of residence time in a physical state (hereinafter designated for sake of simplicity by f(t) although its form might be more complex if distinct states are present in the overall transport volume) and the MRF G(t,t') in (2) of exported mass is:

$$G(t,t') = f(t-t')\, C'(t-t',t') \tag{3}$$

where C'(t,t') is thought of as the Lagrangian concentration of quasi-particles (Rinaldo and Marani [3] p.2109) yielded by the initial value problem:

$$\partial C'(t,t')/\partial t = h\,[\,C_E(t+t') - C'(t,t')\,] \tag{4}$$

where: h , (T^{-1}), is the overall mass-transfer coefficient qualifying the speed of (e.g.) sorption, dissolution or other reaction; t is the contact time between phases in the transport volume; t' is the time of occurrence of the input flow solution or rainfall; $C_E(t)$ is the equilibrium concentration in the mobile phase, which is viewed (Rinaldo and Gambolati [4]) as proportional to the instantaneous fraction $M_s(t)$ of mass sorbed onto immobile regions. This hypothesis is therefore related to the rapid-slow sorption model implied by physical nonequilibrium models of two-component transport in natural media. Hence:

$$C_E(t) = C_E(0) + M_s(t) / K_D \tag{5}$$

where K_D is a distribution coefficient of the sorbed mass. This position postulates an integral equation behavior for (4) in that the fraction of mass is dependent upon the mass actually stored in the mobile region and upon that exportedthrough the exit boundary. This is indeed consistent with the bivariate mathematical form of the transfer function G(t,t'), as already noticed by

Sposito et al. ([5]). The remaining task consists only of the tedious rewriting of (5) in the light of the probabilistic mass balance which yielded both eqs. (1) and (2): nevertheless it seems worth recalling the innovative strength of (2), (3) and (5) for the study of transport phenomena in which the ergodic postulates underlying (2) are valid.
Mass balancing yields:

$$M_s(t) = \text{Inflow mass - exited mass - mass in storage within the mobile phase} =$$

$$= \int_0^{t'+t} i(x)\, C'(0,x)\, dx - \int_0^{t'+t} dx\, i(x) \int_x^{t'+t} dy\, f(y-x)\, C'(y-x,x) - \int_0^{t'+t} dx\, i(x)\, P(T>t-x)\, C'(t-x,x) \quad (6)$$

where: $P(T>x) = \int_0^x f(t)\, dt$; $i(t')$ is thought of (without loss of generality) as solution flow rate or distributed rainfall at the inflow surface bounding the transport volume. An $O(\Delta t^2)$ solution to (4) is given by (let $C'(i,j) = C'(i\Delta t, j\Delta t)$ define a concentration matrix dependent upon the two time scales of contact between phases and of injection of carrier particles; Δt is the partition of time for the discrete solution):

$$C'(i,j) = C'(i-1,j)\exp(-h\Delta t) + [1-\exp(-h\Delta t)][C_E(i+j)+C_E(i+j-1)] / 2 + O(\Delta t^2) \qquad (7)$$

It is interesting to note that the computational problem cast by (5) and (6) relies on the determination of the concentration matrix which describes the history of the concentration of solute particles conditional on the time of injection. Since the current time is the sum of the injection time t' and the contact time t, the computation proceeds diagonally (i.e. i+j = constant for the calculations). It is then understood that any quadrature formula for solution to (6) and (4) leads to the final tools for the applications: and therefore this will not be discussed in the sequel. However, for the sake of completeness, an example will be described in detail with reference to a case study.

SOLUTE NO_3-N IN RIVER WATERS OF THE AI RIVER (JAPAN)

The examples of MRF application described herein are related to the simulation of river quality data in comparison with extensive experimental evidence gathered in a small Japanese catchment (Takeuchi et al. [6]) for basinwide NO_3-N circulation in the hydrological cycle. The study, which presented no modeling effort, focused on the experimental determination of the concentration of solute NO_3-N in all components of the hydrological cycle (rain, surface, subsurface waters). Its findings clearly indicated that the high content of NO_3- load to river water during floods is not brought in by rain or surface discharge, but by the discharge in contact with the humus layer which turned out to be the sole source of NO_3 supply. The hydrologic response is constructed by a two-component geomorphological model (Figure 1), in which a distinction is drawn between surface and subsurface runoff. The relative weight of both mechanisms is established by separating the net inflow rates i(t) into two components namely $i_s(t)$ and $i_p(t)$.
Let: $fO_1(t)$, $fC_1(t)$,..., be the pdf's of residence in the basin states (a superscript s or p eventually denotes surface or subsurface detention); p(1), p(2) the proportion of area in the overland states O_1 and O_2; * denote the convolution operator; water discharge at the control section is given by the following model:

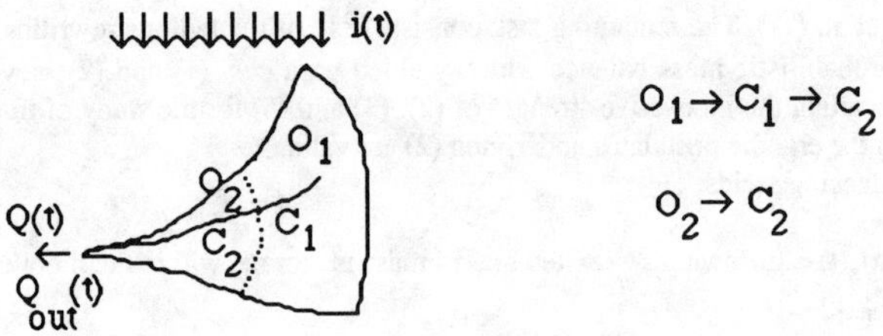

Figure 1. Geomorphological model of the Ai river basin . O are overland states, C are channel states.

$$Q(t) = \int_0^t i_s(x)\ u_s(t-x)\ dx + \int_0^t i_p(x)\ u_p(t-x)\ dx \tag{8}$$

$$u_i(t) = p(1)\ fO_1{}^i * fC_1 * fC_2\ (t) + p(2)\ fO_2{}^i * fC_2\ (t), \quad i{=}s,p$$

The calculation of the lost fraction of rainfall ϕ (t) is obtained by: $\phi(t) = 1 - \exp(-K\ V(t))$, V(t) = instantaneous fraction of water stored in the transport volume computed through the usual probabilistic mass balance (recall that the stored volume S(t) in a state due to a pulse i(0) is i(0) P(T>t) by our usual ergodic postulates). The partition between net rainfall excess into surface and subsurface components is similarly done by a runoff coefficient ϕ_p (t), fitted by the law: $\phi_p(t) = \exp[-\alpha\ i(t) + \beta\ V(t)]$. The parameters α, β, and K have been determined through least-square fitting from the available data.The NO_3-N discharge at the control section is then by (2) given by:

$$Q_{out}(t) = \int_0^t dx\ i_s(x)\ G_s(\ t-x \mid x) + \int_0^t dx\ i_p(x)\ G_p(\ t-x \mid x) \tag{9}$$

$$G_s(t \mid t') = \{p(1)[C'_s(t,t')\ fO_1{}^s(t)]*fC_1 + p(2)[C'_s(x,t')\ fO_2{}^s(x)]\}*fC_2(t)$$

$$G_p(t \mid t') = \{p(1)[C'_p(t,t')\ fO_1{}^p(t)]*fC_1 + p(2)[C'_p(x,t')\ fO_2{}^p(x)]\}*fC_2(t)$$

The concentration at the control section is then defined as the ratio of solute and water discharge:

$$C(t) = Q_{out}(t) / Q(t)$$

The complete description of this computational method for evaluating the solute release needs only specification of the lagrangian concentrations $C'_i(t,t')$, i=s,p through integration of eq. (4).Let:

$$P_1(T>t-x) = 1 - \int_0^{t-x} dx\ fO_1{}^p(x), \quad P_2(T>t-x) = 1 - \int_0^{t-x} dx\ fO_2{}^p(x)$$

(The following formulas will refer for brevity only to the -p- component: analogous expressions characterize the -s- component). Upon trapezoidal quadrature of eq.(6) it is (let $M_{in}(i) = M_{in}(i\Delta t)$, $M_{out}(i)$, $M_{ss}(i)$ denote mass input, exited or instantaneously stored respectively):

$$M_{in}(i) = p(1) \sum_{(k=0,i)} i_p(k)\ C'(0,k)\ \Delta t$$

$$M_{ss}(i) = p(1)\ \Delta t \sum_{(k=0,i)} i_p(k)\ P_1(i-k)\ C'(i-k,k)$$

$$M_{out}(t) = p(1)\ \Delta t^2 \sum_{(k=0,i)} i_p(k) \sum_{(h=k,i)} fO_1{}^p(h-k)\ C'(h-k,k)$$

where the only state 1 is supposed to be supplying solutes. Hence eq.(7) is rewritten with:

$$C_E(i)+C_E(i\text{-}1) = 2\,C_E(0) + \{\, M_{in}(i) + M_{in}(i\text{-}1) + M_{ss}(i\text{-}1) + M_{out}(i\text{-}1) \; +$$
$$+ \Delta t \sum_{(k=1,i)} i_p(k)\,[\,P_1(i\text{-}k)\,p(1)\,]\,C'(i\text{-}k,k) + \Delta t\, i_p(0)\,[P_1(i)\,p(1)]\,C'(i,0) +$$
$$+ \Delta t^2 \sum_{(k=1,i)} i_p(k)\,B_{mas}(i,k) + \Delta t^2\, i_p(0)\,B_{mas}(i\text{-}1,0) + \Delta t^2\, i_p(0)\, fo_1{}^P(i)\,p(1)\,C'(i,0)\}/K_D$$

where: $C'(i,0)=C'(i\text{-}1,0)\exp(-h\Delta t)+[1-\exp(-h\Delta t)]\{C_E(i)+C_E(i\text{-}1)\}/2$, and: $B_{mas}(i,k) = \Sigma_{h=k,i}\, p(1)\, fo_1{}^P(h\text{-}k)\, C'(h\text{-}k,k)$. Upon factorization of the unknown values of C'(i,0) (it is recalled that only once along the diagonal i+j (i.e. the physical current time) has the equilibrium concentration $C_E(i+j)$ to be computed) the updated values are then cast in algebraic form. At the numerator there still exist unknown values of the concentration C'(i,j), i+j=constant. It is chosen for this example to render iterative the calculation of C'(i,j) via the initial prediction of its values by means of a first-order solution: $C'(i,j) = C'(i\text{-}1,j) + \Delta t\, h\, [C_E(i\text{-}1+j) - C'(i\text{-}1,j)]$, where i+j = constant.This position allows a first-order calculation of the fraction of sorbed mass, and hence the prediction of the diagonal concentrations in the preceding equations. This constitutes an iterative adjustment of the concentrations which is stopped as $ERR < 10^{-6}$(mg/l) ($ERR=\Sigma_{k=1,i}\,|\,C'^{\mu}(i\text{-}k,k)-C'^{\mu+1}(i\text{-}k,k)|$) at the μ-th iteration.

The results of the simulations are presented in Figures 2-6.

CONCLUSIONS

MRF's of solute response in hydrologic transport volumes have been examined with reference to the computational structure of the models for practical applications. A set of examples showed the flexibility of the tools for practical applications and the soundness of the theoretical assumptions.

REFERENCES

1. Dagan G. (1986), Statistical Theory of Groudwater Flow and Transport: Pore to Laboratory, Laboratory to Formation, and Formation to Regional Scale, Water Resources Research, 22, pp. 120S-134S.

2. Jury W.A. Sposito G. and White R.E. (1986), A Transfer Function Model of Solute Transport Through Soil 1. Fundamental Concepts, Water Resources Research, 22, pp. 243-247.

3. Rinaldo A. and Marani A. (1987), Basin Scale Model of Solute Transport, Water Resources Research, 23, pp. 2107-2118.

4. Rinaldo A. and G. Gambolati (1987), Basin-Scale Transport of Dissolved Species in Groundwater. Advances in Analytical and Numerical Groundwater Flow and Quality Modelling (Eds. A. Custodio and J.P. Lobo Ferreira), Reidel, Dordrecht, in press.

5. Sposito G. White R.A. Darrah P.R. and Jury W.A. (1986), A Transfer Function Model of Solute Transport Through Soil 3. The Convection-Dispersion Equation, Water Resources Research, 22, pp. 255-262.

6. Takeuchi K. Sakamoto Y. and Hongo Y. (1984), Discharge characteristics of NO_3^- for the analysis of basinwide circulation of water and environmental pollutants in a small river basin, Journal of Hydroscience and Hydraulic Engineering, 2, pp. 73-85.

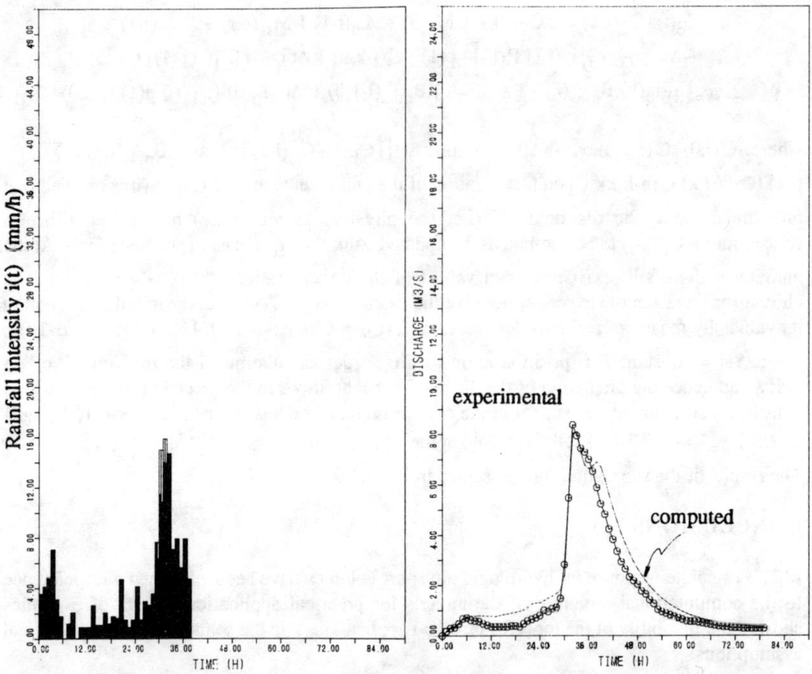

Figure 2. Net rainfall for the event no.1 and computed vs. experimental water discharge.

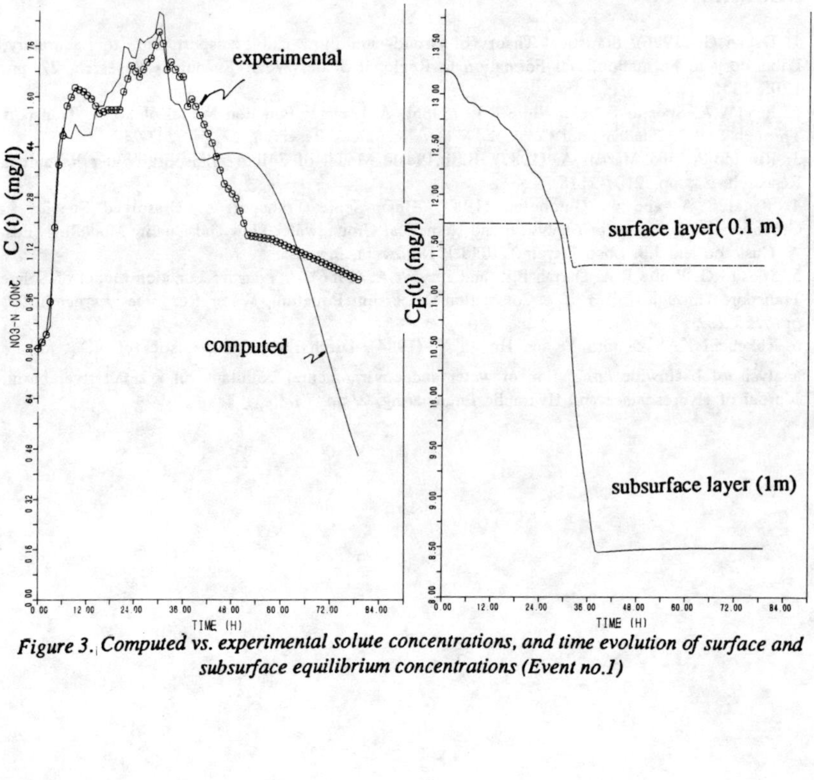

Figure 3. Computed vs. experimental solute concentrations, and time evolution of surface and subsurface equilibrium concentrations (Event no.1)

An Advection Control Method for the Solution of Advection-Dispersion Equations

Ne-Zheng Sun and Wen-Kang Liang
Environmental Science Center, Shandong University, Jinan, Shangdong, The People's Republic of China

INTRODUCTION

A variety of improved numerical methods has been presented in recent years to deal with numerical difficulties arising in the soluition of advection–dominated transport problems, such as Upstream Weighting Methods (UWM) (Heinrich et al., 1977), the Method of Characteristics (MOC) (Konikow and Bredehoeft, 1978), Moving Point Methods (MPM) (Neuman, 1984) and other alternative methods (Prickett et al., 1981; Hwang et al. 1985; Taigbenu and Liggett, 1986). A review paper was given recently by Sun (1988).

UWMs and other Eulerian methods have two disadvantages: the overshoot can be eliminated but at the expense of increasing the numerical dispersion, and upstream weighting coefficients arising in each method require to be designated artificially. The MOC and its modified versions belong to Eulerian–Lagrangian methods that are successful in eliminating the numerical dispersion. However, interpolations of concentrations from moving points to fixed nodes and the other way round may cause significant error in mass conservation. For 3–D complex problems, it is difficult to track moving particles.

Sun and Yeh (1983) presented an upstream weighted Multiple Cell Balance Method (MCBM). In this paper, a modified form of MCBM is presented that is named the Advection Control Method (ACM). It differs from the MOC and UWM in such a way that an advection control term is added onto the right–hand side of the discretized mass balance equation of each node to control the behavior of numerical solutions. The overshoot can be eliminated and the numerical dispersion is smaller than that of UWMs, the boundary conditions and sink or source terms can be treated as in the FEM . Beside that, it is easy to be used to 3–D field problems. However, there is still a undefined coefficient that needs to be designated artificially.

THE ADVECTION CONTROL METHOD

The integral form of 2–D advection–dispersion equation can be written as (Sun and Yeh, 1983):

$$\int_{(L)} \left(D_{xx}\frac{\partial C}{\partial x} + D_{xy}\frac{\partial C}{\partial y}\right)dy - \left(D_{xy}\frac{\partial C}{\partial x} + D_{yy}\frac{\partial C}{\partial y}\right)dx$$

$$+ \int_{(L)} C(V_y dx - V_x dy) = \iint_{(R)} \left[\frac{\partial C}{\partial t} + M\right] dxdy \qquad (1)$$

where C = concentration of solute (M/L^3) ; D_{xx}, D_{xy}, D_{yy} = components of the dispersion tensor (L^2/T) ; V_x, V_y = components of the average velocity (L / T); M = source or sink (M/L^2T) ; (R) = a flow region or a subdomain ; (L) = the boundary of (R).

Assume that the flow region is divided into a number of triangular elements and (e) is any one of them; The vertices i, j, k of the element are real nodes; its center m is taken as an invented node. The concentration values at points i, j, k and m are denoted by C_i, C_j, C_k and C_m, respectively. We define

$$C_m = \gamma\hat{C}_m + (1-\gamma)\overline{C}_m \qquad (2)$$

in which, $0<\gamma<1$ is an undefined weighting coefficient. The first term on the right–hand side of (2) is the regular part, where

$$\hat{C}_m = \frac{1}{3}(C_i + C_j + C_k) \qquad (3)$$

The second term on the right–hand side of (2) is called the advection control part, where

$$\overline{C}_m = C_p^0 \qquad (4)$$

and p is the single–step reverse point (Neuman, 1984) of the invented node, i.e., if, at time t, a particle locates at point p, it just arrives point m at time t + Δt; Δt is the time step; C_p^0 is the known concentration of point p at time t.

By linking the center m with vertices i, j, k, the element (e) is divided into three subelements. Assume that the unknown function C(x,y,t) can be replaced approximately by a linear function in each subelement. For example, in subelement Δijm, we have

$$C(x,y,t) = N_{ki}C_i + N_{kj}C_j + N_{km}C_m \qquad (x,y)\in\Delta ijm \qquad (5)$$

where N_{ki}, N_{kj}, and N_{km} are linear basis functions in Δijm for nodes i , j , k, respectively.

Substituting (2) into (5) yields

$$C(x,y,t) = (N_{ki} + \frac{\gamma}{3} N_{km})C_i + (N_{kj} + \frac{\gamma}{3} N_{km})C_j \tag{6}$$
$$+ \frac{\gamma}{3} N_{km} C_k + (1-\gamma) N_{km} \overline{C}_m \qquad (x,y) \in \Delta ijm$$

There are similar representations for other two subelements. From those, it is easy to obtain representations of $\frac{\partial C}{\partial x}$, $\frac{\partial C}{\partial y}$, and $\frac{\partial C}{\partial t}$ in each subelement. Now consider all elements around node i, by linking the center and the middle point of each side of each element, we obtain a multiangular (R_i)around node i that is referred to as the exclusive subdomain of the node. In equation (1), if the subdomain (R_i) is taken as the region (R), then the equation expresses the mass balance associated with node i. Using (6) and similar representations of other subelements, all integrals arising in equation (1) can be computed directly. We have

$$\sum_{ei} (A^e_{ii} C_i + A^e_{ij} C_j + A^e_{ik} C_k)$$
$$+ \sum_{ei} \left(B^e_{ii} \frac{\partial C_i}{\partial t} + B^e_{ij} \frac{\partial C_j}{\partial t} + B^e_{ik} \frac{\partial C_k}{\partial t} \right) = F_i \tag{7}$$

where $\sum_{ei}$ represents the summation for all elements around node i. A^e_{il}, B^e_{il} (l=i, j, k, m) are easy to be obtained through a simple calculation procedure described in Sun and Yeh (1983). In equation (7),

$$F_i = -\sum_{ei} \left(A^e_{im} \overline{C}_m + B^e_{im} \frac{\partial \overline{C}_m}{\partial t} \right) - Q_i \quad ; \qquad Q_i = \iint_{(R_i)} M\, dx dy \tag{8}$$

By writing (7) for each node where the concentration is unknown and incorporating the given boundary conditions, we obtain a system of equations

$$[A]\{C\} + [B]\left\{\frac{dC}{dt}\right\} = \{F\} \tag{9}$$

that can be used to obtain the unknown concentration distribution. The first term on the right–hand side of (8) is the the advection control term. For dispersion–dominated problems, let $\gamma = 1$, then $A^e_{im} = 0$ and $B^e_{im} = 0$, this term vanishes and the method reduces to general MCBM. For pure advection problems, let $\gamma = 0$, then concentrations of invented nodes are determined almost by pure advection. The effect of the advection control term in (7) now likes to impose an internal boundary condition with given concentration at these invented nodes around each real node. For advection–dominated problems, let $0 < \gamma < 1$, the oscillations of numerical solutions can be eliminated. Therefore, (9) is available for the entire range of Peclet number from 0 to ∞.

In (8), $\overline{C}_m(t + \Delta t)$ is determined by (4) , where $C^o_p = C_p(t)$ can be obtained by an interpolation process using the known concentration distribution at time t . Here, we use following nonlinear interpolation procedure. Let

$$C(x,y,t) = (N_i^{\alpha} C_i + N_j^{\alpha} C_j + N_k^{\alpha} C_k) / (N_i^{\alpha} + N_j^{\alpha} + N_k^{\alpha}) \tag{10}$$

$$(x, y) \in \Delta ijk$$

where N_i, N_j and N_k are linear basis functions of the element; α depends on the slope of concentration front in th element. (10) is able to describe the local shape of the concentration front quite well and the required computational effort is almost equal to that of the linear interpolation method. In numerical experiments, it leads to low numerical dispersion and low overshoot. Furthermore, let

$$\frac{\partial \overline{C}_m}{\partial t} = \frac{\overline{C}_m(t+\Delta t) - \overline{C}_m(t)}{\Delta t} \tag{11}$$

Therefore, in (9), $\{F\}$ is a known right–hand side and unknown nodal concentrations at t +Δt can be obtained by solving the system of equations.

The movement of contaminants in groundwater is an inherently 3–D process. However, for some time past, the practical application of numerical solutions was almost limited in 1–D and 2–D cases. The 3–D upstream weighted FEM and 3–D MOC are more complex than their 2–D forms and more computational effort is needed.

The upstream weighted MCBM was extended to 3–D case (Wang et al., 1986). It is able to deal with advection–dominated field problems in acceptable computational expenses. The disadvantages of the method are: the numerical dispersion is increased and too many weighting coefficients are included.

The ACM mentioned above is easy to be extended to 3–D cases. The triangular prism is used as the 3–D element. Each element has six real and six invented nodes as did in 3–D upstream weighted MCBM (Wang et al., 1986). The concentration at an invented node is considered as the weighted average of a regular part and an advection control part. The latter can be obtained by the 'single–step reverse point' technique. Using the representations of C, $\frac{\partial C}{\partial x}$, $\frac{\partial C}{\partial y}$, $\frac{\partial C}{\partial z}$, and $\frac{\partial C}{\partial t}$ in each subelement, the local mass balance equation associated with each real node is easy to be translated into an algebraic equation but with an advection control term on its right–hand side. The whole procedure is almost same as that of 2–D case mentioned above and the 3–D MCBM. Therefore, it is not necessary to derive it here in detail.

A NUMERICAL EXAMPLE

Considering the 1–D classical problem

$$\frac{\partial C}{\partial t} = D\frac{\partial^2 C}{\partial x^2} - V\frac{\partial C}{\partial x}$$

$$C\ (x,0)\ = 0 \quad C\ (\infty,t)\ = 0 \quad C\ (0,t)\ = C_0 \tag{12}$$

Let D = 0.05, V = 1, x = 5, C_0 = 10, then the Peclet number Pe = 100. 2–D triangular elements are used in the example (Sun and Yeh, 1983). The analytical solution, the numerical solutions of the MCBM and ACM are shown in Fig. 1. The fronts of the two numerical solutions are almost same, but the solution of ACM does not oscillate. The solutions of upstream weighted MCBM and ACM are shown in Fig.2. The overshoot is eliminated in both solutions, but the latter has smaller numerical dispersion than the former.

SUMMARY

A modified numerical method is presented. In which, one or several invented nodes are introduced into each element. The concentration of an invented node is considered as a weighted average of two parts: one is determined by nodal values of real nodes and another by 'single–step reverse particle tracking' technique. The overshoot of numerical solutions is eliminated at the expense of smaller numerical dispersion.

REFERENCES

Heinrich, J. C., P. S. Huyakorn, O. C. Zienkiewicz, and A. R. Mitchell (1977), An "Upwind" Finite Element Scheme for Two–dimensional Convection Transport Equations, Int. J. Numer. Method. Eng., vol. 11: 131–143.

Hwang, J. C., C–J. Chen, M. Sheikhoslami, and B. K. Panigrahi (1985), Finite Analysis Numerical Solution for Two–Dimensional Groundwater Solute Transport, WRR, vol. 21 (9): 1354–1360.

Konikow, L. F., and J. D. Bredehoeft (1978), Computer Model of 2–D Solute Transport and Dispersion in Groundwater, U. S. Geol. Survey Techniques of Water Resources Investigations, Book 7, C2.

Neuman, S. P. (1984), Adaptive Eulerian–Lagrangian Finite Element Method for Advection–Dispersion, Int. J. Numer. Method. Eng., vot. 20: 321–337.

Prickett, T. A., T. G. Namik, and C. G. Lonnquist (1981), A "Random–Walk" Solute Transport Model for Selected Groundwater Quality Evaluations, ISWS / BUL–65 / 81, Bulletin 65, State of Illinois Department of Energy and Natural Resources, Chamaign.

Sun, N–Z. and W. W–G. Yeh (1983), A Proposed Upstream Weighted Numerical Method for Simulating Pollutant Transport in Groundwater, WRR, vol. 19

(6) :1489–1500.

Sun, N–Z. (1988) , Applications of Numerical Methods to Simulate the Movement of Contaminants in Groundwater, submitted to EHP.

Taigbenu, A. and J. A. Liggett (1985) , An Integral Solution for the Diffusion–Advection Equation, WRR, 22 (8) :1237–1246.

Wang, C. C., N–Z. Sun, and W. W–G. Yeh (1986) , An Upstream Weighted Multiple Cell Balance Finite Element Method for Solving 3–D Convection–Dispersion Equations, WRR, 22 (11) : 1575—1589.

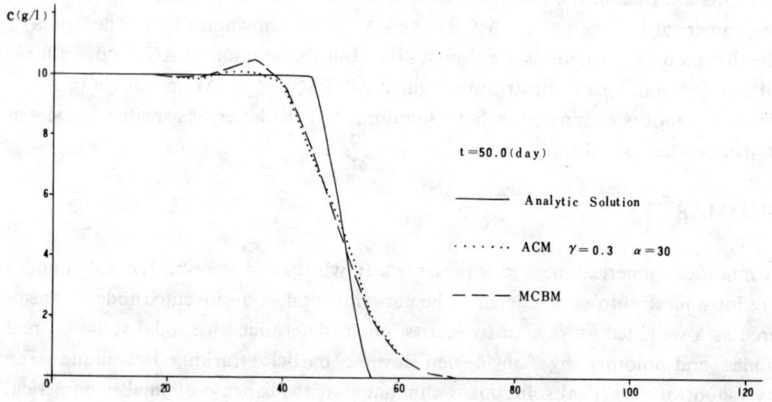

FIG. 1

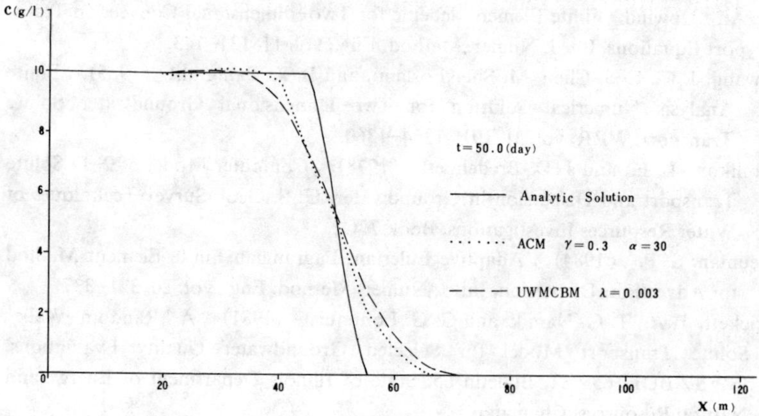

FIG. 2

Non-Diffusive N+2 Degree Upwinding Methods for the Finite Element Solution of the Time Dependent Transport Equation

J.J. Westerink and M.E. Cantekin
Civil Engineering Department, Texas A&M University, College Station, Texas 77843, USA
D. Shea
Department of Civil Engineering, Massachusetts Institute of Technology, Cambridge, MA 02139, USA

INTRODUCTION

The use of standard central type numerical schemes to solve the convection-diffusion equation for convection dominated flows typically results in solutions which exhibit severe spurious oscillations, as well as peak damping and phase lag of the fundamental solution. The use of upwinding methods has been a very popular remedy in both finite difference (f.d.) and finite element (f.e.) methods. However while traditional upwinding methods work well for steady state problems, they lead to over-diffusive solutions for time dependent problems.

For f.e. methods, upwinding has been traditionally applied by using weighting functions which are modified by a function one polynomial degree greater than the basis functions being used [Christie et al.[1]; Heinrich and Zienkiewicz[2]; Christie and Mitchell[3]]. Thus a quadratic biased weighting function is used with linear elements and a cubic biased weighting function is used with quadratic elements. However, as is the case with traditional f.d. upwinding schemes, these so called Petrov-Galerkin f.e. schemes produce overly-diffusive results for time dependent problems due to the appearance of a diffusive type truncation term [Carey and Oden[4]]. A technique which we examine in this paper that overcomes this problem is the use of upwind weighting functions which are modified by functions two polynomial degrees greater than the basis functions. Dick[5] proposed the use of a cubic upwind modification to the weighting function in conjunction with linear elements. We now introduce a quartic biased weighting function for use with quadratic elements. We will detail the use and behavior of these N+2 degree upwind biased weighting functions (where N equals the degree of the basis function) and demonstrate that they lead to excellent solutions with no numerical diffusion.

NUMERICAL DISCRETIZATION OF THE GOVERNING EQUATIONS AND UPWIND WEIGHTING FUNCTIONS

The convection-diffusion equation in one-dimensional form is written as:

$$\Phi,_t + u\,\Phi,_x = D\,\Phi,_{xx} \tag{1}$$

on the interval Γ. Developing a symmetrical weak weighted residual formulation for this equation results in:

$$\int_\Gamma \{\, (\Phi,_t + u\,\Phi,_x)\,w + D\,\Phi,_x\,w,_x \,\}\,dx = D\,\Phi,_x\,w\Big|_{\text{flux b.c.}} \tag{2}$$

where w represents the weighting function.

For traditional f.e. upwinding or Petrov-Galerkin methods, a bias one degree greater than the interpolating functions is added to the basis functions to form the upwinded weighting functions. Thus for linear elements, a quadratic bias is used which has the form [Christie et. al.[1]]:

$$F_2(\xi) = \frac{3}{4}(1+\xi)(1-\xi) \tag{3}$$

N+2 degree upwinding introduces a bias two degrees greater than the basis function. Dick[5] introduced the use of a cubic modification for linear basis of the form:

$$F_3(\xi) = \frac{5}{8}\,\xi\,(\xi+1)(\xi-1) \tag{4}$$

Thus F_2 and F_3 are added to the linear interpolating functions ψ_1 and ψ_2 to produce the following upwinded weighting functions:

$$w_1 = \psi_1 - \alpha\,F_2(\xi) - \beta\,F_3(\xi) \tag{5a}$$

$$w_2 = \psi_2 + \alpha\,F_2(\xi) + \beta\,F_3(\xi) \tag{5b}$$

where α is equal to the N+1 degree (quadratic) upwinding coefficient and β equals the N+2 degree (cubic) upwind bias for linear elements.

For quadratic elements the traditional N+1 upwinding approach modifies the second degree basis functions by a third degree modifying function of the same form as (4) [Heinrich and Zienkiewicz[2]]. We now introduce the use of N+2 degree upwinding on quadratic elements. Thus a quartic upwind modification for quadratic elements is used and is of the form:

$$F_4(\xi) = \frac{21}{16}(-\xi^4 + \xi^2) \tag{6}$$

The upwind biased weighting functions appear as:

$$w_1 = \psi_1 - \alpha_c\,F_3(\xi) - \beta_c\,F_4(\xi) \tag{7a}$$

$$w_2 = \psi_2 + 4\alpha_m\,F_3(\xi) + 4\beta_m\,F_4(\xi) \tag{7b}$$

$$w_3 = \psi_3 - \alpha_c\,F_3(\xi) - \beta_c\,F_4(\xi) \tag{7c}$$

where α_c and α_m and β_c and β_m respectively equal the N+1 and N+2 degree bias for the corner and mid-element nodes for quadratic elements.

The use of any of these weighting functions in conjunction with the symmetrical weak weighted residual formulation (2) will lead to a global system equations of the form:

$$\mathbf{M}\,\mathbf{\Phi}_{,t} + (\mathbf{A} + \mathbf{B})\,\mathbf{\Phi} = \mathbf{P} \qquad (8)$$

where $\mathbf{\Phi}$ = vector of nodal unknowns, $\mathbf{M}$ = mass matrix, $\mathbf{A}$ = convection matrix, $\mathbf{B}$ = diffusion matrix and $\mathbf{P}$ = diffusive boundary flux loading vector. The form and characteristics of these matrices will be significantly influenced by the order of interpolation selected and even more importantly by the degree of upwinding.

Finally (8) is time discretized using a Crank-Nicholson f.d. scheme which results in:

$$[\mathbf{M} + \frac{\Delta}{2}(\mathbf{A}^{n+1}+\mathbf{B}^{n+1})]\mathbf{\Phi}^{n+1} = [\mathbf{M} - \frac{\Delta}{2}(\mathbf{A}^{n}+\mathbf{B}^{n})]\mathbf{\Phi}^{n} + \frac{1}{2}\,\mathbf{P}^{n+1} + \frac{1}{2}\,\mathbf{P}^{n} \qquad (9)$$

where Δ equals the time step and n+1 and n represent the future and current time levels.

NUMERICAL EXAMPLES

Let us compare the performance of the standard, N+1 and N+2 upwinded Galerkin methods for the case of a one-dimensional Gaussian plume of standard deviation σ=264 travelling in pure convection (D=0,u=0.5). A constant node to node distance of h=200 is maintained while time steps are varied although all solutions are compared to the analytical solution at t=9600.

Linear basis with quadratic (N+1 degree) and cubic (N+2 degree) upwinding

Figure 1a shows the standard Galerkin solution at low Courant number, $\mathbb{C}$=0.24. The solution is trailed by wiggles and the peak is depressed. Traditional N+1 upwinding eliminates most of the wiggles at the expense of suppressing the peak, as is shown in figure 1b. N+2 upwinding, figure 1c, results in a much improved solution compared to both the standard and N+1 upwinded solutions. Both amplitude and phase of the plume have dramatically improved and in addition the amplitude of the wiggles has been substantially reduced.

Figure 2a shows the same case computed with a larger time step such that $\mathbb{C}$=0.8. The standard and N+1 upwinded solutions have deteriorated (figure 2a,b) while the N+2 upwinded solution has been improved even further (figure 2c) as compared to the corresponding solution at the lower Courant number. In fact, the overall quality of the N+2 upwinded solution continues to improve as Courant number increases to $\mathbb{C}$=1.0, for which the solution is perfect.

Quadratic basis with cubic (N+1 degree) and quartic (N+2 degree) upwinding

Figure 3 shows the various solutions computed with $\mathbb{C}$=0.24. The N+1 upwinded solution, figure 3b, is slightly worse while the N+2 upwinded solution, figure 3c, is slightly better than the standard Galerkin solution shown in figure 3a. However, at this low Courant number, all three solutions appear very good.

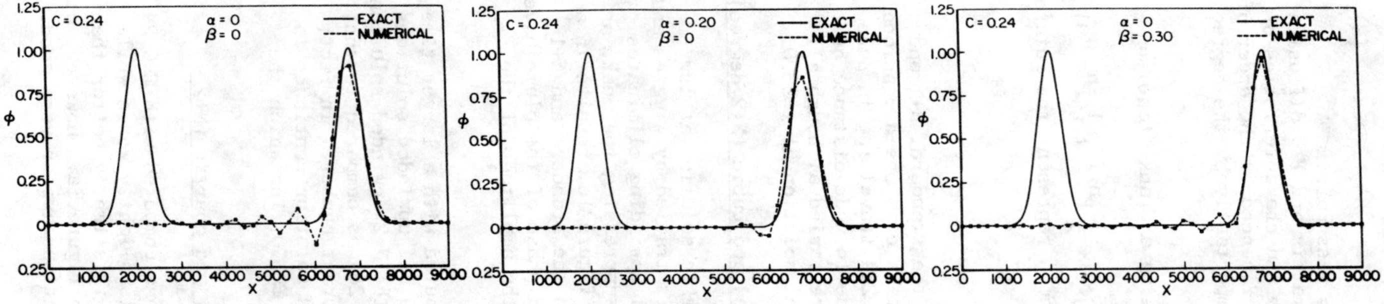

Fig. 1: Pure convection of a plume using linear elements at ℂ=0.24: (a) Standard Galerkin, (b) Quadratic (N+1 degree) upwinded, (c) Cubic (N+2 degree) upwinded.

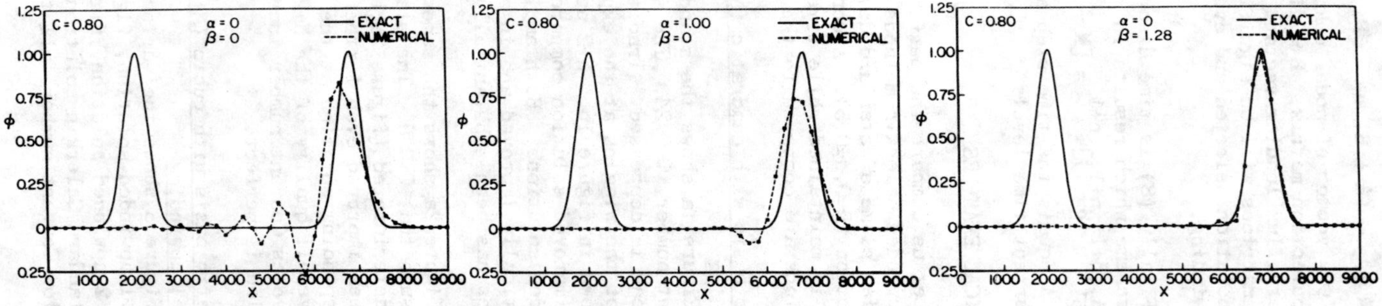

Fig. 2: Pure convection of a plume using linear elements at ℂ=0.80: (a) Standard Galerkin, (b) Quadratic (N+1 degree) upwinded, (c) Cubic (N+2 degree) upwinded.

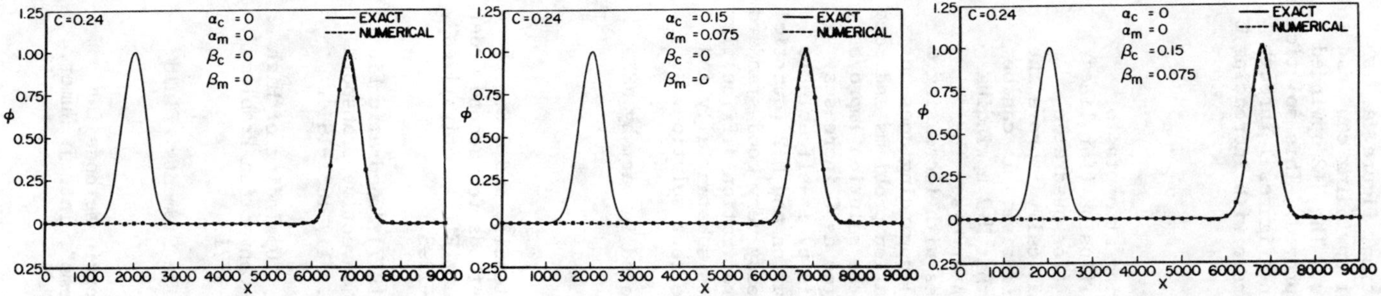

Fig. 3: Pure convection of a plume using quadratic elements at $\mathbb{C}$=0.24: (a) Standard Galerkin, (b) Cubic (N+1 degree) upwinded, (c) Quartic (N+2 degree) upwinded.

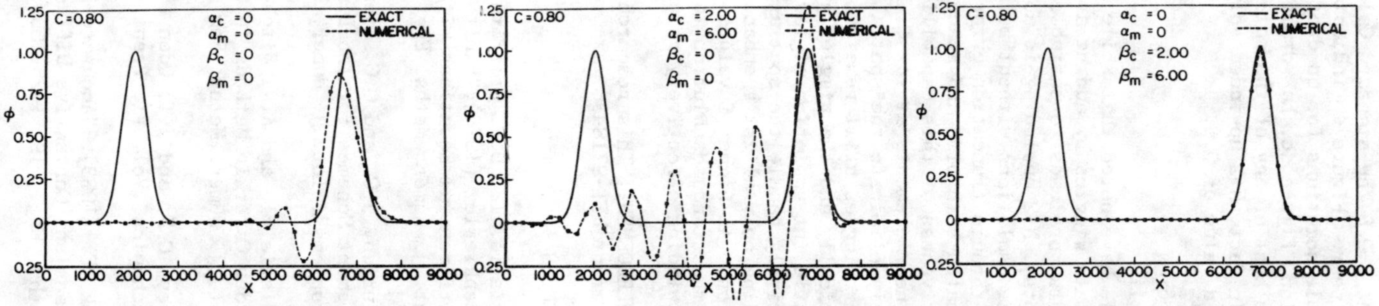

Fig. 4: Pure convection of a plume using quadratic elements at $\mathbb{C}$=0.80: (a) Standard Galerkin, (b) Cubic (N+1 degree) upwinded, (c) Quartic (N+2 degree) upwinded.

At $\mathbb{C}$=0.8 the standard Galerkin solution, figure 4a, is quite poor. Figure 4b indicates that N+1 upwinding can lead to unstable solutions for quadratic elements. The N+2 upwinded solution, figure 4c, has dramatically improved. This solution is the best of any of the solutions shown (figures 1 through 4). In fact, N+2 upwinded solutions improve with increasing $\mathbb{C}$ up to a value of 0.8.

DISCUSSION

We have noted that while N+1 upwinding attempts to eliminate wiggles by adding artificial diffusion (for linear elements) or leads to unstable solutions (for quadratic elements), N+2 upwinding is able to dramatically improve the computed solutions without being over-diffusive. It can be shown through truncation error analysis that N+2 upwinding eliminates both time and space truncation errors. The effectiveness of this capability increases with increasing $\mathbb{C}$. For linear elements at low $\mathbb{C}$, N+2 (cubic) upwinding does not entirely eliminate the spatial discretization problems and some smaller wiggles still remain. However, the solution improves dramatically and the wiggles are eliminated as $\mathbb{C}$ increases. Quadratic interpolation at low $\mathbb{C}$ values is by itself able to effectively handle the spatial discretization and N+2 (quartic) upwinding only slightly enhances the already very good standard solution. At higher $\mathbb{C}$ values, time discretization difficulties cause the standard solution to deteriorate substantially and N+2 upwinding effectively returns an excellent solution.

ACKNOWLEDGEMENT: This research was sponsored in part by NSF under grant EET-8718436.

REFERENCES

1. Christie, I., D.F. Griffiths, A.R. Mitchell, and O.C. Zienkiewicz (1976),"Finite Element Methods for Second Order Differential Equations with Significant First Derivatives", Int. J. Numer. Meths. Eng., 10, 1389-1396.

2. Heinrich, J.C. and O.C. Zienkiewicz (1977),"Quadratic Finite Element Schemes for Two-Dimensional Convective Transport Problems", Int. J. Numer. Meths. Eng., 11, 1831-1844.

3. Christie, I. and A.R. Mitchell (1978), "Upwinding of High Order Galerkin Methods in Conduction-Convection Problems", Int. J. Numer. Meths. Eng., 12, 1764-1771.

4. Carey, G.F. and J.T. Oden (1986), Finite Elements: Fluid Mechanics, Vol. VI, Prentice-Hall, N.J.

5. Dick, E. (1983), "Accurate Petrov-Galerkin Methods for Transient Convective Diffusion Problems", Int. J. Numer. Meths. Eng., 19, 1425-1533.

A Characteristic Alternating Direction Implicit Scheme for the Advection-Dispersion Equation

Yuqun Xue and Chunhong Xie

Department of Geology and Department of Mathematics, Nanjing University, Nanjing, 210008, People's Republic of China

Introduction

The present paper describes a numerical scheme based on combining the utility of a fixed grid in Eulerian coordinates with the computational power of the Lagrangian method. This is followed by a detailed comparison of the simulated concentrations with the analytical solutions. An analytical solution of the three-dimensional advection-dispersion equation is developed in this connection.

Description of the method

The basis of the method will be given for the initial boundary value problem of the advection-dispersion equation in three-space dimensions x,y and z and time t defined as

$$\frac{\partial C}{\partial t}=\text{div}(\underline{\underline{D}}\cdot\text{grad } C)-\text{div}(\underline{V}\, C) \quad \text{in } F \qquad (1)$$

$$C(x,y,z,0)=C_0(x,y,z) \qquad (2)$$

$$C(x,y,z,t)=\varphi(x,y,z,t) \quad \text{on } \Gamma \qquad (3)$$

where C is concentration; $\underline{\underline{D}}$ is dispersion tensor; $\underline{V}$ is seepage velocity vector; C_0 and φ are prescribed functions; Γ is the boundary of the domain F. In steady flow without source and sink, we have div $\underline{V}$ =0. So (1) becomes

$$\frac{\partial C}{\partial t}=\text{div}(\underline{\underline{D}}\cdot\text{grad } C)-\underline{V}\cdot\text{grad } C \qquad (4)$$

we define $d\underline{r}/dt=\underline{V}$, where $\underline{r}=x\underline{i}+y\underline{j}+z\underline{k}$, $\underline{i}$, $\underline{j}$ and $\underline{k}$ are the positive unit vector along x,y and z, respectively. Using Lagrangian coordinates, one can rewrite (4) in Lagrangian form as

$$\frac{dC}{dt}=\text{div}(\underline{\underline{D}}\cdot\text{grad } C) \qquad (5)$$

In unsteady flow we have

$$\frac{dC}{dt}=\text{div}(\underline{\underline{D}}\cdot\text{grad } C)-C\,\text{div } \underline{V} \qquad (6)$$

In order to save computer storage and time, the alternating direction implicit method is used. It is illustrated by a two-dimensional problem as follows: Defining the operator $L_p (p=1,2)$

$$(L_1C)^n_{i,j} = \frac{\Delta t^n}{2(\Delta x)^2} D_{xx}(C^n_{i-1,j} - 2C^n_{i,j} + C^n_{i+1,j})$$

$$(L_2C)^n_{i,j} = \frac{\Delta t^n}{2(\Delta y)^2} D_{yy}(C^n_{i,j-1} - 2C^n_{i,j} + C^n_{i,j+1})$$

we can rewrite (5) in difference scheme as

$$C^{n+\frac{1}{2}}_{i,j} - {}^nC_{i,j} = (L_1C)^{n+\frac{1}{2}}_{i,j} + (L_2C)^n_{i,j} \qquad (7)$$

$$C^{n+1}_{i,j} - {}^{n+\frac{1}{2}}C_{i,j} = (L_1C)^{n+\frac{1}{2}}_{i,j} + (L_2C)^{n+1}_{i,j} \qquad (8)$$

for $i=1,2,\ldots,I-1$; $j=1,2,\ldots,J-1$

where t^n and t^{n+1} denote the beginning and the end of a typical time step n with time increment Δt^n; x_i and y_i are the x- and y-coordinates of the node (i,j), respectively; I+1 and J+1 are number of nodes in x and y direction, respectively; $C^n_{i,j}$ and $C^{n+1/2}_{i,j}$ denote the nodal value of the approximating function to the solution of the problem defined by equation (1)-(3) as follows:

$$C^n_{i,j} = C\{x_i(n\Delta t), y_j(n\Delta t), n\Delta t\} \qquad (9)$$

$$C^{n+\frac{1}{2}}_{i,j} = C[x_i\{(n+\tfrac{1}{2})\Delta t\}, y_j\{(n+\tfrac{1}{2})\Delta t\}, (n+\tfrac{1}{2})\Delta t] \qquad (10)$$

However, $x_i(n\Delta t)$ and $y_j(n\Delta t)$ denote the position of node $p(x_i, y_j)$ at $t = n\Delta t$ and it is recorded by ${}^n\underline{r}_p$; $x_i\{(n+\frac{1}{2})\Delta t\}$ and $y_j\{(n+\frac{1}{2})\Delta t\}$ denote the position of node $p(x_i, y_j)$ at $t=(n+\frac{1}{2})\Delta t$ and it is recorded by ${}^{n+\frac{1}{2}}\underline{r}_p$.

Case 1. Reverse tracking

If (x_i, y_j) is the coordinate of ${}^{n+\frac{1}{2}}\underline{r}_p$, the position of ${}^n\underline{r}_p$ will be determined by (assuming $t^n = n\Delta t$)

$${}^n\underline{r}_p = {}^{n+\frac{1}{2}}\underline{r}_p - \int_{t^n}^{t^n+\frac{1}{2}\Delta t} \underline{V}\, dt \qquad (11)$$

In other words, this means that a particle leaving ${}^n\underline{r}_p$ at t^n will reach the point ${}^{n+\frac{1}{2}}\underline{r}_p$ at $t^{n+\frac{1}{2}}$. Because all $C_{i,j}$ at $t=n\Delta t$ are known and it is not at all certain whether ${}^n\underline{r}_p$ will just move to a grid point or node, $C({}^n\underline{r}_p)$ can be calculated by

$$C({}^n\underline{r}_p, t^n) \simeq \sum_{m=1}^{N} C^n_m \xi_m({}^n\underline{r}_p) \qquad (12)$$

where C^n_m is the concentration of node m at t^n; N is numbers of grid points or nodes surround the ${}^n\underline{r}_p$; ξ_m is a basis function satisfying $\xi_m(\underline{r}_p) = \delta_{mn}$, $\underline{r}_p$ being $\underline{r}$ at point p and δ_{mn} being the Kronecker delta (i.e., $\delta_{mn}=1$, if m=n and $\delta_{mn}=0$, if $m \neq n$). From (12) we have ${}^nC_{i,j} = C({}^n\underline{r}_p, t^n)$. Substituting (12) into (7), one obtains

$$C^{n+\frac{1}{2}}_{i,j} = (L_1C)^{n+\frac{1}{2}}_{i,j} + f_1 \qquad (13)$$

where

$$f_1 = (L_2C)^n_{i,j} + C({}^n\underline{r}_p, t^n)$$

and f_1 is known. The coefficient matrix of (13) is tridiagonal and the whole system can be solved by the highly efficient Thomas algorithm. Equation (8) can be solved by the same way as equation (7).

Case 2: Forward tracking

If the coordinate of ${}^{n}\underline{r}_p$ is just the position of grid point or node $p(x_i,y_j)$, the position of ${}^{n+\frac{1}{2}}\underline{r}_p$ will be determined by

$$^{n+\frac{1}{2}}\underline{r}_p = {}^{n}\underline{r}_p + \int_{t^n}^{t^n+\frac{1}{2}\Delta t} V\,dt \qquad (14)$$

Because the value of ${}^{p}C$ at grid point or node p at $t=n\Delta t$ is known, $C[{}^{n+\frac{1}{2}}\underline{r}_p,(n+\frac{1}{2})\Delta t]$ can be obtained from (7). And $C[{}^{n+1}\underline{r}_p,(n+1)\Delta t]$ will be calculated through the same method. At this stage, it is not at all certain whether ${}^{n+1}\underline{r}_p$ will just move to a original grid point or node. If ${}^{n+1}r_p$ is taken as a new node, it is necessary for changing grids and revising the space increment. Through combining the above-mentioned two methods, computational results may be free of oscillation and reduce the numerical dispersion.

Example

The following examples show preliminary results for one-, two- and three-dimensional dispersion in a uniform or non-uniform steady state velocity field.

Example 1 concerns the one-dimensional problem of solving

$$\frac{\partial C}{\partial t} = D_{xx}\frac{\partial^2 C}{\partial x^2} - V_x\frac{\partial C}{\partial x} \qquad 0 \leqslant x \leqslant x_r \qquad (15)$$

subject to

$$C(x,0)=0, \quad C(0,t)=1, \quad C(x_r,t)=0$$

The physical and grid parameters, in an arbitary system of consistent units are $D_{xx}=0.001$, $V_x=0.005$, $x_r=5.0$ and $\Delta x=0.05$, Δx being the distance between neighbouring grid points. This problem is dispersion-dominated with a Peclet number $Pe_x=V_x x/D_{xx}=0.25$. Fig. 1 shows results at t=2.5, 5.0, 7.5 and 10.0 when $\Delta t=0.1$ (Courant number $\alpha_x=V_x\Delta t/\Delta x=1$). The results agree very well with the analytical solution

$$C(x,t)=\tfrac{1}{2}\text{erfc}\left(\frac{x-V_x t}{\sqrt{4D_{xx}t}}\right)+\tfrac{1}{2}\exp\left(\frac{V_x x}{D_{xx}}\right)\text{erfc}\left(\frac{x+V_x t}{\sqrt{4D_{xx}t}}\right) \qquad (16)$$

which is valid for $x_r \rightarrow \infty$.

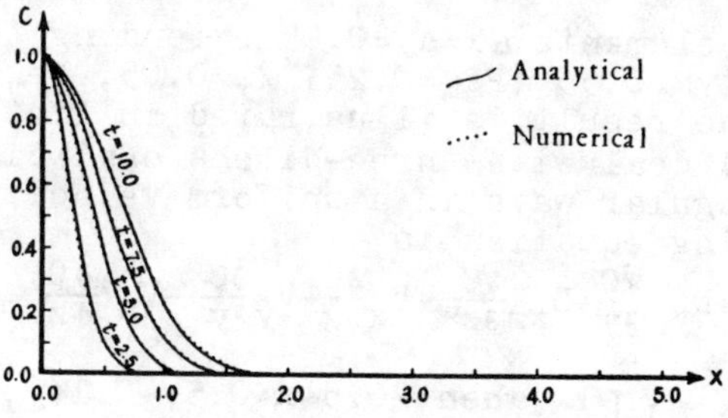

Fig.1 Result of example 1

Example 2 concerns the convection and dispersion of a rectangular wave. The problem is defined by Eq. (15) subject to

$$C(x,0)=\begin{cases}1 & \text{when } -1.0\leqslant x\leqslant 1.0\\ 0 & \text{when } -3.0\leqslant x\leqslant -1.0,\quad 1.0\leqslant x\leqslant 17.0\end{cases}$$

$$C(-3.0,t)=C(17.0,t)=0$$

and its analytical solution is

$$C(x,t)=\frac{1}{2}\left[\operatorname{erf}\left(\frac{b-x+V_x t}{\sqrt{4D_{xx}t}}\right)+\operatorname{erf}\left(\frac{b+x-V_x t}{\sqrt{4D_{xx}t}}\right)\right] \qquad (17)$$

where 2b=2.0 is the width of the rectangle. The results at t=2.5,5.0,7.5 and 10.0 for $D_{xx}=0.013$, $V_x=1.0$, $\Delta x=0.1$ and $\Delta t=0.1$(Peclet number $Pe_x=7.69$, Courant number $\alpha_x=1$) are illustrated in Fig.2.

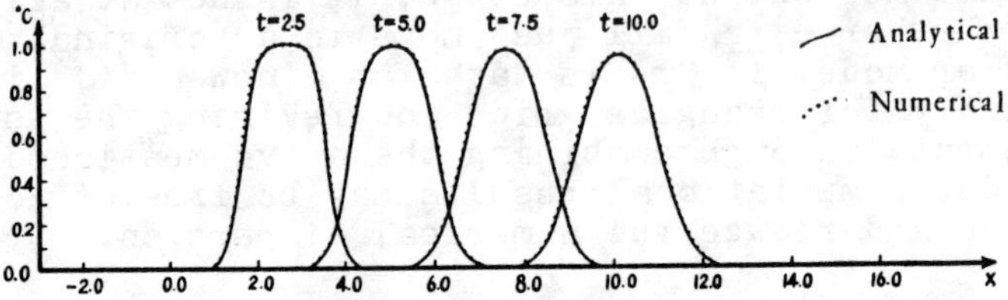

Fig. 2 Result of example 2

Example 3 deals with two-dimensional dispersion of a rectangular wave in a uniform velocity field. The governing equation is

$$\frac{\partial C}{\partial t} = D_{xx}\frac{\partial^2 C}{\partial x^2} + D_{yy}\frac{\partial^2 C}{\partial y^2} - V_x\frac{\partial C}{\partial x} - V_y\frac{\partial C}{\partial y} \qquad (18)$$

subject to initial and boundary conditions

$$C(x,y,0)=\begin{cases}1 & \text{when } -0.5\leqslant x\leqslant 0.5,\quad -0.1\leqslant y\leqslant 0.1\\ 0 & \text{otherwise}\end{cases}$$

$$C(x_l,y,t)=C(x_r,y,t)=C(x,y_b,t)=C(x,y_t,t)=0$$

When $x_l=y_b=-\infty$ and $x_r=y_t=\infty$, the analytical solution is

$$C(x,y,t)=\frac{1}{4}\left[\operatorname{erf}\left(\frac{a-x+V_x t}{\sqrt{4D_{xx}t}}\right)+\operatorname{erf}\left(\frac{a+x-V_x t}{\sqrt{4D_{xx}t}}\right)\right]\left[\operatorname{erf}\left(\frac{b-y+V_y t}{\sqrt{4D_{yy}t}}\right)+\operatorname{erf}\left(\frac{b+y-V_y t}{\sqrt{4D_{yy}t}}\right)\right] \qquad (19)$$

where a=0.5(half the length of rectangle in x direction) and b=0.1(half the length of rectangle in y direction). Exemplifying a case where $D_{xx}=0.005$, $D_{yy}=0.0002$, $V_x=0.025$, $V_y=0.0$, $\Delta t=0.1$ over a grid with equal elements $\Delta x=\Delta y=0.05$ extending to $-1.5\leqslant x\leqslant 3.5$, $-0.55\leqslant y\leqslant 0.55$ ($Pe_x=0.25$, $\alpha_x=0.05$, $Pe_y=0.0$, $\alpha_y=0.0$), the result is illustrated in Fig.3.

Example 4 deals with three-dimensional dispersion of a rectangular wave in a uniform velocity field. The governing equation is

$$\frac{\partial C}{\partial t}=D_{xx}\frac{\partial^2 C}{\partial x^2}+D_{yy}\frac{\partial^2 C}{\partial y^2}+D_{zz}\frac{\partial^2 C}{\partial z^2}-V_x\frac{\partial C}{\partial x}-V_y\frac{\partial C}{\partial y}-V_z\frac{\partial C}{\partial z} \qquad (20)$$

subject to

$$C(x,y,z,0)=\begin{cases}1 & \text{when } -0.5\leqslant x\leqslant 0.5,\ -0.1\leqslant y,z\leqslant 0.1\\ 0 & \text{otherwise}\end{cases}$$

$$C(x_l,y,z,t)=C(x_r,y,z,t)=C(x,y_b,z,t)=C(x,y_t,z,t)=C(x,y,z_f,t)=C(x,y,z_b,t)=0$$

when $x_l=y_b=z_f=-\infty$ and $x_r=y_t=z_b=\infty$, we developed an analytical solution as follows(see Appendix):

$$C(x,y,z,t)=\frac{1}{8}\left[\mathrm{erf}\left(\frac{a-x+V_x t}{\sqrt{4D_{xx}t}}\right)+\mathrm{erf}\left(\frac{a+x-V_x t}{\sqrt{4D_{xx}t}}\right)\right]\cdot\left[\mathrm{erf}\left(\frac{b-y+V_y t}{\sqrt{4D_{yy}t}}\right)+\mathrm{erf}\left(\frac{b+y-V_y t}{\sqrt{4D_{yy}t}}\right)\right]\cdot\left[\mathrm{erf}\left(\frac{c-z+V_z t}{\sqrt{4D_{zz}t}}\right)+\mathrm{erf}\left(\frac{c+z-V_z t}{\sqrt{4D_{zz}t}}\right)\right]$$

where a=0.5, b=c=0.1(half the length of rectangle in x,y and z direction, respectively). The results at t=2.5, 5.0 and 7.5 with $D_{xx}=0.05$, $D_{yy}=D_{zz}=0.0002$, $V_x=0.025$, $V_y=V_z=0$, $\Delta t=0.1$, $\Delta x=\Delta y=\Delta z=0.05$, $x_l=-1.5$, $x_r=2.5$, $y_b=z_f=-0.05$, $y_t=z_b=0.05$ are shown in Fig.4($Pe_x=0.25$, $Pe_y=Pe_z=0$, $\alpha_x=0.05$, $\alpha_y=\alpha_z=0$).

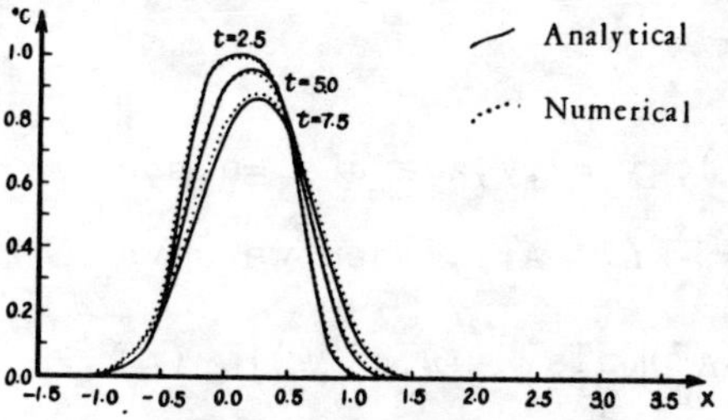

Fig.3 Result of example 3

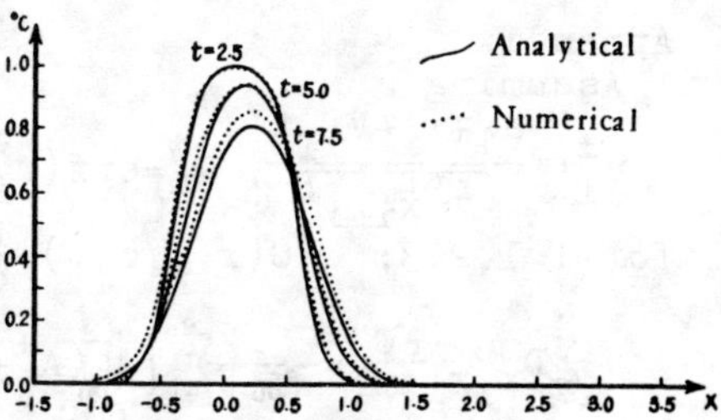

Fig.4 Result of example 4

Example 5 concerns the three-dimensional convection-heat dispersion problem of solving

$$\frac{\partial T}{\partial t}=\frac{1}{c}\,\mathrm{div}(\lambda\ \mathrm{grad}\ T)-\frac{c_w}{c}\underline{V}\ \mathrm{grad}\ T \qquad (x,y)\in F,\quad -\frac{3}{2}M\leqslant z\leqslant\frac{3}{2}M$$

subject to

$$T|_{t=0}=T_0(x,y,z) \qquad (x,y)\in F,\ -\frac{3}{2}M\leqslant z\leqslant\frac{3}{2}M$$
$$T|_{\Gamma}=\varphi(x,y,z) \qquad (x,y)\in F,\ -\frac{3}{2}M\leqslant z\leqslant\frac{3}{2}M$$
$$T(x,y,-\tfrac{3}{2}M,t)=T_1(x,y) \qquad (x,y)\in F$$
$$T(x,y,\tfrac{3}{2}M,t)=T_2(x,y) \qquad (x,y)\in F$$
$$T|_{w_i}=\psi_i(z,t) \qquad w_i\in F$$

where T is temperature; λ is the coefficient of heat dispersion; c and c_w are the heat capacity of the porous medium and water, respectively; $\underline{V}$ is the filtration velocity of groundwater; T_1, T_2, φ and T_0 are prescribed functions; w_i is the surface of injection well i; ψ_i is its temperature; M is the aquifer thickness. A description of the results is given in another paper for this conference.

Conclusion

A new method for the numerical solution of the convection-diffusion equation in one ,two and three dimensions is presented. The method is employed to obtain the numerical solution of some solute transfer and heat transfer problems. The numerical results presented demonstrate that the method is capable of solving advection-dispersion problems without generating significant numerical diffusion when Peclet number is not too large and oscillations. Numerical diffusion is mainly caused by the interpolation between nodes. Also due to the increasing of interpolation and computation, the results of two- and three-dimensional problems are not as good as one-dimensional one.

References

1. Neuman, S. P., Adaptive Eulerian-Lagrangian finite element method for advection-dispersion, Intern. J. Numer. Methods in Eng., Vol.20, 1984.
2. Xue, Yuqun and Xie, Chunhong, Numerical methods in hydrogeology, Coal Ind. Press, Beijing, 1980.

Appendix

Assuming

$$u_i^{\pm}=\frac{a_i \mp x_i \pm V_{xi} t}{\sqrt{4D_{x_i x_i} t}},\quad A_i^{\pm}=\mathrm{erf}(u_i^{\pm}),\quad x_i=x,y,z \text{ and } a_i=a,b,c$$

for i=1,2,3; $C(x,y,z,t)=\frac{1}{8}\prod_{i=1}^{3}(A_i^{+}+A_i^{-})$, then we have

$$\sum_{i=1}^{3} D_{x_i x_i}\frac{\partial^2 C}{\partial x_i^2}=-\frac{1}{8\sqrt{\pi}\,t}\sum_{i=1}^{3}\left\{\left[\prod_{\substack{j=1\\(j\neq i)}}^{3}(A_j^{+}+A_j^{-})\right](u_i^{+}e^{-u_i^{+2}}+u_i^{-}e^{-u_i^{-2}})\right\} \quad (A)$$

$$\sum_{i=1}^{3} V_{x_i}\frac{\partial C}{\partial x_i}=\frac{1}{8\sqrt{\pi}\,t}\sum_{i=1}^{3}\left\{\frac{V_{x_i}}{\sqrt{D_{x_i x_i}}}\left[\prod_{\substack{j=1\\(j\neq i)}}^{3}(A_j^{+}+A_j^{-})\right](e^{-u_i^{-2}}-e^{-u_i^{+2}})\right\} \quad (B)$$

$$\frac{\partial C}{\partial t}=\frac{1}{8\sqrt{\pi}}\sum_{i=1}^{3}\left\{\left[\prod_{\substack{j=1\\(j\neq i)}}^{3}(A_j^{+}+A_j^{-})\right]\left[\frac{V_{x_i}}{\sqrt{D_{x_i x_i}\,t}}(e^{-u_i^{+2}}-e^{-u_i^{-2}})-\frac{1}{t}(u_i^{+}e^{-u_i^{+2}}+u_i^{-}e^{-u_i^{-2}})\right]\right\} \quad (C)$$

Comparing (A) and (B) with (C), we obtain (A)+(B)=(C), so C satisfies the following equation

$$D_{xx}\frac{\partial^2 C}{\partial x^2}+D_{yy}\frac{\partial^2 C}{\partial y^2}+D_{zz}\frac{\partial^2 C}{\partial z^2}-V_x\frac{\partial C}{\partial x}-V_y\frac{\partial C}{\partial y}-V_z\frac{\partial C}{\partial z}=\frac{\partial C}{\partial t}$$

When t=0, we have

$$C(x,y,z,0)=\lim_{t\to 0}\frac{1}{8}\left[\prod_{i=1}^{3}(A_i^{+}+A_i^{-})\right]=1$$

if $-a\leq x\leq a,\ -b\leq y\leq b,\ -c\leq z\leq c$;

otherwise $C(x,y,z,0)=0$

So C satisfies the initial condition. We can prove that C satisfies the boundary conditions:

$$C(x_l,y,z,t)=C(x_r,y,z,t)=C(x,y_b,z,t)=$$
$$C(x,y_t,z,t)=C(x,y,z_f,t)=C(x,y,z_b,t)=0,$$

if $x_l=y_b=z_f=-\infty$ and $x_r=y_t=z_b=\infty$.

Acknowledgement

This work was supported by the National Natural Science Fund of China.

A Zoomable and Adaptable Hidden Fine-Mesh Approach to Solving Advection-Dispersion Equations

G.T. Yeh
Environmental Sciences Division, Oak Ridge National Laboratory, Oak Ridge, Tennessee 37831, USA

ABSTRACT

A zoomable and adaptable hidden fine-mesh approach (ZAHFMA), that can be used with either finite element or finite difference methods, is proposed to solve the advection-dispersion equation. The approach is based on automatic adaptation of zooming a hidden fine-mesh in the place where the sharp front locates. Preliminary results indicate that ZAHFMA used with finite element methods can handle the advection-dispersion problems with Peclet number ranging from 0 to ∞.

INTRODUCTION

Contaminant transport in the subsurface is often modeled with advection-dispersion equations. Many numerical methods have been employed to solve the advection-dispersion equations. Most conventional numerical methods can be classified into two major categories: Eulerian and Lagrangian approaches. In the Eulerian approach, the equation is discretized by a finite difference or a finite element grid system fixed in space. In the Lagrangian approach, either a deforming grid or a fixed grid in deforming coordinate can be used.

Experiments have shown that the Eulerian approach using conventional finite element methods (FEMs) or finite difference methods (FDMs) has performed well for dispersion dominant transport problems. For advection dominant transport problems, oscillation solutions may result when the Eulerian approach is used in conjunction with conventional FEMs or FDMs. The Lagrangian method can also be used to circumvent the problem of oscillations but it is not always easily adapted to deal with complex subsurface media (Neuman[1]).

A third approach which is a mix of the Lagrangian-Eulerian method has been gaining popularity in the past decade (Neuman[1],

Konikow and Bredehoeft[2], Molz et al.[3]). In this mixed method, one adopts a Lagrangian viewpoint when dealing with the advection terms and an Eulerian viewpoint when dealing with all other terms in the transport equations. In the Lagrangian step, either continuous forward particle tracking -- CFPT (Konikow and Bredehoeft[2]), single-step reverse particle tracking -- SRPT (Molz et al.[3]), or the combination of both (Neuman[1]) has been used. The SRPT could introduce a significant amount of numerical dispersion (Yeh and Tripathi[4]). Futhermore, if continuous multi-sources are present in interior nodes, the SRPT would give incorrect solution unless the Courant number is less than or equal to 1. For the CFPT, the treatment of complex boundary conditions and nonlinearities is not straightforward and the constant handling of numerous particles is troublesome and time consuming. The combined SRPT and CFPT approach eliminate some of these deficiencies but still leaves many questions unanswered (Neuman[1]). For example, how the solution quality depends the number of particles and the density of particles around sharp front.

From the above discussions, it is clear that the Eulerian approach is still the simplest and most straightforward way to solve advection-dispersion equations provided numerical oscillations can be eliminated. The easiest way of eliminating numerical oscillations can be achieved by restricting the spatial grid size such that the mesh Peclet number is less than certain critical number, which depends on the numerical scheme used (Jensen and Finlayson[5]). However, it is not always practical to reduce the grid size; and it is certainly impossible to achieve the elimination of oscillations by reducing grid size for the case of pure advection. The alternative is to use upstream FDMs or FEMS that are able to eliminate oscillations for mesh Peclet number ranging from 0 to ∞. However, upstream methods introduce large numerical dispersion coefficient. Numerical dispersion coefficient can be reduced by using fine-grid system or by using higher-order approximations in space, time, or both. Using higher-order finite element techniques may re-introduce oscillations. Hence, higher-order approximations have not proven capable of entirely and efficiently eliminating both numerical oscillation and numerical dispersion. On the other hand, using an extremely fine grid throughout the whole region to reduce the numerical dispersion coefficient may not be practical for many problems.

Since numerical dispersion depends on both the numerical dispersion coefficient and the gradient of concentration, there is no need to reduce numerical dispersion coefficient at the region where the gradient of concentration is very small. Therefore, we propose a zoomable and adaptable hidden fine-mesh approach (ZAHFMA) to solving the advection-dispersion equations. ZAHFMA coupled with upstream methods would entirely eliminate the numerical oscillation and sufficiently reduce the numerical dispersion coefficient at sharp-front regions.

ZOOMABLE AND ADAPTABLE HIDDEN FINE-MESH APPROACH (ZAHFMA)

Let us use a simple linear line finite element (Fig. 1) to illustrate how the ZAHFMA is implemented. First, we discretize the region with M elements (for example M = 3) and N nodes (for example N = 4) in the region (Fig. 1a). Second, we embed a predetermined number of nodes L and elements K (for example L = 3, K = 4) in each element (Fig. 1a). Third, we apply the spatial finite element and temporal finite difference to an advection-dispersion equation for each element to yield an element matrix equation

$$[A^e]\{C^e\} = \{R^e\} \tag{1}$$

where $[A^e]$ is the element coefficient matrix, $\{C^e\}$ is the unknown vector of the concentration, and $\{R^e\}$ is the element load vector. Fourth, we loop over all elements to determine if steep concentration gradient exists within an element. If the element is not a sharp-front element, regular finite element integration is used to obtain $[A^e]$ and $\{R^e\}$. Then $[A^e]$ and $\{R^e\}$ are assembled into the global coefficient matrix and global load vector, respectively.

If the element is a sharp-front element, we zoom the element and renumber the hidden nodes and the global nodes on the boundary of the element consecutively (Fig. 1b). We then obtain $[A^e]$ and $\{R^e\}$ by assembling the fine-mesh element matrix

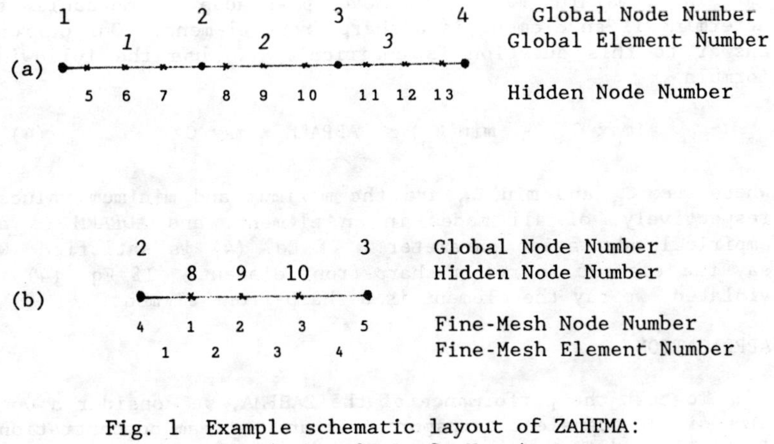

Fig. 1 Example schematic layout of ZAHFMA:
N = 4, M = 3, L =3, K = 4.

and fine-mesh element load vector (both of which are obtained by finite element integration) over K fine-mesh elements. Since $\{C^e\}$ can be subdivided into parts associated with hidden fine- mesh nodes, $\{C_f^e\}$, and others associated with global nodes, $\{C_g^e\}$, we make Gaussian reduction of Eq. (1) up to the L-th equation to yield

$$[U^e]\{C_f^e\} + [V^e]\{C_g^e\} = \{R_f^{*e}\} \tag{2}$$

$$[A^{*e}]\{C_g^e\} = \{R_g^{*e}\} \tag{3}$$

where $[U^e]$ is the upper triangular element coefficient matrix, $[V^e]$ and $[A^{*e}]$ are the reduced element coefficient matrices, and $\{R_f^{*e}\}$ and $\{R_g^{*e}\}$ are the reduced $\{R_f^e\}$ and $\{R_g^e\}$, respectively, after Gaussian reduction. $[A^{*e}]$ and $\{R_g^{*e}\}$ represent the element coefficient matrix and element load vector, respectively, of the zoomed element. Therefore, after the Gaussian reduction, $[A^{*e}]$ and $\{R_g^{*e}\}$ are assembled into the global coefficient matrix and global load vector, respectively.

Fifth, we solve the assembled global matrix equation to yield the concentrations at all global nodes. Finally, we compute the concentrations at all hidden nodes. If the hidden nodes are not in a sharp-front element, we compute the concentrations using a consistent finite element interpolation formula. If the hidden nodes are in a sharp-front element, we can easily solve the concentrations with Eq. (2) because $[U^e]$, $[V^e]$, $\{C_g^e\}$, and $\{R_f^{*e}\}$ are already known. The procedure outlined above completes a one-time step computation.

The remaining task is to develop an adaptive mechanism to determine if an element is a sharp-front element. Our current answer to this question is empirical. We use the following formula

$$(\max C_n - \min C_n) \leq \text{ADPARM} \times \max C_n \tag{4}$$

where $\max C_n$ and $\min C_n$ are the maximum and minimum values, respectively, of all nodes in an element, and ADPARM is an empirical adaptation parameter. If Eq. (4) is satisfied, we say the element is not a sharp-front element. If Eq. (4) is violated, we say the element is a sharp-front element.

APPLICATION

To test the performance of the ZAHFMA, we consider a one-dimensional transient transport from an upstream concentration. Initially, the concentration over the region $0 \leq x \leq 2$ is assumed zero everywhere. Boundary conditions are given as $C = 1$ at $x = 0$ and $C = 0$ at $x = 2$. For ZAHFMA simulation, the region is discretized with 8 elements with element length equal to 0.25. A time step size of 0.1 is used for simulation. The adaptation parameter used is ADPARM = 0.1. Two examples are used for illustration. In the first example, we use a velocity of zero and a dispersion coefficient of 0.01. Thus, the first example represents pure dispersion with Peclet number equal to 0. In the second example, we use a velocity of 0.25 and a dispersion coefficient of zero, which represents a pure advection with Peclet equal to ∞. Figure 2a shows the concentration profile at time equal to 10 and Figure 2b depicts the concentration profile at time equal to 4. It is seen that, for the first example, ZAHFMA yield very close results to the analytical solution whether the hidden nodes are 0, 1, or 3. On the other hand, for the second example, numerical dispersion is greatly reduced with 9 hidden nodes per element. Thus, the number of hidden nodes per element required to reduce numerical dispersion to an acceptable level depends on the nature of problems.

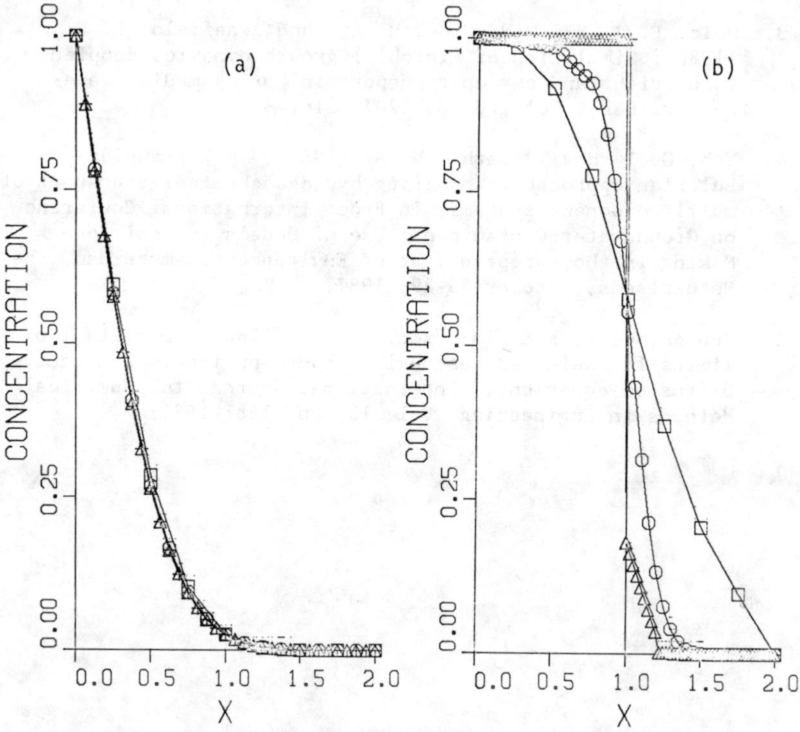

Fig. 2 Concentration profiles for
(a) pure dispersion -
□ = 0 hidden node, ○ = 1 hidden node, and
Δ = 3 hidden nodes, and + = analytical
(b) pure advection -
□ = 0 hidden node, ○ = 4 hidden nodes,
Δ = 9 hidden nodes, and + = analytical.

ACKNOWLEDGEMENTS

This research is supported by the Environmental Sciences Division, Oak Ridge National Laboratory, Oak Ridge, Tennessee. ESD Publication No. 3065.

REFERENCES

1. Neuman, S. P. (1983), Computer Prediction of Subsurface Radionuclide Transport -- An Adaptive Numerical Method., NUREG/CR-3076, Dept. of Hydrology and Water Resources, University of Arizona, Tucson, Arizona.

2. Konikow, L. F. and Bredehoeft, J. D. (1978), Computer Model of Two-Dimensional Solute Transport and Dispersion in Groundwater, Techniques of Water-Resources Investigation of the United States Geological Survey, Chapter C2, Book 7, USGS, Reston, Virginia.

3. Molz, F. J. M., Widdowson, M. A., and Benefield, L. D. (1986), Simulation of microbial growth dynamics coupled to nutrient and oxygen transport in porous media, Water Resour. Res., Vol.22, pp. 1207-1216.

4. Yeh, G. T. and Tripathi, V. S. (1987), A Lagrangian-Eulerian approach to modeling hydrogeochemical transport of multi-component systems, in Proc. International Conference on Groundwater Contaminant: Use of Models in Decision-Making in the European Year of Environment, Amsterdam, The Netherlands, October 26-29, 1987.

5. Jensen, O. K. and Finlayson, B. A. (1980), Oscillation limits for weighted residual methods applied to convective diffusion equations, International Journal for Numerical Methods in Engineering, Vol. 15, pp. 1681-1689.

SECTION 1B - COMPUTATIONAL FLUID DYNAMICS

A Taylor Weak Statement CFD Algorithm for Free Surface Hydromechanical Flows

A.J. Baker and G.S. Iannelli

University of Tennessee, Knoxville, TN, USA

ABSTRACT

A finite element CFD algorithm is established for the two-dimensional, unsteady depth-averaged free surface Navier-Stokes equations for an incompressible isoenergetic flow. The governing equation set is cast in hyperbolic conservation law form, whereupon a temporal Taylor series is utilized to establish truncation error terms. The resulting Taylor weak statement embeds functional expressions dominating dispersion and dissipation error mechanisms of the semi-discrete approximation. Theoretical analyses are developed to characterize these mechanisms for the linear basis implementation. An efficient numerical linear algebra procedure is developed using matrix tensor products. Numerical results are highlighted that document algorithm performance.

INTRODUCTION

Numerical (*CFD*) solution procedures have been under development for the two-dimensional depth-averaged, free surface incompressible Navier-Stokes equations for well over a decade, c.f., Gray, et.al., [1]. The sequence of international conferences on *Water Resources* has served to disseminate topical progress on a regular basis. For finite element methods, recent theoretical issues have focused on conservation law formulations for the governing equation system and efficient code implementation. King and co-workers have developed the RMA code system, written as a Galerkin weak statement on a non-divergence form, and use mixed interpolation in a quasi-Newton implicit procedure, c.f., [2]. In distinction, other research groups, c.f., Werner and Lynch [3], Gray, Drolet and Kinnmark [4], have developed tidal cycle codes using a second order wave form of the mass conservation equation coupled with the non-divergence momentum equations. In distinction, Walters [5] expands the primitive variable description in a temporal Fourier series and develops a Galerkin weak statement that does not utilize mixed interpolation.

In the allied field of computational aerodynamics, recent research the inviscid, compressible Euler equations have focused on CFD algorithms for the associated hyperbolic conservation law form.

Finite element methods have been developed that are robust, stable and efficient, c.f., Baker, et.al [6]. The formulational extension of the classical Galerkin weak statement, via a temporal Taylor series, yields a precise statement of higher order terms dominating algorithm stability. Baker and Kim [7] document this theory, and verify independent establishment of over a dozen dissipative algorithms from various CFD theories.

The free surface hydrodynamics equations can be rearranged into hyperbolic conservation law form, Baker [8, Ch.5]. This paper develops the Taylor weak statement for this system, and highlights associated theoretical analyses for stability and dispersion error. An implicit time integration algorithm yields a large matrix statement, the Newton algorithm for which is efficiently approximated using matrix tensor product factorizations. Numerical results highlight algorithm and code performance for representative test problems.

THEORETICAL DEVELOPMENT

A. Conservation Law Statement

The *shallow-water* form of the incompressible Navier-Stokes equations are obtained by integrating through the fluid depth in the direction parallel to the gravity vector. Denoting h as the free surface elevation above the bathymetry $b(x_i)$, and defining u_i as the depth-averaged velocity vector, the equation set is

$$L(h) = \frac{\partial h}{\partial t} + h\frac{\partial u_i}{\partial x_i} + u_i\frac{\partial h}{\partial x_i} = 0 \tag{1}$$

$$L(u_i) = \frac{\partial u_i}{\partial t} + u_j\frac{\partial u_i}{\partial x_j} + g\frac{\partial h}{\partial x_i} - \frac{\partial \sigma_{ij}}{\partial x_j} + \Omega_{ij}u_j - \tau_i\Big|_b^{h+b} = 0 \tag{2}$$

In (1)-(2), the tensor indices range $1 \leq (i,j) \leq 2 = n$, σ_{ij} is the kinematic stress tensor, Ω_{ij} is the Coriolis coefficient and τ_i is the shear stress evaluated at the channel bed and free surface.

The hyperbolic conservation law form of (1)-(2), following non-dimensionalized by appropriate reference variables, is

$$L(h) = \frac{\partial h}{\partial t} + \frac{\partial}{\partial x_j}(m_j) = 0 \tag{3}$$

$$L(m_i) = \frac{\partial m_i}{\partial t} + \frac{\partial}{\partial x_j}\left(m_j m_i/h + 2Fr^{-1}h^2\delta_{ij} - Re^{-1}h\sigma_{ij}\right)$$
$$+ Fr^{-1}h\frac{\partial b}{\partial x_i} - Re^{-1}\sigma_{ij}\frac{\partial h}{\partial x_j} = 0 \tag{4}$$

where $m_i \equiv hu_i$, $Fr = U^2/Lg$ is the Froude number, $Re = UL/\nu$ is the Reynolds number, and the last two terms in (2) have been deleted for simplicity. For vanishing kinematic viscosity ν, the Reynolds number terms in (4) vanish. It is then easy to prove that (3)-(4) is a hyperbolic system. The necessary and sufficient condition yields

$$\lambda_x^1 = u, \quad \lambda_x^2 = u + \sqrt{gh}, \quad \lambda_x^3 = u - \sqrt{gh} \tag{5}$$

Similar expressions result for λ_y, with u replaced by v, hence (3)-(4) is indeed a hyperbolic conservation law system for large Re.

B. Taylor Weak Statement

Defining $\mathbf{q} = (h, m_1, m_2)$, the matrix form of (3)-(4) is,

$$L(\mathbf{q}) = \frac{\partial \mathbf{q}}{\partial t} + \frac{\partial \mathbf{f}_j}{\partial x_j} + \mathbf{s} = 0 \tag{6}$$

where the flux vector $\mathbf{f}_j$ and source term $\mathbf{s}$ are appropriately defined. In all instances, an evolutionary solution is sought, starting with some initial condition $\mathbf{q}(x_i, t=0) = \mathbf{q}_o$. Thus, the Taylor series is,

$$\mathbf{q}^{n+1} = \mathbf{q}^n + \Delta t \mathbf{q}_t^n + \frac{1}{2}\Delta t^2 \mathbf{q}_{tt}^n + \frac{1}{6}\Delta t^3 \mathbf{q}_{ttt}^n + \cdots \tag{7}$$

where $\Delta t \equiv t^{n+1} - t^n$ and subscript t denotes temporal derivative. Equation (6) allows derivative interchange, specifically,

$$\mathbf{q}_t = \frac{\partial \mathbf{f}_j}{\partial x_j} - \mathbf{s} = -\frac{\partial \mathbf{f}_j}{\partial \mathbf{q}}\frac{\partial \mathbf{q}}{\partial x_j} - \mathbf{s} = -A_j \frac{\partial \mathbf{q}}{\partial x_j} - \mathbf{s} \tag{8}$$

where A_j is the Jacobian (matrix) of the flux vector $\mathbf{f}_j$. The second term in (7) can be formed in two ways, dependent upon derivative exchange; for time independent source term one obtains [7]

$$\mathbf{q}_{tt} = \frac{\partial}{\partial x_j}\left(\alpha A_j \mathbf{q}_t + \beta A_j A_k \frac{\partial \mathbf{q}}{\partial x_k}\right) \tag{9}$$

where α-β=-1. The third derivative can be similarly reexpressed, but is not included herein for brevity.

Substituting (7)-(8), moving $\mathbf{q}^n$ to the left side of (6), dividing through by Δt and taking the approximate limit $\Delta t \to \varepsilon \geq 0$ yields the modified conservation law statement,

$$L^m(\mathbf{q}) = \mathbf{q}_t^* + \frac{\partial \mathbf{f}_j}{\partial x_j} + \mathbf{s} - \frac{\Delta t}{2}\frac{\partial}{\partial x_j}\left(\alpha A_j \mathbf{q}_t + \beta A_j A_k \frac{\partial \mathbf{q}}{\partial x_k}\right) - \ldots = 0 \tag{10}$$

where $\mathbf{q}_t^*$ denotes the approximate limit for $\mathbf{q}_t$.

The finite element algorithm is then established on (10), as the Taylor series augmentation to (3)-(4), following the standard Galerkin procedure. The state variable approximation is,

$$\mathbf{q}(x_i,t) \approx \mathbf{q}^h(x_i,t) = \bigcup_e \{N_k(\eta_i)\}^T\{\mathbf{Q}(t)\}_e \tag{11}$$

where $\{N_k(\eta)\}$ is the kth degree finite element trial space basis and $\{\mathbf{Q}\}$ contains the nodal values of $\mathbf{q}^h$ on the discretization $\cup\Omega_e = \Omega^h \subset \mathbb{R}^2$. The Taylor weak statement (TWS) with Galerkin criteria is thus,

$$TWS(\{\mathbf{Q}\}) \equiv S_e \int_{\Omega_e} \{N_k\} L^m(\mathbf{q}^h)\, d\tau = [M(k,\alpha)]\frac{d\{\mathbf{Q}\}}{dt} + \{R(k,\beta,A_j,\{\mathbf{Q}\},\Delta t)\} = 0 \tag{12}$$

where S_e is the assembly operator. The second form in (12) specifies that the TWS, formed for the semi-discretization (11), produces an ordinary differential equation system on $\{\mathbf{Q}\}$ with the indicated functional dependence. Substituting (12) into a variable-implicit discrete Taylor series [7] yields the fully discrete matrix statement,

$$[J]\{\delta\mathbf{Q}\}_{p+1} = -\left([M]\{\Delta\mathbf{Q}\}^{n+1} + \Delta t\,(\theta\{R\}^{n+1} + (1-\theta)\{R\}^{n})\right) \tag{13}$$

where $[J] \equiv \partial\{rhs\}/\partial\{\mathbf{Q}\}^{n+1}$ is the Newton Jacobian, p is the iteration index, and $\{\Delta\mathbf{Q}\}^{n+1} \equiv \{\mathbf{Q}\}^{n+1} - \{\mathbf{Q}\}^{n}$.

C. Theoretical Analysis

Baker and Kim [7] document the theoretical stability analysis for scalar conservation laws of the form (10). Of the numerous choices available for α and β, the Raymond-Garder [9] definition ($\alpha \equiv \beta$) yields a high-order accurate implicit formulation when using the linear basis specification, $k=1$ in (11). The resultant Fourier stability analysis expresses discrete error mechanisms in terms of modal dispersive and dissipative components, as occur for the even and odd order term expansions in the Fourier symbolic analysis. For the Raymond-Garder definition, these expressions are,

$$\textit{even:}\quad \delta = \frac{a}{m}\left[c\left(\frac{1}{2} - (a\Delta x\,Re)^{-1} - \theta\right)m^2 + \left(c^3\left(\theta^3 - \theta^2 + \frac{1}{2}\theta - \frac{1}{8}\right) - \frac{v}{12}\right)m^4 + O(m^6)\right] \tag{14}$$

$$\textit{odd:}\quad (a - \phi) = \frac{a}{m}\left[\left(c^2 - \theta + \frac{1}{3}\right)m^3 + O(m^5)\right] \tag{15}$$

where $m = \omega\Delta x$ and the Courant number is $c \equiv a\Delta t/\Delta x$ where a is the convection velocity in the model equation, equivalent to u_j in (1)-(2). For large Re, and $\theta = \frac{1}{2}$ in (13), (14) confirms that the (artificial) dissipation level is $O(m^4)$ and controlled by ν, the RG parameter defined via $\alpha \equiv 2\nu\Delta x/a\Delta t \equiv \beta$. The n-dimensional generalization replaces these definitions by $\alpha \equiv 2\nu\, det^{1/n} / |u_j|\Delta t \equiv \beta$, where det is the measure of the finite element domain $\Omega_e \subset \mathbb{R}^2$ and $|\cdot|$ denotes magnitude. The dispersion error is $O(m^3)$ for all θ and c.

NUMERICAL PROCEDURE

The TWS finite element algorithm is code-implemented for bilinear and biquadratic trial space basis formulations and for variable θ, $\frac{1}{2} \leq \theta \leq 1$. It is applicable to general geometries via direct element embedding of the isoparametric coordinate transformation. Acceptable grids include those created by body-fitted coordinate transformations as well as macro-element block-algebraic procedures.

The restriction to the tensor product basis family renders the code amenable to an efficient linear algebra solution procedure using a tensor matrix product factorization of the Newton algorithm Jacobian. Recalling (13), the definition is,

$$[J] \equiv \frac{\partial\{rhs\}}{\partial\{\mathbf{Q}\}} = [M] + \theta\Delta t\,\frac{\partial\{R\}}{\partial\{\mathbf{Q}\}} \tag{16}$$

The lead term in $[M]$ is the assembly of the elemental *mass matrix*.

$$[M]_e \equiv \int_{\Omega_e} \{N_k\}\{N_k\}^T d\tau \tag{17}$$

The tensor product basis admits construction of the n-dimensional matrix $[M]_e$ as the tensor matrix product of n one-dimensional mass matrices $[M_i]_e$, $1 \leq i \leq n$. Since the Jacobian (16) is formed via assembly of $[J]_e$ over $\cup\overline{\Omega}_e$, it can be *approximately* constructed as,

$$[J] \equiv S_e[J]_e = S_e\Big([J_1]_e \otimes [J_2]_e\Big) \tag{18}$$

where,

$$[J_i]_e \equiv [M_i]_e + \theta\Delta t \frac{\partial\{R_i\}}{\partial\{\mathbf{Q}\}}, \quad 1 \leq i \leq 2 \tag{19}$$

which introduces an error in $[J]$ of order $(\theta\Delta t)^2$. The form of (18) suggests replacement of (13) with the numerical linear algebra procedure,

$$\begin{aligned} \gamma[J_1]\{\mathbf{P}\} &= -\gamma^2\{rhs(19)\} - \gamma^2(\theta\Delta t)^2\{error\} \\ \gamma[J_2]\{\delta\mathbf{Q}\}_{p+1} &\equiv \{\mathbf{P}\} \end{aligned} \tag{20}$$

hence $\{\delta\mathbf{Q}\}$ is the approximate solution of (13) and $\gamma<0$. The matrices $[J_i] \equiv S_e([J_i]_e)$ defined in (20) are 3-block, $(2k+1)$-diagonal, for the developed shallow water CFD algorithm, hence are easy to form and require negligible memory.

The execution process for (20) constitutes sweeping on all grid lines parallel (say) to the η_1-curvilinear coordinate system, to compute the intermediate array $\{\mathbf{P}\}$, which serves as the data for the subsequent grid sweepings parallel to the η_2-curvilinear system. For $\|\delta\mathbf{Q}_{p+1}\| \leq \varepsilon$, the convergence requirement, the solution $\{\mathbf{Q}\}^{n+1}$ has converged to satisfy (13) to within the indicated tensor product factorization error. The error size is controlled via the constant γ since the tensor product factorization term is not explicitly evaluated. Hence, it is scaled by γ^2 for all $\gamma<0$ for which (20) is numerically tractible, i.e., not destabilized by round-off error.

DISCUSSION AND RESULTS

The computer code embodying the TWS finite element algorithm statement has been developed for estuarine flow field analyses. Several one-dimensional test case problems confirm robust performance for sub- and super-critical flows over abrupt bed elevation changes. Elementary two-dimensional check cases driven to steady-state include nominally straight channel flows with various bathymetries, to assess application of boundary condition combinations as predicted appropriate by the characteristics analysis. Referring to (5), for $u(v) > 0$, two characteristic lines exist with positive slopes equal to λ^1 and λ^2. Consequently, two Dirichlet boundary conditions are always admissible at an inlet. The relative magnitude of $u(v)$, with respect to the celerity $\sqrt{gh}$, determines the sign of λ^3. Thus, for supercritical flow, $u > \sqrt{gh}$ and a third Dirichlet inlet condition is admissible. Conversely, for $u < \sqrt{gh}$, one Dirichlet boundary condition at the outlet is required. For subcritical flows, code experimentation has verified that numerically stable inlet boundary conditions can be either, a) fixed free surface height and inflow angularity, or b) specified volume flow rate and angularity. The consistent outlet boundary condition is vanishing normal derivative for both h and m_i, wherein constant h is imposed within the momentum equation solutions.

A channel geometry of considerable interest is flow through a sudden enlargement in cross-section, the so-called step-wall diffuser. Figure 1 shows the steady state velocity field for $Re = 50$. The sidewall boundary conditions are no-slip for m_i and vanishing normal derivative for h via the TWS β-term augmentation to (1), see (9). Manuscript length limitations preclude further discussion; the range of numerical test results will be presented at the conference.

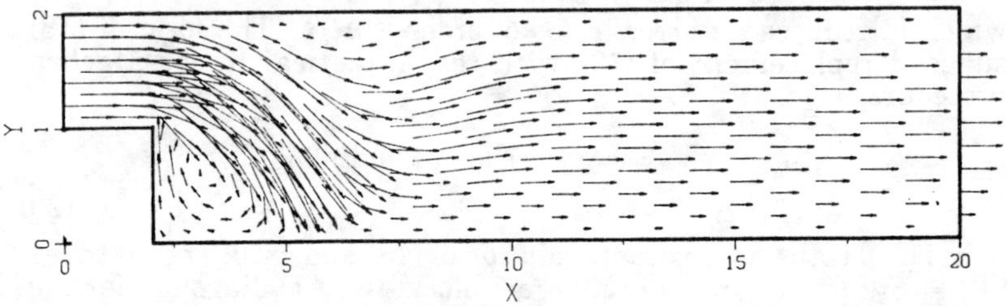

Figure 1. Step Wall Steady State Velocity Field, $Re = 50$.

SUMMARY AND CONCLUSIONS

A Taylor weak statement finite element algorithm is developed for the two-dimensional, shallow-flow Navier-Stokes equations in hyperbolic conservation law form. Theoretical analyses document algorithm accuracy and stability for the linear trial space basis implementation. An efficient numerical linear algebra procedure has been derived and implemented into a computer code. Numerical results verify code performance for prototype estuarine geometries including significantly separated flows.

REFERENCES

1. W.G. Gray, G.F. Pinder and C.A. Brebbia (eds.), *Finite Elements in Water Resources*, Pentech Press, London (1977).
2. I.P. King, Adv. Water Resources, 8, p. 69-76 (1985).
3. F. E. Werner and D. R. Lynch, Adv. Water Resources, 10, p. 115-130 (1987).
4. W.G. Gray, J. Drolet and P.E. Kinnmark, Adv. Water Resources, 10, p. 131-137 (1987).
5. R. A. Walters, Adv. Water Resources, 10, p. 138-148 (1987).
6. A.J. Baker, J.W. Kim, J.D. Freels and J.A. Orzechowski, Int. J. Num Mtd. Fluids, 7, p. 1235-1259 (1987).
7. A.J. Baker and J.W. Kim, Int. J. Num. Mtd. Fluids, 7, p. 489-520 (1987).
8. A. J. Baker, *Finite Element Computational Fluid Mechanics*, Hemisphere/Harper & Row, New York (1983).
9. W.H. Raymond and A. Garder, Mon. Weather Rev., 104, p. 1583-1590 (1976).

Numerical Simulation of the Vortex Shedding Process Past a Circular Cylinder

A. Giorgini and G. Alfonsi
School of Civil Engineering, Purdue University, West Lafayette, Indiana 47906, USA

ABSTRACT

A computational analysis is performed on the two dimensional, time dependent Navier-Stokes equations in their streamfunction-vorticity transport form. The equations are cast in logpolar coordinates and the fields are expanded in Fourier series along the azimuthal coordinate. The numerical techniques used for the computations are finite differences for the time advancement, Fast Fourier Transform to perform the convolutions resulting from the convective terms and matrix inversion method for the radial integration. In order to simulate the vortex shedding process, a perturbation, consisting in a pure rotational field, has been imposed to the initially irrotational flow at time T=0. The perturbation vortex is characterized by two quantities, the strength S and the spread σ, both depending on the values of two parameters, C and n. The results presented in this paper are those corresponding to the couple of values C=0.009 and n=3.44. The results are illustrated by means of computer generated drawings of the flow fields in term of absolute streamlines, relative streamlines and vorticity.

INTRODUCTION

The flow of a viscous fluid past a circular cylinder

has been simulated by integrating the Navier-Stokes equations in two dimensions. The most remarkable feature of the model is the use of a mixed spectral-finite analytic numerical technique which has already given rather accurate results in numerical experiments formerly performed by the authors [1],[2],[3]. In this case the nonsteady nonsymmetric flow past an impulsively started circular cylinder is calculated. The Reynolds number is Re=1000 and the asymmetry is obtained by introducing a perturbation vortex at the nondimensional time T=0; the perturbation, consisting in a pure rotational field, is characterized by two quantities, the strength S and the spread σ, with the following values:

$$S = \frac{C}{(n-2)(n-3)} = \frac{0.8}{56} \tag{1}$$

$$\sigma = \frac{2}{(n-3)} = \frac{32}{7} . \tag{2}$$

The corresponding values of the parameters C and n are:

$$C = 0.009 \quad \text{and} \quad n = 3.44 .$$

The phenomenon is described up to T=40. The CDC-CYBER 205 supercomputer of the Purdue University Computing Center has been used for the calculations.

DESCRIPTION OF THE NUMERICAL EXPERIMENT

Referring to Figure 1 to 3, the number appearing on the low right side of each drawing is the nondimensional time T=Ut/R which can be interpreted as the number of radii travelled by the cylinder since its impulsive start. Up to time T=18 there is almost no evidence of any asymmetry in the fields, despite the fact that the perturbation vortex has been imposed at T=0. At T=20 the vorticity field starts showing a slight asymmetry which will become more evident as time elapses. By looking at the absolute streamlines, vortex 1 is a D (attached) vortex, while vortex 2 is an α (detached) vortex. At T=24 a change takes place: vortex 2 becomes a D (attached) vortex while vortex 1 becomes an α vortex, at T=26 is already detached from the cylinder and separates quickly. The vorticity area corresponding to vortex 1 starts to assume an elongated shape at T=30 and by T=38 and T=40 completely separated vorticity lumps

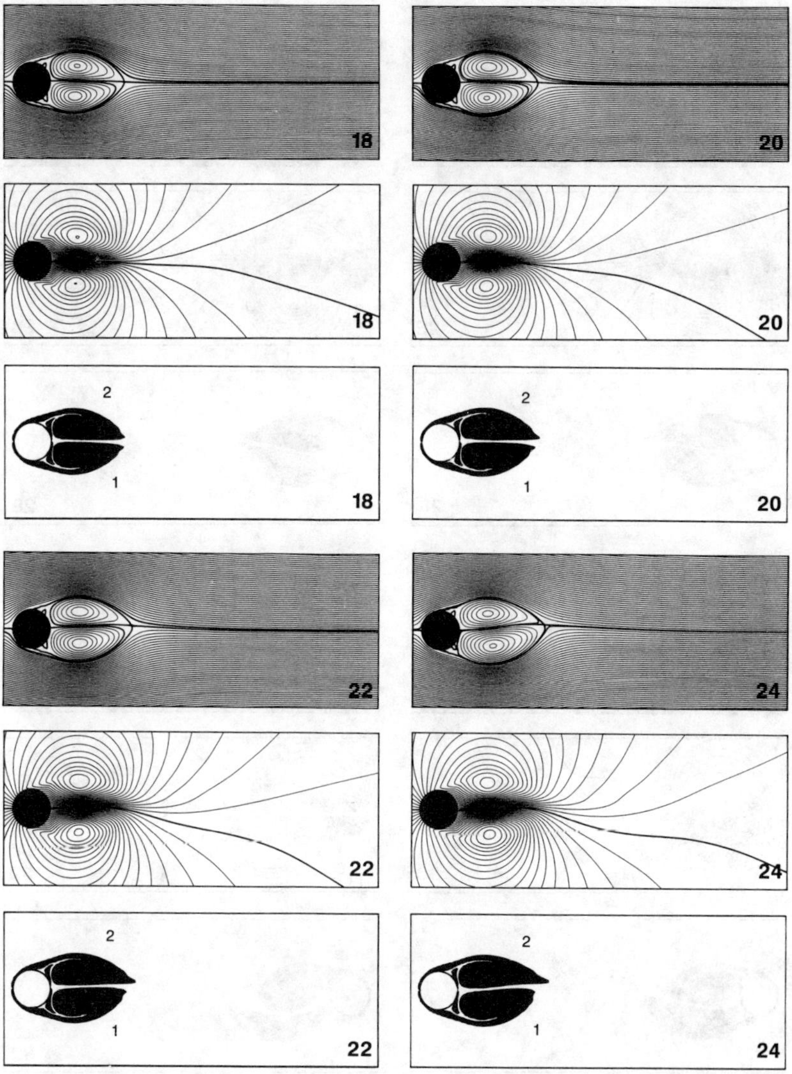

Figure 1. Absolute streamlines, relative streamlines and vorticity field representation for the nondimensional times T=18, T=20, T=22 and T=24.

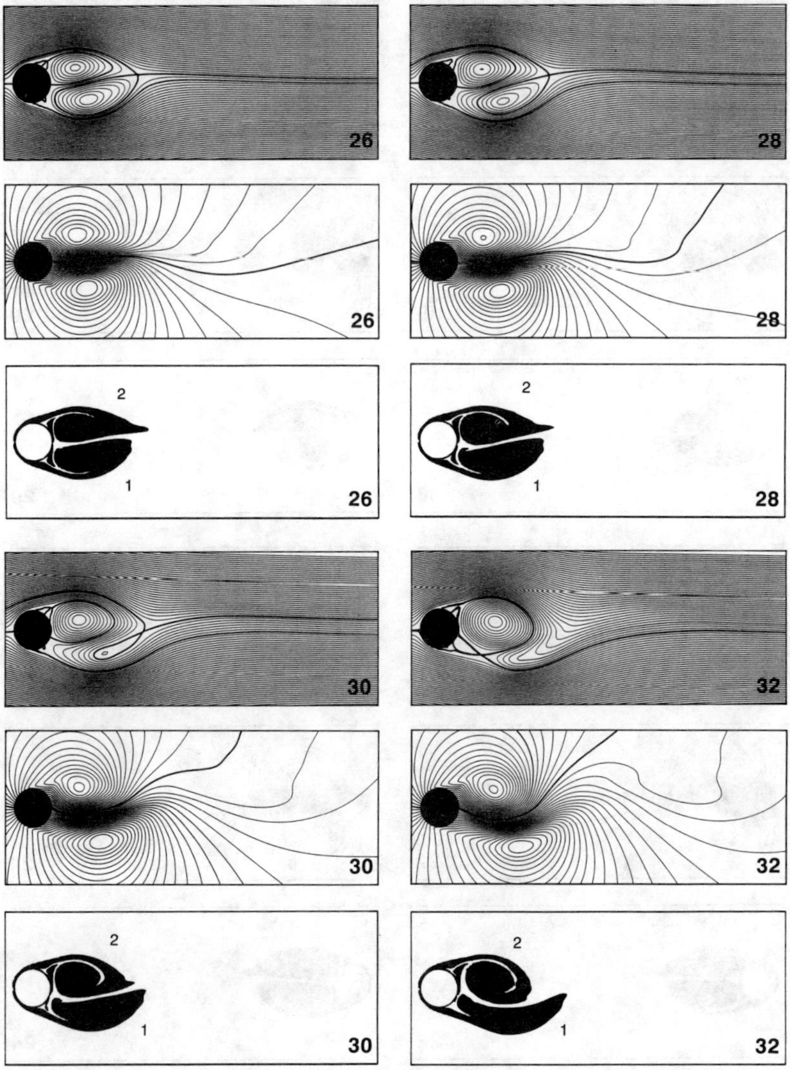

Figure 2. Absolute streamlines, relative streamlines and vorticity field representation for the nondimensional times T=26, T=28, T=30 and T=32.

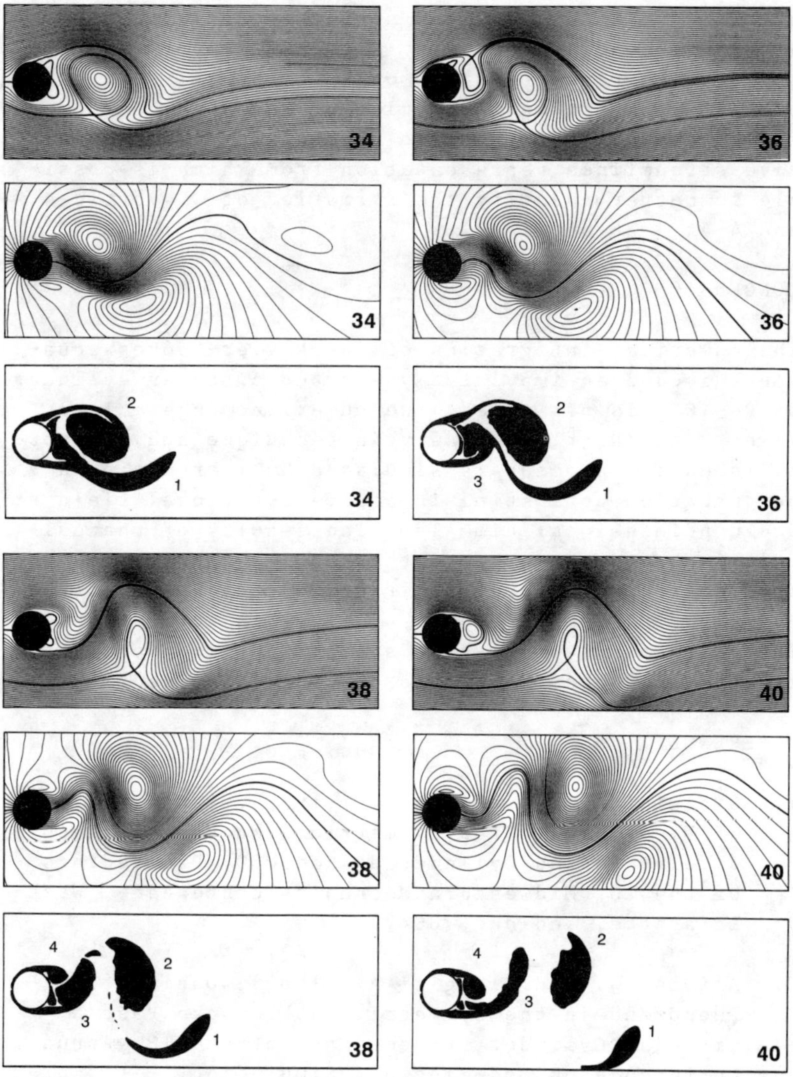

Figure 3. Absolute streamlines, relative streamlines and vorticity field representation for the nondimensional times T=34, T=36, T=38 and T=40.

appear. Vortex 2 grows and separates by T=34 while its corresponding vorticity area has a more rounded shape with respect to the one of vortex 1. All the process of birth and evolution of the vortices downstream the cylinder is also recognizable on the relative streamlines representation from which is possible to better identify the formation of vortices 3 and 4 at T=36 and T=38.

CONCLUSION

The numerical integration of the Navier-Stokes equations around an impulsively started circular cylinder at Re=1000 is performed. The numerical scheme is a mixed spectral-finite analytic technique and the vortex shedding process is simulated by introducing a perturbation consisting in a pure rotational field at the nondimensional time T=0. The results of the calculations are shown up to the nondimensional time T=40 and the sequence of birth and detachment of the primary vortices is clearly visible in term of absolute streamlines, relative streamlines and vorticity from the computer generated drawings provided.

REFERENCES

1. Giorgini A. and Alfonsi G. (1987), Early Stages of Development of a Symmetric Wake Past an Impulsively Started Cylinder at Re=3000, Proceedings of the 20^ Midwestern Mechanics Conference, West Lafayette, Indiana, USA.

2. Alfonsi G. and Giorgini A. (1987), Oscillatory Phenomena in the Symmetric Bubble Wake Past a Circular Cylinder, Proceedings of the 18^ Annual Conference on Modeling and Simulation, Pittsburgh, Pennsylvania, USA.

3. Giorgini A. and Alfonsi G. (1986), Symmetric Wake Development Past a Circular Cylinder, Proceedings of the ASCE Specialty Conference on Advancements in Aerodynamics, Fluid Mechanics and Hydraulics, Minneapolis, Minnesota, USA.

Numerical Investigation of Turbulent Flow Field in a Curved Duct with an Alternating Pressure Difference Scheme

Z.J. Liu, C.G. Gu and Y.M. Miao

Department of Power Mechanical Engineering, Xi'an Jiaotong University, China

ABSTRACT

A finite volume method is presented for the solution of the two-dimensional incompressible, steady Navier-Stocks equations with K- ε turbulence model. A new scheme, Alternating Pressure Difference Scheme (APDS), is proposed to eliminate the pressure oscillation with an ordinary grid arrangement. This numerical method is applied to predicting the turbulent flow field in a curved duct, and the results are compared with the corresponding experimental data. It is found that the calculated mean velocity field agrees well with the experimental one, and the calculated turbulence intensity distributions have the same trend as the experimental ones.

NOMENCLATURE

A_E, A_W, A_N, A_S, A_P	coefficients in the general finite difference equation
c_1, c_2, c_μ, σ_K, σ_ε	constants in the turbulence model equations
G^Φ	diffusion coefficient of the general scalar equation
J	Jacobian
K	turbulence kinetic energy
M_1, M_2	convective terms normal to η, ξ directions
P	pressure
P'	pressure correction
S^Φ	source term of the general scalar equation
$\vec{U}$	mean velocity vector
U, V	mean velocity components in x,y directions
x,y	Cartesian coordinates in physical plane
ξ,η	curvilinear coordinates in physical plane
ε	turbulence energy dissipation
ε_t	turbulence intensity
μ	laminar viscosity

μ_t	turbulent viscosity
ρ	density
Φ	general scalar

Subscripts

E,W,N,S	the east, west, north and south neighbouring grid points of point P
e,w,n,s	the east, west, north and south boundary surfaces of a finite volume
i,j	grid point position

Superscripts

Φ	general scalar
*	preliminary value or the value from last iteration
'	correction

INTRODUCTION

Patankar[5] showed that when an ordinary grid arrangement was employed for a solution of Navier-Stocks equations, pressure oscillation might occur. Harlow and Welch[3] introduced the staggered mesh method with which the pressure oscillation could be suppressed. However, as shown by Rhie and Chow[6], the staggered mesh method is not suitable for general curvilinear coordinates if the Cartesian velocity components are employed to present the govering equations, because the Cartesian velocity components are not related to the general curvilinear grid line orientations. On the other hand, for many practical problems, such as a flow field in turbine passages and over airfoils, the curvilinear coordinates are required to fit the boundaries.

In the present study, in order to develop a computer program which is widely applicable, the general curvilinear coordinates are employed. Instead of using the staggered mesh method, the Alternating Pressure Difference Scheme (APDS) is introduced to eliminate the pressure oscillation.

THEORETICAL FORMULATION AND NUMERICAL COMPUTATION

Governing equations

The governing equations can all be written in a general form:

$$\mathrm{div}(\rho \vec{U}\Phi - G^{\Phi}\,\mathrm{grad}\Phi) = S^{\Phi} \tag{1}$$

The variables Φ, G^{Φ}, S^{Φ} in x,y Cartesian coordinates for each governing equation are shown in Table 1

The near wall region is dealt with by the wall function method. The idea of pressure correction from Chorin[1] is adopted. Therefore, the pressure correction equation is introduced in the solution procedure.

Eq.1 is transformed into (ξ, η) plane, ξ lines and η lines are general body fitted coordinates. After integrating the transformed equation over a finite volume, shaded in Fig.2, and applying a hybrid scheme, the finite-difference equation of Eq.1 is as follows (Gu[2]):

$$A_P\Phi_P = A_E\Phi_E + A_W\Phi_W + A_N\Phi_N + A_S\Phi_S + S^{\Phi}_{new} \quad (2)$$

Table 1 Φ, G^{Φ}, S^{Φ} in x,y Cartesian coordinates

Equations	Φ	G^{Φ}	S^{Φ}
continuity equation	1	0	0
U momentum equation	U	$\mu+\mu_t$	$\frac{\partial}{\partial x}\left(G^{\phi}\frac{\partial U}{\partial x}\right)+\frac{\partial}{\partial y}\left(G^{\phi}\frac{\partial V}{\partial x}\right)-\frac{\partial}{\partial x}\left(P+\frac{2}{3}\rho K\right)$
V momentum equation	V	$\mu+\mu_t$	$\frac{\partial}{\partial x}\left(G^{\phi}\frac{\partial U}{\partial y}\right)+\frac{\partial}{\partial y}\left(G^{\phi}\frac{\partial V}{\partial y}\right)-\frac{\partial}{\partial y}\left(P+\frac{2}{3}\rho K\right)$
K equation	K	μ_t/σ_K	$\mu_t\left[2\left(\frac{\partial U}{\partial x}\right)^2+2\left(\frac{\partial V}{\partial y}\right)^2+\left(\frac{\partial U}{\partial y}+\frac{\partial V}{\partial x}\right)^2\right]-\rho\varepsilon$
ε equation	ε	μ_t/σ_ε	$\mu_t\left[2\left(\frac{\partial U}{\partial x}\right)^2+2\left(\frac{\partial V}{\partial y}\right)^2+\left(\frac{\partial U}{\partial y}+\frac{\partial V}{\partial x}\right)^2\right]c_1\frac{\varepsilon}{K}-c_2\rho\frac{\varepsilon^2}{K}$

where $\mu_t=c_\mu\rho K^2/\varepsilon$, $c_1=1.44$, $c_2=1.92$, $c_\mu=0.09$, $\sigma_K=1.0$, $\sigma_\varepsilon=1.3$ (3)

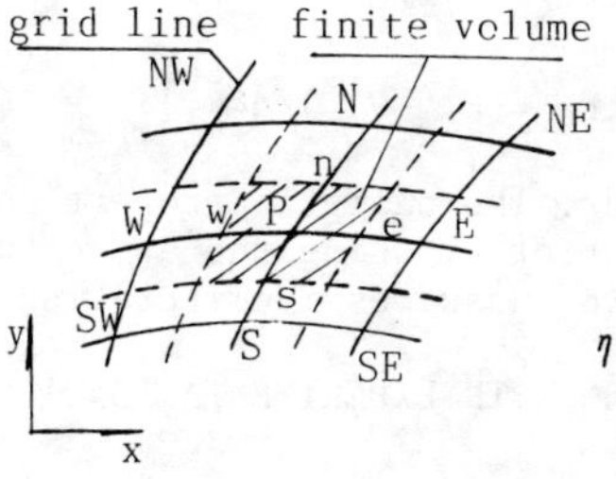

Fig.1 Physical plane

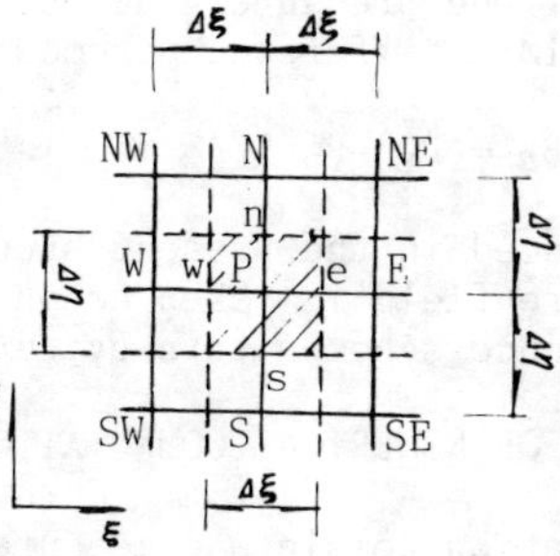

Fig.2 Transformed plane

Alternating pressure difference scheme

As mentioned in the introduction, when the ordinary grid arrangement is adopted, a pressure oscillation may occur, In the governing equations, the only terms which include pressure are S^U_{new} and S^V_{new}.From Table 1 and Eq.2, the finite-difference momentum equations can be written as:

$$A_P U_P = A_E U_E + A_W U_W + A_N U_N + A_S U_S + S^U_r + (y_\xi P_\eta - y_\eta P_\xi) \quad (4)$$

$$A_P V_P = A_E V_E + A_W V_W + A_N V_N + A_S V_S + S^V_r + (x_\eta P_\xi - x_\xi P_\eta) \quad (5)$$

Here S^U_r and S^V_r are residues from S^U_{new}, S^V_{new} after the pressure gradient terms have been extracted from them. Usually the pressure gradient P_ξ , P_η are approximated by centered difference scheme:

$$P_\xi = (P_E - P_W)/(2\Delta\xi) \qquad P_\eta = (P_N - P_S)/(2\Delta\eta) \quad (6)$$

It is seen that for an oscillation pressure field in Fig.3, the pressure gradient P_ξ is equal to zero if the centered $2\Delta\xi$ -difference is adopted. This is because the centered $2\Delta\xi$ -difference scheme relates the pressure values on every alternate point, W and E, it cannot sense the pressure oscillation between them. Therefore, once the oscillation pressure field appears from truncation errors or some other sources, the momentum equations lack the ability to correct the oscillation in succeeding iterations. This is to say the pressure oscillation with the ordinary arrangement originates from the centered $2\Delta\xi$ -difference approximation of the pressure gradient. To eliminate this oscillation, the $1\Delta\xi$ -difference schemes:

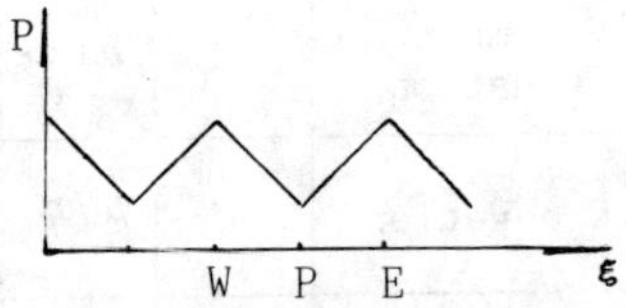

Fig.3 Oscillation pressure field

$$P_\xi = (P_E - P_P)/\Delta\xi \qquad P_\xi = (P_P - P_W)/\Delta\xi \quad (7)$$

are introduced in the present study. The elliptic property of the pressure field is taken account of by using the above two $1\Delta\xi$-difference schemes turn by turn in successive iterations.

COMPARISON OF NUMERICAL COMPUTATION WITH EXPERIMENTAL DATA

The computation domain and measurement stations are shown in Fig.4. The calculation is done on Honeywell DPS8/52. A 31X21 grid is adopted, and the convergence is reached in 100 iterations, the CPU time is about 9 minutes. No pressure oscillation is experienced. The experimental data are from Liu[4]. The inlet Mach number is 0.06. The experiment is done with a TSI 1050 hot-wire anemometer. The comparison of numerical computation with the experiment is shown in Fig.5 and Fig.6.

Fig.5 shows that computed mean velocity field agrees well with the experimental data. In Fig.6, the computed turbulence intensity distributions have the same trend as the experimental ones. Both the numerical computation and the experimental data show that the turbulence intensity near the concave wall is larger than that near the convex wall. However, the computed turbulence intensity changes more rapidly than the experimental near the concave wall. One possible reason for this discrepancy is that the turbulence model has not taken influence of streamline curvature into account.

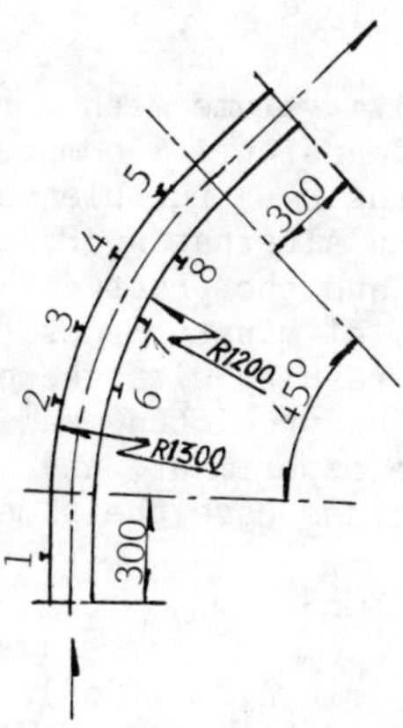

Fig.4 Computation domain & measurement stations

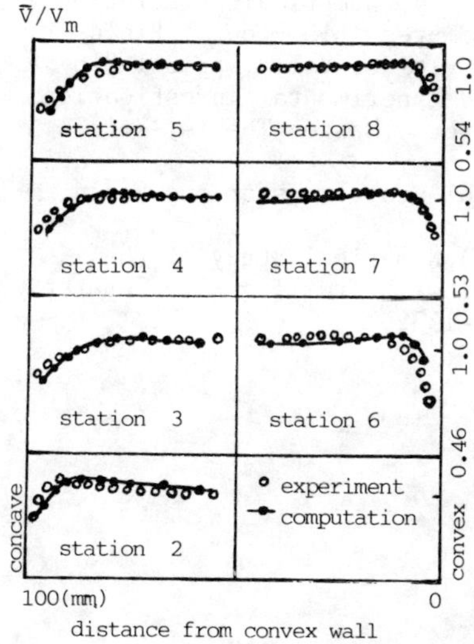

Fig.5 Comparison of mean velocity

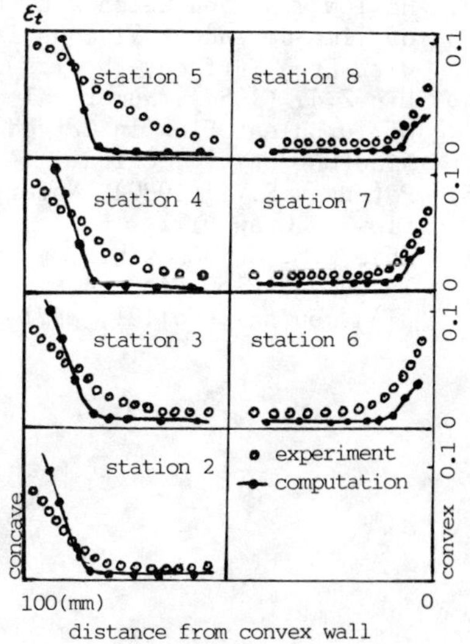

Fig.6 Comparison of turbulence intensity

$\overline{V}$— mean velocity
V_m— mean velocity in center line — · —

CONCLUSIONS

A finite volume method is presented for the solution of the two-dimensional incompressible, steady Navier-Stocks equations with the K-ε turbulence model in general curvilinear coordinates. An Alternating Pressure Difference Scheme method is proposed and the practical computation shows that it can be employed to eliminate pressure oscillation. Comparison of the computed results with the experimental data is made, which shows that the predicted mean velocity field agrees well with that of the experiment, and the computed turbulence intensity distributions have the same trend as the experimental ones.

REFERENCES

1. Chorin A.J. (1967), A Numerical Method for Solving Incompressible Viscous Flow Problems, J. Comp. Physics, Vol.2, pp. 12-26
2. Gu C.G. and Miao Y.M. (1985), Study of Turbulent Flow Through a Plane Cascade at Various Angles of Attack with Separation, Proc. of International Symposium on Refined Flow Modeling and Turbulent Measurement, Vol.1, A25, Univ. of Iowa, Iowa City, Iowa, USA
3. Harlow F.H. and Welch J.E. (1965), Numerical Calculation of Time-dependent Viscous Incompressible Flow of Field with Free Surface, Physics of Fluids, Vol.8, pp.311-346
4. Liu Z.J. (1987), Numerical and Experimental Investigations of Turbulent Flow in Curved Ducts, Master Thesis, Xi'an Jiaotong Univ. , China
5. Patankar S.V. (1980), Numerical Heat Transfer and Fluid Flow, McGraw-Hill
6. Rhie C.M. and Chow W.L. (1983),Numerical Study of the Turbulent Flow Past an Airfoil with Trailing Edge Separation, AIAA Journal, Vol.21, pp.1525-1532

Turbulent Diffusion Simulation by Implicit Factored Solver using K–ε Model

F. Martelli and V. Michelassi

Energy Engineering Department, University of Florence

Via di S. Marta 3, 50139 Firenze, Italy

ABSTRACT

An implicit finite difference procedure for steady two-dimensional incompressible flows with the presence of a pollutant is described. Arbitrary geometries can be solved thanks to the implementation of a curvilinear mesh. Turbulence effects are included by means of Chien[1] version of the k-ε model coupled with the Reynolds averaged Navier-Stokes equations, while the artificial compressibility equation (Chorin[2]) is used to enforce mass conservation. The implicit approximate factorization by Beam and Warming[3] specifically extended to cope with the k-ε non linear source terms is employed in order to avoid stringent stability limitations. Test are then performed for straight channels and hills for various Reynolds numbers, first solving the mean flow equations coupled with the turbulence models, while the pollutant concentration profiles are computed solving a standard transport equation.

I - INTRODUCTION

Due to the availability of powerful computers the numerical simulation of turbulent flows is now possible with formulations that only a few years ago were too heavy to be handled. The present model has been designed to solve inner turbulent two-dimensional flows with the addition of a pollutant that is transported and diffused in the flow field. The choice of an implicit scheme has been done in order to ensure good stability limits. The experience of the author in solving laminar flow fields (Michelassi, Benocci[4]) has driven toward the choice of the implicit approximate factorization of Beam and Warming that was always found convenient with respect to explicit schemes.

Even if the turbulence models with the strongest physical evidence are those based on the Reynolds stresses transport equations, a code based on such model is extremely heavy and could not face complex geometries. According to this, we devoted our attention to the k-ε model in the formulation proposed by Chien[1]. This method has proved its capability in solving inner and outer flows even when recirculations were present (Sahu, Danberg[5]).

Thanks to this formulation the flow field is then used to compute the concentration of a given pollutant solving an additional transport equation using the same scheme imposing that it does not affect the flow field itself.

This hypothesis can be regarded as acceptable for all non-reacting flows and generally valid when the pollutant density and temperature are not strongly different from the main fluid ones (Benocci et al.[6]). One typical result of this procedure for a complex geometry will be shown in the following, together with the turbulence model tests performed on a straight channel.

II - COORDINATE SYSTEMS

In order to ensure the possibility to solve any kind of wall shaped channels the implementation of a curvilinear non orthogonal coordinate system has been carried out by the method proposed by Thompson at al.[7] solving 1 in the physical domain by point S.O.R.; α and β are the transformed plane coordinates while A and B are polynomial stretching functions. With respect to the original method of ref.7, here it is possible to have orthogonal meshing in the neighbourhood of the inlet and outlet sections obtaining simpler expressions for boundary conditions. J is the jacobian of the coordinate transformation and is defined as $J^{-1}=x_{,\alpha}y_{,\beta}-x_{,\beta}y_{,\alpha}$.

$$\begin{cases} \nabla^2 \alpha = A \\ \nabla^2 \beta = B \end{cases} \tag{1}$$

III - CONSERVATIVE FORM OF THE EQUATIONS

Using a curvilinear non orthogonal coordinate system, the Navier-Stokes and the artificial compressibility equations can be written in full conservative form (Pulliam[8]) together with the k-ε model in order to avoid possible instabilities caused by source terms. The system can be conveniently written in vector form, 2, where F, G, F_v and G_v are the convective and viscous flux terms; H stands for the turbulence model source terms. Additionally, since we choose to solve the five equations together in order to couple the mean flow quantities to the turbulence model, it has been possible to obtain a compact form of the differential system. In primitive variables formulation the unknowns vector q is given by 3 for the flow field differential system, while $J^{-1}c$ is the basic unknown for the pollutant transport equation.

$$q_{,t} + F_{,\alpha} + G_{,\beta} = F_{v,\alpha} + G_{v,\beta} + H \tag{2}$$

$$q = J^{-1}(p,u,v,\varepsilon,k)^T \tag{3}$$

(A) Flow field equations.

The artificial compressibility equation is particularly useful when time marching techniques are used to compute steady state for incompressible fluid flows since it avoids the cumbersome solution of a Laplace equation in which no time derivatives are present. Introducing **E** as an artificial compressibility parameter, the continuity equation can be written as

$$p_{,t} + E \operatorname{div}(u,v) = 0 \tag{4}$$

It is clear that equation 4 will ensure mass conservation only at the steady state; this is the main reason why the model is able to compute steady states only. In this case the flux vectors in nondimensional form are;

$$F = J^{-1}\begin{vmatrix} EU \\ uU + \alpha_{,x}p \\ vU + \alpha_{,y}p \\ \varepsilon U \\ kU \end{vmatrix} \quad G = J^{-1}\begin{vmatrix} EV \\ uV + \beta_{,x}p \\ vV + \beta_{,y}p \\ \varepsilon V \\ kV \end{vmatrix} \quad H = J^{-1}\begin{vmatrix} 0 \\ 0 \\ 0 \\ C_1 C_\mu kP - C_2 \varepsilon^2/k - 2/Re/d^2 \varepsilon C_u \\ C_\mu k^2 P/\varepsilon - \varepsilon - 2k/Re/d^2 \end{vmatrix}$$

$$F_v = J^{-1}/Re\begin{vmatrix} 0 \\ \alpha_{,x}\tau_{,xx} + \alpha_{,y}\tau_{,xy} \\ \alpha_{,x}\tau_{,yx} + \alpha_{,y}\tau_{,yy} \\ \alpha_{,x}e_{,x} + \alpha_{,y}e_{,y} \\ \alpha_{,x}h_{,x} + \alpha_{,y}h_{,y} \end{vmatrix} \quad E_v = J^{-1}/Re\begin{vmatrix} 0 \\ \beta_{,x}\tau_{,xx} + \beta_{,y}\tau_{,xy} \\ \beta_{,x}\tau_{,yx} + \beta_{,y}\tau_{,yy} \\ \beta_{,x}e_{,x} + \beta_{,y}e_{,y} \\ \beta_{,x}h_{,x} + \beta_{,y}h_{,y} \end{vmatrix} \qquad (5)$$

where $\tau_{,xx} = 2\nu_{eff}\, U_{,x}$ $\quad e_{,x} = \Gamma\varepsilon_{,x}$ $\quad h_{,x} = \Gamma k_{,x}$ $\quad U = \alpha_{,x}u + \alpha_{,y}v$

$\tau_{,xy} = \nu_{eff}\,(U_{,y} + V_{,x})$ $\quad e_{,x} = \Gamma\varepsilon_{,y}$ $\quad h_{,y} = \Gamma k_{,x}$ $\quad V = \beta_{,x}u + \beta_{,y}v$

$\tau_{,yy} = 2\nu_{eff}\, V_{,y}$

Re is the turbulent Reynolds number based on the wall shear velocity u_*, P is the mean strain production term of the k-ε model while C_μ, C_2, C_u are exponential damping functions as specified by Chien[1]. d here represents the minimum wall distance. The effective viscosity for the momentum equations, ν_{eff}, and the diffusion coefficients for the k-ε equations, Γ, are given by;

$$\nu_{turb.} = Re\ C_\mu\ k^2/\varepsilon$$
$$\nu_{eff} = \nu_{laminar} + \nu_{turb.}$$
$$\Gamma = \nu_{laminar} + \nu_{turb.}/Pr$$

where Pr=1. for k equation and 1.3 for ε equation. This constitutes a system of five partial differential equations in which the mean flow equations are taken fully coupled with the turbulence model that has been chosen since it is valid for low Re too and it allows integration down to the solid wall without the need of any kind of wall function.

(B) Scalar transport equation.

The scalar transport equation is solved using the result of the previous differential system. In this case the flux vectors are simply scalars given by the following expressions.

$$F = J^{-1}\ (cU) \qquad G = J^{-1}\ (cV)$$
$$F_v = J^{-1}/Re\ (\alpha_{,x}C_{,x} + \alpha_{,y}C_{,y}) \qquad G_v = J^{-1}/Re\ (\beta_{,x}C_{,x} + \beta_{,y}C_{,y}) \qquad (6)$$

where $C_{,x} = \Gamma c_{,y}$ and $C_{,y} = \Gamma c_{,x}$. Γ is the diffusion coefficient taking Pr=0.9.

IV - DISCRETIZATION, APPROXIMATE FACTORIZATION

The previous differential equations are discretized in space using centered finite differences for both convective and diffusive terms ensuring second order accuracy. Because of the implementation of the artificial compressibility formulation, in all the equations a time derivative is present; therefore a time marching technique is used to compute the steady state solution for both sets (A) and (B). Since we are not interested in time accuracy, time derivatives are expressed using first order forward differences. Because of the need of resolution of large gradients caused by viscous effects, highly stretched grids are needed close to the solid wall; implicit methods are in general able to avoid stiffness in such problems. Further, time steps much larger than those demanded by explicit schemes can be used. In the present model local time stepping based on the mesh size has been tested finding a considerable gain in convergence speed. Since it is more convenient from an analytical point of view, a solution in δ-form is adopted introducing the following operator giving the correction to the solution at the current time, n, in order to obtain the new value at time n+1:

$$\delta q = q^{n+1} - q^{n}$$

The approximate factorization (Beam, Warming[3]) can be written as it follows:

$$\left[\begin{array}{l}[I+\theta\delta T(H_J/2+(A+R_{,\alpha})_{,\alpha}-R_{,\alpha\alpha})]\delta q^{*}=-\delta T(-F_{,\alpha}-G_{,\beta}+F_{v,\alpha}+G_{v,\beta}+H) \\ {[I+\theta\delta T(H_J/2+(B+S_{,\beta})_{,\beta}-S_{,\beta\beta})]\delta q=\delta q^{*}}\end{array}\right. \qquad (7)$$

in which H_J is the quite complicated jacobian of the source terms H, A and B are jacobians of the convective fluxes F and G, R and S matrices are the jacobians of the diffusive terms obtained in the time linearization of the flux vectors using first order Taylor series expansionand while δT is the local time step. A new parameter $1/2 < \theta \leq 1$ has been introduced in order to weight the implicit and explicit contributions to the space operators: all the calculations have been performed using $\theta=1$. The main advantage of this technique is that the 2-D problem is split in the product of two 1-D problems while in standard ADI we obtain a sum of 1-D operators. In Ref.3 the linear stability analysis is performed showing unconditional stability. Unfortunately the strong nonlinearities peculiar to the turbulence model limit the time step that is, in any case, much larger than what is found for explicit schemes.

V - SOLUTION TECHNIQUE

For both sets of equations (A) and (B) a block tridiagonal matrix has to be solved. This has been done using block Gauss partial pivoting technique. The time marching procedure was carried on for a convenient number of time steps until the residual δq was found to be small enough. Once that set (A) was solved, the obtained solution was used to solve set (B) in the previous hypothesis.

VI - RESULTS

(A) Straight channel.

This first simple test has been performed in order to assess the performances of the adopted turbulence model. Cases with Re=12300 and Re=30800 were investigated for different mesh sizes so that the minimum number of points in the cross stream direction sufficient to resolve viscous effects could be evaluated. Typical mesh points number in the crossflow direction for such calculations is 50, but acceptable results can be obtained even if only 30 points are used. Fig.1, 2 show the adimensional velocity and turbulent kinetic energy for a 30x50 mesh compared with experimental and numerical results. The typical behaviour of almost all the two equations turbulence models is clearly shown in fig. 2 where the peak in turbulent kinetic energy in the viscous layer is not well reproduced. The velocity profile is satisfactorily predicted and the obtained accuracy is considered acceptable for the class of flows under investigation. Nevertheless, it has to be said that this case is particularly simple and it does not ensure good performances in more complicated flow fields.

(B) Hill.
The turbulent flow field in presence of a hill has been computed. Fig.3 shows the 70x31 mesh where a strong point stretching was found necessary to resolve the close to the wall region. As it was expected, two recirculation bubbles are detected in front and after the hill as it is shown in the streamlines pattern in fig.4. Pressure isolines are given in fig.5, showing small pressure wiggles in the past hill region; this clearly indicates that not enough points are used in that region, but limitations in computer storage prevent further refinement.This solution was finally used to compute the distribution of a pollutant c injected in the flow field from the inlet section; the relative isoconcentration profiles are shown in Fig. 6. Particularly interesting is the convergence history plot of the local maximum residual δq for the scalar transport equation shown if Fig. 7. In this case local time stepping has been introduced obtaining a converged solution in less than 400 iterations for the 70x31 mesh that allows the implementation of the scalar transport code on a personal computer.

VII - CONCLUSIONS

An implicit finite difference scheme was developed to solve incompressible two-dimensional steady flows. First tests have shown good convergence characteristics for simple test cases. Thanks to the aforedescribed uncoupled formulation it is possible to compute the diffusion of any kind of pollutant in the flow field. Even if some tests have already been succesful for complex flow field, further investigations and validation are needed in order to assses if the adopted method can be conveniently used to solve problems of practical interest maintaining a two equation turbulence model that, in the opinion of the authors, is the best compromise between accuracy and efficiency.

REFERENCES

1. Kuei-Yuan Chien (1982). Predictions of Channel and Boundary Layer Flows with a Low-Reynolds-Number Turbulence Model, AIAA Journal, Vol.20, No.1, pp. 33-38.
2. Chorin A.J. (1967). A numerical method for solving incompressible viscous flow problems, Journal of Computational Physics, Vol.2, No.1, pp. 12-26.
3. Beam R.M., Warming R.F. (1982). Implicit numerical methods for the compressible Navier-Stokes and Euler equations. von Karman Institute LS 1982-04 "Computational Fluid Dynamics".
4. Michelassi V., Benocci C. (1986). Solution of the steady state incompressible Navier-Stokes equations in curvilinear non orthogonal coordinates. von Karman Institute Technical Note 158.
5. Jubaraj Sahu, Danberg J.E.(1986). Navier-Stokes Computations of Transonic Flows with Two-Equation Turbulence Model. AIAA Journal, Vol.24, No.11, pp. 1744-1751
6. Benocci C., Buchlin J.-M., Michelassi V., Weinacht P. (1985). Numerical Modeling of Gas-Droplet Flows for Industrial applications. Third International Conference on Computational Methods and Experimental Measurements, Porto Carras, Greece.
7. Thompson J.F., Thames F.C., Mastin C.W.(1977). Boundary fitted curvilinear

coordinate systems for solution of partial differential equations on fields containing any number of arbitrary two dimensional bodies. NASA CR 2729.

8. Pulliam T.H., Steger J.L. (1980). Implicit Finite-Differences Simulations of Three-Dimensional Compressible Flow. AIAA Journal, Vol.18, No.2, pp.159-167.

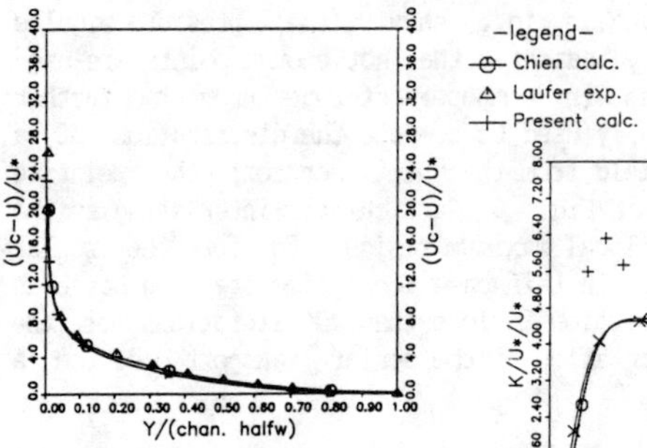

Fig.1. Straight channel Velocity profile

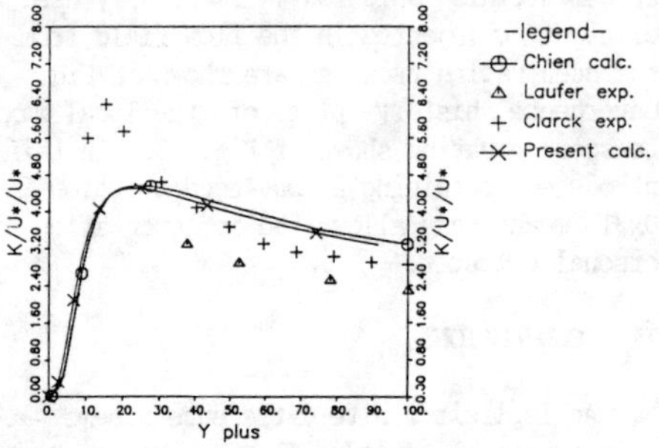

Fig.2. Straight channel Turbulent kinetic energy

Fig.3. Hill Curvilinear grid

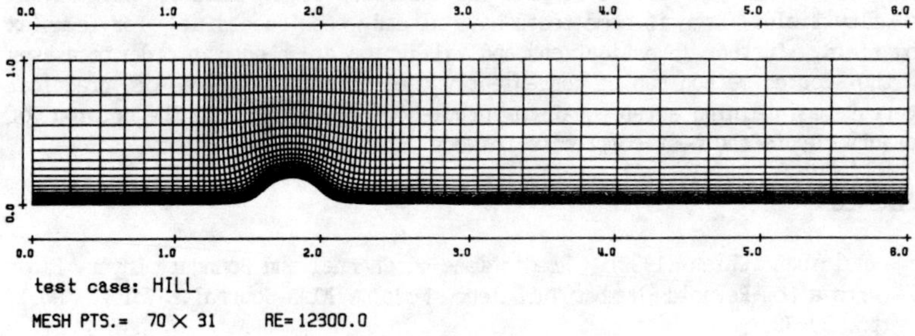

Fig.4. Hill Streamlines pattern

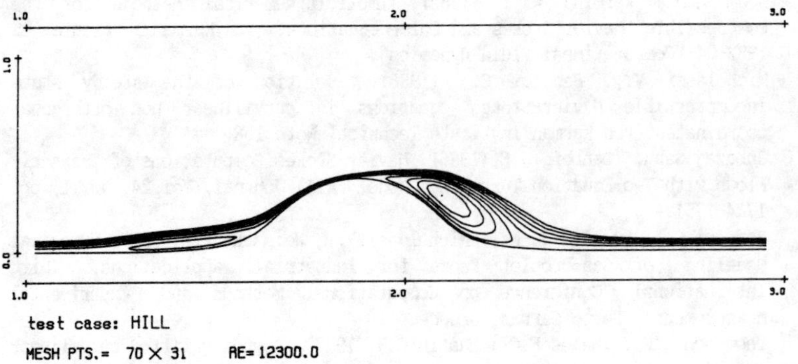

Fig.5. Hill Pressure isolines

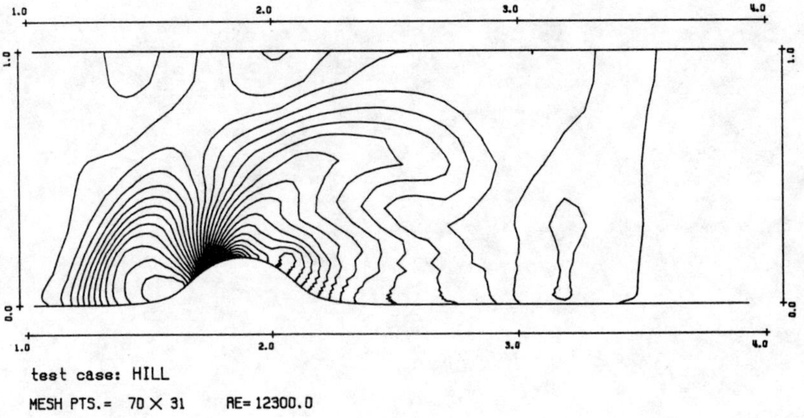

Fig.6. Hill Concentration isolines

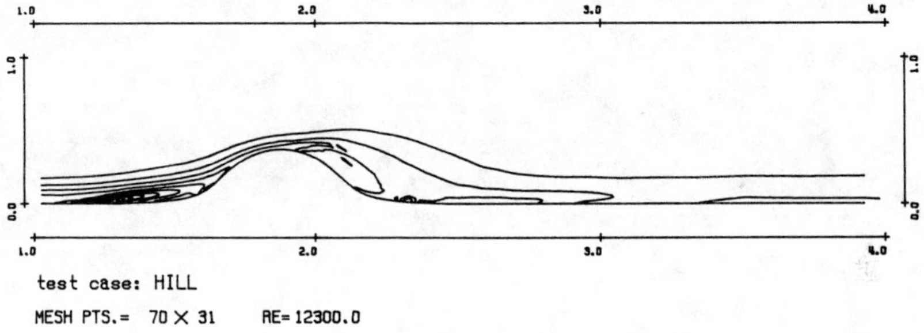

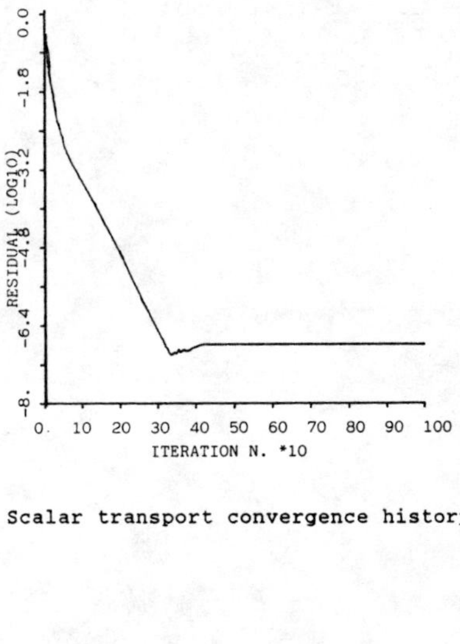

Fig.7. Scalar transport convergence history

INVITED PAPER

A Boundary Element Investigation of Natural Convection Problems

M. Tanaka

Department of Mechanical Engineering, Shinshu University, Nagano, Japan

K. Kitagawa

Consumer Products Engineering Laboratory, Toshiba Corporation, Yokohama, Japan

C.A. Brebbia

Computational Mechanics Institute, Southampton, UK

L.C. Wrobel

COPPE, Federal University of Rio de Janeiro, Rio de Janeiro, Brazil

SUMMARY

This paper presents a boundary element formulation employing a penalty function technique for steady natural convection problems. By regarding the convective and buoyancy force terms as pseudo-body forces, the standard boundary element analysis of elastostatics can be extended to solve the Navier-Stokes equations. In a similar manner, the standard boundary element analysis of potential problems can be applied to the energy transport equation. Emphasis is placed on accurate evaluation of the domain integrals resulting from the pseudo-source terms. A self-adaptive coordinate transformation technique proposed by Telles is used for such purpose. Numerical computation is carried out for some sample problems in two dimensions whereby the usefulness of the proposed solution procedure is demonstrated.

INTRODUCTION

In recent years numerical simulation techniques have been successfully applied to practical design of various engineering products and also investigation of natural phenomena. The boundary element methods (BEM) have been recognized as one of the central techniques of such simulation[1-3]. Generally speaking, it seems to be difficult to develop a general purpose computer code for numerical analysis of the natural convection or viscous fluid flow problems, because of the nonlinearities involved in these problems. However, at present there are available many investigations of these problems using the finite difference methods, e.g. the computer codes named SIMPLE[4] or MAC[5], and also the finite element methods[6]. In the last several years attempts have been made to develop alternative solution procedures for these nonlinear problems by using the boundary element methods[7-12].

In this study the natural convection problems in a steady state are investigated by means of the boundary element method. We propose an integral equation formulation which introduces a penalty function to relate the pressure to the divergence of the velocity field. If we regard the convective terms and the buoyancy forces as pseudo-forces, the Navier-Stokes equations become similar to the governing equations of elastostatics expressed in terms of the displacement. On the other hand, the energy transport equation to be solved simultaneously can be reduced to a steady-state heat conduction equation with a pseudo heat source resulting from the convective terms. The system of differential equations thus obtained can be formulated into the set of integral equations by using the same fundamental solutions as for elastostatics and the Laplace equation. Then, they are solved by means of the boundary element method which employs also the internal cell subdivision. In this paper we discuss in some detail an accurate evaluation of the domain integrals by using a self-adaptive technique proposed by Telles[13]. Finally, numerical computation is carried out for a typical bench-mark problem of the two-dimensional natural convection in a square domain. The results obtained reveal the potential usefulness of the proposed method of solution.

THEORY

Basic Equations in Steady Natural Convection

In this study we assume that the flow is steady and incompressible. The so-called Boussinesq approximation is used, that is, the material constants of the fluid are not temperature-dependent except for the generation of buoyancy force. If we take a rectangular Cartesian coordinates $0\text{-}x_1x_2x_3$ in which the axis x_2 coincides with the opposite direction of gravity, we can express the basic equations of the steady natural convection in the following non-dimensional form:

"continuity equation"

$$v_{i,i} = 0 \tag{1}$$

"Navier-Stokes equations"

$$v_j v_{i,j} = -p_{,i} + Pr(v_{i,j} + v_{j,i})_{,j} \tag{2}$$

"energy transport equation"

$$v_j \theta_{,j} = \theta_{,jj} \tag{3}$$

where v_i , p and θ denote the velocity, the pressure and the excess temperature, respectively. The indices following a comma means the spatial differentiation, i.e. $(\ )_{,i} = \partial(\ )/\partial x_i$. Kronecker's delta is denoted by δ_{ij} , while repeated indices imply the summation convention. We denote by Pr and Ra the Prandtl number and the Rayleigh number, respectively. The notation used is given in more detail in the previous paper[14].

Introducing the penalty number λ we express the relationship between the pressure and the velocity as follows:

$$p = -\lambda v_{i,i} \tag{4}$$

It is clear that the continuity equation (1) can be satisfied as the penalty number approaches infinity. In numerical computation we assume for λ a large positive number, which implies that the incompressibility condition is approximately satisfied.

Substitution of Equation (4) into (3) yields

$$(\lambda + Pr)v_{j,ji} + Pr v_{i,jj} = v_j v_{i,j} - \delta_{2i} Ra Pr \theta \tag{5}$$

On the other hand, the governing equations of elastostatics can be expressed in terms of the displacement as follows:

$$(\lambda' + \mu')u_{j,ji} + \lambda' u_{i,jj} = -b_i \tag{6}$$

where λ' and μ' are Lamé's constants, and b_i the body force. Since Equation (5) is similar to Equation (6), we can apply the standard boundary element method developed so far for elastostatics to numerical analysis of Equation (5).

In a similar manner we can express the energy transport equation (3) in the Poisson equation if we regard the convective term of Equation (3) as a pseudo-heat source. To the solution procedure of this equation we can also apply the standard boundary element method so far developed for the potential problems.

Integral Equation Formulation

Using Kelvin's fundamental solutions with Lamé's constants λ' and μ' replaced with the penalty number λ and the Prandtl number Pr, respectively, we can transform Equation (5) into the following set of integral equations:

$$c_{ij}(y)v_j(y) + \int_\Gamma t^*_{ij}(y,y')v_j(y')d\Gamma(y') + \int_\Gamma u^*_{ij}(y,y')t_j(y')d\Gamma(y') = \int_\Omega u^*_{ij}(y,x)b_j(x)d\Omega(x) \tag{7}$$

where

$$b_j = v_k v_{j,k} - \delta_{2j}RaPr \tag{8}$$

The coefficients $c_{ij}(y)$ in Equation (7) are such that $c_{ij} = \delta_{ij}$ if $y \in \Omega$ (inner domain) and $c_{ij} = \delta_{ij}/2$ if $y \in \Gamma$ (smooth boundary). In Equation (7) the asterisked two-point functions denote Kelvin's fundamental solutions, while t_j stands for the traction on the boundary. Thus, the boundary conditions can be expressed as

$$\begin{aligned} v_i &= \bar{v}_i \quad \text{on } \Gamma_v \\ t_i &= \bar{t}_i \quad \text{on } \Gamma_t = \Gamma - \Gamma_v \end{aligned} \tag{9}$$

where the superimposed bar denotes a prescribed value.

On the other hand, using the well-known fundamental solution for the Laplacian operator, we can transform the energy transport equation (3) into the following integral equation:

$$c(y)\theta(y) + \int_\Gamma q^*(y,y')\theta(y')d\Gamma(y') + \int_\Gamma \theta^*(y,y')q(y')d\Gamma(y') = \int_\Omega \theta^*(y,x)f(x)d\Omega(x) \tag{10}$$

where the asterisked two-point functions are the fundamental solutions. The pseudo-heat source f and the heat flux q on the boundary are defined by

$$\begin{aligned} f &= v_j\theta_{,j} \\ q &= \partial\theta/\partial n \end{aligned} \tag{11}$$

The boundary conditions of this problem are such that

$$\begin{aligned} \theta &= \bar{\theta} \quad \text{on } \Gamma_\theta \\ q &= \bar{q} \quad \text{on } \Gamma_q = \Gamma - \Gamma_\theta \end{aligned} \tag{12}$$

As can be seen, the right hand sides of the boundary integral equations (7) and (10) include the unknown velocity and temperature as well as their gradients in the inner domain. Since these boundary integral equations are nonlinear, we have to employ an iterative solution procedure which will later be discussed in great detail.

Gradients of Velocity and Temperature
To evaluate accurately the domain integrals on the right hand sides of Equations (7) and (10), we can use the following relationships:

$$c_{ij}(y)u_{j,k}(y) + \int_\Gamma t^*_{ijk}(y,y')v_j(y')d\Gamma(y')$$
$$- \int_\Gamma u^*_{ijk}(y,y')t_j(y')d\Gamma(y') = \int_\Omega u^*_{ijk}b_j(x)d\Omega(x) \qquad (13)$$

$$c(y)\theta_{,k} + \int_\Gamma q^*_k(y,y')\theta(y')d\Gamma(y')$$
$$- \int_\Gamma \theta^*_k(y,y')q(y')d\Gamma(y') = \int_\Omega \theta^*_k(y,x)f(x)d\Omega(x) \qquad (14)$$

where

$$u^*_{ijk} = u^*_{ij,k} \;, \quad t^*_{ijk} = t^*_{ij,k}$$
$$\theta^*_k = \theta^*_{,k} \;, \quad q^*_k = q^*_{,k} \qquad (15)$$

Equations (13) and (14) can be obtained by differentiating Equations (7) and (10) with respect to the point y in the inner domain.

METHOD OF SOLUTION

Boundary-Domain Element Discretization
If the boundary is divided into the usual boundary elements and the inner domain into the so-called domain elements, the boundary integral equation (7) can be discretized into the following system of equations:

$$[H]\{v\} = [G]\{t\} + [C]\{b\} \qquad (16)$$

where $\{v\}$ and $\{t\}$ denote the nodal vectors of velocity and traction on the boundary, while $\{b\}$ is the nodal vector of body force in the inner domain. The coefficient matrices $[H]$, $[G]$ and $[C]$ can be computed from the fundamental solutions and the interpolation functions used. If the boundary conditions (9) are taken into consideration, we can rewrite Equation (16) as follows:

$$[A]\{X\} = \{Y\} + [C]\{b\} \qquad (17)$$

where $\{X\}$ denotes the vector of nodal unknowns on the boundary, while $\{Y\}$ is the vector of known components. Provided that $\{b\}$ is given, Equation (17) can be solved for $\{X\}$.

The nodal vector of velocity in the inner domain $\{v_I\}$

and its derivative $\{\partial v_I\}$ can be calculated from Equations (7) and (13). Discretizing these equations by means of the boundary-domain element and taking account of the boundary conditions, we can finally obtain

$$\{v_I\} = [A_I]\{X\} + \{Y_I\} + [C_I]\{b\} \tag{18}$$

$$\{\partial v_I\} = [\partial A_I]\{X\} + \{\partial Y_I\} + [\partial C_I]\{b\} \tag{19}$$

In a similar manner, the relationships corresponding to the energy transport equation yields the discretized sets of equations, which can be summarized as

$$[\hat{A}]\{X_t\} = \{\hat{Y}_t\} + [C]\{f\} \tag{20}$$

$$\{\theta\} = [\hat{A}_I]\{X_t\} + \{\hat{Y}_I\} + [C_I]\{f\} \tag{21}$$

$$\{\partial\theta\} = [\partial\hat{A}_I]\{X_t\} + \{\partial\hat{Y}_t\} + [\partial C_I]\{f\} \tag{22}$$

where

$\{X_t\}$; the nodal unknown vector on the boundary,
$\{\theta\}$; the nodal temperature vector in the inner domain,
$\{\partial\theta\}$; the nodal temperature gradient vector in the inner domain,
$\{f\}$; the nodal heat source vector in the inner domain.

All the coefficient matrices in Equations (20)-(22) can be computed from the fundamental solutions and the interpolation functions used.

Numerical Evaluation of Integrals

For accurate evaluation of the integrals appearing in computation of the coefficient matrices, the inner domain near the boundary must be divided into a finer mesh of domain elements. Since the fundamental solutions in particular their derivatives have strong singularities, some innovative method should be introduced for accurate and also efficient evaluation of these integrals. In this work we employ the self-adaptive method originally proposed by Telles[13].

For two-dimensional problems a general form of the integrals to be evaluated can be expressed as follows:

$$I = \int_{-1}^{1} h(\eta)d\eta \tag{23}$$

The integrand $h(\eta)$ is assumed to be singular at $\eta = \bar{\eta}$. We introduce the following transformation:

$$\eta(\xi) = a\xi^3 + b\xi^2 + c\xi + d \tag{24}$$

where the constants are to be determined by

$$\eta(1) = 1 , \quad \eta(-1) = -1$$
$$[d\eta/d\xi]_{\xi=\bar{\xi}} = \bar{r} , \quad [d^2\eta/d\xi^2]_{\xi=\bar{\xi}} = 0 \tag{25}$$

Then, Equation (23) can re re-written as

$$I = \int_{-1}^{1} h(\eta)[d\eta/d\xi]d\xi \tag{26}$$

It can be expected that if $\bar{r}$ in Equation (25) is selected in an appropriate manner as $0 \leqq \bar{r} \leqq 1$ according to the distance D from the singular point, the integral as shown in Equation (26) can be evaluated accurately under a smaller number of integration points. Telles[13] recently recommended to use the following relations:

$$\begin{aligned} \bar{r} &= 0.85 + 0.24 \ln(D) && 0.05 \leqq D \leqq 1.3 \\ \bar{r} &= 0.893 + 0.0832 \ln(D) && 1.3 < D \leqq 3.618 \\ \bar{r} &= 1 && 3.618 < D \end{aligned} \tag{27}$$

Iterative Solution Procedure
To solve the nonlinear systems of equations mentioned above, we propose in this study an iterative method of solution. The main flow of the proposed method is composed of the following steps:
(1) Assume $\{v_I\}$, $\{\partial v_I\}$ and $\{\partial\theta\}$, and then calculate $\{f\}$.
(2) Solve Equation (20) for $\{X_t\}$.
(3) Compute $\{\theta\}$ and $\{\partial\theta\}$ by using the known values of $\{f\}$ and $\{X_t\}$.
(4) Calculate $\{b\}$ by using the known values of $\{v_I\}$, $\{\partial v_I\}$ and $\{\theta\}$.
(5) Solve Equation (17) for $\{X\}$.
(6) Compute $\{v_I\}$ and $\{\partial v_I\}$ from Equations (18) and (19) with the known values of matrices.
(7) Check convergence with respect to $\{v_I\}$, $\{\partial v_I\}$, $\{\theta\}$ and $\{\partial\theta\}$. Unless convergence is obtained, go to step (1) and repeat the procedure.

It is noted that in this study we adopt an under-relaxation method for iterative computation.

COMPUTATIONAL RESULTS AND DISCUSSION

Now, we show the computational results obtained for a typical benchmark problem of two-dimensional steady natural convection in a square cavity. The boundary conditions and the boundary-domain element division of the cavity are shown in Fig.1 where we use the linear boundary element and the triangular domain element with linear interpolation functions. The total number of boundary nodes is 164 and that of internal nodes 289. The nodal point at each corner of the cavity is considered to be a double node.

In Table 1 comparison is made between the results

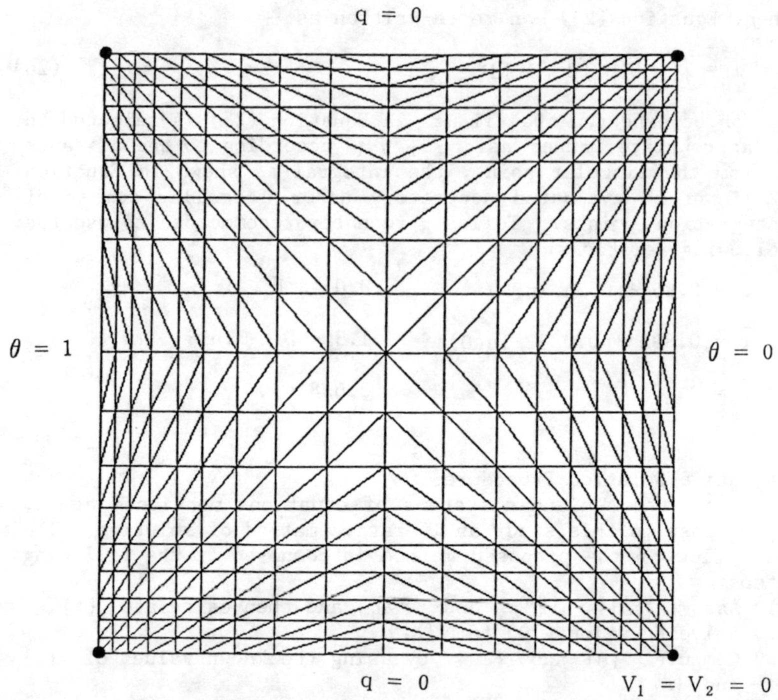

Fig.1 Boundary-domain element discretization

Table 1 Results obtained for square cavity flow problem ($Pr = 0.71$, $\lambda = 10^6$)

Ra	10^4		10^5		10^6	
	BEM	Ref.15	BEM	Ref.15	BEM	Ref.15
$\overline{Nu}$	2.242	2.238	4.438	4.505	8.307	8.903
Nu_{max} x_2 =	3.602 0.135	3.527 0.143	7.657 0.090	7.717 0.082	16.18 0.055	18.562 0.045
Nu_{min} x_2 =	0.596 1.0	0.586 1.0	0.779 1.0	0.729 1.0	1.225 1.0	1.002 1.0
v_{1max} x_2 =	16.429 0.820	16.178 0.823	33.58 0.856	34.77 0.854	58.52 0.820	64.94 0.850
v_{2max} x_1 =	19.727 0.135	19.643 0.119	66.56 0.055	68.25 0.066	205.76 0.055	221.29 0.040

obtained by the present method and the finite difference method with the 41x41 grid[15]. The averaged value of the Nusselt numbers at the nodal points on the higher-temperature wall $\theta = 1$ is denoted by $\overline{Nu}$. It can be seen that both the results are in good agreement.

In Figs.2 and 3 are shown the computational results with respect to velocity distributions and isotherms, respectively.

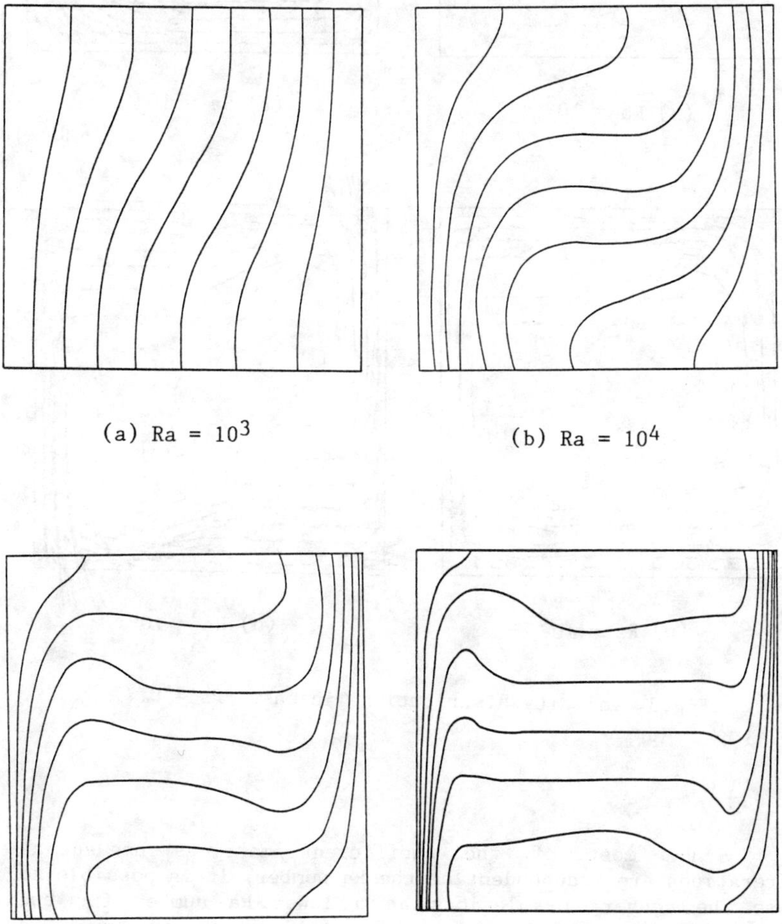

(a) Ra = 10^3 (b) Ra = 10^4

(c) Ra = 10^5 (d) Ra = 10^6

Fig.2 Isotherms in case Pr = 0.71

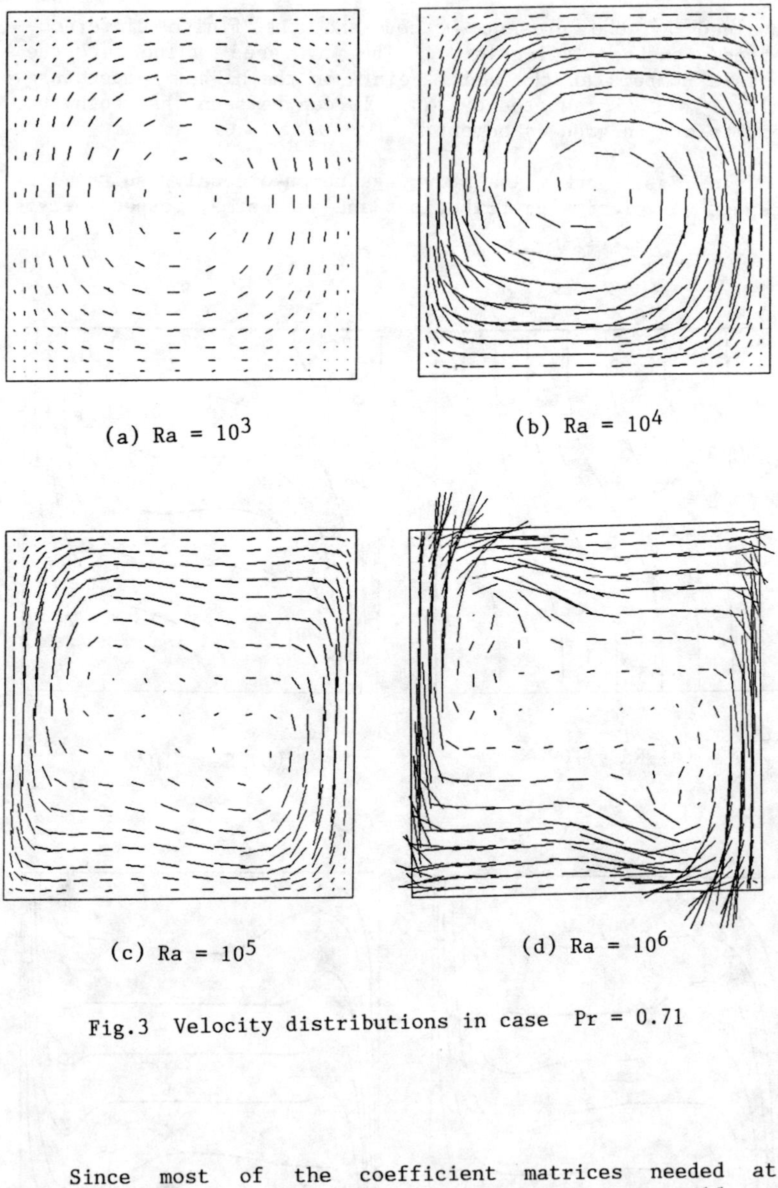

Fig.3 Velocity distributions in case Pr = 0.71

Since most of the coefficient matrices needed at iterations are independent of the Ra number, it is possible to use the converged solutions at a lower Ra number for the initial values of iterative computation for a larger Ra number. It should be noted that no appreciate differences were recognized between the numerical results obtained when the penalty number was changed from 10^3 to 10^6.

CONCLUDING REMARKS

The integral equation formulation and its solution scheme have been presented for the steady natural convection problems. The iterative method of solution was proposed for the resulting systems of nonlinear equations. The self-adaptive method of integration was incorporated into the solution procedure, and its usefulness was demonstrated through the error check of numerical integration. The computational results obtained for a sample problem of the natural convection in a square cavity may reveal the availability of the proposed method of solution.

Although this paper was mainly concerned with the two-dimensional problems, it is straightforward to construct a computer program for the three dimensional problems. It can be recommended as the future research work to develop along the direction of the present work an effective solution procedure for the transient natural convection problems which seem to be more important in practice.

REFERENCES

1. Brebbia C.A., Telles J.C.F. and Wrobel L.C.(1984). Boundary Element Techniques, Springer-Verlag. Berlin-New York-Tokyo.
2. Tanaka M. and Brebbia C.A.(Eds.)(1986). Boundary Elements VIII, ditto, & CM Publications, Boston and Southampton.
3. Brebbia C.A., Wendland W.L. and Kuhn G.(Eds.)(1987). Boundary Elements IX, ditto, & CM Pubs., Boston & Southampton.
4. Patanker S.V. and Spalding D.B.(1972), A Calculation Procedure for Heat-Mass and Momentum Transfer in Three-Dimensional Parabolic Flows, Int. J. Heat & Mass Transfer, Vol.15, pp.1787-1806.
5. Harlor F.H. and Welch J.E.(1965), Numerical Calculation of Time-Dependent Viscous Incompressible Flow of Fluid with Free Surface, Phys. Fluids, Vol.8, pp.2182-2189.
6. Bristeau M.O., Glowinski R., Hauguel A. and Periaux J.(Eds.)(1986). Proc. 6th Int. Symp. on FEM in Flow Problems, Antibes/FRANCE.
7. Kuroki T., Onishi K. and Tosaka N.(1985), Thermal Fluid Flow with Velocity Evaluation Using Boundary Elements and Penalty Function Method, Boundary Elements VII, (Eds. Brebbia C.A. and Maier G.), Springer-Verlag, pp.2/107-114.
8. Tosaka N. and Fukushima N.(1986), Integral Equation Analysis of Laminar Natural Convection Problems, Boundary Elements VIII, (Eds. Tanaka M. and Brebbia C.A.), Springer-Verlag, pp.803-812.
9. Onishi K., Kuroki T. and Tanaka M.(1984), An Application of a Boundary Element Method to Natural Convection, Appl. Math. Modelling, Vol.8, pp.383-390.

10. Onishi K., Kuroki T. and Tanaka M.(1985), Boundary Element Method for Laminar Viscous Flow and Convective Diffusion Problems, Chapter 8, Topics in Boundary Element Research, (Ed . Brebbia C.A.), Springer-Verlag, Vol.2, pp.209-229.
11. Skerget P., Alujevic A. and Brebbia C.A.(1986), BEM for Laminar Motion of Isochoric Viscous Fluid, BETECH 86, (Eds. Connor J.J. and Brebbia C.A.), CM Publications, pp.397-419.
12. Kitagawa K., Wrobel L.C., Brebbia C.A. and Tanaka M.(1986), Modelling Thermal Transport Problems Using the Boundary Element Method, EVVIROSOFT 86, (Ed. Zannetti P.), CM Publications, pp.715-731.
13. Telles J.C.F.(1988), A Self-Adaptive Coordinate Transformation for Efficient Numerical Evaluation of General Boundary Element Integrals, to be appearing on Int. J. Num. Meths. in Eng.
14. Kitagawa K., Wrobel L.C., Brebbia C.A. and Tanaka M.(1988), A Boundary Element Formulation for Natural Convection Problems, to be appearing on Int. J. Num. Meths. in Fluids.
15. Davis G.V. and Jones I.P.(1981), Natural Convection in a Square Cavity - A Comparison Excercise, Numerical Methods in Thermal Problems, Vol.2, (Eds. Lewis R.W. and Morgan K.), Pineridge Press, pp.552-572.

SECTION 1C - NUMERICAL ANALYSIS

A New Family of Shape Functions

S.E. Adeff
Department of Engineering Sciences-Mechanical, The University of Mississippi, POB 3146, University, MS 38677, USA

ABSTRACT

A new family of finite-element shape functions is presented as constructed by using sine and cosine functions. They have the desirable property of sustaining an infinite order of derivatives. The lagrangian family of shape functions requires the addition of new nodes as the order of derivatives to be represented grows. Present family is indifferent to this requirement and consequently is restricted -for the one-dimensional (1-D) subset to two members, for sake of simplicity and computational economy. By obtaining products of 1-D functions, the two- and three- dimensional (2-D and 3-D) subsets are immediately obtained, in the same way corresponding lagrangian functions are built. Although second derivatives appearing in the viscous terms in the Navier-Stokes equations can be reduced to first order derivatives through integration by parts, certain useful techniques like the Dendy version of Petrov-Galerkin weighting functions and the Taylor-Galerkin procedure, avoid the need for computing second derivatives. A quadratic lagrangian function renders constant second derivatives, not allowing enough flexibility to efficiently represent the actual gradients involved. The new functions totally overcome this kind of deficiencies and possess greater generality. Possible economic disadvantages for lower-order derivative cases are easily overcome by tabulating the function values at fixed local abscissas -like gaussian abscissas used in numerical integration. An application to the simple problem of a depth-integrated inviscid-fluid flow is presented. The solution obtained for this inviscid-fluid case is indistinguishable with that resulting from using the lagrangian family. Three-dimensional and depth-integrated turbulent flow problems are currently being solved and are to be reported in the near future. Nevertheless the family is suited for any kind of engineering problems.

INTRODUCTION

The lagrangian family of shape functions is often used in fluid-flow simulations because they provide efficient means of describing complex geometries. The order of derivatives that can be represented is equal to the number of nodes in the basic one-dimensional element, minus one. However, cubic-or-higher-order lagrangian elements are rarely used because of practical reasons. The lagrangian family has been extensively documented in the technical literature and so widely used as to make unnecessary a presentation here. Nevertheless, it should be noted that many of its features are common to any type of shape functions, and should be preserved to guarantee established basic criteria of convergence (e.g. Zienkiewicz[1]). These include the condition that each function has a value equal to one at each node it is referred to and zero at any other node; the sum of all the shape functions at any point in the element should be one,

$$\sum_{i=1}^{n} L_i = 1 \tag{1}$$

to permit a rigid displacement of the element; the sum of the derivatives should similarly be zero,

$$\sum_{i=1}^{n} L_{i,q} = 0 \tag{2}$$

allowing for a constant strain of the element. Also the value of any derivative at the interface between elements should be finite to eliminate any non-valid contribution of the interface to any functional being minimized.

FORMULATION

It is here presented the construction of a new family of shape functions that have many of the features of the lagrangian family but improving it in the sense that it has no bound for the order of derivatives and can be computed without incrementing the number of nodes. It is restricted to two members because these two members are enough for the purpose and more efficient than any possible extension. The first member is a two-node element covering the segment [-1.,+1.] over the local coordinate q. An angle β is defined by

$$\beta = 0.25\,\pi\,(1 + q) \tag{3}$$

and the shape functions are given by:

$$L_1 = +0.5\,(1 + \cos\beta - \sin\beta) \tag{4}$$
$$L_2 = +0.5\,(1 - \cos\beta + \sin\beta).$$

Local derivatives can be computed both in terms of β or q:

$$L_{1,\beta} = -0.5\,(1 + \cos\beta + \sin\beta)$$
$$L_{2,\beta} = +0.5\,(1 + \cos\beta + \sin\beta) \tag{5a}$$
$$L_{1,q} = -0.5\,L_{1,\beta}$$
$$L_{2,q} = +0.5\,L_{2,\beta}$$

$$
\begin{aligned}
L_{1,\beta\beta} &= -0.5\ (1 + \cos\beta - \sin\beta) \\
L_{2,\beta\beta} &= +0.5\ (1 + \cos\beta - \sin\beta) \\
L_{1,qq} &= +0.0625\ L_{1,\beta\beta} \\
L_{2,qq} &= +0.0625\ L_{2,\beta\beta}
\end{aligned} \qquad (5b)
$$

Higher order derivatives can similarly and indefinitely be computed over the same two-nodes element because the expressions contain cosine and sine functions. These functions verify the conditions given by eqs.(1) and (2). In general, for the k-order derivative (up to ∞), we have:

$$\sum_{i=1}^{n} L_{i,q^k} = 0 \qquad (6)$$

Nevertheless, the representation of curved two- and three-dimensional geometries requires the use of a three-node element. Consequently the second member of the new family, again covering the segment [-1.,+1.] of q have three nodes. With the angle β still defined by eq.(3), the three new shape functions are a simple extension of the ones given by eq.(4), since they have the same terms plus an additional one,

$$
\begin{aligned}
L_1 &= +0.5\ [1 + \cos\beta - \sin\beta - \sin(2\beta)] \\
L_2 &= +\sin(2\beta) \\
L_3 &= +0.5\ [1 - \cos\beta + \sin\beta - \sin(2\beta)]
\end{aligned} \qquad (7)
$$

The shape functions given by eqs.(4) and (8) have a value of one at the node to which they are referred and zero at any other node of their respective elements. The local derivatives are:

$$
\begin{aligned}
L_{1,q} &= -0.125\ \pi\ [\cos\beta + \sin\beta + 2\cos(2\beta)] \\
L_{2,q} &= +0.5\ \pi\ \cos(2\beta) \\
L_{3,q} &= +0.125\ \pi\ [\cos\beta + \sin\beta - 2\cos(2\beta)] \\
L_{1,qq} &= -0.03125\ \pi^2\ [\cos\beta - \sin\beta - 4\sin(2\beta)] \\
L_{2,qq} &= -0.25\ \pi^2\ \sin(2\beta) \\
L_{3,qq} &= +0.03125\ \pi^2\ [\cos\beta - \sin\beta + 4\sin(2\beta)]
\end{aligned} \qquad (8)
$$

Higher-order derivatives can be computed indefinitely. Derivatives in both eqs.(5) and (8) compare favourably with their lagrangian counterparts. For a comparative three-node element, lagrangian first derivatives are linear while present ones are nearly quadratic in the sense that are more closely approximated by a quadratic polynomial; second lagrangian derivatives are constant, while present ones are still nearly quadratic, and of course, next lagrangian higher-order derivatives are null. Eqs.(7) and (8) also verify the conditions resumed by eq.(6). Two- and three-dimensional shape functions are obtained by multiplying one-dimensional ones. Despite the numerical differences between the two families, when they are plotted scaled to equate the maximum value of homologous functions a striking similitude may be observed. Figure 1 shows for the node 1 and by means of comparison, at left, functions corresponding to the lagrangian family and at right, those corresponding to the new family. All these functions are shown as perspective views over a reference 2-D element. Only the second derivatives differs notably.

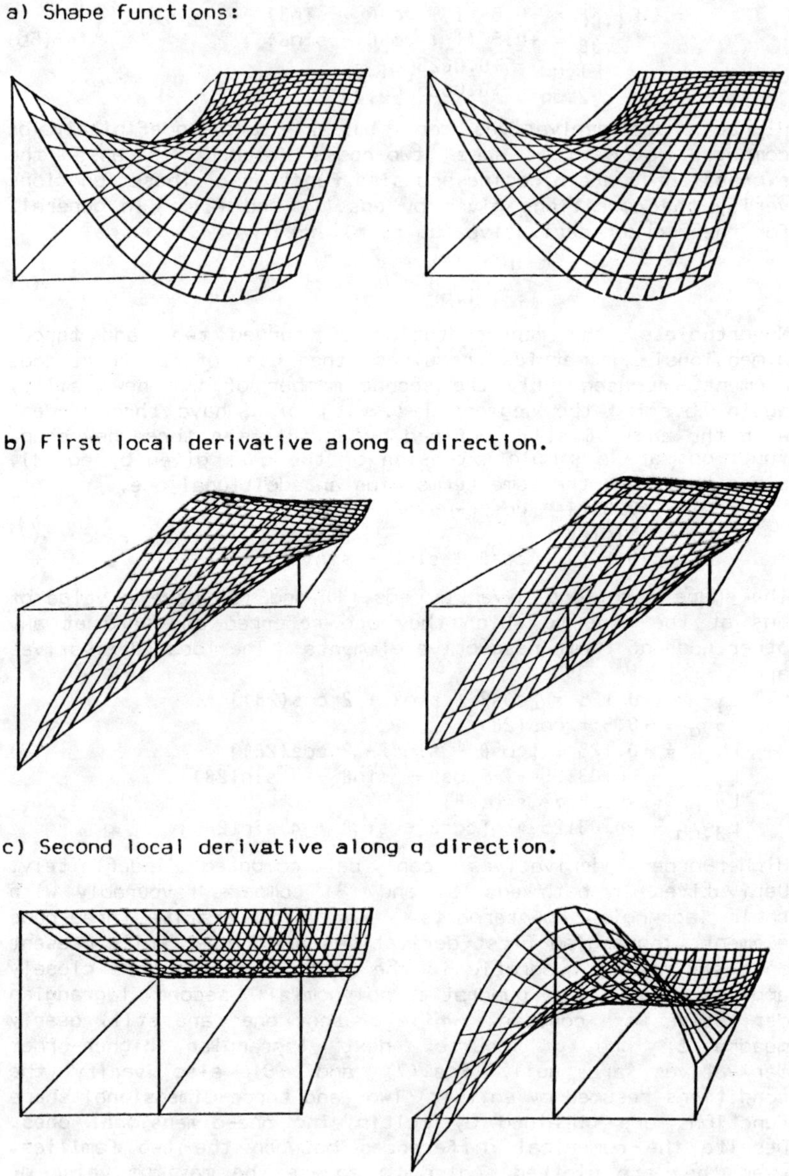

Figure 1 : Shape function and derivatives referred to node 1.
Left: Lagrangian family. Right: Present family.

VERIFICATION

Although the general convergence criteria for shape functions is satisfied, the general applicability of the new family has been tested. The following depth-integrated inviscid flow (DIIF) equation is being used to compute a guess or initial condition for more complex depth-integrated and three-dimensional turbulent-flow models under development (Adeff[2]),

$$\nabla\psi \cdot \nabla h - h\, \nabla^2\psi = 0 \qquad (9)$$

, where h is the depth, a known function $h = h(x,y)$, and ψ is the stream function. Results obtained with DIIF should be regarded as better ones than those obtained with Laplace equation, when applied over a variable-bed channel. Both lagrangian and present family were alternatively used to solve a hypothetical steady flow on the Upper-Chesapeake bay. Plottings done for the finite-element mesh, the bed-contour lines and resulting stream-function lines for both families cannot be distinguished from each other, and hence only those obtained by using the new family are shown here in Figures 2, 3 and 4.

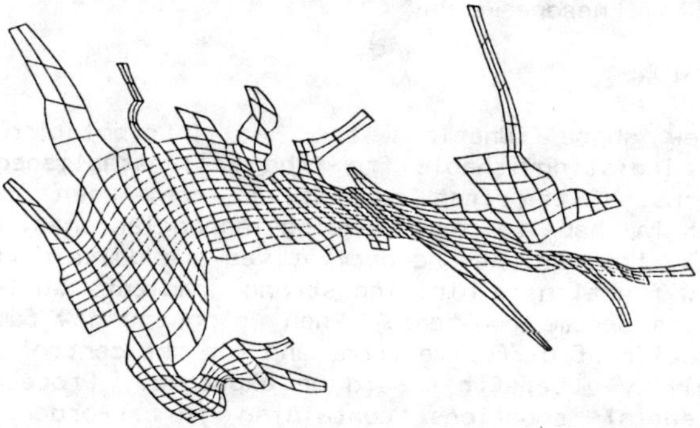

Figure 2 : Finite-element mesh for the Upper-Chesapeake Bay.

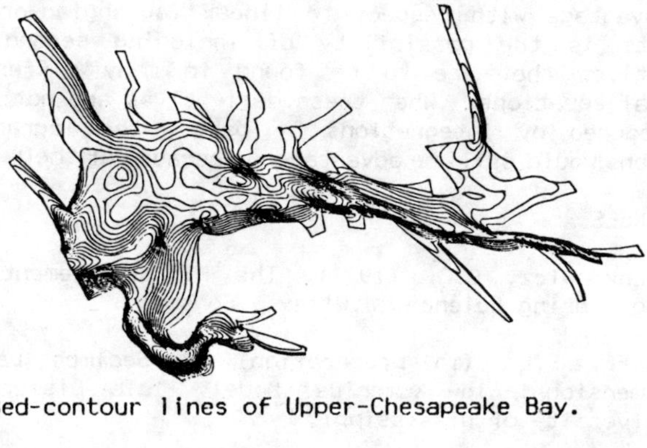

Figure 3 : Bed-contour lines of Upper-Chesapeake Bay.

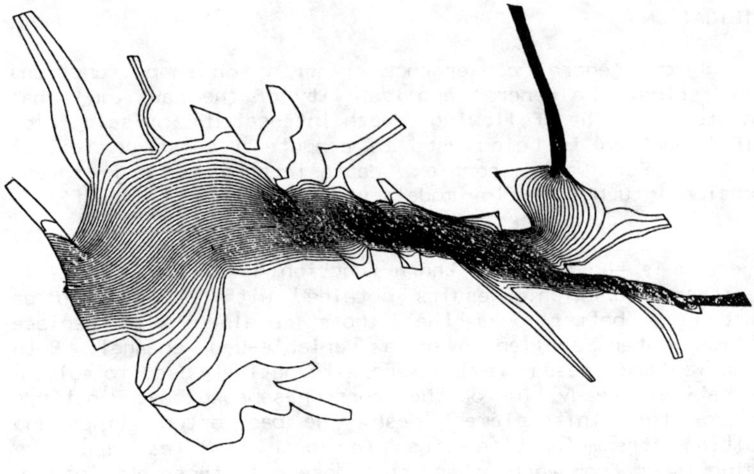

Figure 4 : Streamlines for a hypothetical steady flow in Upper-Chesapeake Bay

CONCLUSIONS

The new shape functions are, as interpolation functions, almost indistinguishable from the well-established lagrangian functions. Its first derivatives are, when scaled, quite similar in shape to corresponding lagrangian ones, although not linear. However, second derivatives are quite different, with the new functions exhibiting strong gradients while lagrangian functions become constant. When using the new family for the computation of diffusive terms and for the control of numerical instability a benefit should be expected. Procedures solving differential equations containing higher-order derivatives would benefit the most. The new two-node element could be approppiate for one-dimensional models, and, for simple geometries, when composing two- or three-dimensional elements. The advantage with respect to linear lagrangian-or-triangular elements is the possibility of including second or higher derivatives that are to be found in many differential and integral equations. When these derivatives are not included or are reduced by integrations by parts, the lagrangian shape functions would still be advantageous because of their simplicity.

REFERENCES

1. Zienkiewicz, O.C. (1971), The Finite Element Method in Engineering Science, J.Wiley & Sons.

2. Adeff, S.E. (In preparation), A Sediment-Laden Three-Dimensional-Flow Numerical Model, Ph.D. Dissertation, The University of Mississippi.

Adaptive Collocation for Burgers' Equation

M.B. Allen III and M.C. Curran
University of Wyoming, Laramie, Wyoming 82071, USA

ABSTRACT

An adaptive gridding algorithm for finite-element collocation improves resolution of steep fronts in advection-dominated flows such as Burgers' equation. The scheme decouples refinement equations from the overall linear system solved at each iteration of each time step. This decoupling preserves the matrix structure associated with coarse-grid calculations and facilitates parallel computation of the unknowns associated with refinement.

1. INTRODUCTION

Many problems in water resources concern flows governed by transient, nonlinear partial differential equations. In transport and multiphase flow problems, these equations have solutions with steep fronts whose resolution requires special numerical consideration. We discuss an adaptive gridding scheme for finite-element collocation that improves the local resolution of steep fronts, using Burgers' equation as a prototype for more complicated flows.

2. COLLOCATION APPROXIMATION TO BURGERS' EQUATION

We consider Burgers' equation in the quasilinear form

$$\frac{\partial u}{\partial t} + u\frac{\partial u}{\partial x} - \frac{\delta}{2}\frac{\partial^2 u}{\partial x^2} = 0, \quad (x,t) \in (0,\infty) \times (0,\infty), \qquad (2-1)$$

assuming the following boundary and initial conditions:

$$u(0,t) = u_L, \quad |u(\infty,t)| \quad \text{bounded}, \quad t > 0;$$

$$u(x,0) = u_I(x), \quad x \geq 0.$$

To discretize Equation (2-1) in time, we use a straightforward implicit Euler scheme: letting k denote the time step, we get

$$\frac{1}{k}(u^{n+1}-u^n)+u^{n+1}\frac{\partial u^{n+1}}{\partial x}-\frac{\delta}{2}\frac{\partial^2 u^{n+1}}{\partial x^2}=0, \qquad (2-2)$$

where $u^n(x)$ represents the approximate value of $u(x, nk)$.

For the discretization in space, we use collocation on Hermite cubic finite elements. We begin with a grid $\Delta = \{0 = x_0, x_1, \ldots, x_N\}$ with $x_0 < x_1 < \cdots < x_N$. Here, x_N is an artificial right boundary chosen so that $\bar{\Omega} = [0, x_N]$ contains the zone of influence of the initial data during the time period of interest. In adaptive refinement schemes Δ depends on the time level n, a feature that we develop in Section 3. We define the trial space $\mathcal{M}_1^3(\Delta)$ of Hermite piecewise cubics on Δ as follows:

$$\mathcal{M}_1^3(\Delta) = \{v \in C^1(\bar{\Omega}) \mid v|[x_{i-1}, x_i] \quad \text{is cubic}\},$$

where $v|[x_{i-1}, x_i]$ denotes the restriction of v to the subinterval $[x_{i-1}, x_i]$. Prenter[1] defines an interpolating basis $\{H_{0,i}, H_{1,i}\}_{i=0}^N$ for $\mathcal{M}_1^3(\Delta)$ having the property that, for any $v \in \mathcal{M}_1^3(\Delta)$,

$$v(x) = \sum_{i=0}^{N}[v(x_i)H_{0,i}(x) + v'(x_i)H_{1,i}(x)].$$

Using this basis, we can also define for later use a projection operator $\pi : C^1(\bar{\Omega}) \to \mathcal{M}_1^3(\Delta)$ by assigning to any function $f \in C^1(\bar{\Omega})$ its Hermite cubic interpolate,

$$(\pi f)(x) = \sum_{i=0}^{N}[f(x_i)H_{0,i}(x) + f'(x_i)H_{1,i}(x)].$$

Standard approximation theory guarantees that, for any $f \in C^4(\bar{\Omega})$, $\|\pi f - f\|_\infty = \mathcal{O}(h^4)$, where $h = \max\{x_i - x_{i-1}\}_{i=1}^N$.

We define the residual for Equation (2-1) as follows:

$$R^{n+1}(x) \equiv \hat{u}^{n+1}(x) - \hat{u}^n(x) + k\hat{u}^{n+1}(x)\frac{d\hat{u}^{n+1}}{dx}(x) - \frac{k\delta}{2}\frac{d^2\hat{u}^{n+1}}{dx^2}(x).$$

The method of collocation for Equation (2-2) determines a sequence $\{\hat{u}^n\}_{n=0}^\infty \subset \mathcal{M}_1^3(\Delta)$ satisfying three conditions:

i. $\hat{u}^n(0) = u_L$ and $\hat{u}^n(x_N) = 0$ for $n = 0, 1, 2, \ldots$;

ii. $\hat{u}^0(x) = (\pi u_I)(x)$ for all $x \in \bar{\Omega}$;

iii. At each point in a collection $\{\bar{x}_k\}_{i=1}^{2N} \subset \Omega$ of collocation points, the pointwise residual $R_k^{n+1} \equiv R^{n+1}(\bar{x}_k) = 0$ for $n = 0, 1, 2, \ldots$.

We choose the points $\bar{x}_k$ to be the abcissae for two-point Gauss quadrature in each element $[x_{i-1}, x_i]$, ensuring the optimal error bound $\|\hat{u}^n - u^n\|_\infty = \mathcal{O}(h^4)$ (Douglas and Dupont[2]).

The algebraic equations $R_k^{n+1} = 0$ are nonlinear, so to solve them at each time level we use Newton's method. In this scheme, each iteration produces a Hermite cubic approximation $\hat{u}^{n+1,m+1}(x)$ to $\hat{u}^n(x)$ by solving a linear system of the form $\mathbf{J}^{n+1,m}\boldsymbol{\delta} = -\mathbf{r}^{n+1,m}$. Here, m stands for the index of the most recently computed iterative level; $\mathbf{r}^{n+1,m} = (R_1^{n+1,m}, \ldots, R_{2N}^{n+1,m})^\top$; and $\mathbf{J}^{n+1,m}$ signifies the Jacobian matrix, whose entries have the form $\partial R_k^{n+1,m}/\partial u_i$ or $\partial R_k^{n+1,m}/\partial u_i'$. The vector $\boldsymbol{\delta} = (\delta u_0', \delta u_1, \delta u_1', \ldots, \delta u_{N-1}, \delta u_{N-1}', \delta u_N')^\top$ contains iterative increments of the nodal coefficients of $\hat{u}^{n+1}$, so that

$$\hat{u}^{n+1,m+1} = \sum_{i=0}^{N} \left\{ \left[(u_i)^{n+1,m} + \delta u_i\right] H_{0,i} + \left[(u_i')^{n+1,m} + \delta u_i'\right] H_{1,i} \right\}.$$

(The boundary values being known, we set $\delta u_0 = \delta u_N = 0$.)

We test for numerical convergence of this iterative procedure by checking the criterion $\|\mathbf{r}^{n+1,m+1}\|_\infty < \tau$ for some small tolerance $\tau > 0$. As soon as this inequality holds, we set $\hat{u}^{n+1} \leftarrow \hat{u}^{n+1,m+1}$ and begin the next time step with initial iterate $\hat{u}^{n+2,0} = \hat{u}^{n+1}$. The scheme converges quadratically, in agreement with theory. This implicit approximation permits large time steps without jeopardizing stability. We have tested the method without instability at global Courant numbers $\|u^n\|_\infty k/h = 100$, a value far exceeding that required for reasonable accuracy. The Newton scheme typically converges with $\tau = 10^{-7}$ in three or four iterations per time step when Δ is uniform and $h = 0.01$.

3. GRID REFINEMENT SCHEME

Adaptive local grid refinement concentrates numerical degrees of freedom in subregions of a problem's domain requiring fine-scale resolution. Such subregions arise for Burgers' equation owing to its tendency to produce shock-like fronts when $\delta << 1$. These subregions may change with time. Therefore, as n increases, we adjust the distribution of nodes x_i to respect the character of the evolving solution. We thus generate a sequence $\{\Delta_n\}$ of grids, each corresponding to a particular time level n. Associated with each Δ_n is a projection $\pi_n : C^1(\bar{\Omega}) \to \mathcal{M}_1^3(\Delta_n)$.

Two desirable features of adaptive local grid refinement schemes are flexibility in regridding and computational efficiency. As a compromise between these often conflicting desiderata, we construct an *elementwise h-refinement* scheme. By "elementwise" we mean that we establish an initial *coarse grid* $\Delta_0 = \{0 = x_0, x_1, \ldots, x_N\}$ and stipulate that $\Delta_n \supset \Delta_0$ at every time level n. This stipulation allows us to add an arbitrary number of new nodes to any coarse-grid element $\Omega_i = [x_{i-1}, x_i]$. Other data structures are possible; as examples we mention the moving-grid approach (Chong[3], Davies[4], Albert and O'Neill[5], Djomehri[6]) and patch refinement (Ewing et al.[7]). By "h-refinement" we mean that the local polynomial degree of the trial space remains constant. In our case, $u^n \in \mathcal{M}_1^3(\Delta_n)$, so the trial function at any time level is Hermite piecewise cubic. This approach stands in contrast to the "p-refinement" schemes , in which the local degree of the approximation can change in time; Mohsen and Pinder[8] discuss such a scheme.

Central to our scheme is an algorithm that decouples the collocation equations for any grid Δ_n into two subsets. One set involves only the nodal unknowns associated with coarse-grid nodes x_i and has the same matrix structure as the collocation system for Δ_0. The other set consists of smaller systems, each involving only those unknowns associated with nodes contained in a single refined coarse-grid element Ω_i. We accomplish this decoupling, or *elementwise condensation*, using row reduction (Allen and Curran[9]). This algorithm allows us to compute coarse-grid variables by solving a system whose size remains constant in time. We then solve for local refinement variables via elementwise condensation and back substitution in each refined Ω_i – procedures that are readily amenable to parallel processing.

To accommodate the nonlinearity of Burgers' equation, we adopt a predictor-corrector strategy that, at a given time level n, determines the refined grid Δ_n only after performing Newton iterations on the equations associated with the coarse grid Δ_0. Thus the iterations on Δ_0 constitute the predictor; those on Δ_n, the corrector. The strategy is as follows: given $u^n(x)$ defined on Δ_n and tolerances $\tau_0, \tau > 0$,

i. Set $u^{n+1,0} \leftarrow \pi_0 u^n$.

ii. Solve $\mathbf{J}^{n+1,m}\boldsymbol{\delta} = -\mathbf{r}^{n+1,m}$ on Δ_0 to get iterates $u^{n+1,m+1} \in \mathcal{M}_1^3(\Delta_0)$. Stop when $\|\mathbf{r}^{n+1,M}\|_\infty < \tau_0$.

iii. Determine Δ_{n+1} according to some refinement strategy.

iv. Set $u^{n+1,M+0} \leftarrow \pi_{n+1}u^{n+1,M}$.

v. Solve $\mathbf{J}^{n+1,M+k}\boldsymbol{\delta} = -\mathbf{r}^{n+1,M+k}$ on Δ_{n+1} to get refined iterates $u^{n+1,M+k+1} \in \mathcal{M}_1^3(\Delta_{n+1})$. Stop when $\|\mathbf{r}^{n+1,M+k+1}\|_\infty < \tau$.

vi. Set $u^{n+1} \leftarrow u^{n+1,M+k+1}$ and $n \leftarrow n+1$.

Step *v* has the following structure:

v.1. Use elementwise condensation to decompose the global system $\mathbf{J}^{n+1,M+k}\boldsymbol{\delta} = -\mathbf{r}^{n+1,M+k}$ into a square system

$$\mathbf{J}_0^{n+1,M+k}\boldsymbol{\delta}_0 = -\mathbf{r}_0^{n+1,M+k}, \qquad (3-1)$$

involving only coarse-grid variables, and nonsquare systems

$$\mathbf{J}_i^{n+1,M+k}\boldsymbol{\delta}_i = -\mathbf{r}_i^{n+1,M+k}, \qquad (3-2)$$

each involving variables associated with one refined element Ω_i.

v.2. Solve the system (3-1) using a standard coarse-grid solver.

v.3. Solve each of the systems (3-2) via back substitution, using values determined in step *v.2.*

4. NUMERICAL RESULTS

To test our method, we solve the N-wave problem (Chong[3]), for which $u_L = 0$ and an appropriate initial condition is

$$u_I(x) = \frac{x}{1 + \exp\left(\frac{4x^2-1}{8\delta}\right)}.$$

The true solution exhibits a boundary layer whose thickness is $\mathcal{O}(\delta)$ and in which $|\partial\hat{u}^n/\partial x| = \mathcal{O}(\delta^{-1})$. Since $|\partial\hat{u}^n/\partial x| = \mathcal{O}(1)$ elsewhere, we can resolve the steep front in the layer with accuracy comparable to that achieved outside it by inserting $\mathcal{O}(\delta^{-1})$ refinement nodes in each coarse-grid element in the layer. Denote by h the coarse-grid mesh, and let $\nu^n = \text{int}(1.5\max\{|(u_i')^n|\}_{i=0}^N)$, where int$(\cdot)$ signifies the integer part. We insert ν^n refinement nodes in each coarse-grid element Ω_i where $(u_i^n - u_{i-1}^n)/h > 2$. We "cushion" any collection of adjacent refined elements by inserting int$(\nu^n/2)$ refinement nodes in neighboring coarse-grid elements. Figure 1 shows the resulting numerical solutions at different time levels, using $h = k = 0.05$, together with a plot of the exact solution for comparison.

5. CONCLUSIONS

The local grid refinement scheme described here offers an efficient method for adding computational degrees of freedom to zones requiring fine-scale numerical resolution. The predictor-corrector iteration strategy gives rise to an algorithm that sidesteps the extra computations associated with refinement during the early Newton steps in each time step of a nonlinear problem. The success of the method in solving Burgers' equation suggests that the approach will be useful in other nonlinear, advection-dominated flows in water resources engineering.

ACKNOWLEDGMENTS

NSF supported this work through grants DMS-8504360 and RII-8610680.

REFERENCES

1. Prenter, P.M. (1975). *Splines and Variational Methods*, John Wiley. New York.

2. Douglas, J., and Dupont, T. (1973), A Finite Element Collocation Method for Quasilinear Parabolic Equations, *Math. Comp.*, Vol. 27, pp. 17-28.

3. Chong, T.H. (1978), A Variable Mesh Finite Difference Method for Solving a Class of Parabolic Differential Equations in One Space Variable, *SIAM J. Numer. Anal.*, Vol. 15, pp. 835-857.

4. Davies, A.M. (1978), Application of the Galerkin Method to the Solution of Burgers' Equation *Comput. Meth. Appl. Mech. Eng.*, Vol. 14, pp. 305-321.

5. Albert, M. R. and O'Neill, K. (1983), The Use of Transfinite Mappings with Finite Elements on a Moving Mesh for Two-Dimensional Phase Change, in *Adaptive Computational Methods for Partial Differential Equations*, (ed. by I. Babuška et al.), pp. 85-110, SIAM, Philadelphia.

6. Djomehri, M. J. (1983), Finite Element Solution of Systems of Partial Differential Equations, Ph.D. dissertation, Dept. of Mathematics, Univ. California, Berkeley, California.

7. Ewing, R. E., McCormick, S., and Thomas, J. (1984), The Fast Adaptive Composite Grid Method for Solving Differential Boundary Value Problems, in *Proceedings, Fifth ASCE Specialty Conference on Engineering Mechanics in Civil Engineering,* Laramie, Wyoming, August 1-3, 1984, pp. 1453-1456. ASCE, New York.

8. Mohsen, M.F.N., and Pinder, G.F. (1984), Collocation with 'Adaptive' Finite Elements, *Int. J. Numer. Meth. Eng.*, Vol. 20, pp. 1901-1910.

9. Allen, M. B. and Curran, M. C. (1988), An Adaptive Gridding Scheme for Solving Advection-Dominated Flows Using Finite-Element Collocation, in *Proceedings, 7th International Conference on Computational Engineering Science*, Atlanta, April 10-14, 1988, (ed. by S. N. Atluri). Springer-Verlag, Berlin.

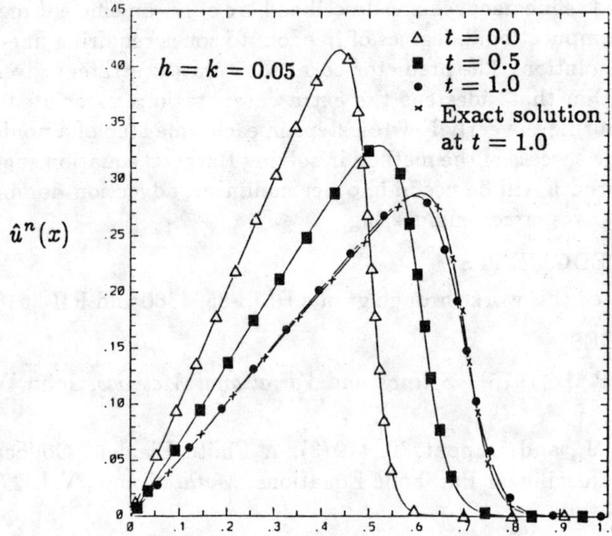

Figure 1. Numerical solution profiles for the N-wave problem.

Alternative Ways of Treating Domain Integrals in Boundary Elements

C.A. Brebbia

Computational Mechanics Institute, Ashurst, Southampton, UK

ABSTRACT

This paper deals with some some alternative techniques to treat domain type integrals present in many boundary element problems. These integrals may originate in body forces distributed throughout the domain, nonlinear effects or initial conditions in the case of time dependent problems.

The success of boundary elements as a generalized numerical technique rests to a large degree on its computationally efficient treatment of domain integrals. The prevalent existing technique of discretizing the domain into cells destroys in great part of the advantages of BEM over FEM. Other approaches tend to be applicable only to a limited range of problems.

This paper describes a more general treatment of domain integrals which preserves the inherent advantages of using boundary elements, i.e. working with unknowns and integrals on the boundary only.

1. INTRODUCTION

The boundary element is popular with practising engineers because of the simplicity of the data required to run a problem and the versatility of boundary element meshes. In addition, it tends to produce more accurate results than finite elements and can be applied to treat problems extending to infinity with little difficulty [1] [2] [3].

All these advantages are important for the solution of water resources topics, many of which have the added complication of being nonlinear or time dependent. It is in these cases when some of the advantages of boundary elements are lost as it is customary to discretize the domain into cells and compute the associated integrals numerically over each of these cells.

The problem of how to treat domain integrals efficiently is of fundamental importance for the further development of the boundary element method and in particular for its application to solve many water resources problems.

Recent efforts by the author and his associates to transform domain into boundary integrals have resulted in two different approaches; they are those using particular solutions and the ones applying higher order fundamental solutions. The latter is generally based on expressing the fundamental solution in terms of Galerkin's functions and in this way defining a higher order solution which can then be integrated. This procedure provides an elegant way of transferring to the boundary terms due to some types of body forces but has severe limitations when the integral varies in a more complex manner. Some applications of these techniques in water resources have been presented in reference [4].

The other possibility is to apply particular solutions which render the problem homogeneous. While this approach is well known and frequently applied for simple cases its has only recently become generalized to account for more complex distributions using the concept of localized functions. The work originated with Nardini and Brebbia [5,6] for solving elastodynamic problems and has now been extended to a wide variety of applications [7,8,9]. The present paper applies the same idea to compute domain integrals due to generalized body forces.

2. BASIC FORMULATION

Domain integrals in boundary elements may arise due to a variety of effects such as body forces, initial states, nonlinear terms and others. In what follows the use of the non-homogeneous differential Poisson's type equation will be studied. The resulting formulation can then be extended to those for which the non-homogeneous terms are function of the potential.

Consider first the use of Poisson's equation, i.e.

$$\nabla^2 u = b \qquad \text{in } \Omega \tag{1}$$

where b is a known function in the Ω space. In addition to (1) the potential has to satisfy certain conditions, such as

$$\left.\begin{array}{lll} \text{Essential conditions,} & u = \bar{u} & \text{on } \Gamma_1 \\ \text{Natural conditions,} & q = \bar{q} & \text{on } \Gamma_2 \end{array}\right\} \Gamma = \Gamma_1 + \Gamma_2 \tag{2}$$

where Γ is the total boundary and q denotes the derivative of potential u with respect to the normal to the boundary n, i.e. $q = \partial u/\partial n$. The bar indicates that the value of the corresponding variable is known.

The problem can be reduced to a boundary integral form using the fundamental solution u*, i.e.

$$\nabla^2 u^* + \Delta^i = 0 \tag{3}$$

where Δ^i is the dirac delta function at a point x_i. One can also define a fundamental solution flux q*, such that $q^* = \partial u * /\partial n$.

The solution u* for three dimensional problems is well known, i.e.

$$u^* = \frac{1}{4\pi^2} \tag{4}$$

and for two dimensional cases,

$$u^* = \frac{1}{2\pi} \ln\left(\frac{1}{r}\right) \tag{5}$$

Applying weighted residual principles or reciprocity [1] [2] between the u and u* fields one obtains an integral equation at any given point 'i'

$$c^i u^i + \int_\Gamma q^* u \, d\Gamma + \int_\Omega b \, u^* \, d\Omega = \int_\Gamma q \, u^* \, d\Gamma \tag{6}$$

This equation is the starting point for the boundary element method in potential problems. Notice that all integrals are on the boundary with the exception of the one corresponding to the b term. The constant c^i depends on the type of boundary point i under consideration (i.e. c $= \frac{1}{2}$ for smooth boundaries or otherwise proportional to its solid angle; $c^i = 1$ for any internal points and $c^i = 0$ for the point i external to the domain).

3. THE USE OF PARTICULAR SOLUTIONS

An obvious way of solving equation (6) without domain integrals is by changing variables in such a manner that these integrals disappear. This can be attempted by adding a particular solution to the new variable.

To illustrate the procedure consider the Poisson equation (1) with boundary conditions as given by (2). Assume now that the potential function u can be written as,

$$u = \tilde{u} + \hat{u} \tag{7}$$

where û is a particular solution of the Poisson equation, such that

$$\nabla^2 \hat{u} = b \tag{8}$$

One can now write the domain term in (6) as,

$$\int_\Omega b\, u^*\, d\Omega = \int_\Omega (\nabla^2 \hat{u}) u^*\, d\Omega \tag{9}$$

Integrating by parts this expression and taking into consideration the special character of the fundamental solution - see equation (3) - one finds the following expression,

$$\int_\Omega b\, u^*\, d\Omega = \int_\Omega (\nabla^2 u) u^*\, d\Omega =$$

$$= \int_\Omega u(\nabla^2 u^*) d\Omega + \int_\Gamma u^* \frac{\partial \hat{u}}{\partial n}\, d\Gamma - \int_\Gamma q^*\, \hat{u}\, d\Gamma \tag{10}$$

$$= -\, c^i \hat{u}^i + \int_\Gamma u^*\, \hat{q}\, d\Gamma - \int_\Gamma q^*\, \hat{u}\, d\Gamma$$

where $\hat{q} = \partial \hat{u}/\partial n$.

Substituting (10) into (6) one finds the following expression,

$$c^i u^i + \int_\Gamma q^*\, u\, d\Gamma - \int_\Gamma q\, u^*\, d\Gamma =$$

$$= c^i \hat{u}^i + \int_\Gamma q^*\, \hat{u}\, d\Gamma - \int_\Gamma \hat{q}\, u^*\, d\Gamma \tag{11}$$

Notice that now all integrals are computed on the Γ boundary. Equation (11) can be written in a more compact form as function of the new variable ũ, i.e.

$$c^i \tilde{u}^i + \int_\Gamma q^*\, \tilde{u}\, d\Gamma = \int_\Gamma \tilde{q}\, u^*\, d\Gamma \tag{12}$$

where ũ = u - û defines the new variable.

4. THE DUAL RECIPROCITY METHOD

This method which was originally proposed by Nardini and Brebbia [5,6] for elastodynamics is essentially a generalized search for particular solutions using localized rather than global functions as was done in section 2. The method can be used to represent body force distributions over the whole boundary or only on parts of it as demonstrated by Niku and Brebbia [9]. A complete description of the technique for the use of elastodynamics has been presented by Nardini and Brebbia in 1985 [10]. The method has also been applied to solve parabolic problems such as transient diffusion [11]. Transient thermal analysis has been studied in detail by Brebbia et al. [7,8,12,13] including extension of the technique to nonlinear diffusion problems.

In order to describe the method let us consider again the domain term in equation (6) and propose the following representation of b, i.e.

$$b = \sum_{j=1}^{N} f^j \alpha^j \tag{13}$$

where f^j represents a set of N known coordinate functions. $f(\xi, x)$ is chosen as a function between the points ξ_j and x, where the 'j' point is usually one of the boundary nodes. α^j are the unknown coefficients associated with each f^j. Notice that the f^j functions are considered to originate at 'j' different points generally - but not normally - on the boundary. They are of the same type for all the points.

One can first define the f^j functions and then find the corresponding particular solutions, i.e. such that

$$\nabla^2 \hat{u}^j = f^j \tag{14}$$

where the $\hat{u}^j$ field and its associated variables, such as $\hat{q}^j$ can be found by integrating the above equation.

Equation (6) can now be written as,

$$c^i u^i + \int_\Gamma q^* \, u \, d\Gamma - \int_\Gamma q \, u^* \, d\Gamma = - \sum_{j=1}^{N} \{\alpha^j \int_\Omega (\nabla^2 \hat{u}^j) u^* \, d\Omega\} \tag{15}$$

Each term on the right hand side can be integrated by parts resulting in only boundary integrals, i.e.

$$c \, u + \int_\Gamma q^* \, u \, d\Gamma - \int_\Gamma q \, u^* \, d\Gamma = \sum_{j=1}^{N} \{\alpha^j (c^j \, \hat{u}^j + \int_\Gamma q^* \, \hat{u}^j \, d\Gamma - \int_\Gamma u^* \, \hat{q}^j \, d\Gamma)\} \tag{16}$$

Notice that this formula involves only boundary integrals but that contrary to an equation such as (11), its right hand side is formed by adding a series of terms, each of them localized at a particular point 'j'.

The superscript 'i' in the first term on the left and right hand side of equation (16) has been eliminated for simplicity's sake but the reader should keep in mind that these terms and the whole equation still refer to a particular position 'i' of the source.

Equation (16) produces the following matrix system after the usual boundary element discretization and integrations (see [1,2,3])

$$\underset{\sim}{H}\,\underset{\sim}{U} - \underset{\sim}{G}\,\underset{\sim}{Q} = \underset{\sim}{S}\,\underset{\sim}{\alpha} \tag{17}$$

where

$$\underset{\sim}{S} = \underset{\sim}{H}\,\underset{\sim}{\hat{U}} - \underset{\sim}{G}\,\underset{\sim}{\hat{Q}} \tag{18}$$

$\underset{\sim}{U}$ and $\underset{\sim}{Q}$ are now square matrices N x N each. (N is the number of points where the function f^j has been applied). The columns of these matrices represent the values of $\hat{u}^j$ and $\hat{q}^j$ at the different nodes for the case of the f^j function acting at a particular 'j' point.

Notice that the α coefficients are different from the values of b at the points under consideration. Both are however related through equation (13) which can be written in matrix form as

$$\underset{\sim}{B} = \underset{\sim}{F}\,\underset{\sim}{\alpha} \tag{19}$$

where the $\underset{\sim}{B}$ vector represents the values of b at these points and $\underset{\sim}{F}$ is a matrix whose elements are the values of f^j at all points for each position of j. The number of coordinate functions ought to be equal to the number of points or 'poles' to be able to invert $\underset{\sim}{F}$, i.e.

$$\underset{\sim}{\alpha} = \underset{\sim}{F}^{-1}\,\underset{\sim}{B} = \underset{\sim}{F}\,\underset{\sim}{B}^{1} \tag{20}$$

Hence equation (17) can finally be written as,

$$\underset{\sim}{H}\,\underset{\sim}{U} - \underset{\sim}{G}\,\underset{\sim}{Q} = (\underset{\sim}{S}\,\underset{\sim}{F})\underset{\sim}{B} \tag{21}$$

The $\underset{\sim}{H}$, $\underset{\sim}{G}$ and $\underset{\sim}{S}$ are only function of the geometry of the problem, the type of discretization and the $\hat{u}^j$, $\hat{q}^j$ field of the particular solution.

5. DISTRIBUTED FORCES ACTING ONLY ON PART OF THE DOMAIN

The previous approach will now be generalized to deal with localized body forces. Consider the case of the b function acting only on a part of the Ω domain called Ω_S (figure 1).

In this case the starting equation is (6) with Ω replaced by Ω_S, i.e.

$$c\,u + \int_\Gamma q^*\,u\,d\Gamma + \int_\Omega b\,u^*\,d\Omega = \int_\Gamma q\,u^*\,d\Gamma \tag{22}$$

The DRM discretization of b now takes place within the subdomain Ω_S and produces the following result,

$$c\,u + \int_\Gamma q^*\,u\,d\Gamma - \int_\Gamma q\,u^*\,d\Gamma = \sum_{j=1}^{N} [\alpha^j \int_{\Omega_S} (\nabla^2\hat{u}^j)u^*\,d\Omega] \tag{23}$$

After substituting formula (13) - which is now valid only in the subdomain - and integrating by parts one can write,

$$c\,u + \int_{\Gamma} q^*\,u\,d\Gamma - \int_{\Gamma} q\,u^*\,d\Gamma = \tag{24}$$

$$= \sum_{j=1}^{N} \alpha^j \left[c_S^j\,\hat{u}^j + \int_{\Gamma_S} q^*\,\hat{u}^j\,d\Gamma - \int_{\Gamma_S} u^*\,\hat{q}^j\,d\Gamma \right]$$

The subscript S in the constant c_S indicates that the value of this coefficient is defined by the boundary of the subdomain Ω_S rather than the external boundary. In practice sources are taken to the external boundary and this term is zero as they are external to Ω_S, i.e. equation (24) becomes,

$$c\,u + \int_{\Gamma} u^*\,q\,d\Gamma - \int_{\Gamma} u\,q^*\,d\Gamma = \tag{25}$$

$$= \sum_{j=1}^{N} \left\{ \alpha^j \left[\int_{\Gamma_S} q^*\,\hat{u}^j\,d\Gamma - \int_{\Gamma_S} u^*\,\hat{q}^j\,d\Gamma \right] \right\}$$

Notice that this formula involves two types of boundary integrals, those on the external Γ boundary and others on the Γ_S internal boundary. The integrals can be evaluated numerically by subdividing both boundaries into elements as shown in figure 2.

Thus equation (25) can be written in matrix form as follows,

$$\underset{\sim}{H}\,\underset{\sim}{U} - \underset{\sim}{G}\,\underset{\sim}{Q} = \underset{\sim}{S}_S\,\underset{\sim}{\alpha} \tag{26}$$

where $\underset{\sim}{S}$ is an N x M matrix, N being the total number of degrees of freedom on the internal boundary Γ_S. The integrals on the Γ_S boundary are non-singular and easy to compute. (They may be singular in the case that part of the boundary of the Ω_S subdomain coincides with part of the Γ boundary). They produce $\underset{\sim}{H}_S$ and $\underset{\sim}{G}_S$ matrices such that,

$$\underset{\sim}{S}_S = \underset{\sim}{H}_S\,\underset{\sim}{U}_S - \underset{\sim}{G}_S\,\underset{\sim}{Q}_S\,\underset{\sim}{\alpha} \tag{27}$$

<u>Special Case of a Triangular Subdomain</u>

To illustrate how the technique can be used to compute domain integrals in complex nonlinear or time dependent problems consider the case of a triangular subdomain Ω_S as shown in figure 3.

The value of b can now be defined in terms of its values at the 3 corners of the subdomain i.e. will be given by a linear relationship,

$$b = \alpha_1 + \alpha_2 x_1 + \alpha_3 x_2 \tag{28}$$

Thus the body force term can be written as,

$$\int_{\Omega_S} b\,u^*\,d\Omega = \int_{\Omega_S} (\nabla^2 u)\,u^*\,d\Omega \tag{29}$$

The corresponding particular solutions for the three terms in (28), i.e.

$$\nabla^2 \hat{u}_1 = \alpha_1 \quad ; \quad \nabla^2 \hat{u}_2 = \alpha_2 x_1 \quad ; \quad \nabla^2 \hat{u}_3 = \alpha_3 x_2 \tag{30}$$

can be proposed as,

$$\hat{u}_1 = \alpha_1\left(\frac{x_1^2 + x_2^2}{4}\right) \quad ; \quad \hat{u}_2 = \alpha_2\left(\frac{x_1^3}{6}\right) \quad ; \quad \hat{u}_3 = \alpha_3\left(\frac{x_2^3}{6}\right) \tag{31}$$

These solutions can easily be applied to transform (25) to the boundary Γ_S, i.e.

$$-\int_{\Omega_S} b\ u^*\ d\Omega = \int_{\Gamma_S} \hat{u}\ q^*\ d\Gamma - \int_{\Gamma_S} \hat{q}\ u^*\ d\Gamma \tag{32}$$

One can then compute the integrals in terms of û and q̂ analytically or numerically.

6. CHOICE OF FUNCTIONS

The most important consideration in the computational implementation of the DRM is the choice of appropriate interpolation functions f^j. After a series of numerical experiments, Brebbia and Nardini [10][11] proposed using 'conical' functions of the type distance between the points of application of the singularity, ξ_i, and any given point x, such as

$$f^j = r(\xi_j, x) \tag{33}$$

This gives a very simple type of $\hat{u}^j$ function for the Laplace operator, namely,

$$\hat{u}^j(x) = \frac{1}{6} r^3 \tag{34}$$

In addition those authors recommended adding a suitably closer constant for completeness, which can then be incorporated through the following function,

$$f^j = 1 \tag{35}$$

This gives for the Laplace's equation

$$\hat{u}^j = \frac{r^2}{2} \tag{36}$$

The introduction of one more equation for the constant will require setting up another equation in terms of α, which can be done by defining an internal degree of freedom or 'pole'. In general the introduction of more degrees of freedom is recommended to obtain better results when the b function is difficult to represent in function of the boundary values only.

Other types of functions proposed by Nardini and Brebbia can be seen in reference [10]. They include localized harmonic functions and polynomials in terms of x_k coordinates, but the best results were always reported using conical functions.

It has already been pointed out that the success of the DRM is strongly dependent on the choice of coordinate functions which ought to satisfy completeness. Further investigation on this was carried out by Tang and Brebbia [14] who proposed using a generalized function to represent the b term. These new functions are given by the power of the distance r, i.e.

$$[f^j(\xi_j, x)]^k = [r_j]^k \tag{37}$$

where k is an integer, k = 0,1,2,3, etc. For the special case k = 0 one has the constant as defined in (34).

For each function f the following equation needs to be satisfied, i.e.

$$\nabla^2 \hat{u}^j = [r]^k \tag{38}$$

which gives as solution,

$$\hat{u}^j = \frac{1}{(k+2)(k+1)} [r]^{k+2} \tag{39}$$

and

$$\hat{q}^j = \frac{\partial \hat{u}^j}{\partial n} = \frac{1}{(k+1)} [r]^{k+1} \frac{\partial r}{\partial n} \tag{40}$$

Notice that the original paper by Nardini and Brebbia considered only the cases k = 0 and k = 1. Tang and Brebbia however found that some results obtained with these functions were unsatisfactory and that it was sometimes necessary to use a more general expression for b(x), such that it involved the various power functions, i.e.

$$b(x) = \sum_{k=1}^{M} \sum_{j=1}^{N} \alpha_j^k [r_j]^k + \alpha_j^\circ \tag{41}$$

Choosing the index k from 0 to M gives a total of M x N unknowns in the resulting system of equations corresponding to the α_j^k coefficients.

The body force term can now be written as follows,

$$- \int_\Omega u^* \, b \, d\Omega = \sum_{j=1}^{N} \alpha_o^j [c^j \, \hat{u}_o^j + \int_\Gamma q^* \, \hat{u}_o^j \, d\Gamma - \int_\Gamma u^* \, \hat{q}_o^j \, d\Gamma]$$

$$+ \sum_{k=1}^{M} \sum_{j=1}^{N} \alpha_k^j [c \, \hat{u}_k^j + \int_\Gamma q^* \, \hat{u}_k^j \, d\Gamma - \int_\Gamma u^* \, \hat{q}_k^j \, d\Gamma] \tag{42}$$

For axisymmetric cases the solution depends not only on r but also on the distance from the source and field points to the axis of revolution. Because of this Nardini, Telles and Brebbia [15] proposed the following function,

$$f^j = r \left(1 - \frac{R_j}{4R} \right) \tag{43}$$

where R_j is the distance from the different poles to the axis of revolution and R is the distance from any point on Γ (or Ω) to the same axis. The û and q̂ functions are the same type as for three dimensional analysis.

CONCLUSIONS

The Dual Reciprocity Method presented in this paper provides a general technique for transforming the BEM domain integrals into boundary integrals. Other prevalent BE treatments discretize the domain into cells which destroys in great part the advantages of BEM over FEM.

The DRM approach has been extended in the paper to deal with distributed forces acting only on part of the domain and in particular to the case of triangular subdomains. A special section deals with the choice of functions for axisymmetric, two and three dimensional applications.

Although the technique was originally developed for elastodynamics [5][6] it has been applied to solve potential and fluid flow problems [7][8][11][12][13][15] and has a wide range of applications in water resources.

REFERENCES

[1] BREBBIA, C.A. (1978) The Boundary Element Method for Engineers. Pentech Press, London.

[2] BREBBIA, C.A., TELLES, J.C.F. and WROBEL, L.C. (1984) Boundary Element Techniques - Theory and Applications in Engineering. Springer-Verlag, Berlin and New York.

[3] BREBBIA, C.A. and DOMINGUEZ, J. (1988) Boundary Elements - An Introductory Course. Computational Mechanics Publications Southampton and Boston. (This book includes a PC diskette with BE Codes)

[4] BREBBIA, C.A. (1982) Some Applications of the Boundary Element Method for Potential Problems, in Finite Elements in Water Resources (Ed. K.-P. Holz et al.), pp.19-3 to 19.28. Proceedings of the 3rd Int. Conf. on F.E.W.R., Hannover, Germany. Computational Mechanics Publications, Southampton and Springer-Verlag, Berlin and New York.

[5] NARDINI, D. and BREBBIA, C.A. (1982) A New Approach to Free Vibration Analysis using Boundary Elements, in Boundary Element Methods in Engineering (Ed. C.A. Brebbia), Springer-Verlag, Berlin and New York.

[6] BREBBIA, C.A. and NARDINI, D. (1983) Int. J. Soil Dynamics and Earthquake Engg., Vol.2, No.4, pp.228-233. Computational Mechanics Publications, Southampton.

[7] WROBEL, L.C., BREBBIA, C.A. and NARDINI, D. (1986) The Dual Reciprocity Boundary Element Formulation for Transient Heat Conduction. In Proc. of the Vth Int. Conf. on F.E.M. in Water Resources. Computational Mechanics Publications, Southampton and Boston.

[8] WROBEL, L.C. and BREBBIA, C.A. (1987) The Dual Reciprocity Boundary Element Formulation for Non-Linear Diffusion Problems. Computer Methods in Applied Mechanics and Engg. Vol.65, pp.147-164.

[9] NIKU, S.M. and BREBBIA, C.A. (1988) Dual Reciprocity Boundary Element Formulation for Potential Problems with Arbitrarily Distributed Forces. Technical Notes, Eng. Analysis, Vol.5, No.1.

[10] NARDINI D. and BREBBIA, C.A. (1985) Boundary Integral Formulations of Mass Matrices for Dynamics Analysis, In Topics in Boundary Element Research, Vol.2, (Ed. C.A. Brebbia) Springer-Verlag, Berlin and New York.

[11] NARDINI, D. and BREBBIA, C.A. (1985) The Solution of Parabolic and Hyperbolic Problems using an Alternative Boundary Element Formulation, in Proc. of VIIth Int. Conf. on B.E.M. in Eng. Computational Mechanics Publications, Southampton and Boston.

[12] WROBEL, L.C., BREBBIA, C.A. and NARDINI, D. (1986) Analysis of Transient Thermal Problems in the BEASY System, in BETECH/86 (Eds. J.J. Connor and C.A. Brebbia), Computational Mechanics Publications, Southampton, Boston.

[13] BREBBIA, C.A. and WROBEL, L.C. (1987) Non-Linear Transient Thermal Analysis using the Dual Reciprocity Method, in Boundary Element Techniques: Applications in Stress Analysis and Heat Transfer (Eds. C.A. Brebbia and W. Venturini), Computational Mechanics Publications, Southampton, Boston.

[14] TANG, W. and BREBBIA, C.A. Critical Comparison of Two Transformation Methods for Taking BEM Domain Integrals to the Boundary. Submitted for publication to Eng. Analysis Journal.

[15] WROBEL, L.C., TELLES, J.C.F. and BREBBIA, C.A. (1986) A Dual Reciprocity Boundary Element Formulation for Axisymmetric Diffusion Problems, in Proc. of VIIIth Int. Conf. on B.E.M. in Engg., Tokyo, 1986. Computational Mechanics Publications, Southampton and Boston.

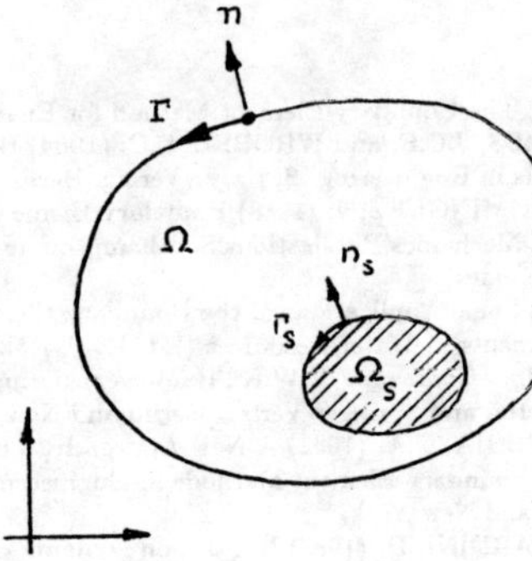

Figure 1 Geometrical Definitions

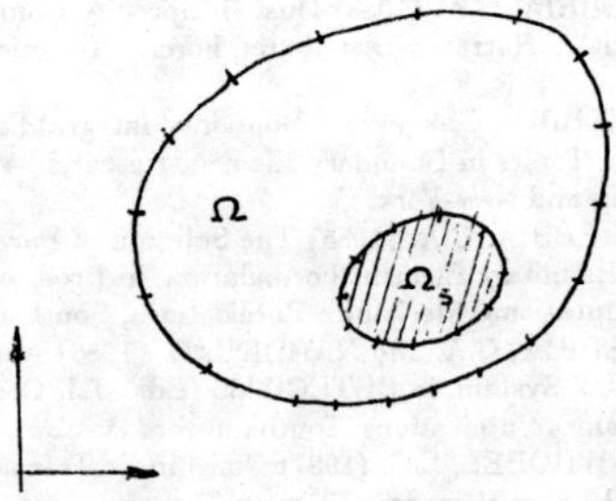

Figure 2 Discretization into Boundary Elements (External and internal elements)

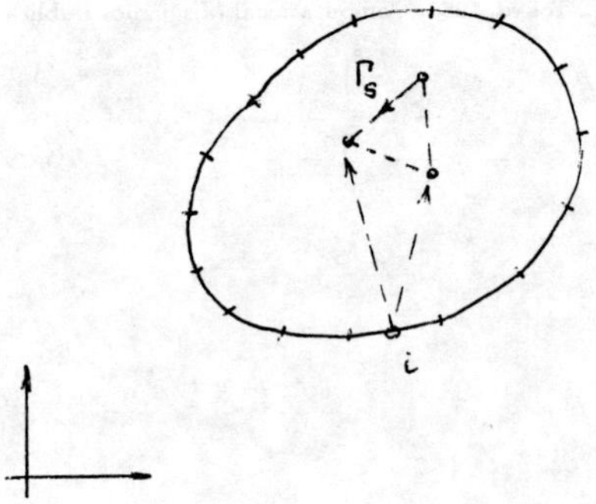

Figure 3 Triangular Subdomain

INVITED PAPER

Advances on the Numerical Simulation of Steep Fronts

I. Herrera and G. Hernández

Geophysics Institute, National University of Mexico, Mexico, D.F.

Summary

Solution of advection-dominated transport problems by discrete interior methods is usually accomplished by employing some type of upstream weighting. Upwinded finite element formulations have also been developed. These are usually based on Petrov-Galerkin formulations. More recently the author has proposed a procedure which produces optimal test functions using approximate solutions of the adjoint equation. Advantages of this method are: a) no arbitrary parameters appear in their definitions, b) the functions vary continuously with the coefficients of the equations, c) the definition of weighting functions results from a systematic and mathematically sound formulation and d) rapidly convergent and accurate solutions are obtained. The only procedure with comparable efficiency is the version of Petrov-Galerkin due to Hughes and Brooks. In this paper a comparison of test functions for these methods is carried out and conclusions drawn from such comparison.

Introduction

The numerical solution of the advective-diffusive transport equation is a problem of great importance because many problems in science and engineering have mathematical representations characterized by sharp fronts. This happens when the process is advection dominated, in which case its numerical treatment is very difficult. Considerable work has been expended in developing discretization formulae for this kind of problems [1-3]. Most have focused on upstream weighting techniques. A fundamental criticism to these methods is the essentially ad-hoc nature of their development. This is manifested through the presence of an arbitrary parameter, the choice of which has to be decided by the analyst. An alternative and very promising approach has been introduced by the author [4-9]. In the past several researchers [2,3], when developing test functions, have considered them optimal when they yield exact values at the nodes. More generally, the author has proposed to consider a system of weighting functions optimal when they yield exact values at inter-element boundaries for arbitrary excitation terms. When this is done, this criterium of optimality reduces to the notion of T(Trefftz)-completeness which has been introduced by the author [4].Herrera's approach consists in using optimal test functions (OTF) systematically.

Numerical comparisons have been carried out between the results obtained using the author's method and other procedures [9]. In general, it was found that the OTF method yields more satisfactory results. The only procedure whose results were very close, is the Petrov-Galerkin method of Hughes and Brooks [3]. In this paper Herrera's OTF method is described. Then, a comparison between the test functions used in these two methods is carried out. They are shown to be quite similar up to fairly large values of the element Peclet number. The optimal test functions used in the author's approach are derived via the solution of the adjoint differential equation. The test functions of Hughes and Brooks are derived in a relatively ad-hoc manner and using very different considerations. However, they turn out to be good approximations of the solutions of the adjoint differential equation (OTF), except for very large values of the element Peclet number or when the source terms are large. This explains the good preformance of Hughes' approach. There are three situations in which the advantages of the author's approach are clear: for large Peclet numbers, when the source term is strong (in a sense which is made clear in the section of conclusions) and when higher order algorithms are required.

<u>Herrera's OTF Method</u>

Let us introduce the approximation for one-dimensional steady-state transport equation with sources, given by

$$\mathcal{L}\, u \equiv \frac{d}{dx}\left(D\,\frac{du}{dx}\right) - V\frac{du}{dx} + Ru = f_{\Omega}(x), \quad 0 \leq x \leq \ell \tag{1}$$

$$u(0) = g_0 \tag{2a}$$

$$u(\ell) = g_{\ell} \tag{2b}$$

first type boundary conditions are chosen for convenience of presentation only. The numerical procedure has been implemented for general differential equations and general boundary conditions [6-8].

The domain $[0,\ell]$ is divided into E subintervals, or elements (not necessarily equal), $\{[x_0,x_1], [x_1,x_2],\ldots,[x_{E-1},x_E]\}$, where $x_0=0$ and $x_E=\ell$. This yields E+1 nodal points $\{x_0,x_1,\ldots,x_E\}$.

We adopt the representation

$$u(0) = \sum_{i=0}^{E} U_i\, \phi_i(x) \tag{3}$$

where U_i are the nodal values of u(x). A test function w(x) will be taken localized in the union of two neighboring subintervals $[x_{j-1},x_j]$ and $[x_j,x_{j+1}]$, where x_j is any interior node (Fig. 1). In addition, the test function w satisfies

$$w(x_{j-1})=0 \tag{4a}$$

$$w(x_{j+1})=0 \tag{4b}$$

$$w(x_j-) = w(x_j+) \tag{4c}$$

condition (4c) states that the limits from the right and form the left agree at the node x_j;i.e. w is continuous at node x_j. However, generally, the derivative of the test function w will have a jump discontinuity at x_j (Fig. 1).

Multiplying $\mathcal{L}u$ by w, integrating from x_{j-1} to x_{j+1}, and applying "generalized Gauss Theorem" for functions with jump discontinuities (see, for example [10]), it is obtained.

$$\int_{x_{j-1}}^{x_{j+1}} w\mathcal{L}u\,dx=-\left[uD\frac{dw}{dx}\right]_{x_{j-1}}^{x_{j+1}} + \left[\!\left[uD\frac{dw}{dx}\right]\!\right]_{x_j} + \int_{x_{j-1}}^{x_{j+1}} u\mathcal{L}^*\,w\,dx \tag{5}$$

Here, the "jump" $[\![\;]\!]$, is defined by

$$\left[\!\left[uD\frac{dw}{dx}\right]\!\right]_{x_j} =u_j(x_j+)D\frac{dw}{dx}\,(x_j+)-u(x_j-)D\frac{dw}{dx}\,(x_j-) \tag{6}$$

while the adjoint operator $\mathcal{L}^*$ is defined by

$$\mathcal{L}^*\, w \equiv \frac{d}{dx}\,(D\frac{dw}{dx}) + \frac{d}{dx}\,(Vw) + Rw \tag{7}$$

In the author's procedure, the optimal test function w satisfies $\mathcal{L}^*\,w=0$. In this case, combining (5) with (1), it is obtained

$$A_{j-}U_{j-1}+A_jU_j+A_{j+}U_{j+1} = \int_{x_{j-1}}^{x_{j+1}} wf_{\Omega}dx \tag{8}$$

where

$$A_{j-}=(D\frac{dw}{dx})_{x_{j-1}};\quad A_j=\left[\!\left[D\frac{dw}{dx}\right]\!\right]_{x_j}\;;\; A_{j+1}=-(D\frac{dw}{dx})_{x_{j+1}} \tag{9}$$

When equation (8) is applied at each one of the interior nodes (i.e., j=1,...,E-1), the unknown values ($U_1,\ldots,U_{E-1}$) of the solution there, can be obtained from the resulting system of E-1 equations. Before closing this Section, we observe that the optimal test function w used in (8), may be thought as defined throughout the whole interval $[0,\ell]$, if its value is identically zero outside $[x_{j-1},x_{j+1}]$. In view of equations (4), such test function is continuous on $[0,\ell]$ but its derivative has jumped discontinuities at interior nodes.

Comparison with Petrov-Galerkin

The procedure explained before has been applied to advection dominated problems using semi-discretization [9]. The results so obtained are quite satisfactory, being oscillation free, to a large extent. The only method whose results are close for a large range of Peclet numbers is the Petrov-Galerkin version of Hughes and Brooks [3]. After a more careful analysis it was found that this is due to the fact that the weighting functions used in both methods are close to each other.

Hughes' test function is

$$w = \phi + \tfrac{1}{2}[(\coth\alpha) - 1/\alpha]\, V\frac{d\phi}{dx} \tag{10}$$

where ϕ is a basis function and α= Vh/2 is the element Peclet number. In Figs. 2 and 3 the test functions for both methods are compared.

In Fig. 3, we have introduced a non-dimensional measure of the strength of the source term

$$\rho = 4\frac{RD}{V^2} \tag{11}$$

From inspection of these figures we conclude that Hughes' test functions yield good approximations except when Peclet number is large, specially if the source strength ρ is not small. An additional point must be made in connection with this comparison, in Herrera's procedure it is possible to produce solutions of any desired order of accuracy [7], something which is not possible when using Hughes' approach.

References

1. Allen, M.B. and Pinder, G.F., Collocation Simulation of Multiphase Porous Medium Flow, Soc. Pet. Eng. J., 23 (1) pp. 135-142, 1983.

2. Christie, I.; Griffiths, D.F. and Mitchell, A.R., Finite Element Methods for Second Order Differential Equations with Significant First Derivatives, Int. J. Num. Method. Engng., 10, 1389-1396 (1976).

3. Hughes, T.J.R. and Brooks, A., A theoretical framework for Petrov-Galerkin methods with discontinuous weighting functions: Applications to the streamline-upwind procedure, in Finite Elements in Fluids. Vol. 4, Gallagher, et al., eds., 47-65, Wiley, N.Y. 1982.

4. Herrera, I., Boundary Methods: An Algebraic Theory, Pitman Advanced Publishing Program, Boston, London 1984.

5. Herrera, I., Unified approach of numerical methods. Part I, Green's formulas for operators in discontinuous fields Numerical Methods for Partial Differential Equations, 1 (1), 12-37, 1985. Part 2, Finite elements, boundary methods and its coupling Numer. Meths. for Partial Differential Equations 1 (3), 159-186. 1985.

6. Herrera, I., Chargoy, L. and Alduncin, G., Unified approach of numerical methods. Part. 3. Finite differences and ordinary differential equations. Numer. Meths. for Partial Differential Equations, 1, (4), 241-258, 1985.

7. Herrera, I., The algebraic theory approach for ordinary differential equation: Highly accurate finite differences. Numer. Meths. for Partial Differential Equations, 3, (3), 199-218, 1987.

8. Celia, M. and Herrera, I., Solution of general ordinary differential equations using the algebraic theory approach, Numer. Meths. for Partial Differential Equations, 3 (2), 117-129, 1987.

9. Celia, M., Herrera, I. and Kindered, S., A new numerical approach for the advective-diffusive transport equation, Comunicaciones Técnicas del Instituto de Geofísica, (UNAM), No. 7, 1987.

10. Allen, M., Herrera, I. and Pinder , G., Numerical Modelling in Science and Engineering, John Wiley, 1987.

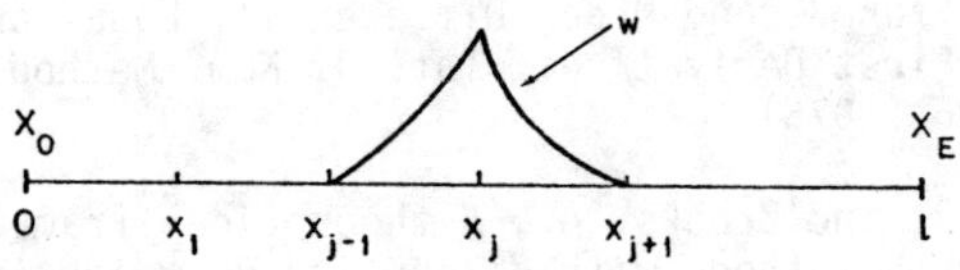

Fig. 1 A typical test function.

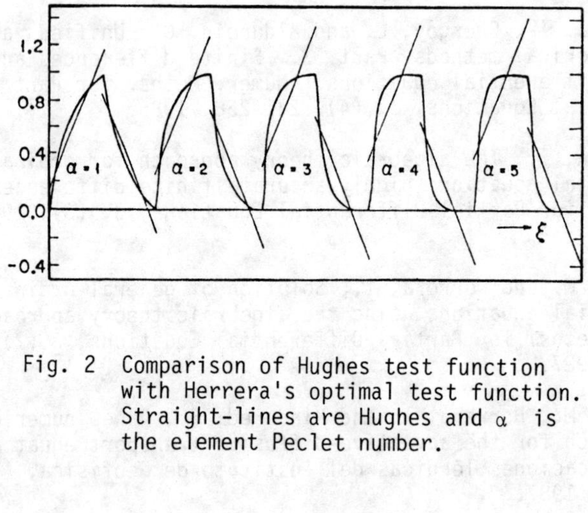

Fig. 2 Comparison of Hughes test function with Herrera's optimal test function. Straight-lines are Hughes and α is the element Peclet number.

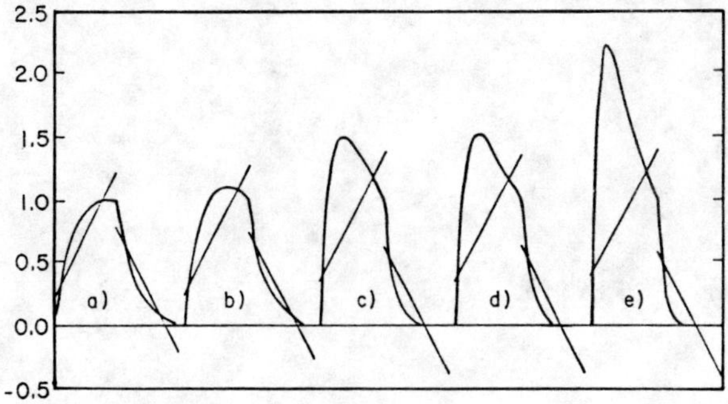

Fig. 3 Comparison of Hughes test function for a) $\rho=.44$, $\alpha=1.5$; b) $\rho=.55$, $\alpha=1.9$; c) $\rho=.064$, $\alpha=3.95$; d) $\rho=.36$, $\alpha=4$; e) $\rho=.36$, $\alpha=6$. 2α is the element Peclet number.

Guidelines for the use of Preconditioned Conjugate Gradients in Solving Discretized Potential Flow Problems

E.F. Kaasschieter

TNO-DGV Institute of Applied Geoscience, PO Box 285, 2600 AG Delft, The Netherlands

1. INTRODUCTION

In the last decade interest has grown in the preconditioned conjugate gradient (cg-)method, because it is a successful iterative method for solving systems of linear equations with a large, sparse and symmetric positive definite coefficient matrix. The method is particularly suitable for solving discretized potential flow problems. Using numerical experiments Gambolati and Perdon[1] have shown that this is particularly the case when modelling stationary groundwater flow in a saturated porous medium. However, to achieve optimal convergence for the preconditioned cg-method in practical modelling, it is necessary to have some insight in the most important aspects of this method. This is the subject of this paper.

After briefly summarizing the potential flow problem, the properties that characterize the cg-method are mentioned. The derivation of a practical termination criterion and a particular preconditioning, namely the well-known incomplete Cholesky decomposition, are discussed.

2. THE POTENTIAL FLOW PROBLEM

The basic laws governing stationary flow of groundwater in a saturated porous medium are

$$\mathbf{q} = -\mathbf{K}\nabla\phi \text{ in } \Omega \text{ (Darcy's law)}, \tag{1}$$

$$\nabla\cdot\mathbf{q} = f \text{ in } \Omega \text{ (continuity equation)}, \tag{2}$$

where $\phi[L]$ is the piezometric head (potential), $\mathbf{K}[LT^{-1}]$ the second rank tensor of hydraulic conductivity (permeability) and $q[LT^{-1}]$ the specific discharge (flux). Here Ω is an open bounded and connected three-dimensional domain with a piecewise smooth boundary $\partial\Omega$. The function f may be used to represent the sources and sinks in Ω. Combining these equations and adding the boundary conditions

$$\phi = g^1 \text{ on } \partial\Omega^1 \text{ (prescribed potential)}, \tag{3}$$

$$\mathbf{n}\cdot\mathbf{q} = g^2 \text{ on } \partial\Omega^2 \text{ (prescribed flux)}, \tag{4}$$

results in the potential flow problem

$$\boxed{\begin{aligned} &-\nabla\cdot(\mathbf{K}\nabla\phi) = f \text{ in } \Omega, \\ &\phi = g^1 \text{ on } \partial\Omega^1,\ -\mathbf{n}\cdot(\mathbf{K}\nabla\phi) = g^2 \text{ on } \partial\Omega^2 \end{aligned}} \tag{5}$$

(for details see Bear[2]). The portions $\partial\Omega^1$ and $\partial\Omega^2$ of the boundary $\partial\Omega$ are such that $\partial\Omega^1 \cap \partial\Omega^2 = \phi$, $\overline{\partial\Omega^1} \cup \overline{\partial\Omega^2} = \partial\Omega$ and the measure of $\partial\Omega^1 \subset \partial\Omega$ is positive. The vector **n** is the normal to $\partial\Omega$ pointing outwards.

3. THE CONJUGATE GRADIENT METHOD

In many practical situations it is impossible to determine an exact solution of Equation 5 and therefore approximations obtained by numerical methods, e.g. the finite difference or the finite element method, have to be used (for a brief introduction to these methods see Bear[2]: Sec. 5.6).
The discretization of Equation 5 by a finite difference or a finite element method leads to a system of linear equations

$$\mathbf{Ax} = \mathbf{b} \tag{6}$$

in which the symmetric positive definite matrix A is large and sparse. It is commonly agreed that for most three-dimensional problems iterative methods (see Golub and Van Loan[3]: Ch. 10, Hageman and Young[4]) are preferable, since direct methods (e.g. Gaussian elimination, see Golub and Van Loan[3]: Ch. 4) usually require an unacceptable amount of computer time and computer storage. Iterative methods require matrix-vector multiplications only and thus avoid fill-in, so that they are cheap per iteration and economical with respect to computer storage.
A very popular iterative method is the conjugate gradient (cg-)method. In this method an arbitrary vector $\mathbf{x}_0$ is chosen and successive approximations $\mathbf{x}_1, \mathbf{x}_2, \ldots$ are computed (for details see Axelsson and Barker[5]: Sec. 1.3, Golub and Van Loan[3]: Sec. 10.2, Hageman and Young[4]: Sec. 7.2, Hestenes[6]) according to

Algorithm 1:

$\mathbf{r}_0 := \mathbf{b} - \mathbf{Ax}_0$
<u>for</u> $i = 0,1,\ldots$
  <u>if</u> $\mathbf{r}_i = \mathbf{0}$ <u>then</u> stop
  $\beta_{i-1} := \mathbf{r}_i^T\mathbf{r}_i/\mathbf{r}_{i-1}^T\mathbf{r}_{i-1}$ $(\beta_{-1} := 0)$
  $\mathbf{p}_i := \mathbf{r}_i + \beta_{i-1}\mathbf{p}_{i-1}$ $(\mathbf{p}_0 := \mathbf{r}_0)$
  $\alpha_i := \mathbf{r}_i^T\mathbf{r}_i/\mathbf{p}_i^T\mathbf{A}\mathbf{p}_i$
  $\mathbf{x}_{i+1} := \mathbf{x}_i + \alpha_i\mathbf{p}_i$
  $\mathbf{r}_{i+1} := \mathbf{r}_i - \alpha_i\mathbf{A}\mathbf{p}_i$.

For the basic relations in the cg-method see Golub and Van Loan[3]: Sec. 10.2, Hestenes[6]: Sec. IV-2, Van der Sluis and Van der Vorst[7]: Sec. 1. The cg-method is characterized by the following property:

$$\|\mathbf{x} - \mathbf{x}_i\|_{\mathbf{A}} = \min\left\{\|\mathbf{x} - \mathbf{y}\|_{\mathbf{A}} \mid \mathbf{y} - \mathbf{x}_0 \in K_i\right\}, \tag{7}$$

where $\|\mathbf{z}\|_{\mathbf{A}} = (\mathbf{z}^T\mathbf{Az})^{1/2}$ for every vector z and $K_i = \mathrm{span}\{\mathbf{r}_0, \mathbf{Ar}_0, \ldots, \mathbf{A}^{i-1}\mathbf{r}_0\}$. Although in principle the cg-method is finite, this property is not exploited, because in practice the approximations $\mathbf{x}_i$ are sufficiently close to the solution **x** of Equation 6 long before exact termination should occur. It can be shown that

$$\|\mathbf{x} - \mathbf{x}_i\|_{\mathbf{A}} \leq 2\left(\frac{\sqrt{\kappa} - 1}{\sqrt{\kappa} + 1}\right)^i \|\mathbf{x} - \mathbf{x}_0\|_{\mathbf{A}}, \tag{8}$$

where κ is the spectral condition number of **A**. Because of its generality and simplicity, the bound given in Equation 8 is very useful. It must be remembered, however, that depending on the distribution of the eigenvalues of **A** Equation 8 might be quite pessimistic, concealing the true rate of convergence. It has also been observed that when i increases the process tends to converge faster. This phenomenon is generally referred to as superlinear convergence. For further reading on this topic see Van der Sluis and Van der Vorst[7,8].

4. A PRACTICAL TERMINATION CRITERION

In practice, exact termination of the cg-method is prevented because of rounding errors. Moreover, an approximation to the solution of Equation 6, obtained long before exact termination should occur, is often sufficient. Therefore, if $\mathbf{x} \neq \mathbf{0}$, the following relative termination criterion is chosen:

$$\|\mathbf{x} - \mathbf{x}_i\| / \|\mathbf{x}\| \leq \epsilon, \tag{9}$$

where $\epsilon > 0$ is a preordained accuracy and $\|\mathbf{z}\| = (\mathbf{z}^T\mathbf{z})^{1/2}$ for every vector $\mathbf{z}$. Unfortunately it is impossible to determine $\|\mathbf{x} - \mathbf{x}_i\| / \|\mathbf{x}\|$ cheaply, because the solution $\mathbf{x}$ is unknown. To solve this problem, note that the criterion can be replaced by

$$\|\mathbf{x} - \mathbf{x}_i\| \leq \|\mathbf{x}_i\| \epsilon/(1 + \epsilon), \tag{10}$$

because Equation 10 implies Equation 9 (for a proof see Kaasschieter[9]: Th. 2). Further, after i iterations of the cg-method:

$$\mathbf{x} - \mathbf{x}_i = \mathbf{A}^{-1}(\mathbf{b} - \mathbf{A}\mathbf{x}_i) = \mathbf{A}^{-1}\mathbf{r}_i \tag{11}$$

and then

$$\|\mathbf{x} - \mathbf{x}_i\| = \|\mathbf{A}^{-1}\mathbf{r}_i\| \leq \|\mathbf{r}_i\| / \mu_1, \tag{12}$$

where μ_1 is the smallest eigenvalue of **A** for which $\mathbf{x} - \mathbf{x}_i$ has a component in the corresponding eigenvector direction.
While iterating it is possible to obtain cheaply an approximation $\mu_1^{(i)}$ to μ_1 from the iteration constants $\alpha_0, \alpha_1, ..., \alpha_{i-1}$ and $\beta_0, \beta_1, ..., \beta_{i-2}$ (for details see Hageman and Young[4]: Sec. 7.5, Kaasschieter[9]). The criterion (see Equation 10) is therefore approximately satisfied if

$$\|\mathbf{r}_i\| \leq \mu_1^{(i)} \|\mathbf{x}_i\| \epsilon/(1 + \epsilon). \tag{13}$$

Since $\mu_1^{(i)}$ tends to be closer and closer to μ_1, for increasing i, this termination criterion is more useful than the commonly used criteria, based on the residual (i.e. $\|\mathbf{r}_i\| \leq \epsilon$) or on the reduction of the residual (i.e. $\|\mathbf{r}_i\| / \|\mathbf{r}_0\| \leq \epsilon$), since these criteria give little information on the error in the solution. As regards to the reliability of the termination criterion (see Equation 13), note that if

$$\mu_1^{(i)} \leq \|\mathbf{r}_i\| / \|\mathbf{x} - \mathbf{x}_i\|, \tag{14}$$

then using Equation 13 guarantees that $\|\mathbf{x} - \mathbf{x}_i\| / \|\mathbf{x}\| \leq \epsilon$. In Kaasschieter[9]:

Sec. 4 it is proved that Equation 14 holds, provided that the cg-process does not terminate prematurely, i.e. when ϵ is not too large. In solving discretized potential flow problems it usually suffices to choose $\epsilon = 10^{-2}$.

5. THE PRECONDITIONED CONJUGATE GRADIENT METHOD

The convergence rate of the cg-method can be significantly improved by replacing Equation 7 by the minimization problem

$$\| \hat{\mathbf{x}} - \hat{\mathbf{x}}_i \|_{\hat{\mathbf{A}}} = \min \left\{ \| \hat{\mathbf{x}} - \mathbf{y} \|_{\hat{\mathbf{A}}} \mid \mathbf{y} - \hat{\mathbf{x}}_0 \in \hat{K}_i \right\}, \tag{15}$$

where $\hat{\mathbf{A}} = \mathbf{C}^{-1}\mathbf{A}\mathbf{C}^{-T}$, $\hat{\mathbf{x}} = \mathbf{C}^T\mathbf{x}$, $\hat{\mathbf{b}} = \mathbf{C}^{-1}\mathbf{b}$, $\hat{\mathbf{x}}_0 = \mathbf{C}^T\mathbf{x}_0$, $\hat{\mathbf{r}}_0 = \hat{\mathbf{b}} - \hat{\mathbf{A}}\hat{\mathbf{x}}_0 = \mathbf{C}^{-1}\mathbf{r}_0$ and $\hat{K}_i = \text{span}\,\{\hat{\mathbf{r}}_0, \hat{\mathbf{A}}\hat{\mathbf{r}}_0, \ldots, \hat{\mathbf{A}}^{i-1}\hat{\mathbf{r}}_0\}$. C is a suitably chosen nonsingular matrix. The preconditioned cg-method (for details see Axelsson and Barker[5]: Sec. 1.4, Golub and Van Loan[3]: Sec. 10.3) can be formulated in two versions, one transformed and one untransformed. The untransformed method directly produces approximations to the solution x of $\mathbf{A}\mathbf{x} = \mathbf{b}$, whereas the transformed method produces approximations to the solution of

$$\hat{\mathbf{A}}\hat{\mathbf{x}} = \hat{\mathbf{b}}. \tag{16}$$

The untransformed method is defined as follows:

Algorithm 2:

$\mathbf{r}_0 := \mathbf{b} - \mathbf{A}\mathbf{x}_0$
<u>for</u> $i = 0, 1, \ldots$
  $\mathbf{z}_i := \mathbf{M}^{-1}\mathbf{r}_i$
  <u>if</u> $\mathbf{r}_i = \mathbf{0}$ <u>then</u> stop
  $\beta_{i-1} := \mathbf{z}_i^T\mathbf{r}_i / \mathbf{z}_{i-1}^T\mathbf{r}_{i-1}$ $(\beta_{-1} := 0)$
  $\mathbf{p}_i := \mathbf{z}_i + \beta_{i-1}\mathbf{p}_{i-1}$ $(\mathbf{p}_0 := \mathbf{z}_0)$
  $\alpha_i := \mathbf{z}_i^T\mathbf{r}_i / \mathbf{p}_i^T\mathbf{A}\mathbf{p}_i$
  $\mathbf{x}_{i+1} := \mathbf{x}_i + \alpha_i\mathbf{p}_i$
  $\mathbf{r}_{i+1} := \mathbf{r}_i - \alpha_i\mathbf{A}\mathbf{p}_i$.

The preconditioning matrix M is given by $\mathbf{M} = \mathbf{C}\mathbf{C}^T$. An analysis of Algorithm 2 shows that the sequence $\hat{\mathbf{x}}_0, \hat{\mathbf{x}}_1, \ldots$, where $\hat{\mathbf{x}}_i = \mathbf{C}^T\mathbf{x}_i$, can be obtained by applying the ordinary cg-method to the system $\hat{\mathbf{A}}\hat{\mathbf{x}} = \hat{\mathbf{b}}$. Since the relations $\hat{\mathbf{x}}_i = \mathbf{C}^T\mathbf{x}_i$ and $\hat{\mathbf{x}} = \mathbf{C}^T\mathbf{x}$ imply that $\| \hat{\mathbf{x}} - \hat{\mathbf{x}}_i \|_{\hat{\mathbf{A}}} = \| \mathbf{x} - \mathbf{x}_i \|_{\mathbf{A}}$, it can be concluded that

$$\| \mathbf{x} - \mathbf{x}_i \|_{\mathbf{A}} \leq 2 \left(\frac{\sqrt{\hat{\kappa}} - 1}{\sqrt{\hat{\kappa}} + 1} \right)^i \| \mathbf{x} - \mathbf{x}_0 \|_{\mathbf{A}}, \tag{17}$$

where $\hat{\kappa}$ is the spectral condition number of $\hat{\mathbf{A}}$. Thus, if M has the property that $\hat{\kappa} < \kappa$, then the preconditioned cg-method has a faster rate of convergence than the ordinary cg-method. The true rate of convergence, however, is strongly determined by the distribution of the eigenvalues of $\hat{\mathbf{A}}$. Clustering of the eigenvalues generally helps to speed up the convergence (see Van der Sluis and Van der Vorst[7,8]).

If $x \neq 0$, then the following termination criterion is chosen:

$$\| \hat{x} - \hat{x}_i \| / \| \hat{x} \| \leq \epsilon. \tag{18}$$

In practice this criterion is replaced by

$$\| \hat{r}_i \| \leq \hat{\mu}_1^{(i)} \| \hat{x}_i \| \epsilon/(1 + \epsilon), \tag{19}$$

where $\hat{r}_i = C^{-1} r_i$ and $\hat{\mu}_1^{(i)}$ is an approximation to the smallest eigenvalue $\hat{\mu}_1$ of $\hat{A}$ for which $\hat{x} - \hat{x}_i$ has a component in the corresponding eigenvector direction (see Sec. 3). Note that $\| \hat{r}_i \| = (z_i^T r_i)^{1/2}$ and $\| \hat{x}_i \| = (x_i^T M x_i)^{1/2}$.
In general a good preconditioning matrix **M** has the following properties:
(i) $\hat{\kappa}$ is significantly less than κ;
(ii) the factors of **M** can be determined quickly and do not require excessive storage in relation to **A**;
(iii) the system $Mz = r$ can be solved more efficiently than $Ax = b$.
The preconditioning matrix **M** may be constructed by an incomplete Cholesky decomposition. There are many variants of incomplete Cholesky decompositions (see Meijerink and Van der Vorst[10]). In one of the simplest, although very successful, variants $M = (D + L)D^{-1}(D + L^T)$, such that
(i) diag (**M**) = diag (**A**),
(ii) **L** is the strictly lower triangular part of **A**.
The entries of the diagonal matrix **D** can be computed recursively from

$$d_{ii} := a_{ii} - \sum_{j=1}^{i-1} a_{ij}^2 / d_{jj}, \; i = 1,\ldots,n. \tag{20}$$

The diagonal entries of **D** are positive, if the symmetric positive definite matrix A is an M-matrix (for details see Meijerink and Van der Vorst[11]). This is generally the case for matrices resulting from potential flow problems discretized by a finite difference method. For finite element methods some extra care is needed (see Axelsson and Barker[5]: Sec. 5.2). Note that the system $Mz = r$ can be solved by successively solving the systems $(D + L)y = r$, $(D + L^T)z = Dy$. Using the preconditioned cg-method with the preconditioning mentioned above, the number of multiplications and additions can be substantially reduced by using an idea of Eisenstat[12].

6. CONCLUSIONS

The conjugate gradient method is a very attractive method for solving discretized potential flow problems, since it is reliable and, when preconditioned, is generally fast. Further, its memory space requirements are modest and it is easy to implement. Moreover, a safe termination criterion based upon the iteration constants can be used.

REFERENCES

1. Gambolati, G. and Perdon, A. (1984), The Conjugate Gradients in Subsurface Flow and Land Subsidence Modelling. Fundamentals of Transport Phenomena in Porous Media, (Eds. Bear, J. and Corapcioglu, M.Y.), pp. 955-984, Martinus Nijhoff, Dordrecht.
2. Bear, J. (1979), Hydraulics of Groundwater, McGraw-Hill, New York.
3. Golub, G.H. and Van Loan, C.F. (1983), Matrix Computations, North Oxford Academic, Oxford.
4. Hageman, L.A. and Young, D.M. (1981), Applied Iterative Methods, Academic Press, New York.
5. Axelsson, O. and Barker, V.A. (1984), Finite Element Solution of Boundary Value Problems, Academic Press, New York.
6. Hestenes, M.R. (1980), Conjugate Direction Methods in Optimization, Springer, Berlin.
7. Van der Sluis, A. and Van der Vorst, H.A. (1986), The Rate of Convergence of Conjugate Gradients, Numerische Mathematik, Vol. 48, pp. 543-560.
8. Van der Sluis, A. and Van der Vorst, H.A. (1987), The Convergence Behaviour of Ritz Values in the Presence of Close Eigenvalues, Linear Algebra and its Applications 88/89, pp. 651-694.
9. Kaasschieter, E.F., A Practical Termination Criterion for the Conjugate Gradient Method, to appear in BIT.
10. Meijerink, J.A. and Van der Vorst, H.A. (1981), Guidelines for the Usage of Incomplete Decompositions in Solving Sets of Linear Equations as They Occur in Practical Problems, Journal of Computational Physics, Vol. 44, pp. 134-155.
11. Meijerink, J.A. and Van der Vorst, H.A. (1977), An Iterative Solution Method for Linear Systems of Which the Coefficient Matrix is a Symmetric M-Matrix, Mathematics of Computation, Vol. 31, pp. 148-162.
12. Eisenstat, S.C. (1981), Efficient Implementation of a Class of Preconditioned Conjugate Gradient Methods, SIAM Journal of Scientific and Statistical Computation, Vol. 2, pp. 1-4.

INVITED PAPER

Non Linear Instability in Long Time Calculations of a Partial Difference Equation

A.R. Mitchell

Department of Mathematics and Computer Science, University of Dundee, Dundee, Scotland

ABSTRACT

The stability of long time ($t \to \infty$) calculations of non linear partial difference equations is examined. The model chosen is a discretisection in space and time of the Korteweg de Vries equation. The standard linearised von Neumann stability analysis, although necessary, is not sufficient, since it ignores the quadratic nature of the non linearity in the problem. Non linear stability is analysed by perturbing a constant solution with a period 3 Fourier mode. This leads to a non linear system of ordinary differential equations, the stability of which is examined by phase plane analysis. The results obtained contrast strongly with those obtained from the linear analysis. A numerical illustration shows how a single Fourier mode initial condition through side band growth, eventually causes blow-up in a long time calculation.

INTRODUCTION

Partial difference equation models of non linear time dependent problems, either obtained by finite difference or finite element methods from differential equations, or as models in their own right, are of use only if they are stable. There is no accepted definition of stability for non linear partial difference equations particularly in long time ($t \to \infty$) calculations with a fixed time step, where lack of blow-up of the solution is necessary for stability but not sufficient. In linear problems there is no such difficulty and many stability theorems exist based on Fourier or energy methods.

If Fourier modes are used to assess the stability of a non linear difference formula, where linearisation has taken place as in the von Neumann method, any Fourier mode with an amplification factor greater than one is enough to make the method unstable. On the other hand, satisfaction of the von Neumann condition, although necessary for stability is not sufficient and non linear effects can still render the difference formula unstable. Alternatively, if the difference formula retains its stability, despite the presence of non linear effects, the solution of the non linear problem may bifurcate into a new stable solution which exists for grid parameter values beyond the linearised stability limits. (Mitchell and Griffiths [1]). These can be illustrated by studying discretisations of two entirely different time dependent partial differential equations involving the non linear term u^2, viz. the Korteweg de Vries equation (Whitham [2])

$$\frac{\partial u}{\partial t} + \frac{1}{2}\frac{\partial}{\partial x}(u^2) + \epsilon\frac{\partial^3 u}{\partial x^3} = 0, \qquad \epsilon > 0 \tag{1}$$

with dispersive wave solutions, and Fisher's [3] equation

$$\frac{\partial u}{\partial t} = \frac{\partial^2 u}{\partial x^2} + u - u^2, \tag{2}$$

a model for reaction diffusion problems in mathematical biology. In the latter case, standard space discretisation of Eq. (2) followed by Euler's method in time, leads to the linearised stability condition

$$r \leqq \tfrac{1}{2} - \tfrac{1}{4}k, \tag{3}$$

where $r = k/h^2$, h and k being the step lengths in space and time respectively. It has been shown that stable periodic solutions of the non linear partial difference equation exist for values of r and k in the range $r > \frac{1}{2} - \frac{1}{4}k$. (Mitchell et al [4]).

Our main objective, however, is the study of non linear instability and so for the remainder of the paper, we shall concentrate on Eq. (1). When $\epsilon = 0$ and Eq. (1) becomes the inviscid Burgers equation

$$\frac{\partial u}{\partial t} + u\frac{\partial u}{\partial x} = 0, \tag{4}$$

several important results exist for non linear instability of difference models. These are gathered together in a clearly written paper by Trefethen [5], where reference is made to fundamental papers on non linear instability by Phillips [6], Fornberg [7], Briggs et al. [8], and others.

THE KORTEWEG DE VRIES EQUATION

Useful analytical properties of the KdV equation are;

(i) the initial value problem with compact support initial data and solutions which tend to zero rapidly as $|x|$ tends to infinity is completely integrable by inverse scattering, and
(ii) the equation has an infinite number of conservation laws. Consequently it might be thought that numerical solutions of the KdV equation will present few problems, in contrast to the case when $\epsilon = 0$, when considerable difficulties are known to arise. This is true in some cases e.g. solitons emerging from compact support, positive initial data, but not in others.

The problem we actually consider involves Eq. (1) in the x-range [0,L] with <u>periodic</u> boundary conditions and an initial condition $u(x,0) = f(x)$, $0 \leq x \leq L$. Initially we discretize Eq. (1) in space only to give

$$\dot{U}_j = -\frac{\theta}{4h}\left(U^2_{j+1} - U^2_{j-1}\right) - \frac{1-\theta}{2h} U_j\left(U_{j+1} - U_{j-1}\right) - \frac{\epsilon}{2h^3}\left(U_{j+2} - 2U_{j+1} + 2U_{j-1} - U_{j-2}\right), \quad 0 \leq j \leq J-1 \tag{5}$$

where $x = jh$, $Jh = L$, a dot denotes differentiation with respect to time, and θ is a parameter in the range $0 \leqq \theta \leqq 1$. This conserves $\sum_{j=0}^{J-1} U_j^2$ exactly if $\theta = 2/3$.

For the purpose of comparing linear and non linear stability, we consider <u>leap frog</u> discretisation in time of Eq. (5) giving

$$U_j^{n+1} = U_j^{n-1} - r\left(U^n_{j+1} - U^n_{j-1}\right)\left[\tfrac{1}{2}\theta\left(U^n_{j+1} + U^n_{j-1}\right) + (1-\theta)U^n_j\right] - \frac{\epsilon r}{h^2}\left(U^n_{j+2} - 2U^n_{j+1} + 2U^n_{j-1} - U^n_{j-2}\right), \quad 0 \leq j \leq J-1,\ n = 1,2,\ldots \tag{6}$$

where $t = nk$ and $r = k/h$. The extra starting condition required by Eq. (6) is obtained from Eq. (5), with Euler discretisation in time. The non linear partial difference formula Eq. (6) is second order accurate in space and time, and for $\theta = 2/3$ conserves $\sum_{j=0}^{J-1} (U_j^n)^2$ to order k^2.

LINEAR STABILITY

We now linearise Eq.(6) by making the substitution

$$U_j^n \sim C + \delta_j^n,$$

where δ is a small perturbation about the constant solution C. This gives

$$\delta_j^{n+1} = \delta_j^{n-1} - Cr\left(\delta_{j+1}^n - \delta_{j-1}^n\right) - \frac{\epsilon r}{h^2}\left(\delta_{j+2}^n - 2\delta_{j+1}^n + 2\delta_{j-1}^n - \delta_{j-2}^n\right), \tag{7}$$

a result <u>independent of θ and the initial condition</u>, and where $u_{min} \leq C \leq u_{max}$.

Standard von Neumann analysis with

$$\delta_j^n \sim e^{\alpha nk} e^{i\beta jh}, \qquad \alpha, \beta \text{ real},$$

leads to <u>neutral</u> stability $(|e^{\alpha k}| = 1)$ for Eq.(7) if

$$\max_{\phi} \left| r \sin\theta\left(C - \frac{4\epsilon}{h^2}\sin^2\frac{\phi}{2}\right)\right| \leq 1, \tag{8}$$

where $\phi = \beta h$. If it is assumed that $u_{min} = 0$, and so the constant about which the linearisation is performed is never negative, Vliegenhart [9] derives from Eq. (8) the stability condition

$$r \leq \left(u_{max} + \frac{4\epsilon}{h^2}\right)^{-1},$$

and Sanz Serna [10], the more realistic linearised condition

$$r \leq \frac{2}{3\sqrt{3}} \frac{h^2}{\epsilon}, \tag{9}$$

provided $u_{max} \leq 1/r$.

NON LINEAR STABILITY OF THE SEMI-DISCRETE SYSTEM

Non linear stability is analysed by considering a solution of Eq.(5) in the form

$$U_j(t) = A_1(t)e^{i\frac{2\pi}{3}j} + (*) + C, \qquad j = 0,1,2. \tag{10}$$

where A_1 is complex and (*) denotes complex conjugate. This represents a period 3 perturbation about the constant solution C. The non linear terms in Eq. (5) acting on the $2\pi/3$ Fourier mode give

$$e^{i\frac{2\pi}{3}j\,2} \cdot e^{i\frac{2\pi}{3}j} = e^{i\frac{4\pi}{3}j} = e^{-i\frac{2\pi}{3}j},$$

and

$$e^{-i\frac{2\pi}{3}j} \cdot e^{-i\frac{2\pi}{3}j} = e^{-i\frac{4\pi}{3}j} = e^{i\frac{2\pi}{3}j},$$

respectively, and so the quadratic interactions are closed. Hence no new Fourier modes are created. This phenomenon is known as <u>aliasing</u>.

If we substitute (10) into (5) and put

$$A_1(t) = V_1(t) + iV_2(t) , \quad V_1, V_2 \in \underline{R} ,$$

we obtain the system of ordinary differential equations

$$\begin{aligned} \dot{V}_1 &= 2\lambda V_1 V_2 - \beta V_2 \\ \dot{V}_2 &= \lambda(V_1^2 - V_2^2) + \beta V_1 \\ \dot{C} &= 0 , \end{aligned} \tag{11}$$

with $\lambda = \frac{\sqrt{3}}{4h}(2 - 3\theta)$ and $\beta = \frac{\sqrt{3}}{2h}\left(\frac{3\epsilon}{h^2} - \tfrac{1}{2}(2 - 3\theta)C\right)$. Phase plane analysis in the $(V_1 V_2)$ plane reveals four critical points with (0,0) stable, provided β and λ are not zero. Special cases are

(i) $\lambda = 0$, $(\theta = 2/3)$ when the system is linearised and integrates to give the conservation law

$$V_1^2 + V_2^2 = \text{constant}, \tag{12}$$

and

(ii) $\beta = 0$, giving $C = \frac{6\epsilon}{(2-3\theta)h^2}$ $(\theta \neq 2/3)$, when the system reduces to

$$\begin{aligned} \dot{V}_1 &= 2\lambda V_1 V_2 \\ \dot{V}_2 &= \lambda(V_1^2 - V_2^2) . \end{aligned} \tag{13}$$

This system is also obtained for perturbations about the zero solution when $\epsilon = 0$ and $\theta \neq 2/3$ (Sloan and Mitchell [11]), where it is shown that the only critical point is a centre at (0,0) and the integral curves satisfy

$$V_1(V_1^2 - 3V_2^2) = \text{constant}. \tag{14}$$

The initial condition determines the solution curve all of which eventually lead to infinity. Thus the system Eq. (13) is completely unstable.

In a similar manner, a period 4 perturbation about the constant solution C given by

$$U_j(t) = A(t)e^{i\frac{\pi}{2}j} + (*) + B(t)\cos \pi j + C(t), \quad j = 0,1,2,3 \tag{15}$$

results in the system

$$h\dot{V}_1 = [(1 - 2\theta)B + \beta]V_2$$

$$h\dot{V}_2 = [(1 - 2\theta)B - \beta]V_1 \qquad (16)$$

$$h\dot{B} = 4(1 - \theta)V_1V_2$$

$$\dot{C} = 0$$

where $\beta = (C - \frac{26}{h^2})$. Special cases of Eq. (16) are

(i) $\theta = 1$. the system is linearised and integrates to give

$$V_1^2 - \gamma V_2^2 = \text{constant} \qquad (17)$$

where $\gamma = (B - \beta)(B + \beta)^{-1}$. Hence depending on the sign of γ, the orbits in the phase plane are either ellipses or hyperbolae. In the latter case, the system is unstable.

(ii) $\theta = 2/3$. Here the system integrates to give the energy conservation

$$V_1^2 + V_2^2 + \tfrac{1}{2}B^2 = \text{constant} \qquad (18)$$

and so the solution cannot blow-up.

We conclude from the two cases $J = 3,4$ that non linear analysis of the semi-discrete system is very dependent on θ, in sharp contrast to the linearised analysis for the fully discretised system, where the stability condition is independent of θ.

LEAP FROG DISCRETISATION IN TIME AND A NUMERICAL EXPERIMENT

We now return to Eq. (6) which is the leap frog discretisation in time of Eq. (5) and put $\theta = 2/3$. This results in the partial difference formula

$$U_j^{n+1} = U_j^{n-1} - \frac{1}{3}r\,(U_{j+1}^n - U_{j-1}^n)(U_{j+1}^n + U_j^n + U_{j-1}^n)$$
$$- \frac{\epsilon r}{h^2}(U_{j+2}^n - 2U_{j+1}^n + 2U_{j-1}^n - U_{j-2}^n),\quad 0 \le j \le J-1,\; n = 1,2,\ldots \qquad (19)$$

which was introduced first by Zabusky and Kruskal [12] as a means of solving the K.d.V. equation.

A numerical experiment is now set up in order to study the growth of non linear instabilities. It is based on a long time calculation of Eq.(19) with a fixed time step. The initial condition is chosen from Eq. (10) to be

$$U_j(0) = 2\sigma(\cos\frac{2\pi}{3}j - \sin\frac{2\pi}{3}j) + C,\quad j = 0,1,2, \qquad (20)$$

and since periodic boundary conditions require J to be a multiple of 3, we choose $J = 120$. The additional starting values U_j^1, $j = 0,1,\ldots,N-1$, are calculated using Euler's rule. Although the quantity $\sum (U_j^n)^2$ is conserved by Eq.(5) with $\theta = 2/3$, leap frog discretisation reduces the conservation to order k^2, leaving blow-up a possibility using Eq.(19). The other parameters in Eqs. (19) and (20) are $r = 1$, $h = 1/120$, $\epsilon = .2315 \times 10^{-4}$ and $C = 1$, values which satisfy the linearised stability condition Eq.(9) but marginally violate the upper limit condition on u_{max}. It is shown that for this choice of parameters, round-off errors in the numerical calculation induce non linear instability, during which the relative spread and growth of Fourier modes can be studied prior to blow-up.

We define the energy in the perturbation term of the initial condition Eq.(20) by

$$E = h \sum_{j=0}^{J-1} (U_j^o - C)^2 = 4\sigma^2 \tag{21}$$

and blow up time by the first value of n for which $\max_j |U_j^n| > 10^5$. In fact in our numerical experiment, with $E = 0.01$, blow-up occurred after 16,774 time steps. The distribution of Fourier modes in the solution as time increases is best seen in Fourier space where the Fourier transform of U_j^n, $j = 0,1,..,J-1$ is obtained from the formula

$$\hat{U}_k = \frac{1}{\sqrt{J}} \sum_{j=0}^{J-1} U_j^n \, e^{\frac{2\pi ijk}{J}}, \qquad k = 0,1,..,J-1, \tag{22}$$

and the initial condition

$$|\hat{U}_k^o| = \begin{cases} 0 & k \neq 40 \\ \sqrt{\dfrac{JE}{2}} \,. & k = 40 \end{cases} \tag{23}$$

In x-space for the initial stages the solution is $U_j^n = U_j^o$. As the calculation proceeds in time, the first sign of instability is a modulation of the constant envelope and then the solution begins to focus at local patches on the grid. Blowup occurs within approximately one hundred steps after focusing. A fuller account of the numerical experiment can be found in Herring[13].

REFERENCES

1. Mitchell A.R. and Griffiths D.F. (1986) Beyond the linearised stability limit in non linear problems. Pitman Research Notes in Mathematics 140, pp. 140-156.

2. Whitham G.B. (1974) Linear and Nonlinear Waves, John Wiley and Sons, Chichester.
3. Fisher R.A. (1950) Gene frequencies in a cline determined by selection and diffusion, Biometrics, Vol. 6, pp.353-361.
4. Mitchell A.R., John-Charles P., and Sleeman B.D. (1987), Long time calculations and non linear maps. Numerical Methods for Transient and Coupled Problems, pp. 199-211, John Wiley and Sons, Chichester.
5. Trefethen L.N., (1983), Nonlinear instability of difference models of $u_t = -uu_x$, (Private Communication).
6. Phillips N.A., (1959), An example of non-linear computational instability, The Atmosphere and the Sea in Motion (Ed. Bolin B) pp. 501-504, Rockefeller Institute, New York.
7. Fornberg B. (1973), On the instability of Leap Frog and Crank-Nicolson Approximations of a Nonlinear Partial Differential Equation, Math. Comp., Vol. 27, pp. 45-57.
8. Briggs W., Newell A.C., and Sarie T., (1983), Focusing: A mechanism for instability of nonlinear finite difference equations. J. Comp. Phys., Vol. 51, pp. 83-106.
9. Vliegenhart A.C. (1971), On finite difference methods for the Korteweg de Vries equation. J. Engrg. Math., Vol. 5, pp. 137-155.
10. Sanz-Serna J.M., (1982), An explicit finite difference scheme with exact conservation properties. J. Comp. Phys., Vol. 47, pp. 199-210.
11. Sloan D.M. and Mitchell A.R. (1986), On non linear instabilities in leap frog finite difference schemes. J. Comp. Phys. Vol. 67, pp. 372-395.
12. Zabusky N.J. and Kruskal M.D. (1965), Interaction of Solitons in a Collisionless Plasma and the Recurrence of Initial States, Phys. Rev. Lett. Vol. 15, pp. 240-243.
13. Herring S. (1987), Nonlinear instability and the numerical solution of the Korteweg de Vries equation. Ph.D. Thesis, University of Dundee.

The Numerical Treatment of Partial Differential Equations by the Parallel Application of a Hybrid of the Ritz-, Galerkin-Product Integral Methods

N.L. Petrakopoulos

Mathematics Department, Concentration of Natural and Applied Sciences, The University of Wisconsin-Green Bay, 2420 Nicolet Drive, Green Bay, WI 54311-7001, USA

ABSTRACT

Recently the utility of the transcendental functions of matrices, and the multiplicative integral methods (Volterra-Petra) has been demonstrated for the simulation of systems modeled by ordinary differential equations[1], and by the semi-discrete Navier-Stokes equations[2], which are derived from the exact Navier-Stokes equations using Galerkin methods. It was demonstrated that the Volterra-Petra method is direct and simple to use from the point of view of the simulationist, notwithstanding the complexity of the key theoretical results which form the basis of the method. It is now proposed to demonstrate the utility of the Volterra-Petra method to the simulation of systems modelled by partial differential equations which are Stäckel-separable. Laplace's equation, Poisson's equation, the diffusion equation, the wave equation, the damped wave equation, and the transmission line equation are examples of Stäckel-separable partial differential equations [3, 4, 5].

INTRODUCTION

Finite element methods are formulated in several ways, some involving variational principles, others not. Ritz methods, based on the approximate solution of variational problems, appear most frequently in structural analysis and in the theoretical literature[6,7]. Galerkin methods form the basis of most nonstructural finite element calculations[8,9].

In either case, an approximate solution to the partial differential equation is required in the form $\sum_i U_i \rho_i(x)$ where the U_i's are the nodal parameters, and the ρ_i's are the basis functions constructed element-by-element from a variety of shape functions (triangular, rectangular, quadrilateral, tetrahedral, and hexahedral elements with straight or curved boundaries). The approximation in a finite-element method requires a certain amount of inter-element continuity. For an elliptic equation of order 2, or 4, this requirement is either C° or C^1 for Ritz and Galerkin, and C^1 or C^3 for least squres. Considering that polynomials of degree 9 are required to obtain C^1 continuity between tetrahedral elements, it is clear that the application of elements with the required amount of continuity for the fourth order elliptic equation is a difficult task. From the perspective of simulation it is necessary to use elements with less continuity than it is mandated by the theory.

Straight edge networks of elements are not adequate for curved boundaries and surfaces, usually inhabiting the real problems which need simulation. Basis functions for networks composed of elements with curved edges and surfaces must be used. The computations involved in the use of a finite element method are too numerous even when simulating small problems. Computers with large memories are needed to solve complicated problems. Even though large computers exist, and several commercial and governmental organizations have developed extensive computer programs, nevertheless, the large computer and corresponding programs are readily available to a very small fraction of potential users and this characteristic of the finite element method is still a problem, in an alleviated form.

Several properties of the finite element method have contributed to its extensive use for most systems modeled by partial differential equations:

1. System properties in adjacent elements could be different.
2. The size of the elements can be varied.
3. Boundary conditions such as discontinuous surface loadings present no difficulties to the method, and mixed boundary conditions can be handled.

THE HYBRID METHOD: AN ALTERNATIVE TECHNIQUE

Simulation through the use of the Hybrid Finite-Elements/-Volterra-Petra method is carried out as follows:

1. An orthogonal coordinate system (u^1, u^2, u^3) suggested by the system to be simulated is chosen with metric coefficients g_{11}, g_{22}, g_{33}.
2. A finite number of nodal points in the domain of the system is identified: (u^1_i, u^2_j, u^3_k) $(i,j,k=1,\ldots,N_H<<N_{FE})$. In the hybrid method the number of nodal points N_H is much smaller than the number of nodal points usually required in the finite-element method, N_{FE}.

3. The value of the continuous quantity to be simulated is denoted as a variable at each nodal point; this variable is to be determined.
4. The domain of the system is divided into a finite number of elements connected at common nodal points. These elements are orthogonal hexahedrons with curved bounding surfaces.
5. The nodal values of the continuous quantity are determined using either variational methods or Galerkin methods (whenever the equations of the system do not depend on a true variational principle). For real-time/on-line simulations the nodal values of the continuous quantity are determined by direct measurement/input.
6. The continuous quantity is determined over each element by the Volterra-Petra method through the parallel computation of the solution of at most three two-point boundary-value problems. The solution over each element is an approximation of the exact solution over each element, and the required continuity of the solution along the element boundaries is automatically maintained.

Example

Consider any system modeled by a Stäckel-separable second order PDE, reducible to a series of two-point boundary-value problems.

The second order linear boundary value problem of form:

$$L(y) = a_2(t)y'' + a_1(t)y' + a_0(t)y = f(t),\ y(a) = y_a,\quad y(b) = y_b,\ a_2(t) \neq 0 \tag{1}$$

is transformed to a first order linear system with boundary values as follows: We define

$$X = \begin{pmatrix} X_1 \\ X_2 \end{pmatrix} = \begin{pmatrix} y \\ y' \end{pmatrix};\ A(t) = \left(\begin{array}{c|c} 0 & 1 \\ \hline -\frac{a_0}{a_2} & -\frac{a_1}{a_2} \end{array}\right);\ f(t) = \begin{pmatrix} 0 \\ f(t)/a_2 \end{pmatrix};$$

$$X(a) = \begin{pmatrix} y_a \\ y'_a \end{pmatrix};\ X(b) = \begin{pmatrix} y_b \\ y'_b \end{pmatrix}$$ where y'_b and y'_a remain unspecified.

Equations (1) are then equivalent to the system:

$$X' = A(t)X + f(t);\ X(a) = X_a;\ X(b) = X_b \tag{2}$$

To solve (2) we find the solutions X_1, X_2 of two initial value problems:

$$X_1' = A(t)X_1 + f(x);\ X_1(a) = \begin{pmatrix} y_a \\ 0 \end{pmatrix}$$

$$X_2' = A(t)X_2 \qquad ;\ X_2(a) = \begin{pmatrix} 0 \\ 1 \end{pmatrix} \tag{3}$$

This yields two values $X_1(b)$ and $X_2(b)$ at $t = b$.

Because of the linearity of the problem, the solution of the system with initial condition $X(a) = (y_a, y'_a)^T$ is given by $X(t) = X_1(t) + y'_a X_2(t)$. We calculate y'_a by insisting that $X(b) = (y_b, y'_b)^T = X_1(b) + y'_a X_2(b)$. Since y'_b need not be specified, we find that $y'_a = [y_b - y_1(b)]/y_2(b)$. Thus the boundary condition at $x = b$ is also satisfied, and we have the required solution to the boundary value problem (1) by means of the application of the product integral algorithm to initial value problems (3).

THE PRODUCT INTEGRAL ALGORITHM

The system is $\dot{X} = A(t)X$, $X(t_0) = X_0$, where $A(t)$ is an nxn matrix with continuous elements $a_{ik}(t)$ in some interval (a,b) of the argument t. The solution to the system is of the form:

$$X(t) = \Omega_{t_0}^{t} X_0; \quad \Omega_{t_0}^{t} = E + \int_{t_0}^{t} A(\tau)d\tau + \int_{t_0}^{t} A(\tau) \int_{t_0}^{\tau} a(\sigma)d\sigma d\tau + .. \quad (4)$$

The series for the matricant $\Omega_{t_0}^{t}$ converges absolutely and uniformly in every closed interval in which $A(t)$ is continuous. For the inhomogeneous system

$$\dot{X} = A(t)X + f(t) \quad (5)$$

$$X(t) = \Omega_{t_0}^{t} X(t_0) + \int_{t_0}^{t} \Omega_{t_0}^{t} [\Omega_{t_0}^{\tau}]^{-1} f(\tau)d\tau \quad (6)$$

Let us now divide the basic interval (t_0, t) into n parts by introducing intermediate points $t_1, t_2, \ldots, t_{n-1}$ and set $\Delta t_k = t_k - t_{k-1}$ $(k = 1,2,\ldots,n;\ t_n = t)$. Then by the properties of the matricant

$$\Omega_{t_0}^{t} = \Omega_{t_{n-1}}^{t} \cdots \Omega_{t_1}^{t_2} \Omega_{t_0}^{t_1} \quad (7)$$

In the interval (t_{k-1}, t_k) let $\tau_k (k=1,2,\ldots,n)$ be an intermediate point.

By regarding the Δt_k as small quantities of the first order we can take for the computation of $\Omega_{t_{k-1}}^{t_k}$ to within small quantities of the second order, $A(t) \approx \text{const.} = A(\tau_k)$. Then

$$\Omega_{t_{k-1}}^{t_k} = e^{A(\tau_k)\Delta t_k} + (**) = E + A(\tau_k)\Delta t_k + (**) \quad (8)$$

The symbol (**) denotes the sum of terms beginning with terms of the second order. From the transitional property of the matricant (7) and (8) we find:

$$\Omega_{t_0}^{t} = \exp(A(\tau_n)\Delta t_n)\ldots\exp(A(\tau_2)\Delta t_2)\exp(A(\tau_1)\Delta t_1) + (*) \quad \text{and}$$

$$\Omega_{t_0}^{t} = [E + A(\tau_n)\Delta t_n]\ldots[E + A(\tau_2)\Delta t_2][E + A(\tau_1)\Delta t_1] + (*) \qquad (9)$$

When we pass to the limit by increasing the number of intervals indefinitely and letting the length of these intervals tend to zero, the exact limit formulae are obtained:

$$\Omega_{t_0}^{t} = \lim_{\Delta t_k \to 0} [\exp(A)(\tau_n)\Delta t_n)\ldots\exp(A(\tau_2)\Delta t_2)\cdot\exp(A(\tau_1)\Delta t_1)] \qquad (10)$$

$$\Omega_{t_0}^{t} = \lim_{\Delta t_k \to 0} [E + A(\tau_n)\Delta t_n]\ldots[E + A(\tau_2)\Delta t_2][E + A(\tau_1)\Delta t_1] \qquad (11)$$

The expression under the limit sign of (11) is the product integral--an analogue to the sum integral for the ordinary integral. The limit of the product integral is the multiplicative integral of Volterra. The multiplicative integral was first introduced by Volterra in 1887. On the basis of this concept, Volterra developed an original infinitesimal calculus for matrix functions.

Formulae (8) and (9) are crucial to the numerical treatment of a linear system with variable coefficients. Their practical computational significance is strengthened by the exact theoretical statements reflected by formulae (10) and (11). The multiplicative integral algorithm has been tested on specific linear systems whose coefficients exhibited considerable variation. The numerical results obtained exhibited relative errors of the order of 10^{-6} percent.

REFERENCES

1. Petrakopoulos, N.L. (1986). The Numerical Treatment of Initial- and Boundary-value Problems in Differential Equations by the Volterra-Petra methods in The Proceedings of the 1986 Summer Computer Simulation Conference, Reno, Nevada.

2. Petrakopoulos, N.L. (1986). The Numerical Treatment of the Vertically-Average PDE's for a Well-Mixed Estuary by the Multiplicative Integral and Transcendental Functions of Matrices Method in the Proceedings of the 2nd European Simulation Congress, Antwerp, Belgium.

3. Stäckel, P., Uber die Integration der Hamilton-Jacobischen Differentialgleichung mittels Separation der Variabelen, Habil-Schr. Halle, 1891. Sur une classe de problemes de dynamique, C. R. Acad. Sci., Paris 116, 485 (1893). Uber die integration der Hamiltonschen Differentialgleichungen mittels Separation der Variabelen, Math. Ann. 49, 145, (1897).

4. Moon, P. and D. E. Spencer, Separability Conditions for the Laplace and Helmholtz Equations, J. Franklin Inst., 253, 585 (1952). Theorems on Separability in Riemmannian n-space, Amer. Math. Soc. Proc. 3, 635 (1952). Recent Investigations of the Separation of Laplace's Equation, Amer. Math. Soc. Proc. 4, 302 (1953). Separability in a Class of Coordinate Systems, J. Franklin Inst., 254, 227 (1952).

5. Moon, P. and D. C. Spencer (1974), Field Theory Handbook, Springer-Verlag, Berlin.

6. Courant, R. and D. Hilbert (1966). Methods of Mathematical Physics, Interscience Publishers, New York.

7. Necas, J (1967). Les Methodes Directes en Theorie des Equations Elliptiques, Academia, Prague.

8. Wait, R. and A. R. Mitchell (1985). Finite Element Analysis and Applications, John Wiley and Sons, New York.

9. Segerlind, L. J. (1976). Applied Finite Element Analysis, John Wiley and Sons, New York.

Fractional Steps and Process Splitting Methods for Industrial Codes

J.M. Usseglio-Polatera
CEFRHYG, BP172X, 38042 Grenoble, France
M.I. Chenin-Mordojovich
IMAG/TIM3, Université de Grenoble, 38031 Grenoble, France

ABSTRACT

The method of fractional steps is a remarkable and cost-efficient technique introduced by N.N. Yanenko for computational mechanics. Among the various fractional step methods, process-splitting is particularly attractive for 2-D and 3-D simulations in Water Resources. The authors applied these methods in various 2-D and 3-D industrial codes in the field of Computational Hydraulics; they try to define a concise state-of-the art and to establish the formal limitation of the method for 2-D shallow water applications.

INTRODUCTION TO SPLITTING

Splitting reduces a complex and multidimensional problem to a sequence of simpler elementary ones. Since it tends to minimise the magnitude of the computational task, this apparently simple technique looks attractive and is widespread in the field of Computational Hydraulics. However, "splitting" is used loosely and this term covers various numerical procedures. There is quite a mix-up and neither the theoretical background nor advantages and limitations of the relevant splitting technique are clearly established.

Operator splitting is the basis of fractional step methods introduced by N.N. Yanenko and G. Marchuk (Yanenko[17]). The governing equations are split up into elementary parts subsequently solved in succession within each time step. The combination produces a "weak" approximation to the original operator. Various splitting criteria are used: space-splitting, time-splitting and process-splitting must be distinguished.

Space-splitting is customary when 2-D or 3-D modelling based on finite differences is involved (Benqué et al.[2]). Space-splitting consists in separating partial space derivatives according to the relevant space directions. This reduces the multidimensional problem to a sequence of one-dimensional equations and leads to tri-diagonal matrices, even with implicit schemes. This directional separation may cause spurious polarisations along the axes of discretization when the time step is too large, leading to stringent limitations making iterative procedures necessary (Benqué et al.[3]).

On the other hand, a series of numerical techniques uses fractional steps based not on physical reasons (space directions, physical processes), but on numerical statements (stabilisation of the scheme, alleviation of sparse matrix inversion). Within each fractional step, the complete equations are solved using different discretization schemes. Well-known methods like Predictor-Corrector schemes or Alternating Direction Implicit schemes (Leendertse[11]) fall in this category which may be termed "time-splitting" methods.

Process-splitting is based on a recognition of the physical processes described by the governing equations. Following Yanenko (Yanenko[17]), the governing equations are split up into fractional steps corresponding to the basic physical processes involved. Thus numerical methods best fitted to their specific physical and numerical behaviour can be used within each fractional step. Combination of process-splitting and space-splitting is customary.

PROCESS-SPLITTING IN COMPUTATIONAL HYDRAULICS

Process-splitting is particularly adapted to equations involving several physical processes of different numerical nature. A typical example is hyperbolic advection and parabolic diffusion. Hence the most common application of process-splitting is to <u>scalar transport</u> (contaminant, temperature, kinetic energy...), leading to methods often referred to as "Eulerian-Lagrangian" (Baptista et al.[1], Glass et al.[7], Holly et al.[9], Neuman et al.[13], Sobey[15]).

In the particular case where water depth and diffusion coefficient do not vary in space, splitting introduces no errors provided that at each time step advection is solved first by an explicit technique (characteristics) and diffusion next by an implicit technique. In the general case, hyperbolic components will remain in the diffusion equation leading to numerical diffusion when advection is dominant (Baptista et al.[1]). However the use of characteristics-based methods for advection coupled with high order interpolation schemes leads to almost "wiggle-free" and remarkably accurate and cost-efficient solutions even for high Peclet numbers.

Process-splitting methods have been widely implemented in France in the field of Computational Fluid Dynamics by

Electricité de France, the French Electricity Board (Benqué et al.[2], Hauguel[8]) and SOGREAH Consulting Engineers (Holly et al.[9]). In the field of Computational Hydraulics, the major benefit of process-splitting is for non-linear shallow water wave equations. Applications have been extensively developed for: 2-D depth-averaged equations (Benqué et al.[3]), 2-D width-averaged equations (Sauvaget et al.[14]), 3-D equations coupled with scalar transport (Donnars et al.[6]), coupled 2-D/3-D equations (Usseglio-Polatera et al.[16]).

In addition to the two basic fractional steps involved in scalar transport (horizontal advection and diffusion), shallow water equations involve a wave propagation step (including non-linear bed friction) and vertical advection and diffusion steps whenever non depth-averaged equations are solved. Due to non-linearities (advection, bed friction) and the number of fractional steps, it is difficult to establish formally the accuracy of the fully discrete fractional step algorithm. The discretization errors associated with approximations for the elementary operators interact with the intrinsic error involved with operator splitting in a complex form. Nevertheless provided that implicit schemes are used for every fractional step, the overall scheme is unconditionally stable. Furthermore the order of accuracy is at least the lowest order of the elementary solvers (generally first order in time and second order in space). Finally the method can be justified even for non-commuting operators (Yanenko[17]).

ADVANTAGES OF PROCESS-SPLITTING FOR INDUSTRIAL CODES

Process-splitting is not a numerical method in itself but a reliable framework for more physics-based schemes. Within each elementary fractional step, the cost of using stable implicit procedures or specifically adapted schemes is small especially when space-splitting is combined with process-splitting. This leads to accurate, powerful and cost-effective schemes with no formal limitations. Furthermore, this basic framework makes possible a combination of competing numerical techniques (characteristics, finite differences, finite elements) when they are really complementary (Baptista et al.[1], Neuman et al.[13]).

A further advantage concerns coupling. Take the example of coupling 2-D depth-averaged to fully 3-D shallow water equations (Usseglio-Polatera et al.[16]). Coupling each of the common physical processes involved within each time step (horizontal advection, horizontal diffusion, wave propagation) has been made possible. It has been combined with a dependable solution to the coupling of bed friction formulations in spite of their inconsistent expressions in 2-D and 3-D. This physics-based coupling technique allows definition of a 3-D block inside any 2-D computational domain and accounts for the feedback hindered by conventional nested models.

Finally, when real world applications do not clearly fulfil the basic assumptions and when variant equations are locally necessary (tidal flat procedure described by Benqué et al.[3]) or when a simplification of the governing equations is adequate (suppression of inertia terms in the initial computation towards the steady state for instance), process-splitting is obviously suitable.

LIMITATION OF PROCESS-SPLITTING TECHNIQUES

The differences in the numerical nature of the basic physical processes involved in the governing equations also influence boundary conditions requirements. On the one hand, upstream conditions are predominant for advection leading to a polarisation in the flow directions. On the other hand, whatever the wave celerity, wave propagation or momentum diffusion do not create such a space polarisation. Since the necessary boundary conditions are not the same, the process-splitting technique, which requires specific boundary conditions for each time step, should be advantageous.

Conversely, since only the conditions corresponding to the complete differential equations are known, the treatment of boundary conditions is a major drawback to process-splitting as it is to every fractional step technique. In spite of unconditional stability, the maximum allowable time step is constrained by the treatment of intermediate level boundary conditions. Due to the lack of information, trade-off is necessary and evaluation of the induced error is difficult.

FORMAL LIMITATION OF THE TIME STEP IN THE CASE OF SHALLOW WATER EQUATIONS

Although the formal limitation looks uncertain, the maximum possible time step can be clearly estimated. This is the case for shallow water wave equations when momentum diffusion is neglected. The following statements proceed from theoretical and numerical investigations recently carried out by M.I. Chenin-Mordojovich. They are developed in more detail in J.A. Cunge's paper (Cunge[5]).

Consider the elementary procedure described by Benqué et al.[3]. For the sake of clarity, consider a 1-D equation and a process-splitting based solution with two steps: horizontal advection and wave propagation. Consider the application to a canal reach AB with subcritical flow. Discharge is imposed at upstream boundary A and water elevation at downstream B. Assume the solution known at every grid point at time t_n and consider the calculation of solution at $t_n + \Delta t$. Solving the propagation step discretized implicitly requires knowledge of the advection discharge increment at every point except at boundary points. Thus the intermediate level (after advection) upstream boundary value is necessary only when the Advection Courant number is greater than one within the first upstream grid cell because it then requires a time interpolation to estimate the foot value of the backward characteristics line. Actually, due to the conventional space-staggered grid, the Courant number limitation for a dependable solution is 0.5. The generalisation to 2-D applications is obvious. This limitation is allowable for most coastal engineering applications but it may be stringent for some river applications of 2-D codes.

Furthermore, the solution of split-up advection and propagation steps does not allow an estimate of water elevation at the upstream boundary! The solution proposed is to use the characteristic line coming from inside the domain and related to the complete equation. In 2-D the solution is similar although characteristic surfaces are not so easy to handle. Nevertheless, space-splitting is possible.

REFERENCES

1. Baptista A.M., Adams E.E and Stolzenbach K.D. (1984). Eulerian-Langrangian analysis of pollutant transport in shallow water. Ralph M. Parsons Laboratory, MIT, Report No 296, June 1984.

2. Benqué J.P., Hauguel A. and Viollet P.L. (1982), Engineering Applications of Computational Hydraulics, Vol. II, (Ed. Abbott M.B. and Cunge J.A.), Part 1, Tidal Flows, Pitman Advanced Publishing Program, Boston-London-Melbourne.

3. Benqué J.P., Cunge J.A., Feuillet J., Hauguel A. and Holly F.M. Jr (1982), New method for tidal current computation, Journal of the Waterway, Port, Coastal and Ocean Division, ASCE, Vol. 108, No WW3, pp. 396-417.

4. Chenin-Mordojovich M.I. (1980), Numerical solutions for wave equations in a characteristics network, PhD Thesis (in French), University of Grenoble, France.

5. Cunge J.A. (1988), Interaction of numerical and physical aspects in algorithms of the free surface mathematical modeling, Proceedings of the VII International Conference on Computational Methods in Water Resources, Cambridge, USA.

6. Donnars Ph. and Péchon Ph. (1987), 3-D numerical model, for thermal impact studies, Proceedings of the 18th International Liège Colloquium on Ocean Hydrodynamics, Liege, Belgium, to be published by Elsevier, Amsterdam.

7. Glass J. and Rodi W. (1982), A higher numerical scheme for scalar transport, Computer Methods in Applied Mechanics and Engineering, Vol. 31, pp. 337-358.

8. Hauguel A. (1985), Numerical Modeling of Complex Industrial and Environmental Flows, Proceedings of the 3rd Symposium 'Turbulence Measurements and Flow Modeling', (Ed. Chen C.J., Chen L.D. and Holly F.M. Jr), Springer-Verlag, Berlin.

9. Holly F.M. Jr and Preissmann A. (1977), Accurate calculation of transport in two dimensions, Journal of the Hydraulics Division, Vol. 103, No HY11, pp. 1259-1277.

10. Holly F.M. Jr and Usseglio-Polatera J.M. (1984), Dispersion in two dimensional tidal flow, Journal of the Hydraulics Division, Vol. 110, No 7, pp. 905-926.

11. Leendertse J.J. (1970), A Water Quality Simulation Model for Well-Mixed Estuaries and Coastal Seas, Principles of Computation, Rand Corporation Memorandum, Vol. 1, RM-6230-RC.

12. Mc Rae G.J., Goodin W.R. and Seinfeld J.H. (1982), Numerical Solution of the Atmospheric Diffusion Equation for Chemically Reacting Flows, Journal of Computational Physics, Vol. 45, pp. 1-42.

13. Neuman S.P. and Sorek S. (1982), Eulerian-Lagrangian methods for Advection-Dispersion, Proceedings of the 4th International Conference "Finite Elements in Water Resources", Hannover, Germany, pp. 14-41, 14-68.

14. Sauvaget P. and Usseglio-Polatera J.M. (1987), Numerical simulation of stratified flows in estuaries and reservoirs, Proceedings of the 3rd International Symposium on Density Stratified FLows, Pasadena, USA, February 1987.

15. Sobey R.J. (1983), Fractional step algorithms for estuarine mass-transport, International Journal for Numerical methods in fluids, Vol. 3, pp. 567-581.

16. Usseglio-Polatera J.M. and Sauvaget P. (1987), A coupled 2-D/3-D modeling system for tidal and wind-induced currents computation, Proceedings of the 18th International Liège Colloquium on Ocean Hydrodynamics, Liège, Belgium, to be published by Elsevier, Amsterdam.

17. Yanenko N.N. (1971), The method of fractional steps, English translation edited by M. Holt, Springer-Verlag, Berlin.

On the Construction of N-th Order Functions for Complete Interpolation

S.Y. Wang, K.K. Hu, P.G. Kramer and S.E. Swartz

School of Engineering, The University of Mississippi, University of MS 38677 and The Kansas State University, Manhattan, KS 66502, USA

ABSTRACT

A general methodology for constructing n-th order shape functions over 2-d and 3-d elements is presented. These shape functions are capable of accurately representing the physical phenomena modeled by the finite element analysis. The development of this methodology and the demonstration of its application are given.

INTRODUCTION

In the application of Finite Element Modeling to scientific research, there is a need of interpolation functions which are capable of accurately representing physical phenomena including rigid body displacement, uniform strain and stress distributions, and/or uniform values of high order partial derivatives. Interpolating functions having these capabilities are called a complete set and the element adopting these interpolating functions is sometimes called the complete element.[1,2]

A general methodology for constructing n-th order shape functions over 2- and 3-d elements bounded by polygonal areas which leads to complete elements is presented. It is an extension of the early works of generalized Lagrangian interpolation[3] and the n-th order Hermitian shape function[4] developed by the authors.

GENERALIZED LAGRANGIAN INTERPOLATING FUNCTION

Generalized shape functions of the Lagrangian family over 2- or 3-d elements bounded by quadrilateral boundaries with pre-selected nodal points can be constructed by the use of products of the ratios of properly selected areas or volumes. If all nodal points are on the surfaces or on the edges of an element, the interpolation formulae obtained through the generalized

Lagrangian interpolation technique will be the same as those of the Serendipity family[5]. A typical case of 2-d quadrilateral elements, having both interior and boundary nodes, shown in Fig. 1 is used to illustrate the method. Interpolation functions of the generalized Lagrangian family for side node Q_2 and interior node I_1 are given as follows:

$$L\{P,Q_2\} = \left[\frac{\Delta PQ_3Q_5}{\Delta Q_2Q_3Q_5} \frac{\Delta PQ_4Q_7}{\Delta Q_2Q_4Q_7} \frac{\Delta PQ_7Q_{10}}{\Delta Q_2Q_7Q_{10}} \frac{\Delta PQ_{10}Q_1}{\Delta Q_2Q_{10}Q_1} \right] \left[\frac{\Delta PI_1I_2}{\Delta Q_2I_1I_2} \frac{\Delta PI_1I_5}{\Delta Q_2I_1I_5} \frac{\Delta PI_3I_4}{\Delta Q_2I_3I_4} \right] \quad (1)$$

$$L\{P,I_1\} = \left[\frac{\Delta PQ_1Q_4}{\Delta I_1Q_1Q_4} \frac{\Delta PQ_4Q_7}{\Delta I_1Q_4Q_7} \frac{\Delta PQ_7Q_{10}}{\Delta I_1Q_7Q_{10}} \frac{\Delta PQ_{10}Q_1}{\Delta I_1Q_{10}Q_1} \right] \left[\frac{\Delta PI_2I_5}{\Delta I_1I_2I_5} \frac{\Delta PI_3I_4}{\Delta I_1I_3I_4} \right] \quad (2)$$

With the shape function for every node of an element constructed, one can find the interpolation of the function of interest:

$$f(X) = \sum_{i=1}^{N} f(X_i)\, L(X,R_i) \quad (3)$$

where R_i is the set of all nodal points of an element, and X is any point of the element.

THE N-TH ORDER HERMITIAN INTERPOLATION

The n-th order Hermitian interpolation formulae[4] depends on the construction of a Hermitian distribution function $H(L(P,R_i),n)$ where n is the highest order of the partial derivatives used in the interpolation and the argument $L(P,R_i)$ is the Lagrangian interpolation function described previously. This function has the following properties.

a) $H(L(R_j,R_i),n) = 0$, if $R_j \neq R_i$,

b) $H(L(R_i,R_i),n) = 1$,

c) all partial derivatives of H at every node in all directions are zero. (4)

The product of H_i and the terms of the Taylor series expansion of order n will form a set of n-th order Hermitian interpolating functions at the i-th node of the element. Some examples are listed below.

$$T_i^{o} = H_i, \quad T_i^{,x} = (x-x_i)\, H_i, \quad T_i^{,y} = (y-y_i)\, H_i,$$

$$T_i^{,xy} = (x-x_i)\,(y-y_i)\, H_i, \quad T_i^{,xx} = (x-x_i)^2\, H_i/2$$

$$T_i^{,yy} = (y-y_i)^2\, H_i/2, \ldots, \quad T_i^{,(x\ldots x)} = (x-x_i)^n\, H_i/n! . \tag{5}$$

Superscripts used in each shape function represent partial derivatives of the function at the node identified. The interpolated result in a 2-d domain using second order Hermitian function is given below.

$$f(x,y) = \sum_{i=1}^{N} [f_i T_i^{o} + f_{,xi}\, T_j^{,x} + f_{,yi}\, T_i^{,y} + f_{,xxi}\, T_i^{,xx}$$

$$+ f_{,xyi}\, T_i^{,xy} + f_{,yyi}\, T_i^{,yy}\,] . \tag{6}$$

It is interesting to note that in the case of a rectangular element the above interpolating functions form a complete set of the second order as defined by Felippa/Clough[1]. For the general case, one can construct a complete set of interpolating functions of order n by evaluating the coefficients, C_i, of the following polynomial,

$$p_n(x) = 1 - x^{n+1}\,[c_o + c_1\, x + c_2\, x^2 + \ldots + c_n\, x^n] \tag{7}$$

such that $p_n(1)$ and all its derivatives are equal to zero, at $x = 1$. Substituting x by $[1 - L(P,R_i)]$ into $p_n(x)$, one obtains the desired distribution function, H_i.

$$H(L(P,R_i),n) = 1 - \sum_{r=0}^{n} c_r [1 - L(P,R_i)]^{(n+1+r)} \tag{8}$$

MODIFIED HERMITIAN INTERPOLATING FUNCTION

A general formula for the construction of a complete set of modified Hermitian interpolation functions for finite element analysis is given below. The modifications should be carried out systematically, beginning with zero-th order in an ascending order, using the modified lower ordered interpolating functions to obtain the modified higher order interpolating functions.

Let the function to be completed by the Hermitian interpolation be a general term in the Taylor expansion,

$$g(x) = \frac{x^p\, y^q\, z^r}{p!\; q!\; r!} \tag{9}$$

The deviation between the given $g(x)$ and interpolated result obtained from some imcomplete set of interpolating functions may be expressed as:

$$E'[(x)^p(y)^q(x)^r] = x^p y^q z^r - \sum_{j=1}^{N} T_j{}'[(x)^p(y)^q(z)^r]$$

$$- \sum_{m=1}^{N} \sum_{i=0}^{p-1} \sum_{j=0}^{q-1} \sum_{k=0}^{r-1} \frac{x^{p-1} y^{q-j} z^{r-k}}{(p-i)!(q-j)!(r-k)!} S_m{}'[(x)^i(y)^j(z)^k] \qquad (10)$$

where the superscripts after a common (,) designate partial derivatives. After distributing this error term to each of the N interpolating functions of the set $\{T_j{}'[(x)^p(y)^q(z)^r]\}$ one has the modified or corrected shape function:

$$S_j{}'[(x)^p(y)^q(z)^r] = T_j{}'[(x)^p(y)^q(z)^r] + E'[(x)^p(y)^q(z)^r]/N. \qquad (11)$$

Now, one can use these formulae, starting from zero-th order, to modify each of the interpolating functions until a complete set of functions from 0 to n-th order polynomials of spatial variables can be reconstructed.

AN EXAMPLE: THE USE OF TRIGONOMETRIC FUNCTIONS

Trigonometric functions are harmonic functions and are very useful in engineering computations. However, one could not construct a complete set of interpolation functions for an element using only a few terms of these functions. The following simple example demonstrates that the construction of a complete set of interpolation functions for a line element of three nodal points can be achieved by using the methodology presented.

Let x_1, x_2 and x_3, be the three nodes at (-1,0,1) and L_i, H_i, T_i and S_n be their corresponding Lagrangian interpolation, Hermitian distribution, Taylor-Hermitian interpolation, and shape functions for complete element at the i-th node, respectively. The Lagrangian interpolation functions are constructed as:

$$L_1 = \cos\left[\frac{(1+x)\pi}{2}\right] \cos\left[\frac{(1+x)\pi}{4}\right]$$

$$L_2 = \sin\left[\frac{(1+x)\pi}{2}\right]$$

$$L_3 = \cos\left[\frac{(x-1)\pi}{2}\right] \cos\left[\frac{(x-1)\pi}{4}\right] \qquad (12)$$

They satisfy the following conditions

$$L_i(x_j) = \begin{cases} 1, & \text{if } i = j \\ 0, & \text{if } j \neq i \end{cases} .$$

Using the conditions: $p_1(1) = p_1'(1) = 0$, one obtains: $C_o = 3$ and $C_1 = -2$. This implies that

$$H(L_j,1) = 1 - 3[1 - L_j]^2 + 2[1 - L_j]^3 \ . \quad (13)$$

Substituting (12) into (13) yields the first order hermitian distribution function in terms of trigonometric functions.

The Taylor-Hermitian interpolating function can be obtained by substituting (13) into (6):

$$T_j^o = H_j, \quad T_j^{,x} = (x - x_j) H_j \quad (14)$$

Finally, the shape functions and derivatives, S_j^0 and $S_j^{,x}$, can be obtained by substituting (14) into (10) and (11). These results are shown in Fig. 2 and Fig. 3. The deviations E^o and $E^{,x}$ are shown in Fig. 4. To verify the completeness of the element, the function $f(x) = (2x - 1) / 4$ is reconstructed by

$$f(x) = -0.75S_1^o - 0.25S_2^o + 0.25S_3^o + 0.5(S_1^{,x}+S_s^{,x}+S_3^{,x}). \quad (15)$$

Each of the shape functions used in (15) and the reconstructed function are plotted in Fig. 5 to illustrate the completeness of the interpolating functions obtained by the newly developed method. Two-d cases are to be presented at the Conference.

CONCLUDING REMARKS

A methodology for constructing complete elements of n-th order by modifying the interpolation functions of Taylor-Hermitian family in 1-, 2- or 3-d space is presented. It can be applied to elements enclosed by polygonal areas. The interpolated results from complete elements should be smoother than those of incomplete Hermitian interpolation. The shape functions derived are in terms of the spatial coordinates without having to transform the element into some normalized spaces. This allows direct calculations of partial derivatives without the complication of the Jacobian matrices. It allows the numerical integration over the domain to be computed more efficiently. If collocation methods are used in the finite element analysis, the use of this direct interpolation will be one of the most convenient ways to establish the algebraic equations.

ACKNOWLEDGEMENT

The research reported here was sponsored in part by the US Army Engineer Waterways Experiment Station, Contract No. DACW 39-87-K-0088, The Mississippi Water Resources Research Institute, Grant No. G1234-06, and by the National Science Foundation under Grant No. MSM-8513982.

REFERENCES

1. Felippa and Clough (1970). The Finite Element Method in Solid Mechanics, pp. 210-252, SIAM-AMS Proc., Vol. 2, American Mathematical Society, Providence, RI.

2. Oliveira (1968). Theoretical Foundations of the Finite Element Method, pp. 929-952, Int. J. Solid & Structures, V. 4.

3. Hu, Swartz, Kirmser and Wang (1984) On a Generalized Shape-Function for Two and Three Dimensional Elements Bounded by Quadrilateral Boundaries, pp. 561-564, Proc. of the 4th Int. Conf. on Applied Numerical Modeling, National Cheng-Kung University, Tainan, Taiwan, R.D.C., Dec.

4. Hu, Kirmser, Swartz and Wang (1985) On the Construction of N-th Order Hermitian Shape Functions for 2- and 3-dimensional Elements, pp. 561-564, Collected Technical Papers, Part I, AIAA CP851, 26th Structures, Structural Dynamics and Materials Conf., Orlando, FL, April 15-17.

5. Zienkiewicz, Irons, Ergatoudis, Ahmad and Scott (1963) Iso-Parametric and Associated Element Families for Two and Three Dimensional Analysis, pp. 383-342, Finite Element Methods in Stress Analysis, I. Holand & K. Bell (eds), Tapir Press, Trendheim, Norway.

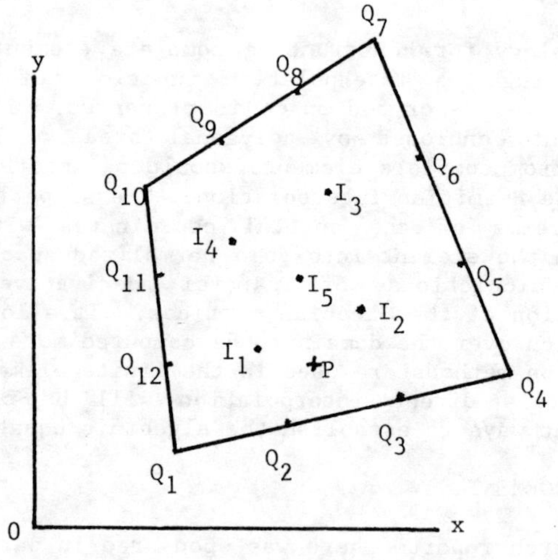

Figure 1. A typical quadrilateral element

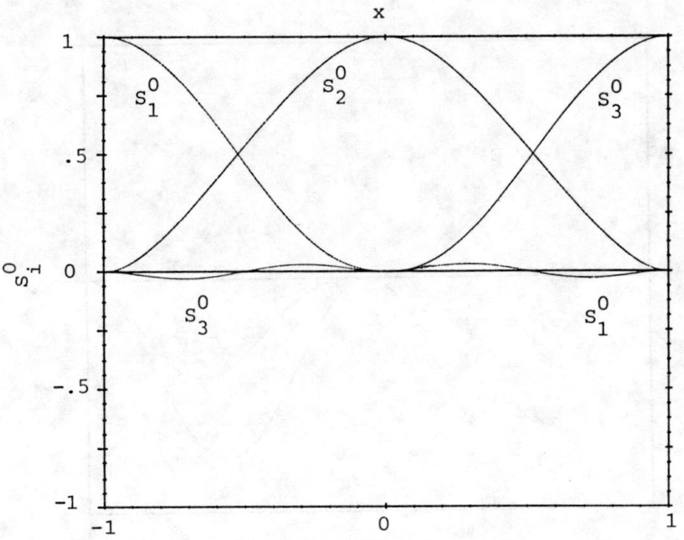

Figure 2. Modified (Complete) Hermite Interpolation of Order 1 for F(x)

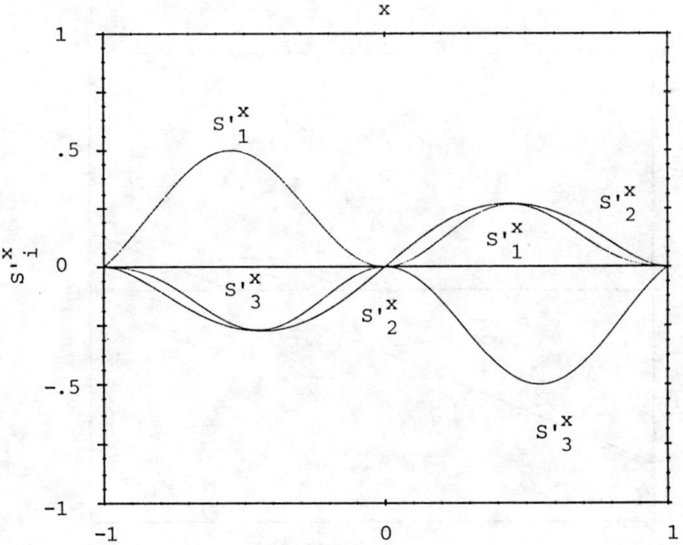

Figure 3. Modified (Complete) Hermite Interpolation of Order 1 for F'(x)

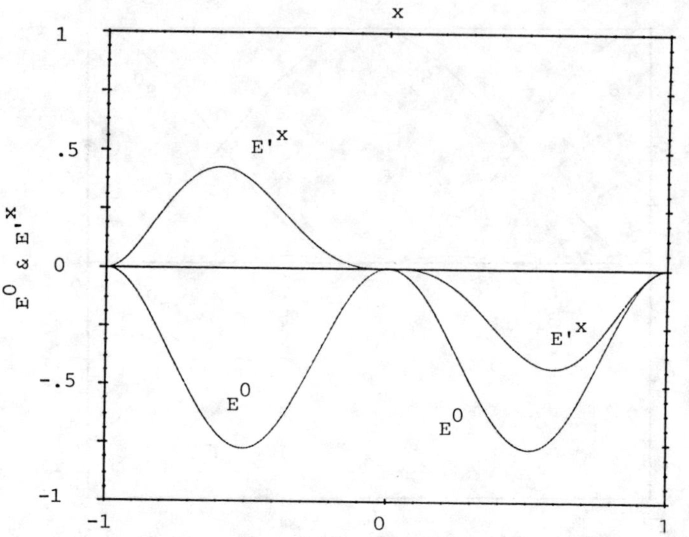

Figure 4. Errors due to First Order Taylor-Hermite Interpolation

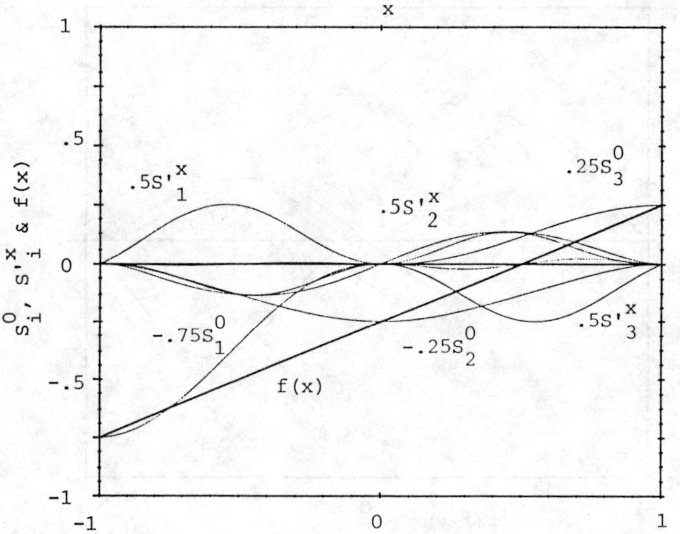

Figure 5. Evaluation of Completeness

SECTION 2 - TRANSPORT

SECTION 2A - SOLUTE TRANSPORT IN SATURATED POROUS MEDIA

INVITED PAPER

Three-Dimensional Adaptive Eulerian-Lagrangian Finite Element Method for Advection-Dispersion

R. Cady

Conservation and Survey Division, University of Nebraska, Lincoln, Nebraska 68588-0517, USA

S.P. Neuman

Department of Hydrology and Water Resources, University of Arizona, Tucson, Arizona 85715, USA

ABSTRACT

The adaptive Eulerian-Lagrangian finite element method of Neuman[1] is improved and extended to three-dimensional domains. Accurate results are obtained for the entire range of Peclet numbers from very small to infinity and Courant numbers at least up to 15. Difficulties resolved include oscillations, numerical smearing, the clipping of peaks, and grid orientation effects.

INTRODUCTION

Numerical methods for the advection-dispersion equation may be classified according to the emphasis they place on the parabolic and/or hyperbolic nature of the problem. In Eulerian methods, the equation is usually solved on a fixed grid in space. Lagrangian methods utilize either a deforming grid or a fixed grid in deforming coordinates. Eulerian-Lagrangian methods combine aspects of both approaches so as to merge the simplicity of a fixed Eulerian grid with the computational power of the Lagrangian method. Reviews and comparisons can be found, among other sources, in the papers of Neuman[2] and Smith and Hutton[3].

Neuman[2] proposed a formal decomposition of the concentration field into two parts, the first best suited for Lagrangian treatment and the other amenable to an Eulerian solution. The Eulerian problem is parabolic and can be formulated in terms of time-implicit symmetric finite element matrices. For the Lagrangian problem Neuman[1] proposed an adaptive approach that relies on the backward or modified method of characteristics away from steep fronts but switches automatically to forward particle tracking in the vicinity of such fronts. Preliminary tests against analytical solutions of one- and two-dimensional dispersion in a uniform steady-state velocity field suggested that the adaptive method can generate accurate solutions over the entire range of Peclet numbers from 0 to ∞ with a high efficiency characterized by Courant numbers well in excess of 1.

In this paper we modify and improve some aspects of Neuman's[1] adaptive Eulerian-Lagrangian method and illustrate its extension to three spatial dimensions.

THEORY

Consider the advection-dispersion equation

$$R\frac{\partial c}{\partial t} = \nabla\cdot(\mathbf{D}\nabla c - \mathbf{v}c) - \lambda Rc + q \tag{1}$$

where c is concentration, R is retardation coefficient, t is time, ∇ is gradient operator, $\mathbf{D}$ is dispersion tensor, $\mathbf{v}$ is velocity vector, λ is radioactive decay constant, and q is source term. The equation is to be solved for c, subject to initial and boundary conditions

$$c(\mathbf{x},0) = C_0(\mathbf{x}) \tag{2}$$

$$(-\mathbf{D}\nabla c + \mathbf{v}c)\cdot\mathbf{n} + \alpha(c-C) = Q \qquad \text{on } \Gamma_i \tag{3}$$

where $\mathbf{x}$ is position vector; Γ_i is inflow or noflow boundary; $\mathbf{n}$ is unit vector normal to Γ and pointing outward; C_0, C and Q are prescribed functions; and α controls the type of boundary condition: Dirichlet if $\alpha \rightarrow \infty$, Neumann if $\alpha = 0$, and Cauchy otherwise. Along outflow boundaries, Γ_0, it is common to assume (without obvious physical justification) that

$$\mathbf{D}\nabla c\cdot\mathbf{n} = 0 \qquad \text{on } \Gamma_0. \tag{4}$$

We will discuss another way to handle outflow boundaries later in the text.

Defining a hydrodynamic derivative in terms of retarded velocity,

$$\frac{D}{Dt} = \frac{\partial}{\partial t} + \frac{\mathbf{v}\cdot\nabla}{R}, \tag{5}$$

we can rewrite (1) in Lagrangian form as

$$R\frac{Dc}{Dt} = \nabla\cdot(\mathbf{D}\nabla c) - c\nabla\cdot\mathbf{v} - R\lambda c + q. \tag{6}$$

Next, we decompose c into two parts,

$$c(\mathbf{x},t) = \tilde{c}(\mathbf{x},t) + \hat{c}(\mathbf{x},t) \tag{7}$$

and require that $\tilde{c}$ satisfy the nonhomogeneous differential equation

$$R\frac{D\tilde{c}}{Dt} = -\,\tilde{c}\nabla\cdot\mathbf{v} - R\lambda\tilde{c} \tag{8}$$

subject to the initial condition

$$\tilde{c}(\mathbf{x},0) = C_0(\mathbf{x}), \tag{9}$$

and the Cauchy condition

$$\mathbf{v}\tilde{c}\cdot\mathbf{n} + \alpha(\tilde{c}-C) = Q \qquad \text{on } \Gamma_i \tag{10}$$

on inflow and noflow boundaries. Conditions along outflow boundaries are irrelevant to the "advection problem" defined by (8) - (10) which can be solved for $\tilde{c}$ independent of $\hat{c}$.

Subtracting (8) from (6) leads to the "residual dispersion problem" for c, defined by

$$R\left[\frac{Dc}{Dt} - \frac{D\tilde{c}}{Dt}\right] = \nabla\cdot(\mathbf{D}\nabla c) - (c-\tilde{c})\nabla\cdot\mathbf{v} - R\lambda(c-\tilde{c}) + q \tag{11}$$

subject to homogeneous initial conditions on $\hat{c}$ together with (3) and (4), or a modification of the latter as discussed below.

NUMERICAL APPROACH

Our adaptive numerical method can be described briefly as follows. Suppose that c is known at time t_k, and we wish to compute it at $t_{k+1} = t_k + \Delta t$. First we set

$$\tilde{c}(\mathbf{x},t_k) = c(\mathbf{x},t_k) \qquad \hat{c}(\mathbf{x},t_k) = 0. \tag{12}$$

We then solve the advection problem for $\tilde{c}(\mathbf{x},t_k)$ by reverse or forward particle tracking (depending on local concentration gradients) combined with an analytical solution of (8) over Δt. Finally, we solve the residual concentration problem for $c(\mathbf{x},t_{k+1})$ by finite elements. For this, c is approximated by

$$c(\mathbf{x},t) \cong c^N(\mathbf{x},t) = \sum_{n=1}^{N} c_n(t)\, \xi_n(\mathbf{x}) \tag{13}$$

where N is number of nodes in the finite element grid, c_n is concentration at node n, and $\xi_n(\mathbf{x})$ is basis function satisfying $\xi_n(\mathbf{x}_m) = \delta_{nm}$ where δ_{nm} is the Kronecker delta. A similar finite element approximation is employed for $\tilde{c}$. Currently we use triangular elements in two dimensions and tetrahedra forming layered prisms with triangular or quadrilateral cross-sections in three dimensions.

SINGLE-STEP REVERSE PARTICLE TRACKING

Consider a fictitious particle that moves from a location ${}^k\mathbf{x}_n$ at t_k to a new location $\mathbf{x}_n{}^{k+1}$ at t^{k+1} which coincides with node n. Its initial location is then given by

$${}^k\mathbf{x}_n = \mathbf{x}_n{}^{k+1} - \int_{t_k}^{t_{k+1}} \frac{\mathbf{v}}{R}\, Dt. \tag{15}$$

If v and Δt are constant, this calculation needs to be performed only once. In nonuniform and/or transient velocity fields, the above integral is evaluated in small increments by a fourth-order Runge-Kutta-Fehlberg algorithm (Thomas[4]).

The initial particle concentration, ${}^k c_n$, is identical to $\tilde{c}_n{}^k$, the initial $\tilde{c}$ value of the particle. In the absence of a particle cloud cover of the type described below, it is computed from (13). The final $\tilde{c}$ value of the same particle upon reaching node n, $\tilde{c}_n{}^{k+1}$, is obtained by solving (8) analytically over Δt,

$$\tilde{c}_n^{k+1} = {}^k c_n \exp\left[- \int_{t_k}^{t_{k+1}} \left[\frac{\nabla \cdot \mathbf{v}}{R} + \lambda \right] Dt \right]. \tag{16}$$

CONTINUOUS FORWARD PARTICLE TRACKING

Steep concentration fronts are tracked with the aid of particle clouds that cover the fronts until their gradients dissipate. A particle, p, moves from its initial location at t_k, x_p^k, to a new location

$$x_p^{k+1} = x_p^k + \int_{t_k}^{t_{k+1}} \frac{\mathbf{v}}{R} Dt \tag{17}$$

at t_{k+1} where the integration is performed by the same method as in (15). If the particle concentration at t_k is c_p^k, its concentration at t^{k+1} is, in analogy to (16),

$$\tilde{c}_p^{k+1} = c_p \exp\left[- \int_{t_k}^{t_{k+1}} \left[\frac{\nabla \cdot \mathbf{v}}{R} + \lambda \right] Dt \right]. \tag{18}$$

The introduction of front-tracking particles at strategic locations is done on the basis of criteria discussed in the section on "Adapative Mechanism." The initial concentrations of these particles are determined by interpolation between nodal concentrations according to (13). The algorithm also allows for particles to be continually emitted along inflow boundaries. Any new particle, r, introduced at (x_r, t_k) along such a boundary is assigned the initial concentration

$$\tilde{c}_r^k = \frac{\alpha C + Q}{\mathbf{v} \cdot \mathbf{n} + \alpha} \tag{19}$$

in accord with (10).

Suppose that during single-step reverse tracking a particle reaches node n at t_{k+1} by starting from ${}^k x_n$ at t_k. If the starting point ${}^k x_n$ is covered by a cloud of front-tracking particles, the corresponding concentration, ${}^k c_n$, is calculated from the concentrations of the cloud particles. This is done by optimally triangulating (or, in three dimensions, tetrangulating) the cloud particles within the element according to a method described by Sloan and Houlsby[5]; locating ${}^k x_n$ within a triangle (or tetrahedron) of neighboring cloud particles and/or nodes; and interpolating the concentration linearly onto ${}^k x_n$ between the vertices of the triangle or tetrahedron.

RESIDUAL DISPERSION BY FINITE ELEMENTS

Applying the Galerkin orthogonalization procedure to (11) yields

$$\int_{\Omega}\left\{R\left[\frac{Dc}{Dt} - \frac{D\tilde{c}}{Dt}\right] - \nabla\cdot(D\nabla c) + (\nabla\cdot\mathbf{v} + R\lambda)(c-\tilde{c}) - q\right\}\xi_n \, d\Omega = 0$$

$$n = 1,2,\ldots,N \quad (20)$$

where Ω is the region bounded by Γ and N is the number of nodes in the grid. Use of Green's first identity together with (3) and mass lumping of the terms involving time derivatives, velocity divergence, and decay yields

$$\left[\frac{Dc_n}{Dt} - \frac{D\tilde{c}_n}{Dt}\right]\int_{\Omega} R\xi_n \, d\Omega + \int_{\Omega} \mathbf{D}\nabla c\cdot\nabla\xi_n \, d\Omega - \int_{\Gamma_i}[\mathbf{v}c\cdot\mathbf{n} + \alpha(c-C) - Q]\xi_n \, d\Gamma_i$$

$$+ (c_n-\tilde{c}_n)\int_{\Omega}(\nabla\cdot\mathbf{v} + R\lambda)\xi_n \, d\Omega - \int_{\Omega} q\xi_n \, d\Omega = 0 \quad (21)$$

Note that when $\alpha = \infty$ near some node, n, the latter represents a Dirichlet boundary and the corresponding equation drops out. Along outflow boundary segments satisfying condition (4) there is no boundary contribution to (21). However, instead of using (4), we prefer to approximate c by $\tilde{c}$ on outflow boundaries which adds the following term to (21),

$$\int_{\Gamma_o} \mathbf{D}\nabla\tilde{c}\cdot\mathbf{n}\xi_n \, d\Gamma_o. \quad (22)$$

A more accurate representation of conditions along outflow boundaries might involve replacing $\tilde{c}$ in (22) by c but this would lead to a nonsymmetric matrix which we wish to avoid.

Each node is treated as a particle that has reached $\mathbf{x}_n$ at t_{k+1}. This leads naturally to a backward difference formulation of the Lagrangian time derivatives,

$$\frac{Dc_n}{Dt} \cong \frac{c_n^{k+1} - {}^k c_n}{\Delta t}, \quad \frac{D\tilde{c}_n}{Dt} \cong \frac{\tilde{c}_n^{k+1} - {}^k c_n}{\Delta t}. \quad (23)$$

With this and (13), it is a matter of routine to develop algebraic finite element equations for (21) that involve only symmetric and diagonal matrices as do (31) - (38) of Neuman[1].

The final step is to project c^{k+1} onto moving particles if such particles exist in the flow field. Neuman[1] (his eq.39) did this by writing

$$c_p^{k+1} = \tilde{c}_p^{k+1} + (c^N(\mathbf{x}_p^{k+1}, t_{k+1}) - \tilde{c}^N(\mathbf{x}_p^{k+1}, t_{k+1})) \quad (24)$$

but this may violate the maximum principle in elements containing sharp fronts. In such elements, we obtain greatly improved results by triangulating (or, in three dimensions, tetrangulating) the nodes and particles according to the algorithm of Sloan and Houlsby[5]. This yields a local grid superimposed on the major element, the nodes of which coincide with particles and major grid nodes. We then solve a system of residual dispersion finite element equations over this local grid by treating the surrounding edges of the major element as Dirichlet boundaries (the effect of residual dispersion along these boundaries being known from the earlier solution of similar equations over the major grid). The approach is akin to the telescopic mesh refinement idea described by Ward *et al.*[6].

ADAPTIVE MECHANISM

To decide whether front-tracking particles are needed in elements surrounding a node, n, where such particles are currently not present, the algorithm evaluates the maximum and minimum components of the concentration gradient along all coordinates in these elements. If the difference between the maximum and minimum values of any component exceeds a prescribed fraction tolerance of the maximum, particles are evenly distributed over all elements in the vicinity of n at a density specified by the user. As mentioned earlier, the user can opt for particles to be continually emitted from sources in the interior or on the boundary.

To decide whether existing particles can be eliminated from a cloud-covered element, we use the ratio between the maximum values of $|\tilde{c}_p^{k+1} - \tilde{c}^N(x_p^{k+1}, t_{k+1})|$ and $|c_n^{k+1} - \tilde{c}_n^{k+1}|$ within the element as a criterion. The element cloud cover is eliminated whenever this ratio falls below a user specified tolerance.

EXAMPLES

Example 1 concerns the two-dimensional dispersion of a rectangular wave in a uniform velocity field over an infinite domain at relatively low Peclet numbers. The governing equation, initial and boundary conditions, and analytical solution are (with the exception of specific parameter values) similar to those in (45) - (47) of Neuman[1]. The background concentration is zero and the wave has an initial concentration of one. It is centered at (0.15,0.02) and has a length of 0.05 parallel to the x-coordinate and 0.01 parallel to the y-coordinate. The grid consists of 2800 rectangular elements (each divided into two triangles) oriented parallel to the x and y axes. Dispersion coefficients, velocities, and grid spacings along the x and y directions, respectively, are (in arbitrary consistent units) $D_x = 1$, $D_y = 0.1$, $v_x = 1000$, $v_y = 400$, $\Delta x = 0.01$, and $\Delta y = 0.004$. These correspond to directional Peclet numbers of $Pe_x = v_x \Delta x / D_x = 10$ and $Pe_y = v_y \Delta y / D_y = 16$.

Figure 1 shows sections through the wave center parallel to the coordinates at three consecutive times. These results were obtained without front-tracking by using a time step, $\Delta t = 0.0000545$, which corresponds to Courant numbers, $C_x = v_x \Delta t / \Delta x$ and $C_y = v_y \Delta t / \Delta y$, of 5.45 in both coordinate directions. In all cases, the peaks are seen to be clipped. Figure 2 demonstrates that peak clipping can be virtually eliminated by using a maximum of 770 particles in an adaptive mode. As the velocity is inclined at approximately 22^0 to the x axis, we see that front tracking eliminates adverse grid orientation effects. When a smaller time step was used corresponding to Courant numbers of 1.87, clipping became slightly more pronounced and required 870 particles to correct.

Example 2 is similar to the first except that dispersion is reduced by an order of magnitude and v_y is reduced to 300 so that the Peclet numbers are larger, $Pe_x = 100$ and $Pe_y = 120$. Figures 3 and 4 are analogous to 1 and 2, showing how both peak clipping and numerical smearing (which is more pronounced here than in Example 1) can be greatly reduced by an adaptive use of front-tracking particles with Courant numbers as large as 5.45 (4.09 in the y direction). The velocity being inclined at about 17^0 to the x axis, we see that there is virtually no grid orientation effect even at these relatively high Peclet numbers. With smaller Courant numbers ($C_x = 1.87$, $C_y = 1.41$) without front-tracking, our calculations show that both clipping and numerical dispersion become very severe as time progresses; at t = 0.0008, the peak reduces to nearly half its correct height. The introduction of particles greatly improves such results which then remain only slightly inferior to those in Figure 4. We don't illustrate this phenomenon as it has been demonstrated earlier by Neuman[1].

Example 3 deals with a pair of injection and discharge wells in an extensive aquifer without background flow. The problem is based on a laboratory tracer experiment conducted by Hoopes and Harleman[7] in which the wells are 61 cm apart and operate at a rate of 2.339 cm^3/sec each. The aquifer is 8.9 cm thick, has a porosity of 0.374 and transverse dispersivity within it is zero. The situation was simulated by Huyakorn *et al.*[8] using a traditional Galerkin finite element scheme on the rectangular grid shown in Figure 5. Their results oscillated at moderate Peclet numbers, a condition they were able to alleviate by redesigning their grid parallel and normal to the streamlines and using upstream weighting. Our adaptive Eulerian-Lagrangian finite element method can solve this problem on the original rectangular grid without oscillations and with a substantial increase in time step size (*i.e.*, Courant number) for Peclet numbers ranging from very small (nearly pure dispersion) to infinity (pure advection). Figure 6 depicts our results at the discharge well obtained with time steps of 2000 sec and a maximum Courant number over the grid of about 15. Figure 6a corresponds to a maximum Peclet number (near the wells) of about 1.7 (longitudinal dispersivity of 2.94 cm); Figure 6b corresponds to a maximum Pe of about 17 (longitudinal dispersivity of 0.294 cm); and Figure 7 shows a comparison with the analytical solution of Hoopes and Harleman for $Pe = \infty$. The results are seen to be quite satisfactory.

Example 4 demonstrates the three-dimensional capability of our current adaptive code. We consider the problem presented by Huyakorn *et al.*[9] of a well injecting contaminated water into a 4 m thick aquifer confined between two identical aquitards having a thickness of 50 m each. Transport in the aquifer is controlled by a porosity of 0.25 and a longitudinal dispersion coefficient of 0.608 m^2/day. In the aquitards these parameters are 0.025 and 6.08 m^2/day, respectively. The well has a radius of 0.1 m and injects solute-free water into the aquifer at a rate of 10 m^3/day. At reference time t = 0, the concentration of the water is raised to 1 ppm and is maintained so indefinitely. Due to symmetry, it was sufficient to simulate transport in a wedge-shaped prism that includes the upper half of the aquifer and the upper aquitard. The prism extends radially from the edge of the injection well (at a radius of 0.1 meter) an additional distance of 20 m. It is subdivided into 600 quadrilateral prismatic elements arranged in 2 aquifer and 13 aquitard layers (Figure 8). Each quadrilateral element consists of five tetrahedra and the radial distance between nodes is 0.5 m.

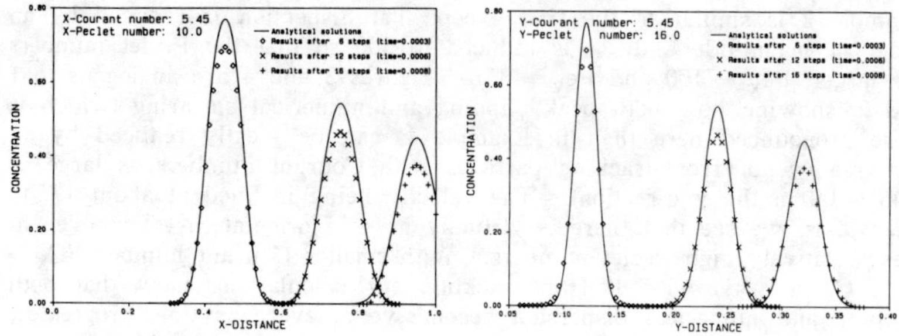

Figure 1. Example 1 concentrations without particles.

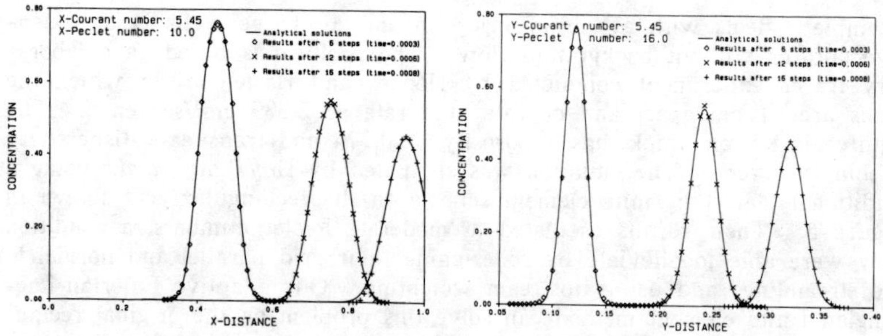

Figure 2. Example 1 concentrations with particles.

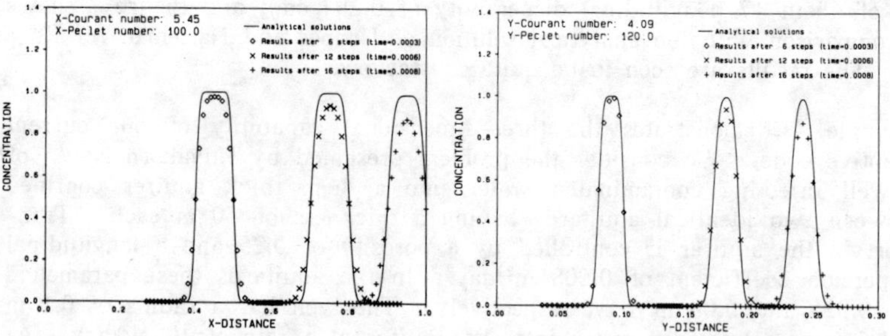

Figure 3. Example 2 concentrations without particles.

An analytical solution exists for the aquifer which assumes that the system has infinite radial extent, the aquitards are infinitely thick, advection in the aquitards is insignificant, dispersion in these units is vertical, the aquifer longitudinal dispersion coefficient is constant, and there are no vertical concentration gradients in the aquifer. For this, we had to temporarily disable horizontal diffusion in all aquitard elements within our code. Results for the aquifer are shown in Figure 9a, and for the aquitard at a radial distance of 6 m from the well in Figure 9b. The Courant and Peclet numbers attain their maximum values of about 4.5 and 3.7, respectively, in the immediate vicinity of the well. At one time during the simulation 415 particles were automatically introduced to adequately define the concentration front but their number diminished to 168 toward the end of the simulation.

REFERENCES

1. Neuman, S. P. (1984), Adaptive Eulerian-Lagrangian finite element method for advection-dispersion, *Intern. Jour. Num. Meth. Engin.*, Vol. 20, pp. 321-337.

2. Neuman, S. P. (1981), A Eulerian-Lagranginan Numerical Scheme for the Dispersion-Convection Equation Using Conjugate Space-Time Grids, *Jour. Comp. Phys.*, Vol. 41, pp. 270-294.

3. Smith, R. M., and Hutton, A. G. (1982), The Numerical Treatment of Advection: A Performance Comparison of Current Methods, *Num. Heat Transfer*, Vol. 5, pp. 439-461.

4. Thomas, B. (1986), The Runge-Kutta Methods, *BYTE Magazine*, Vol. 11, pp. 191-210.

5. Sloan, S. W., and Houlsby, G. T. (1984), An Implementation of Watson's Algorithm for Computing 2-Dimensional Delaunay Triangulations, *Adv. Engin. Software*, Vol. 6, pp. 192-197.

6. Ward, D. S., Buss, D. R., Mercer, J. W., and Hughes, S. S. (1987), Evaluation of a Groundwater Corrective Action at the Chem-Dyne Hazardous Waste Site Using a Telescopic Mesh Refinement Modeling Approach, *Water Resour. Res.*, Vol. 23, pp. 603-617.

7. Hoopes, J. A., and Harleman, D. R. F. (1967), Wastewater Recharge and Dispersion in Porous Media, *Jour. Hydr. Div. ASCE*, Vol. 93, pp. 51-71.

8. Huyakorn, P. S., Andersen, P. F., Güven, O., and Molz, F. J. (1986), A curvilinear Finite Element Model for Simulating Two-Well Tracer Tests and Transport in Stratified Aquifers, *Water Resour. Res.*, Vol. 22, pp. 663-678.

9. Huyakorn, P. S., Jones, B. G., and Andersen, P. F. (1986), Finite Element Algorithms for Simulating Three-Dimensional Groundwater Flow and Solute Transport in Multilayer Systems, *Water Resour. Res.*, Vol. 22, pp. 361-374.

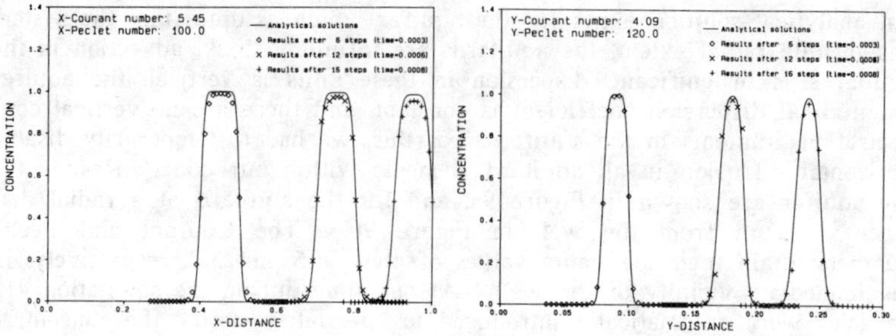

Figure 4. Example 2 concentrations with particles.

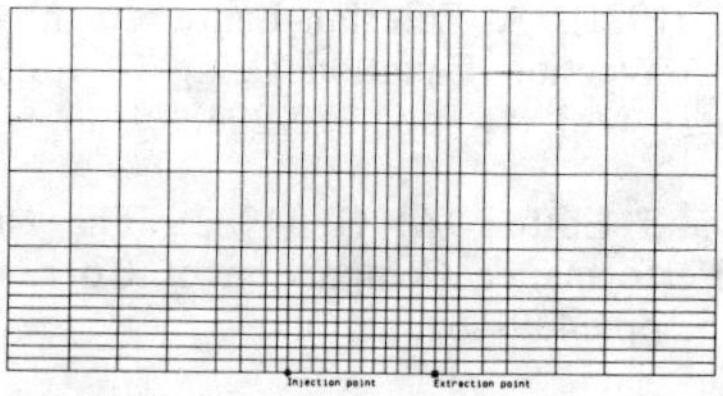

Figure 5. Grid for Example 3.

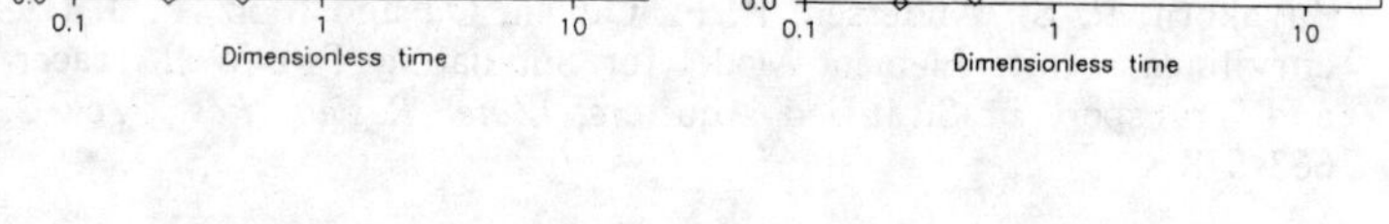

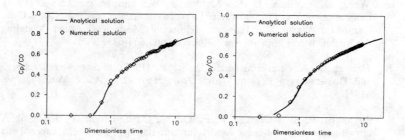

Figure 6. Example 3 concentrations with dispersivity (a, left) 2.94 cm and (b, right) 0.294 cm.

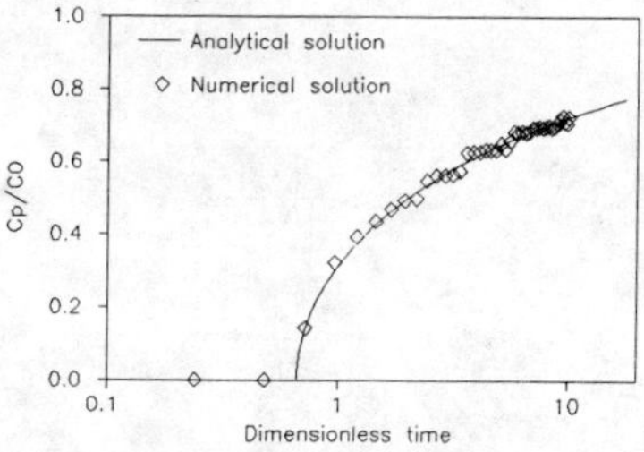

Figure 7. Example 3 concentrations with zero dispersivity.

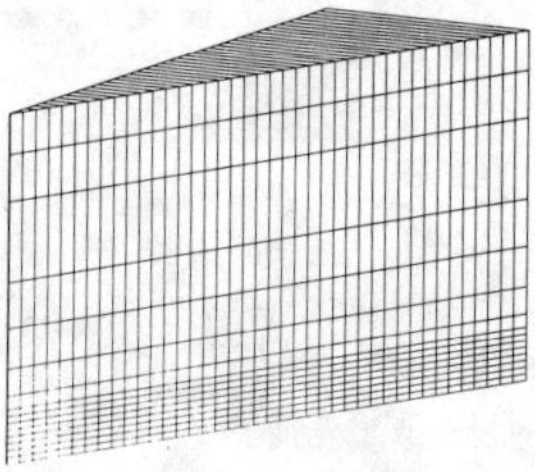

Figure 8. Grid for Example 4.

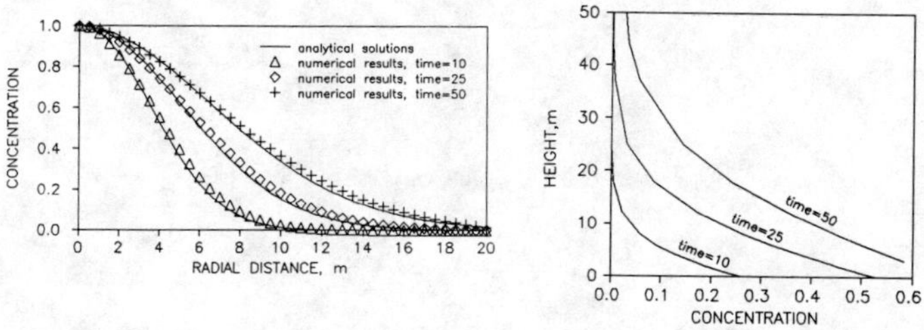

Figure 9. Example 4 concentrations in (a, left) aquifer and (b, right) aquitard at 6 m from well.

Computer Modeling of Groundwater Flow Through Porous Media using a Monte-Carlo Simulation Technique

J.S. Loitherstein

TGG Environmental, Inc., Needham, MA 02194, USA

ABSTRACT

The purpose of this paper is to discuss a computer model developed for simulating the movement of groundwater through porous media. The movement is simulated in terms of its two components: advection and hydrodynamic dispersion. Advection is a purely translational movement directly related to the magnitude and direction of the groundwater velocity. The groundwater's movement due to dispersion is too complex to analyze numerically and instead is modeled using a Monte-Carlo simulation technique. This permits calibrating the dispersion component by varying the statistical parameters in the model.

CONCLUSIONS

The original intention of this paper was to determine whether varying the distribution of the random numbers used to simulate dispersion would have a significant effect on the resultant groundwater plume. It became apparent very early in the project that indeed the random numbers have a major effect on the groundwater plume. Although this can be accomplished somewhat by varying the longitudinal and transverse dispersivities, the random number distribution adds another degree of sensitivity that previously was unavailable.

In addition, the following specific observations were made:

1. **The shape and extent of the plume is indeed sensitive to manipulation of the random numbers used to simulate the dispersion. These parameters, and their effects on the plume should not be overlooked by modelers in the future.**
2. **The particle cloud method of presenting the data is more discernable than a group of lines showing the paths of the particles. The use of color graphics enhances the results.**
3. **There are two types of parameters that are input into the code: those that are related to the groundwater's movement and those that are related to the graphical presentation of the data. The graphics parameters are determined somewhat**

by the aquifer parameters. For example, fewer particles could be used with a simulation involving a small longitudinal or transverse dispersivity. The advantage of using fewer particles is that it would reduce the amount of time required to perform each simulation.

EXPLANATION OF GROUNDWATER MOVEMENT

Groundwater takes a tortuous path as it passes between particles of sand, clay, silt, or gravel. The path which the groundwater follows, and its velocity through the porous medium depend upon physical properties of both the groundwater and the aquifer. The path of the groundwater can be divided into two components: advection and hydrodynamic dispersion. These are vector quantities that describe the groundwater's movement during a fixed period of time. Advection is proportional to the velocity of the groundwater and occurs in the general direction of the groundwater gradient. The rate of advection is equal to the average linear groundwater velocity, V_L where $V_L = v/n$, v = specific discharge, and n = porosity (1) (see Equation 1 below).

Hydrodynamic dispersion is used to denote the spreading (at the macroscopic level) resulting from both mechanical dispersion and molecular diffusion. Mechanical dispersion is due to the interaction of the groundwater with the porous medium. If the groundwater hits an impermeable boundary such as a sand particle, it reflects. Molecular diffusion takes place in the absence of motion (at the macroscopic level). Molecular diffusion, often called Brownian motion, is independent of macroscopic motion. Its effects on the overall dispersion will be more significant at low flow velocities. It is molecular diffusion which makes the phenomenon of hydrodynamic dispersion in purely laminar flow irreversible.(3)

As described by Bear "Obviously it is impossible to carry out experiments and tests in the aquifer itself in order to determine its response...to activities proposed in the future...to make comparisons among responses...to determine the most desirable one...or to incorporate the responses in some decision making procedure. Like in all branches of science and engineering, whenever the treatment of real systems or phenomena is impossible...models of the considered systems or phenomena are introduced. Instead of treating the real system, we manipulate its model and use the results of these manipulations in order to make decisions regarding the operation of the real system."(3)

The two dimensional form of the advection-dispersion equation for nonreactive dissolved constituents in saturated, homogeneous, isotropic, materials under steady-state, uniform flow is:

$$D_L*(d^2C/dl^2) - (V_L)*(dC/dl) = dC/dt \qquad (1)$$

where l is a curvilinear coordinate direction taken along the flowline, V_L is the average linear groundwater velocity (L/T), D_L is the coefficient of hydrodynamic dispersion

(L^2/T) in the longitudinal direction (i.e. along the flow path), and C is the solute concentration (M/L^3).(2)

SIMULATION OF DISPERSION USING COMPUTER MODEL

If dispersion did not exist, it would be possible to predict precisely where a particle of groundwater would be located after a fixed period of time if its velocity and direction were known. Dispersion introduces an element of "unpredictability" into the groundwater movement. In theory, if the exact position of every sand grain were known, their exact sizes and shapes, and exactly how the particles of water will reflect off these grains, it would be possible to predict exactly how a particle of water would move due to dispersion. However, not only would the mathematics of this type of solution reach beyond the capabilities of even the largest computers, the technology for obtaining these types of data on subsurface strata does not currently exist. Therefore, it is necessary to simulate the unpredictability of dispersion using statistical means.

The following equations have been developed by Smith and Schwartz (4) to statistically simulate dispersion through the use of random numbers:

$$d_L = (24*D_L*(t_2-t_1))^{0.5}*(0.5-R) \quad (2)$$
$$d_T = (24*D_T*(t_2-t_1))^{0.5}*(0.5-R) \quad (3)$$

Where:

d_L = the dispersion in the longitudinal direction
d_T = the dispersion in the transverse direction
D_L = the longitudinal dispersion coefficient
D_T = the transverse dispersion coefficient
t_2 = the time at the end of the time step
t_1 = the time at the beginning of the time step
R = a random number between 0 and 1.0

The components of the net displacement during a time step are then:(4) (see Figure 1)

$$\text{del } X = V_L(\cos\Theta)(t_2-t_1)+d_L(\cos\Theta)+d_T(\sin\Theta) \quad (4)$$
$$\text{del } Y = V_L(\sin\Theta)(t_2-t_1)+d_L(\sin\Theta)+d_T(\cos\Theta) \quad (5)$$

Where:

V_L = the velocity in the longitudinal direction
Θ = the angle between the longitudinal flow direction and the X-axis

It should be noted that for the axes selected $\Theta = 0$ and $\cos\Theta = 1$ while $\sin\Theta = 0$. Therefore:

$$\text{del } X = V_L(t_2-t_1) + d_L \quad (6)$$
$$\text{del } Y = d_T \quad (7)$$

As stated before, dispersion is a seemingly random movement that is the sum of two components: molecular diffusion and mechanical dispersion. At low flow velocities, molecular diffusion is dominant, and at higher velocities, mechanical dispersion becomes dominant. The sum of these two components is known as hydrodynamic dispersion. For purposes of this report, an estimate of the total hydrodynamic dispersion was used to compute the groundwater movement due to dispersion.

There are several computer models that use basic mathematical formulae to simulate the movement of groundwater. Most programs employ a normal distribution for the random numbers with a mean of 0.5 and limits of 0 and 1.0. One objective of this research was to develop a model to determine the effect of changing the random number distribution.

MODEL CALIBRATION

After initial best-estimates of the input data are made, model development is an iterative process in which results of previous simulations are interpreted to make modifications and adjustments to the model. The testing process of adjusting input data and comparing the calculated results to field observations allows for a better understanding of the flow system and an improvement of the conceptual model.(6)

The hydrogeological parameters used in the model should be from actual soil samples, are therefore known with a reasonable degree of accuracy, and probably should not be changed merely for calibration purposes. In contrast, the statistical parameters are only being used to simulate dispersion in the field and may change from site to site due to interactions at the microscopic level within the groundwater and between the groundwater and the soil that are too complex to model with existing numerical techniques. Model calibration with the current program is concerned with varying the statistical parameters to achieve a more accurate representation of the actual situation.

An example of how the calibration would be performed is as follows: Figure 2 shows a hypothetical contaminant site with a source at a concentration of 500 ppm and several monitoring wells downstream. Hypothetical contaminant concentrations were developed and are shown in Table 1. After inputting typical hydrogeological and statistical parameters into the model, predicted concentrations were calculated which also appear in Table 1:

TABLE 1

Well No.	Hypothetical Concentration	Predicted Concentration
1	11	0
2	3	0
3	14	91
4	20	51
5	10	0

Performing a linear regression analysis on these data resulted in a correlation coefficient of 0.10 which is extremely low. This means that the model must be recalibrated if it is to better represent the specified conditions.

To perform the calibration, differences between the hypothetical and predicted conditions must be determined. Apparently, the plume should be wider to include Wells 1, 2, and 5 in the plume and to reduce the concentration along the middle of the flowline. This can be accomplished in several ways. Among them are increasing the standard deviation of the normally distributed random numbers or by changing from a normal distribution to a uniform distribution.

Actually, these two operations are similar. If the standard deviation of a normal distribution is increased, in effect the probability for numbers farther away from the mean to occur is increased. For a normal distribution with an extremely high standard deviation, the probability of any number occurring would be almost equal. In a uniform distribution, the probabilities are equal.

First, an attempt was made to calibrate the model with a normal distribution and a standard deviation of 1.50. Although the concentrations were closer to the ones specified, they were still not representative of the hypothetical situation. After several permutations, it was found that a standard deviation of 2.50 yielded the best results with a normal distribution. The calibration was also performed by changing from a normal to a uniform distribution.

The predicted contaminant concentrations at each well for both calibrations are presented in Table 2.

TABLE 2

Well Number	Hypothetical Observed Concentration (ppm)	Predicted Concentration Normal Dist'n (S.D.=2.50) (ppm)	Predicted Concentration Uniform Dist'n (ppm)
1	11	12	5
2	3	3	4
3	14	13	11
4	20	19	20
5	10	9	8

The correlation coefficient for these data was 0.99 for the normal distribution and 0.91 for the uniform distribution. This close correlation indicates that manipulation of statistical parameters is a viable method of calibrating this, or possibly any statistically based model, to actual field situations.

BIBLIOGRAPHY

1. Freeze, R. A., and Cherry, J. A.; 1979; Groundwater; Prentice-Hall.
2. Bouwer, H.; Groundwater; 1978; McGraw-Hill.
3. Bear, J.; Hydraulics of Groundwater; 1979; McGraw-Hill.
4. Smith, L., and Schwartz, F. W.; Mass Transport-A Stochastic Analysis of Macroscopic Dispersion; 1980; Water Resources Research
5. Lecture notes from Prof. Hemond; M.I.T.; 1982.
6. Devore; Probability and Statistics for Engineers; 1982; Brooks/Cole
7. Wilson, J. R. III; BSCE Lecture Notes, 1981.

FIGURE 1

COMPONENTS OF
ADVECTION AND DISPERSION

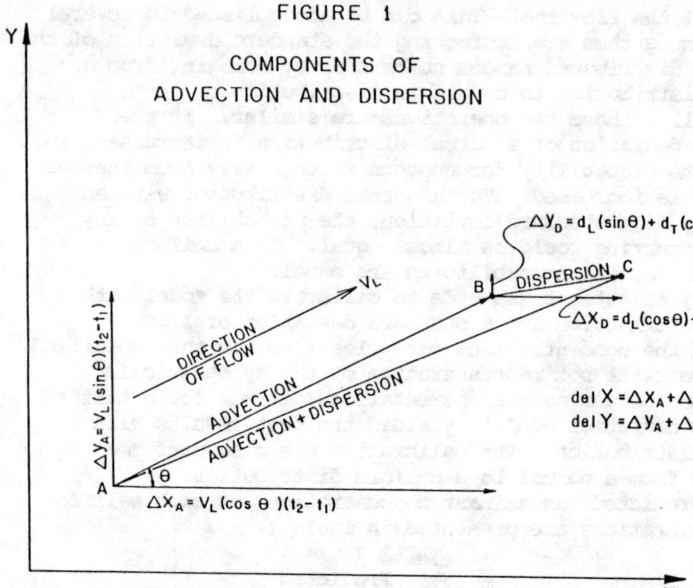

FIGURE 2

HYPOTHETICAL
CONTAMINANT
SITE

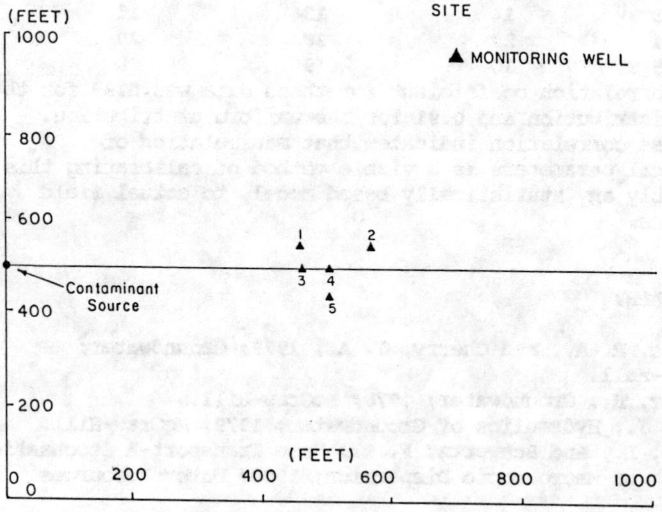

Dispersion of Contaminants in Saturated Porous Media: Validation of a Finite-Element Model

G.L. Moltyaner

Atomic Energy of Canada Limited, Chalk River Nuclear Laboratories, Chalk River, Ontario, KOJ 1JO, Canada

ABSTRACT

The main objective of this paper is to outline the experimental and theoretical investigations performed in an attempt to validate the applicability of finite element based numerical models for the prediction of the behaviour of a conservative tracer (radioiodine) at the Twin Lake aquifer, Chalk River Nuclear Laboratories, Chalk River, Ontario. The essential point is that the 3/4 of a million data points obtained at the Twin Lake site from a 40 m natural gradient tracer test provide a unique opportunity for quantifying the system variability and for testing finite element model of the dispersion process.

The subject of this discussion is the advection-dispersion model of radioiodine transport - its equation and solution by the standard Galerkin finite element method.

The developed coarse-mesh (460 nodes, 830 element) finite element model of the radioiodine transport poorly describes the overall behaviour of the tracer plume and lacks the capability to simulate the fingerlike spreading of the plume due to the fact that the grid does not have an adequately fine space discretization. A refinement of the grid spacing (7,128 nodes, 13,920 elements) has significantly improved the simulation. It was concluded that for the advection-dominated transport, as that encountered at the Twin Lake aquifer, the failure to satisfy fine mesh requirement causes numerical dispersion. It was also concluded that the conventional finite element model may produce accurate simulation of the tracer cloud provided that the adequately fine space discretization of the grid compatible with the support scale of measurements and the adequately fine time discretization are made. This demands large computing resources. The paper describes a state-of-the-art methodology for calibration, validation and sensitivity analysis of a finite-element model of nonreactive solute transport.

INTRODUCTION

The objective of this work is to outline experimental and theoretical investigations performed in an attempt to validate a two-dimensional finite-element model describing groundwater flow and mass transport at the Twin Lake aquifer and to demonstrate a methodology proposed for validating the model. The approach adopted in our model validation program is based upon the interconnections between modelling, controlled experiments and observational studies of the existing contaminated sites at CRNL (Chalk River Nuclear Laboratories).

A field-oriented experimental program has been established at CRNL with the objective of evaluating the validity of existing models of the dispersion process. This is achieved through two natural gradient tracer tests performed at the Twin Lake aquifer using ^{131}I and ^{3}H as non-reactive tracers and numerous laboratory tests (Moltyaner[1], Taylor et al.[2]). The essential point is that the 3/4 of a million data points collected from the tracer tests site provide a unique opportunity for validating numerical models of the dispersion process. Model validation was defined as the comparison of model results with experimental data derived from the tracer experiments. The success of the model was assessed by validation criteria which included visual inspection of the goodness of fit of observed and simulated data, estimation of local deviation between observed and simulated data using subjective criteria for acceptance (the difference between observed and calculated data should not be larger than measurement error), estimation of global deviation using statistical measures to check agreement between observed and simulated tracer distribution and sensitivity analysis with respect to model parameters, grid spacing, time steps, initial and boundary conditions.

The validation criteria also give evidence for the validity or otherwise of the assumptions made in dispersion model formulation including model concepts, equations and their finite element solutions under specified initial and boundary conditions.

In using field data for validating finite-element based models of radioiodine migration, the emphasis was on sampling scale (sampling size and monitoring grid spacing); sampling frequency; methods used to interpolate between points of measurements in time and space; methods used to interpret the dispersion parameter that cannot be directly measured (inverse problem solution); and discretization in time and space used in the predictive groundwater flow and transport models.

A sensitivity analysis with respect to the space discretization was performed by reducing grid spacing by a factor of ≈ 15 in order to evaluate the effect of grid spacing on the numerical dispersion. Limited sensitivity analysis was also performed with regard to time steps, transverse and longitudinal dispersion, initial and boundary conditions. The aim of the performed analysis was to achieve a reasonable assurance that the model and code give a good representation of the naturally occurring process that was measured at the scale of the order of centimetres.

Twin Lake Aquifer

The Twin Lake aquifer is relatively homogeneous, being composed of highly uniform medium to fine-grained sands of aeolian and fluvial origin. Minor silts, situated at the vicinity of the water table, and locally preserved tills at the base of the sequence may be found within the aquifer. The sand is approximately 10 to 15 m thick at the site. Geologic data from 82 boreholes combined with the ground probing radar data were used to map the interface between the sandy sediments and essentially impermeable underlying bedrock. Groundwater flow occurs predominantly in the horizontal direction. The spatial variability of hydraulic conductivity at the Twin Lake aquifer was examined in detail using close-interval grain-size analysis, indirect estimates from natural gradient tracer tests and permeameter tests. An average ratio of horizontal to vertical hydraulic conductivities is 3. Average linear flow velocities were estimated from the tracer test data and porosities from the core sections used for the permeameter tests. The estimated dispersivities were 1 cm and 0.1 cm for longitudinal and transverse values respectively (Moltyaner [1]). Hydraulic head was measured in all piezometers immediately before tracer injection. The tracer was injected into a fully-penetrating well and monitored over 40 m distance over a period of 45 days in a network of 82 boreholes.

The hydrostratigraphic cross sectional scale-model of the aquifer system along the mean direction of tracer flow was used to construct a steady state two-dimensional mathematical model for flow. The boundary conditions were selected to represent the real hydrologic boundaries and to approximate as closely as possible the artificially imposed boundary conditions. The right and the left hand side boundaries were drawn approximately as equipotential lines with specified water fluxes. The water table elevations are refined by applying a zero flux boundary condition in the series of computer iterations and thus obtained values were kept constant. The bedrock surface was simulated as a no-flow boundary. The finite element discretization of the hydrostratigraphic scale model was performed using three node triangular elements. The two dimensional linear interpolation functions were used to approximate the unknown hydraulic head distribution. The same interpolation functions were used in the weighted residuals procedure. The model was first calibrated against the hydraulic head only. The calibrated model showed a relatively poor agreement between simulated velocities and those estimated from the tracer test. The significant improvement in the model performance was gained from calibrating the model against both hydraulic head and tracer velocity measurements.

The velocity field was generated using the calculated hydraulic head and Darcy's law. The generated velocities were used in the transport part of the finite element model. The two-dimensional advection-dispersion equation describing tracer transport was solved under specified initial and boundary conditions using Galerkin finite element approximating in space and finite-difference approximation in time, (Pinder and Gray[3]). The model of tracer transport simulated the changes in the tracer concentration injected into the system along the right hand side boundary of the specified mass flux. The distribution of concentration observed 15 hours after the injection was taken as the initial concentration distribution. A scenario using observed after 4.44 days concentration distribution as the initial one was also considered. The top and left side boundaries were treated through Dirichlet boundary condition. The bottom boundary conditions were Neumann type with a zero concentration gradient.

For the advection-dominated problem with small dispersivities, the finite element model must meet a number of requirements to avoid oscillation, numerical dispersion and negatively of concentration. The effect of this phenomena on the solute transport may be controlled with a proper choice of grid Peclet and Courant numbers, (Frind and Hokkanen[4]). Unfortunately refinement of the grid spacing is limited by the restrictive size of computer memory and by the high computational costs. The optimum irregular grid accommodated by our site computer consisted of 452 nodes and 819 triangles. The average horizontal spacing between nodal points is approximately 1.25 m and the vertical spacing is 0.68 m. The time step is 0.1 days. The model is run until plume reaches the boundary. After the first iteration, the model immediately produces small values of negative concentration. With the evalution of time, this numerical error accumulates and eventually results in the creation of discontinuous small plumes. In the vicinity of bedrock surface this leads to the development of a long dense tail. The effect of numerical dispersion was significant. The simulations based on the calibration of flow model against both hydraulic head and velocity measurements show that the finite element model produces less spreading and the centroids of time concentration profiles are much better predicted as compared with the simulations based on the calibration against hydraulic head only. The results can be anticipated if one recalls that the velocity field affects the advective transport through the advective term and through the coefficients of hydrodynamic dispersion. A sensitivity analysis pertaining to the dispersivity showed that increase in dispersivity causes a decrease in negative values of concentration and in concentration gradients. The same quantitative effect was produced by increasing time steps. Increasing the value of transverse dispersivity causes the tracer to move towards the water table. A sensitivity analysis of the model with respect to space discretization of the domain was performed by refining finite element grid to 7128 nodes and 13 920 elements. This finite element code was executed on a super minicomputer by Floating Point Systems Incorporation. The computer simulations begin with an initial tracer concentration distribution at 4.44 days after injection. The results are illustrated in Figure 1.Numerical dispersion is reduced significantly and the observed variations in the tracer plume are predicted. The capability of the finite element model to simulate the observed data is greatly enhanced as it is clear from Figure 1 (only the well-behaved features are plotted in this figure).

CONCLUSIONS

The finite element model is appropriate for simulating the steep concentration gradients and the fingerlike spreading of the tracer plume provided the requirements to avoid oscillation, numerical dispersion and negativity of concentration are met. It is essential to calibrate the flow model against measured hydraulic head and velocities. Numerical dispersion is reduced significantly following a refinement of the mesh. A drawback, more accurate simulation of the velocity field is required to simulate the fingerlike spreading of the plume.

Given a specific mesh, the solute transport model is sensitive to the choice of dispersivity and time steps. The simulations can be conveniently summarized and compared with observed data by assessing integrated measures of solute transport based on spatial moments calculations.

REFERENCES

1. Moltyaner, G.L. (1987). Mixing Cup and Through-The-Wall Measurements in Field-Scale Tracer Tests and their Related Scales of Averaging. J. Hydrol, No. 89 pp. 281-302.

2. Taylor, S.R., Moltyaner, G.L., Howard, K.W.F. and Killey, R.W.D. (1987). A Comparison of Field and Laboratory Methods for Determining Contaminant Flow Parameters, Ground Water, Volume 25, No. 3, pp. 321-330.

3. Pinder, G.F. and Gray, W.G. Finite Element Simulation in Surface and Subsurface Hydrology. Academic Press, New York.

4. Frind, E.O. and Hokkanen, G.E. (1987). Simulation of the Borden Plume Using the Alternating Direction Galerkin Technique. Water Resour. Res., Vol. 23, No. 5, pp. 918-930.

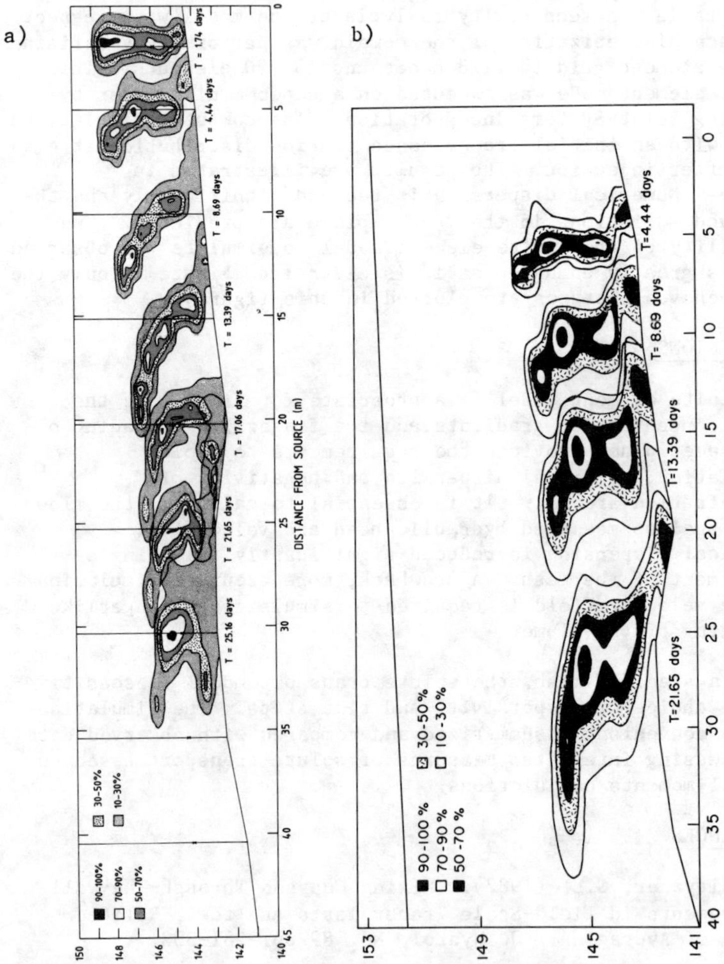

Figure 1. Observed (a) and Simulated (b) concentration distribution (Concentration distribution observed at 4.44 days is taken as initial one).

Modeling Water and Contaminant Transport in Unconfined Aquifers

G. Pantelis

Environmental Science Division, Australian Nuclear Science and Technology Organisation, Private Mail Bag 1, Menai 2234, NSW, Australia

Introduction

There are many available computer programs for investigating saturated-unsaturated water flow in porous media. These are usually based on a Galerkin finite element method which allows for such generalities as anisotropic and heterogeneous soil properties, general boundary conditions and complicated boundary domains [1]. However the use of these computer codes to simulate long-time three-dimensional water transport in unconfined aquifers can take so much computer time that they become impractical to use. This is largely due to the fact that even under homogeneous and isotropic conditions the equations describing the flow in the unsaturated zone are highly nonlinear. Therefore it is desirable to seek simpler models which retain much of the complexity and detail of the problem.

Here we shall consider the flow on the large timescale $D/(\varepsilon K_s)$, where D is the vertical length scale of the aquifer, K_s the characteristic saturated hydraulic conductivity of the aquifer, $\varepsilon=(D/L)^2 \ll 1$ and L is the horizontal length scale of the aquifer. We shall divide the flow domain into the saturated zone $\Omega_1(t')$, and an unsaturated zone $\Omega_2(t')$, which both may vary with time t'. (All dimensional variables will be primed and nondimensional variables will be unprimed). The two zones are separated by a surface $\Gamma(t')$, and the unconfined aquifer will be bounded below by an impermeable bed $\Gamma^{(1)}$, and above by the ground surface $\Gamma^{(2)}$ (see Figure 1). For $T>0$ and $0 \le t' \le T$ let

$$\Omega(t')=\bigcup_{i=1}^{2}\Omega_i(t'),\ \Omega_T=\bigcup_{0<t'\le T}\Omega(t'),$$
$$\Gamma_T=\bigcup_{0<t'\le T}\Gamma(t'),\ \Gamma_T^{(k)}=\Gamma^{(k)}\times(0,T],\ k=1,2.$$

If θ is the moisture content, K' is the hydraulic conductivity and ψ' is the water matric head then

$$\frac{\partial\theta}{\partial t'}= div\,[K'\,grad\,(\psi'+z')] \quad in\ \Omega_T \tag{1}$$

The usual Cartesian coordinates will be denoted by (x',y',z') with z' pointing vertically upwards. Since both θ and K' are functions of the matric head in the unsaturated zone Ω_2, and are equal to the constants θ_s, K_s (resp.) in the saturated zone Ω_1, it is seen that in Ω_1 the left hand side of (1) vanishes. On the upper and lower boundaries of the aquifer are imposed the boundary conditions

$$\underline{n}^{(k)}\cdot K'\,grad\,(\psi'+z')=Q^{(k)} \quad on\ \Gamma_T^{(k)},\ k=1,2, \tag{2}$$

where $\underline{n}^{(k)}$ is the upward unit normal vectors to the sufaces $\Gamma^{(k)}$, k=1,2. The quantity $Q'^{(2)}=Q'^{(2)}(x',y',t',\psi')$ can be thought of as the recharge rate and $Q'^{(1)}=Q'^{(1)}(x',y',t',\psi')$ can be thought of as the amount of flux of water across the surface $\Gamma^{(1)}$ if it is considered significant. In most problems this can be set to zero. On the surface $\Gamma(t')$ the matric head is set to zero and the matric head and normal flux are continuous, i.e.

$$[\psi']_\Gamma=[\![\psi']\!]_\Gamma=\underline{n}\cdot[\![K'grad(\psi'+z')]\!]_\Gamma=0, \tag{3}$$

where $[\![\;]\!]_\Gamma$ denotes the discontinuity across the surface Γ and $\underline{n}$ is the unit normal vector on Γ. While it can be assumed that θ and K' are analytic functions of the matric head in Ω_2 and continuous functions across Γ, i.e $[\![\theta]\!]_\Gamma=[\![K']\!]_\Gamma=0$, both θ and K' may have discontinuous derivatives with respect to ψ' across Γ.

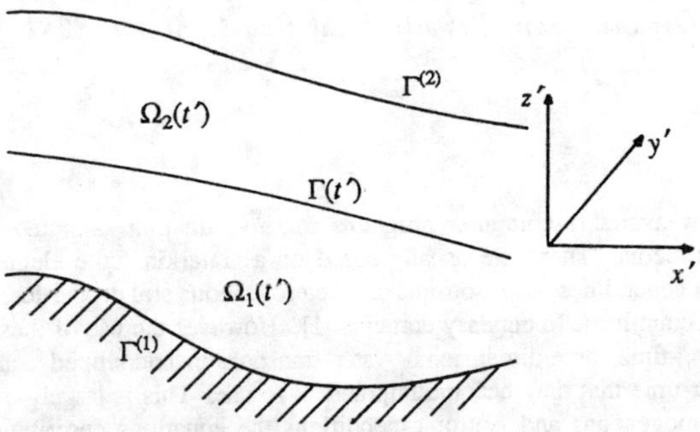

Figure 1

The following nondimensional variables are introduced:

$$x=x'/L\,,\ y=y'/L\,,\ z=z'/D\,,\ K=K'/K_s\,,\ \psi=\psi'/D\,,$$
$$b^{(k)}=b'^{(k)}/D\,,\ f=f'/D\,,\ t=t'/[D/(\varepsilon K_s)],\ Q^{(k)}=Q'^{(k)}/(\varepsilon K_s), \tag{4}$$

where $z'=b'^{(k)}(x',y')$ are the equations of the surfaces $\Gamma^{(k)}$, k=1,2, and $z'=f'(x',y',t')$ is the equation of the surface Γ. The last scaling in (4) is appropriate if $Q'^{(k)}$, k=1,2 are taken as some time averages over the timescale $D/(\varepsilon K_s)$. In terms of these nondimensional variables (1)-(3) become (resp.)

$$\varepsilon\frac{\partial\theta}{\partial t}=\varepsilon\nabla_{xy}\cdot(K\nabla_{xy}\psi)+\frac{\partial}{\partial z}[K(\frac{\partial\psi}{\partial z}+1)] \quad in\ \Omega_T \tag{5}$$

$$-\varepsilon K\nabla_{xy}b^{(k)}\cdot\nabla_{xy}\psi+K(\frac{\partial\psi}{\partial z}+1)=\varepsilon Q^{(k)} \quad on\ \Gamma_T^{(k)},\ k=1,2 \tag{6}$$

$$[\psi]_\Gamma=[\![\psi]\!]_\Gamma=[\![-\varepsilon K\nabla_{xy}f\cdot\nabla_{xy}\psi+K(\frac{\partial\psi}{\partial z}+1)]\!]_\Gamma=0, \tag{7}$$

where $\nabla_{xy}\equiv(\partial/\partial x,\partial/\partial y)$. In [2] it is demonstrated that

$$\psi(x,y,z,t;\varepsilon)=\phi(x,y,t)-z+O(\varepsilon),\quad f(x,y,t;\varepsilon)=\phi(x,y,t)+O(\varepsilon), \tag{8}$$

where the quantity $\phi(x,y,t)$ is obtained from the solution of the parabolic equation

$$\frac{\partial k}{\partial t}+\nabla_{xy}\cdot\underline{q}=Q \tag{9}$$

where $Q=Q^{(1)}(x,y,t,\phi-z)+Q^{(2)}(x,y,t,\phi-z)$ and

$$k=\int_{b^{(1)}}^{b^{(2)}}\theta(\phi-z)dz\,,\quad \underline{q}=-\int_{b^{(1)}}^{b^{(2)}}K(\phi-z)dz\ \nabla_{xy}\phi. \tag{10}$$

The expansions (8) state that on the long timescale the pressure head is essentially at equilibrium along the aquifer verticals. This is reasonable since a separate analysis shows that, for any isolated rainfall event, the vertical equilibration time occurs on the timescale $D/K_s \ll D/(\varepsilon K_s)$.

The modification of the analysis in [2] to take account of the case where the entire aquifer vertical is unsaturated is obvious. In such a case (8)-(10) are still valid except that the water table height is not defined. For the case where the entire aquifer vertical is saturated the flux boundary condition on the upper surface $\Gamma^{(2)}$ is replaced by the boundary condition $\psi=0$ *on* $\Gamma_T^{(2)}$. Following the procedure outlined in [2] with this modification reveals again the expansions of (8), but (9) is no longer required as ϕ is directly given as $\phi=b^{(2)}(x,y)$. Thus any solution of (9) must be accompanied by the constraint $\phi(x,y,t)\le b^{(2)}(x,y)$, $t\ge 0$, if full vertical saturation of the aquifer is expected.

Specific discharge field

It has so far been shown that on the long timescale, $D/(\varepsilon K_s)$, the system (1)-(3) governing saturated-unsaturated flow in unconfined aquifers can be reduced, in the first approximation, to the single parabolic equation (9). In fact it is shown in [2] that in the limit of decreasing capillarity (9) reduces to the Boussinesq equation which is used to describe only saturated flow and in which the water table is treated as a free surface [3].

The horizontal and vertical specific discharges $\underline{v}_{xy}'$ and v_z' (resp.), can be expressed in terms of the above nondimensional variables through Darcy's Law

$$\underline{v}_{xy}'=\varepsilon^{1/2}K_s(-K(\psi)\nabla_{xy}\psi),\quad v_z'=K_s[-K(\psi)(\frac{\partial\psi}{\partial z}+1)]. \tag{11}$$

Substitution of the expansion for ψ of (8) into the first expression of (11) and then expanding gives

$$\underline{v}_{xy}=\underline{v}_{xy}'/(\varepsilon^{1/2}K_s)=-K(\phi-z)\nabla_{xy}\phi+O(\varepsilon)\quad in\ \Omega_T. \tag{12}$$

To get v_z' an expression for $K(\psi)[\partial\psi/\partial z+1]$ is first obtained by integrating both sides of (5) with respect to z, $b^{(1)}\le z<b^{(2)}$, and using (6)-(7). Substituting this into (11) and recalling the expansion for ψ of (8) is obtained

$$v_z=v_z'/(\varepsilon K_s)=-\frac{\partial}{\partial t}[\int_{b^{(1)}}^{z}\theta(\phi-\zeta)d\zeta]+\nabla_{xy}\cdot[\int_{b^{(1)}}^{z}K(\phi-\zeta)d\zeta\nabla_{xy}\phi]$$
$$-Q^{(1)}(x,y,t,\phi-z)+O(\varepsilon)\quad in\ \Omega_T. \tag{13}$$

In (13) the Leibniz rule for the differentiation of integrals has been employed.

The advantage of these asymptotic approximations is that while ϕ is determined from (9), which only involves two horizontal space coordinates, the knowledge of ϕ is sufficient to provide the first approximation of the three-dimensional specific discharge field in both the unsaturated and saturated zones (through (12)-(13)). This discharge field can then be accessed by the contaminant transport component of the model which involves the solution of the convection-dispersion equation.

Convection-dispersion equation

In dimensional form the convection-dispersion equation is given by

$$\frac{\partial(\theta c')}{\partial t'}+\frac{\partial}{\partial x_i'}(v_i'c'-\theta D_{ij}'\frac{\partial c'}{\partial x_j'})=S',\quad in\ \Omega_T, \tag{14}$$

where c' is the concentration, S' is a source/sink term and D_{ij}' is the hydrodynamic dispersion coefficient which is defined by

$$\theta D_{ij}'=a_T'v'\delta_{ij}+(a_L'-a_T')v_i'v_j'/v'. \tag{15}$$

Here $i,j=1,2,3$ with $(x_1,x_2,x_3)\equiv(x,y,z)$ and summation of repeated indices is assumed. The quantities $a_T'(\theta)$ and $a_L'(\theta)$ are the transverse and longitudinal dispersivities (resp.) which are functions of the moisture content θ and $v'=(\underline{v}'\cdot\underline{v}')^{1/2}$. The additional nondimensional quantities are introduced:

$$c=c'/c_0,\ a_T=a_T'/D\,,\ a_L=a_L'/D\,,\ S=S'/(\varepsilon K_s c_0/D), \tag{16}$$

where c_0 is some characteristic concentration of the system. Using the nondimensional variables of (4) and (16) and recalling the expansions (8),(12) and (13) the convection-dispersion equation becomes

$$\varepsilon^{1/2}\frac{\partial[(\theta(\phi-z)+O(\varepsilon))c]}{\partial t}+\frac{\partial J_i}{\partial x_i}=\varepsilon^{1/2}S,\quad in\ \Omega_T \tag{17}$$

where J_i is the sum of the advective and dispersive fluxes (nondimensional) given by

$$J_i=\varepsilon^{1/2}v_i c-D_T\delta_{i3}\frac{\partial c}{\partial z}-\varepsilon D_{ij}\frac{\partial c}{\partial x_j},\quad (i,j=1,2,3) \tag{18}$$

and

$$D_T=a_T(\theta(\phi-z))K(\phi-z)(\nabla_{xy}\phi\cdot\nabla_{xy}\phi)^{1/2} \tag{19}$$

In (18) $(v_1,v_2,v_3)\equiv(v_x,v_y,v_z)\equiv(\underline{v}_{xy},v_z)$ are the components of the nondimensional specific discharges given in (12) and (13). The quantity D_T is an expression for the first approximation of the hydrodynamic dispersion coefficient (15) involving the transverse dispersivity a_T. The higher order terms of $\theta D_{ij}'$ are contained in the elements of D_{ij} appearing in (18) ($D_{ij}=D_{ij}(x,y,z,t;\varepsilon)=O(1)$ for $\varepsilon\to 0$). It is these higher order terms ($O(\varepsilon)$ terms of (17)) which contain expressions involving the longitudinal dispersivity a_L. Thus it is expected that any solution of (17) will not be sensitive to longitudinal dispersion. A most important feature which has been overlooked too often is that while the horizontal discharge is much larger than the vertical discharge (by $O(\varepsilon^{1/2}K_s)$) the vertical advection term is as important as the horizontal advection term, i.e. both are contained in terms $O(\varepsilon^{1/2})$ of (17). Finally, the dominant term in (17) is the transverse dispersion.

It must be understood that the relative importance of these processes applies only to the order that they appear in the convection-dispersion equation which has been nondimensionalised to treat the problem on the long-time $D/(\varepsilon K_s)$ and over the regional scale, i.e. over the horizontal length scale L. Of course on the local scale the dominance of these physical processes appear in a different order.

While these approximating equations for the water and contaminant transport have significantly simplified the original formulation, nonlinearity remains and numerical techniques must be employed. There are many standard techniques for the numerical solution of (9), e.g. Galerkin method [4]. However, as already pointed out the solution of (9) should be accompanied by the additional constraint $\phi(x,y,t)\le b^{(2)}(x,y)$, $t\ge 0$, if full saturation of the aquifer vertical is expected. If (9) were linear parabolic the problem could be set up as a variational inequality [5]. Unfortunately the literature on the analysis of constrained problems involving nonlinear parabolic equations such as (9) is limited and therefore a rigorous examination of the numerical solution of this problem is needed.

Having obtained ϕ and $\nabla_{xy}\phi$ numerically it is then straightforward to calculate the specific discharges from (12) and (13) which can then be used in the convection-dispersion equation. It appears that a weak formulation of (17) would be appropriate, seeking a solution for c in a class satisfying $[\![c]\!]_\Gamma=[\![n_iJ_i]\!]_\Gamma=0$. Having developed a weak formulation of (17) it would then seem reasonable to omit terms $O(\varepsilon)$ thus retaining terms up to $O(\varepsilon^{1/2})$ which do not involve longitudinal dispersion. This procedure, apart from simplifying the mathematical formulation of the problem, removes the necessity of obtaining values for the longitudinal dispersivity which is often difficult to measure in the field.

References

[1] Neuman, S.P.(1975) - Galerkin approach to saturated-unsaturated flow in porous media, in Finite Elements in Fluids - Vol 1 (R.H. Gallagher. J.T. Oden, C. Taylor and O.C. Zienkkiewicz, Eds.), John Wiley, London.

[2] Pantelis, G.(1985) - Saturated-unsaturated flow in unconfined aquifers, Journal of Applied Mathematics and Physics (ZAMP), 36, 648-657.

[3] Bear, J. (1972) - Dynamics of Fluids in Porous Media, American Elsevier, New York.

[4] Luskin, M.(1979) - A Galerkin method for nonlinear parabolic equations with nonlinear boundary conditions, SIAM J. Numer. Anal., 16, 285-299, 1979.

[5] Glowinski, R., Lions, J.L., Tremolieres, R. (1981) - Numerical Analysis of Variational Inequalities, North Holland, New York.

Accurate Fine-Grid Simulations to Derive Coarse-Grid Models of Fine-Scale Heterogeneities in Porous Media

T.F. Russell

Department of Mathematics, University of Colorado at Denver, Denver, CO 80204, USA

ABSTRACT

In the context of a model of miscible transport in a porous medium, a Galerkin modified method of characteristics and a mixed finite-element method are introduced. Previous theoretical and numerical studies demonstrating the accuracy of these methods are summarized. Future studies, to be presented in detail elsewhere, are outlined; these will examine the question of how to represent heterogeneities of widely varying length scales in an averaged model. Unstable fronts, which occur when a fluid is displaced by another fluid of lower viscosity, are emphasized; in this situation, averaging should take fluid properties as well as rock properties into account.

INTRODUCTION

One of the greatest difficulties in modeling of subsurface flows is the treatment of heterogeneity. Heterogeneities in permeability and porosity exist at all length scales ranging over several orders of magnitude.[1] Even if we had precise knowledge of permeability and porosity everywhere in a flow, which we never do, current modeling technology would not tell us how to simulate variations of smaller size than the discrete grid.

For simple flows, some theory is available as a guide. Homogenization can be applied to the Navier-Stokes equations to obtain Darcy's law.[2,3] For the microscopic convection-diffusion equation, a similar derivation is possible, leading to an equation of the same form in averaged variables.[4] Macroscopic dispersion (mechanical mixing due to tortuous flow paths) has been shown to be representable by a diffusion-like term with coefficient proportional to flow velocity when the permeability distribution is log-normal.[5]

As a practical matter, however, the theory does not yield the values of the relevant coefficients for an actual simulation. The theory also does not address unstable displacements, in which a resident fluid is pushed by a fluid of lower viscosity, leading to viscous fingering. This type of flow has been of interest to petroleum engineers for a long time,[6] and there are indications of interest in the hydrological community as well.[7] Such flows are controlled by interactions of heterogeneities, dispersion, and viscosity, and we do not see a theory on the horizon that will account for all of these effects. Accordingly, we pursue an empirical approach to this problem. This will use the methods to be described next.

METHODS

The governing equations for incompressible miscible transport in a porous medium with no-flow boundary conditions can be written in the form

$$\nabla\cdot\mathbf{v} = q, \qquad \mathbf{x}\in\Omega,\ 0<t<T, \tag{1}$$

$$\mathbf{v}\cdot\mathbf{n} = 0, \qquad \mathbf{x}\in\partial\Omega,\ 0<t<T, \tag{2}$$

$$\phi\ \partial c/\partial t - \nabla\cdot(D(\mathbf{v})\nabla c - \mathbf{v}c) = \tilde{c}q, \qquad \mathbf{x}\in\Omega,\ 0<t<T, \tag{3}$$

$$D(\mathbf{v})\nabla c\cdot\mathbf{n} - c(\mathbf{v}\cdot\mathbf{n}) = 0, \qquad \mathbf{x}\in\partial\Omega,\ 0<t<T, \tag{4}$$

$$c(\mathbf{x},0) = c_0(\mathbf{x}), \qquad \mathbf{x}\in\Omega. \tag{5}$$

In Equations (1) through (5), $c(\mathbf{x},t)$ is the concentration of a fluid that mixes with the resident fluid (e.g., a solute in water), and the Darcy velocity vector $\mathbf{v}(\mathbf{x},t)$ is given by Darcy's law

$$\mathbf{v} = -\ (K/\mu(c))\ (\nabla p - \rho g\nabla z). \tag{6}$$

We take the viscosity μ to be concentration-dependent in order to account for the effects of viscous fingering; in hydrology, this is of more interest in immiscible multiphase transport, and the methods to be described here have been extended to that case.[8] Inside the parentheses in Equation (6) are a pressure gradient and a gravity term. Other symbols are the porosity $\phi(\mathbf{x})$, the permeability tensor $k(\mathbf{x})$, the tensor $D(\mathbf{x},\mathbf{v})$ accounting for molecular diffusion and anisotropic hydrodynamic dispersion, the source density $q(\mathbf{x},t)$, and the source concentration $\tilde{c}$ (specified at injection locations, equal to c for production). The forms of μ and D (in two dimensions) are

$$\mu(c)^{-1/4} = c\mu(1)^{-1/4} + (1-c)\,\mu(0)^{-1/4}, \tag{7}$$

$$D(\mathbf{v}) = D_m I + \frac{\alpha_\ell}{|v|}\begin{bmatrix} v_x^2 & v_x v_y \\ v_x v_y & v_y^2 \end{bmatrix} + \frac{\alpha_t}{|v|}\begin{bmatrix} v_y^2 & -v_x v_y \\ -v_x v_y & v_x^2 \end{bmatrix}. \tag{8}$$

The usual primary variables are c and the pressure p(x,t).

Modified method of characteristics

This procedure was introduced in the mathematical literature by Douglas and Russell[9] for convection-dominated parabolic equations and by Russell[10] in the context of miscible displacement. Similar schemes have gone by the names of Eulerian-Lagrangian,[11] transport-diffusion,[12] Lagrange-Galerkin,[13] and combined finite-element method and method of characteristics.[14] It is convenient to replace Equation (3) by its nondivergence form

$$\phi\, \partial c/\partial t + \mathbf{v}\cdot\nabla c - \nabla\cdot(D(\mathbf{v})\nabla c) = (\tilde{c}-c)q, \tag{9}$$

obtained by expanding $\nabla\cdot(\mathbf{v}c)$ and applying Equation (1). The essential idea is to treat the hyperbolic portion (the first two terms) of Equation (9) as a directional (or total) derivative and to "time"-difference along the characteristics. The remaining diffusion-dispersion problem is solved by an implicit Galerkin procedure. The equation to advance the solution from time step n-1 to n can be written as

$$\int_\Omega \phi(c^n(\mathbf{x})-c^{n-1}(\hat{\mathbf{x}}))/\Delta t\; w\, dx + \int_\Omega D(\mathbf{v}^n)\nabla c^n\cdot\nabla w\, dx + \int_\Omega c^n q^n w\, dx = \int_\Omega \tilde{c}^n q^n w\, dx, \tag{10}$$

where c^n is the unique function in the finite-element space S satisfying Equation (10) for all test functions w in S. The point $\hat{\mathbf{x}}$ is obtained by tracking <u>backward</u> in time along the characteristics of the velocity field **v**; if the tangent to the characteristic at time step n is used, this yields

$$\hat{\mathbf{x}} = \mathbf{x} - (\mathbf{v}^n(\mathbf{x})/\phi(\mathbf{x}))\Delta t. \tag{11}$$

The tracking can be done more accurately by closely following the curvature of the characteristics.[15]

Backward tracking makes the method easy to implement in two and three dimensions. All occurrences of the new time step in Equation (10) are of standard form, so contorted meshes are not necessary. The only unusual term, $c^{n-1}(\hat{\mathbf{x}})$, involves evaluation of the previous solution at an easily determined point and goes to the right-hand side. If the integral of this term is computed accurately, the procedure is not numerically diffusive; it amounts to characteristic tracking, least-squares fitting of the function $\hat{c}^{n-1}(\mathbf{x}) = c^{n-1}(\hat{\mathbf{x}})$ by a member of S, and finally incorporation of physical diffusion. If the integral is computed inaccurately, for example by the trapezoidal rule, numerical diffusion results.

Assuming an accurate integral, the method empirically resolves fronts accurately with about 3 intervals across the range $c = 0.9$ to $c = 0.1$; this mesh of order $Pe^{-1/2}$, where Pe is the Peclet number, contrasts favorably with the order of Pe^{-1} required by the usual finite-difference and finite-element methods. Because the time step follows the flow, the time-truncation error of order $\partial^2 c/\partial \tau^2 \, \Delta t$ (where τ is the characteristic direction) is much less than the order of $\partial^2 c/\partial t^2 \, \Delta t$ for standard methods, so that large time steps are appropriate; in particular, the CFL condition can be ignored. An additional advantage of the scheme is that the implicit system in Equation (10) is symmetric, which is favorable for iterative solution techniques.

The principal disadvantage is that mass is not conserved exactly. This is readily apparent; fluid movement is determined directly by the velocity, which is known only approximately, and any errors in tracking will also affect conservation. Accuracy in conservation demands accurate velocities, motivating our other method.

Mixed finite-element method

A standard procedure would combine Equations (1) and (6) into an elliptic pressure equation, solving for p. Sharp discontinuities in the permeability would lead to inaccuracies in **v**, because the product of a rough permeability and a rough pressure gradient would attempt to approximate the smooth velocity. It is better to obtain an approximation of **v** directly from a mixed method as follows. Equations (1) and (6) are respectively discretized by

$$\int_\Omega \nabla \cdot \mathbf{v}^n \, w \, d\mathbf{x} = \int_\Omega q w \, d\mathbf{x}, \qquad (12)$$

$$\int_\Omega (\mu(c^n)/k) \, \mathbf{v}^n \cdot \chi \, d\mathbf{x} - \int_\Omega p^n \, \nabla \cdot \chi \, d\mathbf{x} = \int_\Omega \rho g \nabla z \cdot \chi \, d\mathbf{x}, \qquad (13)$$

where $\mathbf{v}^n$ and p^n are the unique functions in certain finite-element spaces W and X satisfying Equations (12) and (13) for all test functions w in W and χ in X. Details of the spaces and of a specialized formulation to handle point sources are given elsewhere.[15]

Note that there is a circular dependence between c^n and $\mathbf{v}^n$ in Equations (10) and (13). We have circumvented this by using an extrapolated $\mathbf{v}^n$, based on time steps n-1 and n-2, in Equation (10). The overall sequential time-stepping procedure proceeds by solving Equation (10) for c^n (with extrapolated $\mathbf{v}^n$) and then Equations (12) and (13) for $\mathbf{v}^n$ and p^n (using the new c^n). Because **v** changes relatively slowly in time, extrapolating **v** should not incur serious errors.

SUMMARY OF RESULTS

Convergence theorems have been proved for these methods.[17,18] Assuming sufficiently smooth solutions, the theorems show that the difference between the computed concentration and the true concentration, in the $L^2(\Omega)$-norm, is of the order of $h^{k+1} + \Delta t$, where h is the spatial mesh size and k is the degree of finite elements used. The coefficient of Δt depends on derivatives of c with respect to τ, not t, and of **v**, not p. These convergence rates are optimal. When no lower bound is assumed for D, the theorems yield $h^k + \Delta t$, so that one order of spatial convergence is lost.

Numerical studies have yielded very accurate results.[15,16] As expected theoretically, meshes can be coarser than those of standard methods and time steps can be one to two orders of magnitude larger. For high-mobility-ratio problems ($\mu(0)/\mu(1)$ ranging from 41 to 1000) with severe tendencies toward viscous fingering, answers converge as grids and time steps are refined. For homogeneous media (permeability independent of **x**) with finite dispersion, the answers converge to stable unfingered fronts, as they should; dispersion prevents the growth of short-wavelength disturbances arising from small numerical errors. For slightly heterogeneous media, answers converge to fingered fronts; the homogeneous tests show that these results reflect physical fingering and not effects of numerical errors.

OUTLINE OF FUTURE STUDIES

We outline here the thrust of further work, which we cannot discuss in detail for lack of space. We would like to understand how heterogeneities, dispersion, and mobility ratio combine to determine displacement behavior. We believe that our numerical methods have demonstrated sufficient accuracy to be able to model the subtleties of the physics. The idea is to use these methods on very fine grids and then to empirically match the fine-grid results with coarse-grid simulations, thereby finding effective averaged parameters for coarse grids.

Fine-grid computations will use different log-normal permeability distributions with the same variance to see whether they lead to qualitatively similar transport and fingering behavior; if so, then it makes sense to use a deterministic model. Different variances will show how the results depend on the variance. The dependence should be influenced by the values of the mobility ratio $\mu(0)/\mu(1)$, the longitudinal dispersivity α_l, and the transverse dispersivity α_t; these will also be varied systematically. We hypothesize that an effective coarse-grid dispersion term of some form can account for the fingering caused by fine-scale heterogeneities; this will be tested by attempting generalizations of Equation

(8) on coarse grids. It is conceivable that the physical instability is such that fine-scale fingering leads to widely varying behavior at the coarse scale, in which case this program is doomed to failure; we are optimistic enough to believe that this is not so.

REFERENCES

1. Claridge E.L. (1972), Soc. Pet. Eng. J., Vol.12, pp. 352-361.
2. Gray W.G. and O'Neill K. (1976), Water Resour. Res., Vol.12, pp. 148-154.
3. Bensoussan A. Lions J.L. and Papanicolaou G. (1978). Asymptotic Analysis for Periodic Structures. North-Holland, Amsterdam.
4. Gray W.G. (1975), Chem. Eng. Sci., Vol.30, pp. 229-233.
5. Warren J.E. and Skiba F.F. (1964), Soc. Pet. Eng. J., Vol.4, pp. 215-230.
6. Russell T.F. and Wheeler M.F. (1984). Chapter 2, The Mathematics of Reservoir Simulation (Ed. Ewing R.E.), pp. 35-106, Society for Industrial and Applied Mathematics, Philadelphia.
7. Pinder G.F. and Abriola L.M. (1986), Water Resour. Res., Vol.22, pp. 109S-119S.
8. Dahle H. Espedal M. and Ewing R.E. (1988). In Proceedings of Symposium on Numerical Simulation in Oil Recovery (Ed. Wheeler M.F.), Springer-Verlag, Berlin.
9. Douglas J. Jr. and Russell T.F. (1982), SIAM J. Numer. Anal., Vol.19, pp. 871-885.
10. Russell T.F. (1980), Ph.D. Thesis, University of Chicago; SIAM J. Numer. Anal., Vol.22(1985), pp. 970-1013.
11. Neuman S.P. (1981), J. Comp. Phys., Vol.41, pp. 270-294.
12. Pironneau O. (1982), Numer. Math., Vol.38, pp. 309-332.
13. Morton K.W. Priestley A. and Süli E. (1986), Report No. 86/14, Oxford University Computing Laboratory, U.K.
14. Baptista A.M. Adams E.E. and Stolzenbach K.D. (1984). In Proceedings of 5th International Conference on Finite Elements in Water Resources (Ed. Laible J.P. et al.), Burlington, VT.
15. Ewing R.E. Russell T.F. and Wheeler M.F. (1983). In Proceedings of 7th SPE Symposium on Reservoir Simulation, Society of Petroleum Engineers, Dallas.
16. Russell T.F. Wheeler M.F. and Chiang C. (1986). Mathematical and Computational Methods in Seismic Exploration and Reservoir Modeling (Ed. Fitzgibbon W.E.), Society for Industrial and Applied Mathematics, Philadelphia.
17. Ewing R.E. Russell T.F. and Wheeler M.F. (1984), Comp. Meth. Appl. Mech. Eng., Vol.47, pp. 73-92.
18. Dawson C.N. Russell T.F. and Wheeler M.F. Some Improved Estimates for the Modified Method of Characteristics. SIAM J. Numer. Anal., submitted.

Numerical Experiment with Euler-Lagrange Method for a Pair of Recharge-Pumping Wells

S. Sorek

Faculty of Bio-Medical Engineering, The Technion, Haifa 32000, Israel

ABSTRACT

The EL (Euler-Lagrange) method is used to numerically solve the transport problem of a pair of recharge-pumping wells. For the sake of comparison with a different numerical scheme, physical parameters (e.g. Peclet numbers of approximately 18 near the injection well) and discretization data (refined rectangle elements at the wells locations) are kept the same. The EL scheme deploys at each time step, rays of particles commencing from the injection well and carrying the advective part of the concentration. The EL method proves to be converging consistently and superior in less deviation from the analytical solution together with higher Courant numbers (i.e. CPU efficiency).

INTRODUCTION

The mixed Eulerian-Lagrangian method has been gaining popularity in recent years. The method combines the simplicity of the fixed Eulerian grid with the Lagrangian approach being especially effective in regions with high Peclet numbers.

Following Neuman and Sorek for transport problems,[1,2,3,4,5,6] Sorek for flow problems,[3,4,7] Sorek and Braester for multiphase flow and transport[8] a technique consisting of the following two steps is used:

(a) Formal decomposition of the dependent variable into two parts, one controlled by pure Lagrangian advection and a residual governed by a combination of Euler-Lagrange approaches.

(b) Solution of the resulting advection problem, by the method of characteristics for forward particle tracking. The remaining problem is solved by an implicit finite element scheme on a fixed grid.

Information is projected back and forth between the Eulerian-Lagrangian and the Lagrangian schemes.

To understand the concept of such a decomposition, consider a governing equation of the form

$$Lg = f \tag{1}$$

where L is the partial differential operator; g is the dependent variable, and f may be regarded as a source term.

The approach now is to decompose g into a translative term g_T and a residual term g_R

$$g = g_T + g_R \quad . \tag{2}$$

Decouple the differential operator into

$$L = L_T + L_R \tag{3}$$

where L_T and L_R are the translative and residual differential operators respectively.

In view of eqs. (1)-(3) we write

$$(L_T + L_R)\,(g_T + g_R) = f \tag{4}$$

One way to perform the decomposition is to allow

$$L_A\, g_T = 0 \tag{5}$$

The term g_T is then solved along characteristic lines defined by

$$L_T \underline{X} = \underline{V} \tag{6}$$

where $\underline{X}$ is the vector of spatial position, and $\underline{V}$ is the velocity vector tangent to the characteristic path. The location $\underline{X}$ is known only if $\underline{V}$ is known, otherwise iterations are required.

By virtue of (5), the next step is to solve the equation

$$L g_R = f - L_R\, g_T \tag{7}$$

However, we may expect the condition

$$L g_R \gg L_R\, g_T \tag{8}$$

In view of (7) and (8), we obtain

$$L g_R \cong f \tag{9}$$

which is similar to a Poisson equation and is not expected to suffer from stability difficulties typical to advection dominant flow regimes. In cases with dominant source, terms one may perform the decomposition, expressed in (4), by allowing g_T to obey

$$L_T\, g_T = f \tag{10}$$

By virtue of (7) and (8), we then obtain the Laplacian form

$$L g_R \cong 0 \tag{11}$$

which is more stable than equation (9).

The major problem of the decomposition strategy is the interpolation between the residual and translation solutions (as these two solutions are obtained at different spatial locations).

The interpolation (coupling between the solutions) should be performed in a mass conservation fashion to allow a conservative solution.

The following paragraphs will describe the theory of adopting the above mentioned method to the mass transport problem and its exemplification a pair of recharge-pumping wells.

THEORY

Consider the advection-dispersion equation for an effluent patch in saturated domain and incompressible flow

$$(l+s)\,\frac{\partial(\varepsilon C)}{\partial t} = \nabla\cdot(\varepsilon \underline{\underline{D}}\nabla C - \varepsilon \underline{q} C) - \varepsilon\lambda C + Q_R C_R + Q_{Rech} C_{Rech} - Q_P C \;;\; \underline{X}\,\varepsilon B x(0,\tau] \quad (12)$$

where $(l+s) > 0$ is the retardation factor; ε - porosity coefficient; t - time starting at $t=0$ ending at $t=\tau$ ((0,τ]); C - solute concentration; ∇ - gradient operator; $\underline{\underline{D}}$ - tensor of hydrodynamic dispersion (symmetric and semi-definite positive); $\underline{V}$ - fluid velocity vector; $\lambda > 0$ - radioactive decay coefficient; Q_R, Q_{Rech} are rain and recharge flux terms respectively with C_R and C_{Rech} as their associative concentrations; Q_P is the pumpage flux term and B - time and space domain.

In equation (12) the velocity of flow is governed by Darcy's Law

$$\underline{q} = \varepsilon \underline{V} = -\underline{\underline{K}}\nabla(P+\gamma Z) = -\underline{\underline{K}}\nabla(\gamma h) \quad (13)$$

Here, $\underline{q}$ - Darcy's specific flux vector; $\underline{\underline{K}}$ - tensor of hydrodynamic permeability; P - pressure; γ - fluid's specific weight; Z - elevation; h - $(=\dfrac{P}{\gamma}+Z)$ hydraulic head.

The hydrodynamic dispersion in dimension x and y can be written as

$$D_{xx} = D_m + \frac{\alpha_L V_x^2 + \alpha_T V_y^2}{|\underline{V}|} \quad (14.1)$$

$$D_{yy} = D_m + \frac{\alpha_L V_y^2 + \alpha_T V_x^2}{|\underline{V}|} \quad (14.2)$$

$$D_{xy} = (\alpha_L - \alpha_T)\,\frac{V_x V_y}{|\underline{V}|} \quad (14.3)$$

Here, D_m - molecular diffusion; α_L, α_T - longitudinal and transverse dispersivities.

Equation (12) is to be solved for C, subject to Cauchy initial condition and boundary conditions, respectively,

$$C_{(\underline{x},o)} = C_{o(\underline{x})} \qquad \underline{X}\,\varepsilon R \;;\; t=0 \quad (15)$$

$$\begin{aligned}(-\varepsilon \underline{\underline{D}}\nabla C + \underline{q} C)\cdot \underline{n} + \alpha_\Gamma(t)\cdot(C - C_\Gamma(t)) &= \\ = (Q_R(t)C_R)_\Gamma + (Q_{Rech}(t)C_{Rech})_\Gamma - Q_{P_\Gamma} C & \qquad \underline{X}\,\varepsilon\Gamma x(0,\tau]\end{aligned} \quad (16)$$

where, R - spatial domain at t=constant; Γ - boundary line surrounding R; $\underline{n}$ - unit vector perpendicular to Γ directed outward; $C_o, C_\Gamma, (Q_R C_R)_\Gamma, (Q_{Rech} C_{Rech})_\Gamma, Q_{P_\Gamma}$ are prescribed functions, and α_Γ $(0 \le \alpha_\Gamma \le \infty)$ controls the type of boundary condition prevailing along Γ.

Let us now write fluid's equation of continuity

$$\frac{\partial \varepsilon}{\partial t} + \nabla\cdot\underline{q} = Q_R + Q_{Rech} - Q_P \quad (17)$$

multiplying by C, $(C \neq 0)$ and subtracting from (12), we obtain

$$\varepsilon\,\frac{\partial C}{\partial t} + s\,\frac{\partial(\varepsilon C)}{\partial t} = \nabla\cdot(\varepsilon \underline{\underline{D}}\nabla C) - \underline{q}\cdot\nabla C - \varepsilon\lambda C + Q_R(C_R - C) + Q_{Rech}(C_{Rech} - C) \quad (18)$$

Note that terms involving Q_P disappeared, also when the field concentration equals the source concentrations, the source terms in (18) are omitted.

We define the hydrodynamic derivative

$$\frac{D}{Dt} = \frac{\partial}{\partial t} + \underline{V}^* \cdot \nabla \tag{19}$$

with a velocity term of the form

$$\underline{V}^* = \underline{V} + \underline{V}' = \underline{V} - \frac{1}{\varepsilon}\nabla(\varepsilon \underline{\underline{D}}) \tag{20}$$

and rewrite (18)

$$\left[\frac{D}{Dt} + \frac{S}{\varepsilon}\frac{\partial(\varepsilon C)}{\partial t} \underline{\underline{D}}\nabla^2 + \lambda + \frac{1}{\varepsilon}(Q_R + Q_{Rech})\right]C = \frac{1}{\varepsilon}(Q_R C_R + Q_{Rech} C_{Rech}) \tag{21}$$

The additional velocity term $\underline{V}'$ in (20) is expected to drive particles out of stagnation points.

In equivalence to (2)-(6) the Lagrangian operator reads

$$L_T \equiv \frac{D}{Dt} \tag{22.1}$$

the characteristic path is described by

$$D\underline{X} = \underline{V}Dt \tag{22.2}$$

and concentration is decomposed into an advection part $\overline{C}$ and a residual $\overset{o}{C}$ regarded as the dispersion part

$$C = \overline{C} + \overset{o}{C} \tag{23}$$

In view of (5), (15), (16) and (22.1) $\overline{C}$ satisfies

$$\frac{D\overline{C}}{Dt} = 0 \tag{24}$$

subject to initial condition

$$\overline{C}_{(\underline{X},o)} = C_{o(\underline{X})} \tag{25}$$

and conditions along boundaries

$$q\overline{C}\cdot\underline{n} + \alpha_\Gamma(\overline{C} - C_\Gamma) = (Q_{Rech} C_{Rech})_\Gamma + (Q_R C_R)_\Gamma - Q_{P_\Gamma}\overline{C} \tag{26}$$

Condition along outflow boundaries have no effect on $\overline{C}$ and are thus irrelevant. The $\overline{C}$ value remains constant in time, thus the "advection problem" as given by (24)-(26) is first solved within each time step independently from the "dispersion problem" denoted by $\overset{o}{C}$.

The residual concentration $\overset{o}{C}$, is then evaluated by (21) and the already known $\overline{C}$ field which may be regarded as prescribed source terms.

By virtue of (15)-(16), (25)-(26) initial condition for $\overset{o}{C}$ is

$$\overset{o}{C}_{(\underline{X},o)} = 0 \tag{27}$$

conditions along boundaries are as in (16) with boundary source functions generated by the solved $\overline{C}$ field.

After sharp fronts of C were flattened, we solve adaptively (21), (22.2) in its Lagrangian form without decomposing C into $\overline{C}$ and $\overset{o}{C}$.

It is beyond the scope of this article to present the numerical implementation of the theory. Detailed description of the numerical schemes employed to the advective and dispersive parts may be found in Sorek[5].

In the following example the $\underline{V}'$ of (20) was not generated, its influence will be described in an imminent paper.

EXAMPLE

The power of the Eulerian-Lagrangian method is exemplified on a numerical solution of transport between two wells. The pumping and injection wells are situated in a 2-D infinite aquifer. For reasons of comparison, we follow the case presented by Huyakorn et al.[9]

The governing transport equation is given by

$$\frac{\partial}{\partial X}\left(D_{XX}\frac{\partial C}{\partial X}+D_{XY}\frac{\partial C}{\partial Y}\right)+\frac{\partial}{\partial Y}\left(D_{YY}\frac{\partial C}{\partial Y}+D\,XY\frac{\partial C}{\partial X}\right)+ \\ -\frac{V_x}{\varepsilon}\frac{\partial C}{\partial X}-\frac{V_y}{\varepsilon}\frac{\partial C}{\partial Y}=\frac{\partial C}{\partial t} \tag{28}$$

subject to initial condition and boundary condition at the injection well, respectively

$$C(X,Y,0)=0 \tag{29.1}$$

$$C(X,Y,t)=1 \tag{29.2}$$

For homogeneous and isotropic aquifer with a fully penetrating well doublet operating at constant flow rate Q, we obtain a flow regime of the form (see Huyakorn et al.[9]).

$$V_x=-\frac{Q}{2\pi b}\left[\frac{X-X_o}{(X-X_o^2)+Y^2}-\frac{X+X_o}{(X+X_o)^2+Y^2}\right] \tag{30.1}$$

$$V_y=-\frac{Qy}{2\pi b}\left[\frac{1}{(X-X_o)^2+Y^2}-\frac{1}{(X+X_o)^2+Y^2}\right] \tag{30.2}$$

where b is the aquifer thickness and X_o is half the well spacing assumed as the origin of the coordinate axes with X axis along the line joining the center of the two wells.

Physical and grid data are given in table 1. The values of flow parameters and the longitudinal dispersivity correspond to data described by Hoopes and Harleman[10].

By virtue of (30) and table 1, the local Peclet number ($P_i=\frac{|V_i|\Delta X_i}{D_{ii}}$) at the node adjacent to the injection well is, $P_X=18.7,\ P_y=17.$

In this simulation we have deployed particles in a 9 ray configuration depicted in figure 2. Each ray holds 15 equally spaced particles starting at the injection well and extending to a radius of 36.6 cm. The 9×15 particles were deployed at the beginning of each time step for the 15 first time steps from the total simulation time $T_f=75609$ Sec. which was divided to $K=17$ equal time steps, ($\Delta t=\frac{T_f}{K}=4447.6Sec$).

The time step scheme used by Huyakorn et al.[9] was
$\Delta t_1=50sec.\ \Delta t_k=1.2\,\Delta t_{k-1}\le 300Sec,\ k=2,3,\cdots K'$.

From the final non dimensional time $t_d = 9.09$ ($= Qt/2\pi\phi bX_o^2$) and table 1, we evaluate T_f from which we conclude that the maximum time steps used by Huyakorn et al.[9] was $K' \cong 300$.

The comparison of the Eulerian-Lagrangian numerical solution (ADVDSP), an analytical solution by Hoopes and Harleman[10] and the numerical solution (SEFTRAN) by Huyakorn et al.[9], is presented in figure 3.

This figure demonstrates the pronounced difference in terms of deviations from the analytical solution and the required number of time steps.

CONCLUSIONS

A EL method was developed to rigorously decompose the transport problem into advective and dispersive (residual) problems. It was demonstrated how to incorporate an additional velocity term in the Lagrangian scheme. The effectiveness of this additional particle velocity in cases with velocity stagnation points will be reported in the future.

The EL method proved to be superior in terms of solution reliability and efficiency of computer resources.

REFERENCES

1. Neumann, S.P. and Sorek, S. (1982), Eulerian-Lagrangian Methods for Advection-Dispersion, Proc. 4th Inter. Conf. F.E.W.R. FR Germany, 14.41-14.68.
2. Neuman, S.P. (1984), Adaptive Eulerian-Lagrangian finite element method for advection-dispersion, Int. J. Num. Meth. in Eng., 20, 321-337.
3. Sorek, S. (1985a), Eulerian-Lagrangian Formulation for Flow in Soils, Adv. W. Res. 8, 118-120.
4. Sorek, S. (1985b), Adaptive Eulerian-Lagrangian Method for Transport Problems in Soils, Scientific basis for water resources management, IAHS Pub. 153, 393-403.
5. Sorek, S. (1987a), 2-D Adaptive Eulerian-Lagrangian Method for Mass Transport with Spatial Velocity Distribution, (submitted for publication).
6. Sorek, S. (1987b), Eulerian-Lagrangian Method for Solving Transport in Aquifers, (submitted for publication).
7. Sorek, S. and Braester, C. (1986), An Adaptive Eulerian-Lagrangian Approach for the Numerical Simulation of Unsaturated Flow, Proc. 6th Inter. Conf. F.E.W.R. Portugal, 87-100.
8. Sorek, S. and Braester, C. (1987c), Eulerian-Lagrangian formulation of the equations for groundwater denitrification using Bacterial Activity, (submitted for publication).
9. Huyakorn, Peter S., et al. (1986), A Curvilinear Finite Element Model for Simulating Two-Well Tracer Tests and Transport in Stratified Aquifers, W. Resour. Res. 22, 5, 663-678.
10. Hoopes, J.A. and Harleman, D.R.F. (1967), Waste Water Recharge and Dispersion in Porous Media, J. Hydraulics Div. Am. Soc. Civ. Eng. 93, 51-71.

TABLE 1: Values of Physical Parameters and Discretization Data

Parameter	Value
Well flow rate, Q	2.339 cm^3/s
Well spacing, $2Xo$	61.0 cm
Thickness of aquifer model, b	8.9 cm
Porosity, ϕ	0.374
Molecular diffusion, D_m	0
Lateral dispersivity, α_T	0
Longitudinal dispersivity, α_L	0.294 cm
x_i (cm), i=1,2,...,29	-145., -120, -100, -80., -60., -50., -40 -35., -30.5., -25., -20., -15., -10., -5., 0., 5., 10., 15., 20., 25., 30.5, 35., 40., 50., 60., 80.., 100., 120., 145
y_j (cm) j=1,2,...,15	0., 5., 10., 15., 20., 25., 30., 35., 40., 50., 60., 80., 100., 120., 145

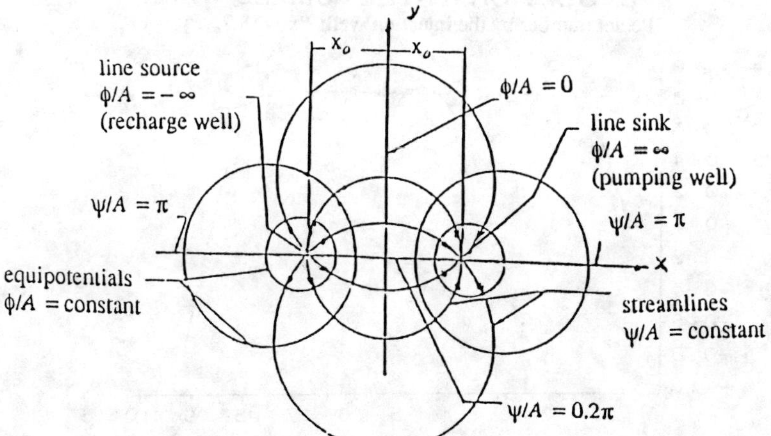

Figure 1: Stream and equipotential lines of injection and pumping wells

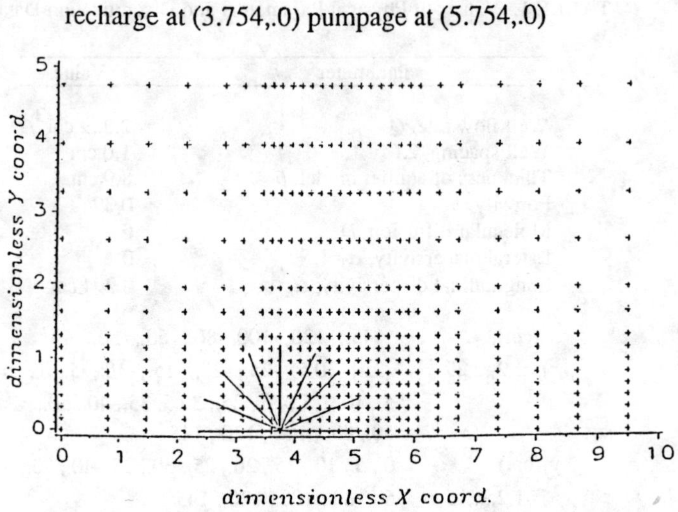

Figure 2: Finite element grid and particles configuration

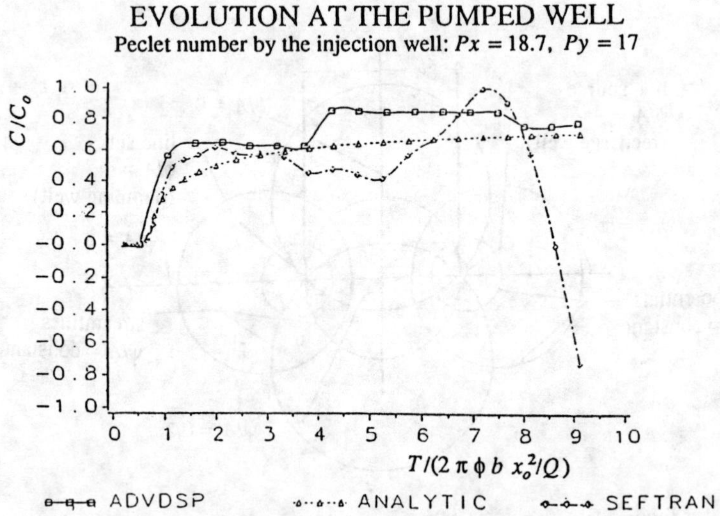

Figure 3: Comparisons between the proposed method (ADVDSP), Huyakorn's scheme (SEFTRAN) and an analytical solution

On the Use of Particle Tracking Methods for Solute Transport in Porous Media

A.F.B. Tompson

Earth Sciences Department, Lawrence Livermore National Laboratory, Livermore, CA 94550

D.E. Dougherty

Department of Civil Engineering, University of California, Irvine, CA 92717, USA

1. INTRODUCTION

The numerical solution of the advection-diffusion (-reaction) equation in any of its forms, such as

$$a\frac{\partial c}{\partial t} + \nabla \cdot (\mathbf{v}c) - \nabla \cdot (\mathbf{D} \cdot \nabla c) = r, \tag{1}$$

remains the object of much research after decades of work [Gray, 1988]. In this paper we review briefly the application of particle methods for the solution of (1). We point out many of the advantages, disadvantages, difficulties, and pleasures of using the technique.

Particle tracking methods have been used to solve partial differential equations in a variety of application areas [Harlow, 1964; Hockney and Eastwood, 1981; Birdsall and Langdon, 1985]. They are based on representing the distribution of an extensive quantity, such as the mass of a particular solute species, as a large collection of particles. At every point in time, a particle is associated with a position in the flow domain, and perhaps other attributes such as mass, charge, or species type. In addition, each particle may be associated with a velocity, certain diffusion characteristics, or other type of forcing vector, each of which may be space dependent. Individual particles are displaced in space over discrete increments of time by these forces. In many cases, a particle distribution at an advanced time level may serve to modify or update the displacement forces. In general, a particle technique is a numerical method for the solution of a time-dependent problem on a region Ω in which the unknown function is approximated at each time t by a linear combination of a finite set of Dirac measures in space [Raviart, 1986]. In the case of (1), we may approximate the solute concentration c by

$$c(\mathbf{x},t) \approx \hat{c}(\mathbf{x},t) = \sum_{j \in J} m(\mathbf{X}_j(t),t)\delta(\mathbf{x} - \mathbf{X}_j(t)). \tag{2}$$

If j is a particle index taken from the set of particle indices J, we denote by $\mathbf{X}_j(t)$ the position of the j-th particle at time t which was initially located at $\mathbf{X}_j(0)$, $m(\mathbf{X}_j)$, the mass of particle j, and $\delta(\mathbf{x})$, the Dirac delta "function" at an arbitrary point $\mathbf{x}$ in Ω. By inspection, (2) is a Lagrangian description of c. As such, the method has the advantage of being self-adaptive in that that the discretization (the particles) moves with the flow field to naturally accomodate sharp fronts and other phenomena that are difficult to handle with conventional finite difference or element methods.

2. DEVELOPMENT OF A PARTICLE METHOD

Several particle methods are currently used to solve solute transport problems. They differ in the treatment of the dispersive term and in the meaning of a particle. We next consider one of these in detail, presuming the velocity field $\mathbf{v}$ is known in some manner.

2.1 Random Walk Method. One of the more popular particle methods that has been applied in solute transport problems in hydrology is based on a *random walk* approach. It involves the use of both deterministic and random displacements in each time step. The magnitudes of these steps depend on the velocities and the dispersion properties of each particle. An individual particle is moved in three-dimensional space via a *step equation*, such as

$$\mathbf{X}^n = \mathbf{X}^{n-1} + \mathbf{A}(\mathbf{X}^{n-1}, t_{n-1})\Delta t + \mathbf{B}(\mathbf{X}^{n-1}, t_{n-1}) \cdot \delta\mathbf{W}(t_n), \tag{3}$$

where $\mathbf{X}^n$ is its position at time level t_n, $\mathbf{A}$ is a deterministic forcing vector, $\delta\mathbf{W}$ is a random forcing vector, and $\mathbf{B}$ is a deterministic, square scaling tensor. The vector $\delta\mathbf{W}$ may be written as $\mathbf{Z}\sqrt{\Delta t}$, where $\mathbf{Z}$ contains three independent random numbers with mean zero and variance one, and $\Delta t = t_n - t_{n-1}$. Equation (3) is a discrete representation of the more general *Langevin equation*,

$$\frac{d\mathbf{X}}{dt} = \mathbf{A}(\mathbf{X}, t) + \mathbf{B}(\mathbf{X}, t) \cdot \boldsymbol{\xi}(t), \tag{4}$$

originally used to describe Brownian motion [Van Kampen, 1981; Gardiner, 1985]. Molecular movements were thought to occur from a set of coherent forces $\mathbf{A}$, and a rapidly shifting (perceptibly uncorrelated) set of forces, $\boldsymbol{\xi}$. The reduction of (4) to (3) involves integration over an increment of time Δt and a specific interpretation of the integral of the second term of (4) [Haken, 1983].

Consider an experiment where one particle, located initially at $(\mathbf{X}, t_0)$, is moved many steps using (3) to a final position $(\mathbf{X}_1, t)$. Because of the random nature of (3), a second experimental particle starting at the same point may end up at a different point $(\mathbf{X}_2, t)$, and so on. In the infinite limit, the spatial distribution of particles at time t can be described by a probability density function $f(\mathbf{x}, t)$. For small Δt, equations (3) and (4) can be used [Haken, 1983] to formulate a balance equation for $f(\mathbf{x}, t)$:

$$\frac{\partial f}{\partial t} + \nabla \cdot (\mathbf{A} f) - \nabla\nabla : \left(\frac{1}{2}\mathbf{B} \cdot \mathbf{B}^T f\right) = 0, \tag{5}$$

where $f(\mathbf{x}, t_0) = \delta(\mathbf{x} - \mathbf{X})$. This relationship conserves the probability density function f and is known as a *Fokker-Planck* equation.

2.2 Analogy to Field Equations. If a finite number of particles, N, are moved simultaneously in one experiment by (3), the density of particles lying in some small volume V_s centered at $\mathbf{x}$ and time t can be estimated by $f(\mathbf{x}, t) \approx N_s/(NV_s)$, where N_s is the number of particles in V_s. This becomes an exact solution of (5) in the limit as $N \to \infty$. If each particle has constant mass m, then $mNf(\mathbf{x}, t)$ is a dimensional measure of concentration, c. A particle-based estimate of f is thus one for c. Solute distributions satisfying (1) with $a = 1$ and $r = 0$ can be found using (3) if $\mathbf{A}$ and $\mathbf{B}$ are chosen from

$$\mathbf{A}(\mathbf{X}^n, t_n) \equiv \mathbf{v}(\mathbf{X}^n, t_n) + \nabla \cdot \mathbf{D}\left(\mathbf{v}\left(\mathbf{X}^n, t_n\right)\right) \tag{6}$$

$$\mathbf{B}(\mathbf{X}^n, t_n) \cdot \mathbf{B}^T(\mathbf{X}^n, t_n) \equiv 2\mathbf{D}\left(\mathbf{v}\left(\mathbf{X}^n, t_n\right)\right), \tag{7}$$

where $\mathbf{v}$ is the groundwater velocity and $\mathbf{D}$ is the solute dispersion tensor. The random step is clearly associated with the dispersion effect, while the deterministic step is associated with both the velocity and dispersion effects. Insofar as the particles are not "lost" from observation, the method conserves mass exactly. Strictly speaking, a grid is not necessary for this computation. In many applications, however, a grid may be employed for the definition and computation of the forcing terms [Hockney and Eastwood, 1981]. Furthermore, interpretation of the particle distribution as a concentration field may require a mesh or some other means of interpolation from the particles. Boundary conditions, finite domains, and source terms are discussed later. Particle models based on this random walk method have been used by several authors [Ackerer and Kinzelback, 1985; Uffink, 1983, 1987; Tompson *et al.*, 1987]. Earlier random walk models (e.g., Prickett *et al.* [1981]) were developed more intuitively and do not include the effects of the $\nabla \cdot \mathbf{D}$ term, so these models will not be correct when $\nabla \cdot \mathbf{D}$ is important.

2.3 Other Things You Can Do. Equation (1) is viewed as the fundamental mathematical statement of the physical phenomena to be modeled. Use of (3), (6), and (7) is a means to approximate its solution. This contrasts with the original Brownian motion studies, which regard (3) and (4) as a fundamental relationship and (5) as a resulting macroscopic balance. The random walk analogy can also be used to solve unsaturated transport problems [Tompson *et al.*, 1987], and heat flow problems [Ahlstrom *et al.*, 1977]. Moreover, it can also be used in multispecies or fracture flow applications. Torney and Warnock [1987] use a random walk model to study chemical reactions between species represented as particles moving on a lattice. Schwartz *et al.* [1983] use a particle method to study solute movement in fractures.

2.4 Other Particle Methods. The random walk model described above is not the only particle method available for solute transport simulations. For instance, rather than use a random walk approach, Raviart [1986] discusses a particle method for (1) based on deterministic steps for both the advective and dispersive displacements. Another widely-known particle technique is a particle-in-cell (PIC) method which has come to be known as the "method of characteristics" (MOC). Introduced by Garder *et al.* [1964] in the petroleum engineering literature, Pinder and Cooper [1970], Bredehoeft and Pinder [1973], and Konikow and Bredehoeft [1978] applied a splitting algorithm to solve (1) in two stages. In the first stage, particles are employed to advect solute in a manner similar to the first part of the random walk. Because each particle is assigned a concentration (and not a mass!), the concentrations associated with the particles are interpolated to a grid and the advection-free dispersion equation is solved. The two-stage process is then repeated. When too many grid cells contain no particles, the grid concentrations are interpolated to a new set of particles to carry out the procedure. The two interpolations lead to numerical error. Practice has shown it does not conserve mass and that the cost of bookkeeping can be quite high.

3. IMPLEMENTATION & APPLICATION ISSUES OF RANDOM WALK

3.1 Determination of Point Displacement Forces. To apply the random walk step equation (3) in practice, one must adopt a method for determining $\mathbf{v}$ and $\mathbf{D}(\mathbf{v})$ so that $\mathbf{A}$ and $\mathbf{B}$ can be determined from (6) and (7). If the velocity field and $\mathbf{D}(\mathbf{v})$ are known analytically, then they (and their spatial derivatives) can be computed exactly at the individual points $\mathbf{X}_j$ where the particles are located. The vector $\mathbf{A}$ can be determined directly from (6). Since $\mathbf{D}$ is real and symmetric in most cases, the tensor $\mathbf{B}$ in (7) can be evaluated using a diagonalization procedure [Tompson *et al.*, 1987]; a Cholesky decomposition method [Forsythe and Moler, 1977] can also be used. If the velocity field is calculated numerically, then $\mathbf{v}$ and its gradients at the particle locations $\mathbf{X}_j$ must be approximated. Nodal values of $\mathbf{v}$ can be interpolated within discrete flow cells or elements. The interpolation method may also allow for the velocity gradient to be estimated. If, for example, velocities are known on a three-dimensional rectilinear grid, then C^0 estimates of c within a cell can be found through trilinear interpolation. Velocity derivatives may also be estimated within the cell, but they will be discontinuous across cell boundaries. The velocity derivatives within the $\nabla \cdot \mathbf{D}$ term are usually significant only near stagnation zones [Uffink, 1987]; in many cases, crude approximations may be sufficient. Computational savings may be achieved by defining velocities and their derivatives as cell constants. This approximation may be more acceptable in problems whose length scale is much larger than a typical grid scale [Tompson *et al.*, 1987].

3.2 Time Stepping. In moving particles through space with the discrete relationship (3), the time step, Δt, must be chosen carefully. Too large a value will generally lead to overshoot errors and will be inconsistent with the derivation of (5). The time step must be small enough that errors due to the discrete step $\mathbf{v}\Delta t$ can be tolerated; it should be bounded in some sense by $\|\nabla \mathbf{v}\|^{-1}$. Higher order schemes for integrating the coherent force term of (4) may also be used as long as the form of (5) is unchanged. When dispersion is present, overshoot may occur from both the deterministic and random steps. If D is a typical dispersion coefficient, then the average dispersive step will be $\sqrt{2\mathsf{D}\Delta t}$. If D is small enough and the effects of $\nabla \cdot \mathbf{D}$ are ignored, then a limiting choice of Δt can be made by observing advective errors only. Two kinds of time stepping procedures are generally used. A single value of Δt can be chosen so that all particles are moved together during each time step. It may be adjusted if the velocities vary significantly from one time step to another. Alternatively, when cell- or element-wise constant velocities are used, a variable time step may be associated with each particle, chosen so that the particle is always advected to the nearest cell boundary and then dispersed. Each particle is moved in this manner to several common *rendezvous times* where the spatial mass distributions can be analyzed.

3.3 Interpretation of Results. Frequently, concentrations of solute are desired for analysis, rather than particle distributions. In this case, a smoothing approximation can be used to estimate c at some point $\mathbf{x}$ and time t. Given the representation $\hat{c}$ defined in (2), this approximation can be built via a regularization or weighting function, $\zeta(\mathbf{x})$, as

$$c(\mathbf{x},t) \approx \tilde{c}(\mathbf{x},t) = \int_{\Omega} \hat{c}(\mathbf{x},t)\zeta(\mathbf{x}'-\mathbf{x})d\mathbf{x}' = \sum_{j\in J} m(\mathbf{X}_j(t),t)\zeta(\mathbf{x}-\mathbf{X}_j(t)), \tag{8}$$

where $m(\mathbf{X}_j)$ is the mass associated with particle j. In many conservative transport problems all the particles will have the same mass, m. Comparing (8) and (2), we see that, in effect, each point particle in (2) is replaced by a particle of finite size. The size and shape of each finite particle is defined by ζ.

Spatial moments of the mass (particle) distribution are alternative measures for gauging the overall behavior of the solute and for the experimental determination of transport parameters, such as velocity and dispersion tensor. This method is very useful in the experimental verification of asymptotic, analytical solutions (see Tompson *et al.* [1987]).

3.4 Auxiliary Conditions. The analogy between the Fokker-Planck equation and the solute mass balance was made under an implicit assumption of an infinite domain. Solution of an equation of the form (1) is, however, usually associated with a finite domain Ω, a set of initial conditions for c, and a set of conditions on the boundary for c or ∇c. Statement of any condition on a boundary is tantamount to an arbitrary specification of physical behavior outside Ω that obeys (5). Familiar functional conditions for c must be phrased in terms of particle distributions or motions. As an example, an initial condition of $c_0 = c(\mathbf{x},0)$ may be specified by uniformly distributing c_0V_s/m particles of constant mass m in some small volume V_s surrounding $\mathbf{x}$. Note that this is equivalent to using (8) with $\zeta = 1/V_s$ (a constant) for $\mathbf{x}$ in V_s and $\zeta = 0$ otherwise. Specification of concentrations in a subregion Ω_i of Ω can be accomplished by dividing Ω_i into several nonoverlapping volumes V_s and repeating the procedure. Constant concentration boundary conditions can be applied similarly by dividing up a thin region Ω_b against the boundary and using an analogous procedure at each time step. The validity of this method will be limited by Δt. Constant flux conditions can be achieved by ensuring a certain number of particles enter the domain through Ω_b during each time step. No-flux boundaries can be maintained by bouncing particles off of a boundary in a "billiard-ball" fashion. Free or "absorbing" boundaries can be defined where the particles are allowed to naturally flow out of Ω, say, at a downstream location, never to reappear. This is approximate in that exited mass could actually disperse upstream into Ω. In the same sense, portions of an initial particle distribution near an upstream boundary could disperse past it and be mistakenly removed from the system. As this method is evolutionary in nature, steady state solutions can only be achieved in an asymptotic manner by fixing steady boundary conditions and letting transients due to initial conditions decay.

3.5 Nonconservative Problems. Nonconservative solute transport is of interest in many situations. A species may adsorb on the solid matrix, undergo biotransformation, decay, or react chemically with other species. In the special case of linear adsorption, the velocity and dispersion may be simply adjusted so the conservative random walk step equation may still be used. On the other hand, nonlinear reactions, including chain processes, typically require some sort of "correction" algorithm to be applied after each time step to account for the loss or gain of mass. Techniques of this sort are not well developed and remain an active area for future research (see §5.1).

3.6 Odds and Ends. How many particles, N, are needed for solute transport simulations? There is no fixed number. The answer really depends on the kind of analysis needed by a user. Spatial moments are less sensitive to N than are the smoothed estimates $\tilde{c}$ determined from (8). In general, the use of a finite number of particles will lead to a noisy approximation of c via (8). The noise can be reduced by simply increasing the number of particles N used. Because the reduction is proportional to $1/\sqrt{N}$, less expensive approaches may involve the use of modified weighting functions ζ or application of filtering techniques (as in image processing) to reduce the noisiness of the approximation. Experience in conservative systems indicates a minimum of about 1000 particles are needed in homogeneous flow systems and upwards of 10,000 are needed in inhomogeneous flow fields. If the total mass in the system is M, the sample volume for interpretation as concentrations is V_s, ζ is as described in §3.4, and the particles are of the same mass, then the resolution of the simulation can be no better than M/V_sN. In practice it will not be so good because of the stochastic nature of the particle method. When M varies in time, the accuracy of the method will also vary in time.

Fortunately, when greater accuracy is required in conservative solute transport simulations, the linearity of (1) can be used. Superposition of additional particles will lead to increased simulation accuracy (see §3.6), and is simply performed. For particles, the computational complexity is roughly linear in N. On the other hand, conventional approaches will require the solution of N_g equations, where N_g is the number of grid points, and a computational effort that is of order N_g^2. The selection of a method based on work estimates is, therefore, dependent on estimates of the values of N and N_g. These, in turn, are problem dependent and are not always easily estimated (see §4.1). In coupled multispecies problems, the particle computational effort will (minimally) increase linearly with the number of species considered, as compared with (at least) a quadratic increase with conventional methods. This apparent advantage may, in fact, be diminished when the reaction coupling effects are taken into account. These latter phenomena will increase the cost of the particle method per time step, but they are also likely to impart nonlinearity to the conventional discretization methods, further adding to their cost.

Finally, we again note that the movement of each particle is independent of all other particles in the random walk method (see §3.6 on superposition). This is important in mapping the particle method onto advanced computer architectures. For parallel computers, it means that each particle can be assigned to a separate processor and the only common communication is interrogating the velocity field. For vector machines, this observation means that the vector registers can be filled with no conflicts, except, possibly, for interrogation of the velocity field [Martin and Brown, 1987]. Other particle methods, such as that of Raviart [1986], include particle-particle interactions in each step, which lead to higher communications costs. However, certain procedures from plasma physics may be employed to reduce these costs.

4. APPLICATIONS

4.1 Example 1. Tompson *et al.* [1987] consider a simulation of a conservative solute in a large heterogeneous flow system. As described by Ababou [1987], a steady three-dimensional velocity field was developed from a synthetic heterogeneous hydraulic conductivity distribution on a 51 × 51 × 51 rectilinear grid. It was desired to analyze the dispersive behavior of expanding solute plumes in such a variable flow system for comparison with theoretical predictions [Tompson *et al.*, 1987]. Conventional finite element approaches would involve the repeated solution of 125,000 algebraic equations to move the plume forward in time; a particle method merely involved the iterated advancement of about 10,000 particles. The potential for grid Peclet numbers to reach values on the order of 10 to 100 was another disadvantage of using a conventional method. In Figure 1, the evolution of an 8000 particle plume over

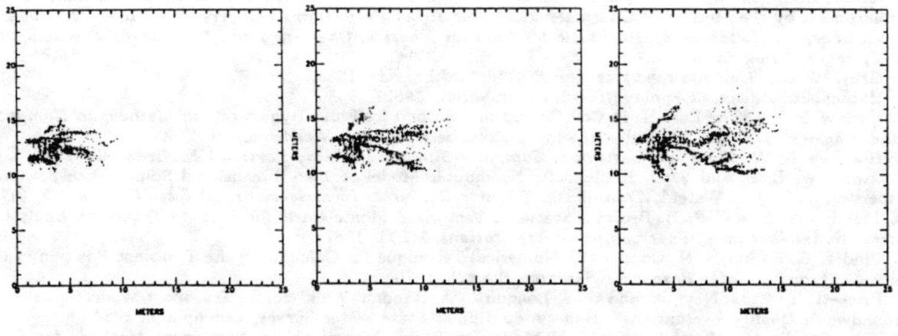

Figure 1.

1250 time steps is shown in three distinct vertical-projection snapshots. The solute mass was released into the flow field from a small cube located midway along the vertical (at left).

4.2 Example 2. Dougherty and Sovich [1988] describe an application of a particle method to the study of the effects of heterogeneous adsorption in an otherwise homogeneous porous medium. The objective was to examine the large scale effects of "chemical" dispersion, rather than "hydrodynamic" dispersion. A reversible, linear adsorption or retardation process was considered for simplicity. Systems layered with respect to retardation factor were examined. Initial distributions of particles were subjected to advection parallel, perpendicular, and diagonal to the bedding of these layers. Two dimensional simulations using 4,000 to 16,000 particles and up to 7000 timesteps were performed on a Cray X-MP to evaluate the effects of these heterogeneities on a larger scale. Spatial moments of the particle distributions were used to estimate megascopic dispersion coefficients and to provide diagnostics for the simulations.

5. FUTURE DIRECTIONS

5.1 Reactions and Multiple Species. Important areas for future improvements of particle transport models lie in the treatment of reactions and multicomponent chemical problems. General strategies for dealing with source terms in a post-processing sense must initially be developed. This includes deciding whether to address the problem in a continuum (e.g., Ahlstrom *et al.*, [1977]) or particle (e.g., Torney and Warnock [1987]) sense. Effective, mass conservative smoothing algorithms will be necessary for the former approach, because errors associated with concentration measurements will propagate into the particle solution in the form of inexact reaction "corrections". The latter method can be very complicated, as evidenced by the restriction of particles to move on a lattice used by Torney and Warnock [1987]. Particle methods offer the potential for addressing coupled multispecies problems in a more efficient manner than conventional methods. Issues regarding the computational effort for these problems are complex, as discussed in §3.6.

5.2 Concentration Dependent Flow. In many practical problems, such as coastal aquifers and waste storage in briny formations, the velocity or external forcing function may be concentration dependent. In this case, the time-stepping procedure must be modified so that changes in solute mass distributions can be fed back into the flow equation. Effective methods to estimate continuum concentration values will be required, as above, to modify the velocity. This kind of feedback algorithm is regularly used in other applications [Hockney and Eastwood, 1981].

6. REFERENCES

1. Ababou, R., "Three-Dimensional Flow in Random Porous Media", Ph.D. Dissertation, Department of Civil Engineering, MIT, Cambridge, 1987.
2. Ackerer, Ph., and W. Kinzelbach, "Modelisation du Transport de Contaminant par la Methode de Marche au Hasard: Influence des Variations du Champ d'Ecoulement au Cours du Temps sur le Dispersion", *Proceedings of the International Symposium on the Stochastic Approach to Subsurface Flow*, Montvillargene, France, 1985.
3. Ahlstrom, S., H. Foote, R. Arnett, C. Cole, and R. Serne, "Multicomponent Mass Transport Model: Theory and Numerical Implementation", Report BNWL 2127, Batelle Pacific Northwest Laboratories, Richland, WA, 1977.
4. Birdsall C., and A. Langdon, *Plasma Physics via Computer Simulation*, McGraw Hill, 1985.
5. Bredehoeft, J. and G. F. Pinder, "Mass Transport in Flowing Groundwater", *Water Resources Research*, **9**:1:194, 1973.
6. Dougherty, D. E. and T. J. Sovich, "Effects of Spatially Varying Adsorption in Saturated Porous Media: Layered System", in review, 1988.
7. Forsythe G. and C. Moler, *Computer Solution of Linear Algebraic Systems*, Prentice Hall, 1977.
8. Garder, A. O., D. W. Peaceman, and A. L. Pozzi, "Numerical Calculation of Multidimensional Miscible Displacement by the Method of Characteristics", *Society of Petroleum Engineers Journal*, **4**:1:26, 1964.
9. Gardiner, C., *Handbook of Stochastic Methods for Physics, Chemistry and the Natural Sciences*, 2nd ed., Springer Verlag, 1985.
10. Gray, W. G., "Can you read this size, Bill?", Hard Pressed, 1988.
11. Hacken, H., *Advanced Synergetics*, Springer Verlag, 1983.
12. Harlow, F. H., "The Particle in Cell Computing Method for Fluid Dynamics", in *Methods in Computational Physics*, B. Alder, S. Fernbach, and M. Rotenberg (eds.), **3**, Academic, 1964.
13. Hockney, R. W. and J. W. Eastwood, *Computer Simulation Using Particles*, McGraw-Hill, 1981.
14. Konikow, L. F. and J. D. Bredehoeft, "Computer Model of Two-Dimensional Solute Transport and Dispersion in Ground Water", *Techniques of Water-Resources Investigations, Chapter C2, Book 7*, 1978.
15. Martin, W. R. and F. B. Brown, "Status of Vectorized Monte Carlo for Particle Transport Analysis", *International Journal of Supercomputer Applications*, **1**:2:11, 1987.
16. Pinder, G. F. and H. H. Cooper, "A Numerical Technique for Calculating the Transient Position of the Saltwater Front", *Water Resources Research*, **6**:3:875, 1970.
17. Prickett, T., T. G. Naymik, and C. G. Lonquist, "A 'Random-Walk' Solute Transport Model for Selected Groundwater Quality Evaluations", Bulletin 65, Illinois State Water Survey, Champaign, 1981.
18. Raviart, P. A., "Particle Numerical Models in Fluid Dynamics", in *Numerical Methods for Fluid Dynamics II*, K. W. Morton and M. J. Baines (eds.), 1986.
19. Schwartz, F. W., L. Smith, and A. S. Crowe, "A Stochastic Analysis of Macroscopic Dispersion in Fractured Media", *Water Resources Research*, **19**:5:1253, 1983.
20. Tompson, A. F. B., E. G. Vomvoris, and L. W. Gelhar, "Numerical Simulation of Solute Transport in Randomly Heterogeneous Porous Media: Motivation, Model Development, and Application", Report UCID-21281, Lawrence Livermore National Laboratory, Livermore, CA, 1987.
21. Torney D. and T. Warnock, "Computer Simulation of Diffusion-Limited Chemical Reactions in Three Dimensions", *International Journal of Supercomputer Applications*, **1**:2:33, 1987.
22. Uffink, G.,"A Random Walk Model for the Simulation of Macrodispersion in a Stratified Aquifer", *IUGG 18th General Assembly, Hamburg, Proceedings of the IAHS Symposia, Vol HS2*, 1983.
23. Uffink, G.,"Modeling of Solute Transport with the Random Walk Method", *NATO Advanced Workshop on Advances in Analytical and Numerical Groundwater Flow and Quality Modeling*, Lisbon, June 2-6, 1987.
24. Van Kampen, N., *Stochastic Processes in Physics and Chemistry*, North Holland, 1981.

SECTION 2B - SOLUTE TRANSPORT IN UNSATURATED POROUS MEDIA

Mass Exchange Between Mobile Fresh Water and Immobile Saline Water in the Unsaturated Zone

H. Gvirtzman and M. Magaritz

Isotope Department, The Weizmann Institute of Science, 76100 Rehovot, Israel

ABSTRACT

A profile of tritium concentrations measured in the unsaturated zone in loessial sediments in a semiarid area is interpreted in terms of mobile and immobile water domains, according to a physical nonequilibrium transport model. The mobile domain is represented by percolating fresh water from both rain and irrigation, and the immobile one is represented by isolated fossil saline water pockets. The two domains are connected by partially-saturated narrow passages within dispersed clay minerals. The transport of the mobile water is described by convective–dispersive flow and mass exchange between the two water domains takes place simultaneously. The relevant equations with the given initial – boundary conditions are solved numerically, and the simulated profile is adjusted to fit the measured profile. By taking into account variations of the mass exchange coefficient in relation to matrix characteristics, we were able to obtain an adequate reconstruction of the measured profile. Temporal changes in matrix characteristics are attributed to dispersion kinetics of clays at the interface between fresh and saline waters.

INTRODUCTION

During the past few decades, several models have been developed for describing the dynamics of water flow in the unsaturated zone [Nielsen et al.[4]]. Most of the models focused on conceptual and mathematical aspects and were based on laboratory experiments. However, there appears to be a lack of studies of natural systems which enable confirmation or rejection of these models.

One of the techniques for quantitative study of water transport in the unsaturated zone is the use of environmental tritium as a tracer. The principle is to date the water molecules along the sediment column according to their tritium content, using background tritium concentration data in the applied water – rainwater and irrigation [Gvirtzman et al.[1]]. Using this method, a 14-year record of water flow in the unsaturated zone was reconstructed, and the existence of an immobile water domain within the medium was hypothesized [Gvirtzman and Magaritz[2]]. The objective of this study is to utilize these field data together with the appropriate two-domain water transport model in order to evaluate numerically the hydrological parameters of the natural system. It was suggested that the mass exchange law between mobile and immobile water domains should be modified based on geochemical considerations and numerical simulation.

THE TRITIUM PROFILE

The profile was sampled in the northern Negev of Israel [Gvirtzman et al.[1]], in an area where precipitation averages 200 mm/winter and irrigation is 650 mm/summer. The unsaturated zone consists of loessial sediments containing large quantities of salts. A 14-year record of vertical flow was reconstructed (Fig. 1). Alternate layers with high and low tritium concentration, corresponding to winter rains and summer irrigations, were detected down to 8.5 m. Based on the peak separation, the vertical water velocity was calculated to be 0.66 m year^{-1}.

This tritium profile corresponds to a period during which the tritium content in the applied water was decreasing (1971–1983). However, the actual profile shows the opposite trend: an excess of tritium in the upper part of the section, and a deficiency of tritium in the lower part. This discrepancy led to interpretation of the tritium profile in terms of mobile and immobile water domains. It was hypothesized that the mobile domain was represented by percolating fresh water of both rain and irrigation, and the immobile one was represented by isolated, saline water pockets related to the fossil saline inclusions [Magaritz et al.[3]]. The immobile water domain has a memory of the large atmospheric tritium pulse of the "high tritium period" (1950's and 1960's). During that period, when the mobile water contained very high concentrations of tritium, the diffusive exchange process caused an increase in the tritium content of the immobile water, which was not completely leached out back to the mobile domain during the "low tritium period" (1970's and 1980's). This transient dynamic process hypothesis needs a verification by an appropriate numerical simulation.

A MATHEMATICAL MODEL

Transport of water in porous media has traditionally been described by the convective–dispersive equation. Often, however, local equilibrium may not be assumed. Consequently, the presence of sources and sinks has been hypothesized, i.e., a physical nonequilibrium transport model. According to this model the water in the porous media is partitioned into two domains, namely mobile (flowing) and immobile (stagnant). The pore water velocity is treated as bimodal: convective–dispersive flow occurs only in the mobile domain, whereas the remainder of the pores contain stagnant water. A diffusion-controlled mass exchange between the two water domains occurs concurrently, and is assumed to be a first-order mass transfer process [Van Genuchten and Wierenga[7]]. Making the assumptions of steady, uniform flow in a homogeneous medium, it leads to the equations in one dimension:

$$\frac{\partial C_m}{\partial t} = D_m \frac{\partial^2 C_m}{\partial x^2} - \frac{q}{\phi_m \theta} \frac{\partial C_m}{\partial x} - \alpha (C_m - C_{im}) - \lambda C_m \qquad (1)$$

$$\frac{\partial C_{im}}{\partial t} = \frac{\phi_m}{1-\phi_m}\alpha(C_m - C_{im}) - \lambda C_{im} \tag{2}$$

where C_m and C_{im} are the solute concentration (M L^{-3}) in the mobile and immobile water domains, respectively, D_m is the dispersion coefficient in the mobile domain (L^2 T^{-1}), q is the water flux (L T^{-1}), θ is the total water content (L^3 L^{-3}) of the porous medium, θ_m and θ_{im} are the water content (L^3 L^{-3}) of the porous medium in the mobile and immobile water domains, respectively ($\theta = \theta_m + \theta_{im}$), ϕ_m is the fraction of the water which is mobile ($\phi_m = \frac{\theta_m}{\theta}$), α is the exchange rate coefficient of solute between the mobile and immobile domains (T^{-1}), λ is the rate coefficient of radioactive decay (T^{-1}), t is time (T), and x is distance (L) along the flow direction. These equations represent a slight modification of those of Van Genuchten and Wierenga[7] to include solutes which decay radioactively such as tritium.

Tritium concentration along the profile until 1957 (prior to cultivation) was assumed to be negligible, owing to the effect of radioactive decay [Gvirtzman and Magaritz[2]]. Accordingly, the initial conditions in conjunction with Eqs. (1) and (2) are:

$$C_m(x,0) = 0 \quad ; \quad 0 < x < 15 \quad meters \tag{3}$$

$$C_{im}(x,0) = 0 \quad ; \quad 0 < x < 15 \quad meters \tag{4}$$

The conditions at upper boundary depend on the tritium concentration in the applied water and were determined according to the historical record of precipitation and irrigation and their tritium content, $f(t)$, for the period 1957–1983:

$$C_m(0,t) = f(t) \quad ; \quad 0 < t < 26 \quad years \tag{5}$$

At the lower boundary, 15 m below the surface, no variation of tritium content with depth was assumed (based on the observed profile of tritium content), i.e.

$$\left.\frac{\partial C_m}{\partial x}\right|_{x=15} = 0 \quad ; \quad 0 < t < 26 \quad years \tag{6}$$

A computer program incorporating NAG Routine D03PGF [Numerical Algorithms Group[5]] was written to numerically integrate the above system of two partial differential equations, (1) and (2), subject to the initial-boundary conditions eqs. (3)–(6). The method of solution is to discretize the space derivatives using finite differencing, and to solve the resulting system of ordinary differential equations using Gear's method.

The following values were used [Gvirtzman and Magaritz[2]]: $\lambda = 0.0558$ year^{-1} (as tritium has a half life of 12.43 years), $D_m = 10^{-3}$ m^2 year^{-1} in the unsaturated zone, and $\theta = 0.2$ along the whole profile and during time. We looked for appropriate values for α and ϕ_m which would yield a curve with pattern and characteristics of the field observations. For each set of fixed α and ϕ_m, Eqs. (1) and (2) were solved numerically, subject to eqs. (3)–(6), and the resulting curve was compared with the observed data. Using an optimization program, the calculated (for specific α and ϕ_m) and the measured profiles were compared by summing the squares of differences. Unfortunately, the best fit α and ϕ_m that could be obtained produced a profile that had no resemblance to the field observations, i.e. the assumption of a constant mass exchange coefficient caused depletion of tritium in the immobile domain during the "low tritium period", such that it did not retain a memory of the "high tritium period". We looked, therefore, for a conceptual model that would yield rapid tritium enrichment of the immobile water domain during the 1960's and slow depletion thereafter.

THE MASS EXCHANGE LAW

Van Genuchten and Wierenga[7] observed in laboratory experiments that the mass exchange rate, α, in Eqs. (1) and (2) varied with velocity of displacement, time, and the water content. In addition, Rao et al.[6] observed in other laboratory experiments that α depends mainly on the geometry of the medium. It seems that in our case, α changed in the course of time due to geochemical processes which could cause modifications in the geometry of the medium and thereby affect the rate of mass transfer between the mobile and immobile water domains.

The evolution of this system was explained as follows: Until 1957, before cultivation of the area, the sediments solution contained relatively high solute concentration. In 1957 fresh water first started to percolate downward due to the addition of irrigation water during the summers. The percolating water moved through preferential conduits of inter-granular width and formed a "finger-shape" interface with the original immobile water. An assembly of isolated saline micro-scale water pockets with different chemical (and isotopical) compositions was created due to the evolution of the "finger-shape" interface. Clay minerals tend to disperse in dilute sodic solutions according to the diffuse double layer theory. Therefore, clay minerals (mainly Montmorillonite) disperse at the interface between the two water types and the resulting gels of dispersed clays and swollen clay tactoids caused continuous clogging of some pores and narrowing of passages. The thickness of the passages between the flowing fresh water and the stagnant saline waters changes due to changes of fluid chemistry. Therefore, a necessary modification of the existing model is to assume that the mass exchange rate coefficient, α, depends on the kinetics of dispersion of the clay minerals, and changes in the course of time. The kinetics of the dispersion process depends on the amount of fresh water which penetrates through the profile. It should be emphasized that the overall recharge during 26 years of percolation was less than half the pore volume of the 10 m sediment column. Consequently, it is proposed that the exchange rate between the two water domains follows on exponential decay with time, i.e.

$$\alpha = \alpha_o e^{-\beta t} \tag{7}$$

where α_o is a constant which represents the initial exchange rate, and β is another constant which expresses the relative rate of decay of the exchange rate, and should be related to the change in the rate of the "clogging" process.

This new expression for α (Eq. (7)) was substituted into Eqs. (1) and (2) which, along with the appropriate initial - boundary conditions, eqs. (3)–(6), were run on the computer. Fig. 2 shows the profile obtained with the best fit parameters (using a least–squares optimization procedure for ϕ_m, α and β). We found that $\alpha = 0.49e^{-0.12t}$ and $\phi_m = 0.36$. Accordingly, α decreased from 0.5 to

0.02 year^{-1} during the 26 years. The modified model fits the data much better than the model with the constant mass exchange rate.

REFERENCES

1. Gvirtzman H., Ronen D. and Magaritz M. (1986), Anion exclusion during transport through the unsaturated zone, *J. Hydrol., Vol.87*, pp. 267–283.
2. Gvirtzman H. and Magaritz M. (1986), Investigation of water movement in the unsaturated zone under irrigated area using environmental tritium, *Water Resour. Res., Vol.22*, pp. 635–642.
3. Magaritz M., Gvirtzman H. and Nadler A. (1988), Salt accumulation in the loessial sequence in the Be'er-Sheva Basin, Israel, *Environ. Geol. Water Sci., Vol.11*, in Press.
4. Nielsen, D.R., Van Genuchten M.Th. and Biggar J.W. (1986), Water flow and solute transport processes in the unsaturated zone, *Water Resour. Res., Vol.22*, pp. 89S–108S.
5. Numerical Algorithms Group (1984), Routine D03PGF, *NAG FORTRAN Library Manual, Mark 11, Vol.2*, pp. 1–15, Numerical Algorithms Group, Oxford.
6. Rao P.S.C., Jessup R.E., Rolston D.E., Davidson J.M. and Kilcrease D.P. (1980), Experimental and mathematical description of nonadsorbed solute transfer by diffusion in spherical aggregates, *Soil Sci. Soc. Am. J., Vol.44*, pp. 684–688.
7. Van Genuchten M.Th. and Wierenga P.J. (1976), Mass transfer studies in sorbing porous media, 1, Analytical solutions, *Soil Sci. Soc. Am. J., Vol.40*, pp. 473–480.

Figure 1: Tritium concentration versus depth, with horizontal bars representing the errors. Peak dates (R=Rain) and velocity (V) are indicated.

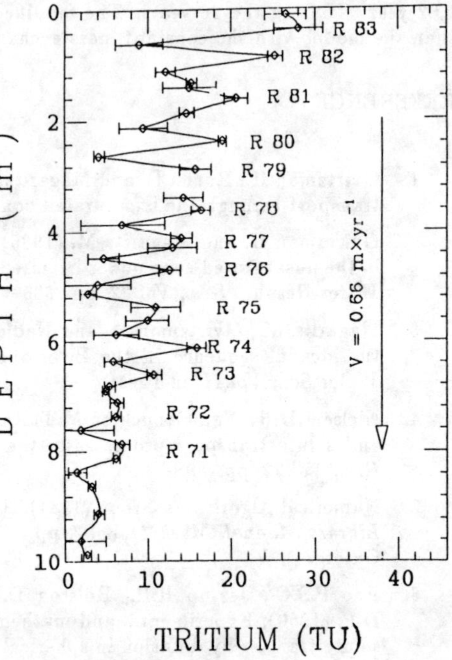

Figure 2: Calculated tritium content in (a) the mobile water, (b) immobile water, and (c) a weighted combination of these components. The best-fit parameters, obtained by using optimization procedure, were: $\alpha_o = 0.5$, $\beta = 0.12$, and $\phi_m = 0.36$. The dotted-dashed line joins the measured data points and agrees with the calculated curve.

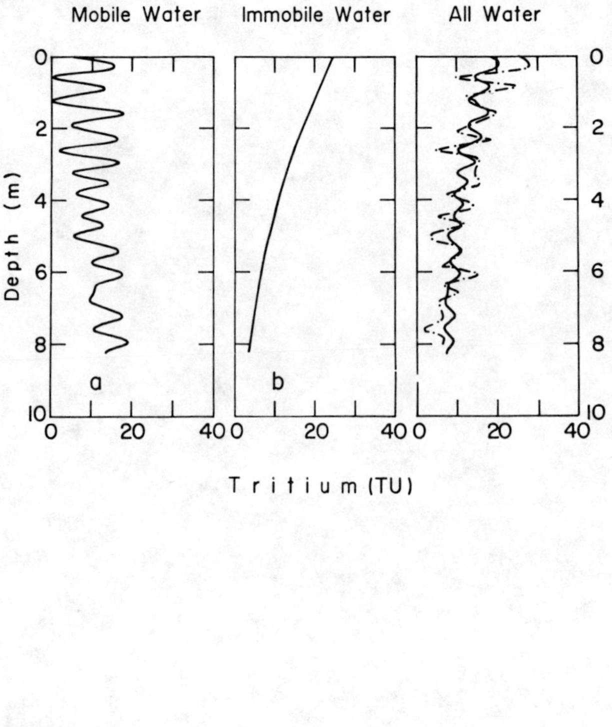

Solution of Saturated-Unsaturated Flow by Finite Element or Finite Difference Methods Combined with Characteristic Technique

Kang-Le Huang

Department of Geology and Mining Engineering, Fuzhou University, China

ABSTRACT

Besides the high nonlinearity, the hyperbolic feature of flow equation due to a predominantly gravitational term is another important factor leading to the numerical oscillation and diffusion in the simulation of vertically saturated - unsaturated flow problems. In this paper, tow schemes, the characteristic finite elememt (CFE) and characteristic alternating direction difference (ADCI) ,are developed for 1 and 2 - D problems respectively to eliminate the numerical difficulties and ensure the high accuracy and efficiency. Several examples are simulated to demonstrate the advantages of the proposed methods.

INTRODUCTION

Saturated - unsaturated flow in vertical plane has greatly interested researchers in hydrology, agronomy and environmental hydraulics. Although many successes have been achieved on numerical simulation of these problems, conventional schemes in certain degrees still suffer from artifical diffusion and oscillation,especially in cases with sharp wetting front. Besides high nonlinearity, the hyperbolic feature of flow equation due to the predominantly gravitative term is another major factor leading to the numerical diffusion and oscillation in the simulation of vertical saturated - unsaturated flow. To overcome such difficulties, the Eulerian - Lagrangian approach based on characteristic method, which has been proved to be more effective on the convective - dispersive problems (Neuman[4]) , is extended to the simulation of vertical saturated - unsaturated flow problems.

PROBLEM STATEMENT

The governing equation for transient saturated - unsaturated flow in vertical plane of homogeneous media is described as

$$(s\theta/\varepsilon + \partial\theta/\partial h)\,\partial h/\partial t = \nabla\cdot(K\nabla h) - \partial K/\partial y \qquad (1)$$

where h, K, θ, ε and s are pressure head, hydraulic conductivity, water content, porosity and specific storage respectively, and y is positive downward.

The initial and boundary conditions are

$$h|_{t=0} = h_i \quad (2)$$

$$[-K\partial h/\partial x n_x - (K\partial h/\partial y - K)n_y]|_{\Gamma_2} = q_0 \quad (3)$$

$$h|_{\Gamma_1} = h_0 \quad (4)$$

$$h|_F = 0 \quad (5)$$

where $\vec{n} = (n_x, n_y)$ is normal unit outward vector; $\Gamma = \Gamma_1 + \Gamma_2 + F$ the entire boundary of considered domain; F is the seepage surface. h_1, h_0 and q_0 are prescribed functions.

NUMERICAL SCHEME

With the Eulerian-Lagrangian approach and $\partial K/\partial y = \partial K/\partial h \cdot \partial h/\partial y$, Eq. (1) may be transformed into a purely parabolic equation

$$(\theta s/\varepsilon + \partial\theta/\partial h)\,dh/dt = \nabla\cdot(K\nabla h) \quad (6)$$

by definition of the hydrodynamic derivative

$$dh/dt = \partial h/\partial t + V^{\star}\partial h/\partial y \quad (7)$$

along the characteristic path line described as

$$dy/dt = V^{\star} = \begin{cases} \partial K/\partial\theta & \text{for unsaturated zone} \\ 0 & \text{for saturated zone.} \end{cases} \quad (8)$$

Eqs. (2) to (6) form the parabolic problem, while following equations

$$d\bar{h}/dt = \partial\bar{h}/\partial t + V^{\star}\,\partial\bar{h}/\partial y = 0 \quad (9)$$

$$\bar{h}|_{t=0} = h_i \quad (10)$$

$$\bar{h}|_{\Gamma_1} = h_0 \quad (11)$$

constitute the hyperbolic problem [for prescribed flux condition, it is incorporated into the solution by Eq. (7) and $dh/dt \approx (h^{n+1} - \bar{h}^n)/\Delta t$.] If the pressure head h^n at time t_n is known, h^{n+1} at $t_{n+1} = t_n + \Delta t$ can be solved by following procedures.

The hyperbolic problem is first solved independently along the characteristic line by an effective characteristic technique developed by the writer for the simulation of transport problem (Huang[2]). The technique can be briefly stated that only one moving particle is needed to track each wetting front along the characteristic line, and then high order and linear interpolation is used for region near the front and remaining parts of the domain respectively to gain the convective contribution of each grid. The parabolic problem can then be solved for h^{n+1} on fixed grids by FE or FD with $\bar{h}^n$ combined by $dh/dt \approx (h^{n+1} - \bar{h}^n)/\Delta t$, which treats each grid as a moving particle will reach the position of the grid at t_{n+1}.

Adopting θ as the independent variable when unsaturated flow only is involved, the flow equation can be treated similarly as follows

$$d\theta/dt = \partial\theta/\partial t + V^{\star}\partial\theta/\partial y \quad (12)$$

$$d\theta/dt = \nabla\cdot(D\nabla\theta) \quad (13)$$

where D is the soil diffusity.

It is worthy to note that there are identical characteristic lines whether θ or h is taken as the variable and furthermore the characteristic line is always unidirection along the gravitational direction even for 2-D problem. For 2-D flow, the alternating direction implicit (ADI) is combined through time step splitting as follows

$$c^{n+1/2}(h^{\star} - \bar{h}^n)/\Delta t/2 = \partial(K^{n+1/2}\partial h^{\star}/\partial x)/\partial x + \partial(K^{n+1/2}\partial h^{n+1,L-1}/\partial y)/\partial y + \mathrm{I\bar{K}}(h^{\star} - h^{n+1,L-1}) \quad \ldots \quad (14a)$$

$$c^{n+1/2}(h^{n+1,L}-\bar{h}^{n})/\Delta t/2=\partial(K^{n+1/2}\partial h^{*}/\partial x)/\partial x+$$
$$\partial(K^{n+1/2}\partial h^{n+1,L}/\partial y)/\partial y+I\bar{K}(h^{n+1,L}-h^{*}) \qquad \ldots (14b)$$

where $c=s\theta/\varepsilon+\partial\theta/\partial h$, h^{*} represents the immediate value of ADI. The third term in right hand side of Eq. (14) is added as Luthin[3] did to achieve a faster convergence of iteration, in which $I=R^{s}(s=1,\ldots,6)$ is the convergence factor and $\bar{K}=(K_{i+1/2,j-1/2}+K_{i+1/2,j+1/2}+K_{i-1/2,j-1/2}+K_{i-1/2,j+1/2})/4$ the average hydraulic conductivity. The Picard's scheme is used in the iterate process with the absolute criterion $\varepsilon_1=0.1$cm (pressure head) and the relative one $\varepsilon_2 = 2\%$.

APPLICATION

As the applications of the proposed methods, following examples are simulated.

One - dimensional

Exam.1 The field experiment of infiltration under ponding in Panoch clay loam (Warrick[7]) is modelled. The properties of soil and the initial and boundary conditions are demonstrated in Fig.1. Using $\Delta z=2$cm and varing time steps: $\Delta t=2, 3, 5, 10$min and $\Delta t_{n+1} = 1.05\times\Delta t_n, \Delta t_1 = 1, 5, 10$ and 60sec., solutions of all discretions are very close to the 'correction', i. e. numerical solution attained by van Genuchten[6] through Philip's quasi - analytical scheme and high order Hermite FE with iteration and fine discretions and have good agreement with observed data (shown in Fig.1).

Exam.2 Suppose that the infiltration as in example 1 occurred under raining condition of intensity $q_0 = 37.8$cm/d which is equal to permeability of the soil, the results simulated by the same discreted steps as above also of good consistency with the 'correction' (Fig.2).

Solutions, however, by FD or linear FE for both of examples above suffer from certain numerical diffusion and oscilation near the wetting front using $\Delta t=2$min and divergence occur at time about t= 86min when $\Delta t>5$min is adopted.

Exam.3 The evaporation experiment in Ida soil which was performed in cylinders with diameter of 7.5cm and height of 38cm under various intensities of illumination by bulbs (Fritton et al[1]) is simulated. The cumulative evaporation process $\varepsilon(t)$ of each soil core is illustrated in Fiq.3. The initial and boundary conditions are $\theta|_{t=0} = 0.463(cm^3/cm^3)$ and $(-D\partial\theta/\partial z+K)|_{z=0}=\varepsilon(t)$, $(-D\partial\theta/\partial z+K)|_{z=38cm} = 0$. The properties of the soil are $D(\theta)=3.33*10^{-13}e^{29.34\theta}cm^2/min$ and $K(\theta)=3.33*10^{-13}e^{54.11\theta}cm/min$ (Selim et al[5]). Using $\Delta z=0.5$cm, $\Delta t=2$min$\sim$4hr increasing with the process of solution, high agreement between the numerical solutions and experimental data is shown in Fig.4

It is noted that no iteration was used during the simulation of above examples by the proposed scheme.

Two - dimensional

An experiment of partially ponding wastewater disposal under symmetrically spaced drainages into loam having intially a water table at depth of y= 130cm and hydrostatic equilibrium thoughout the flow domain was conducted in a polymethyl slab of $3\times2\times0.3m^3$. Adopting $\Delta x=5$cm, $\Delta y=2$cm and $\Delta t_{n+1}=1.05\times\Delta t_n, \Delta t_1=1$min (the maximum Δt is limited to 30min), the free surfaces of some times modelled by the present method (ADCI) with determined properties of the soil (see Fig.5a) exhibit well consistent with those observed.

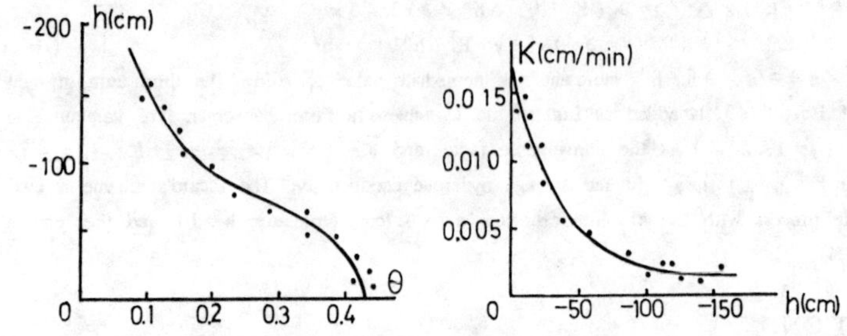

(5a) characteristic curve and hydraulic conductivity of loam soil

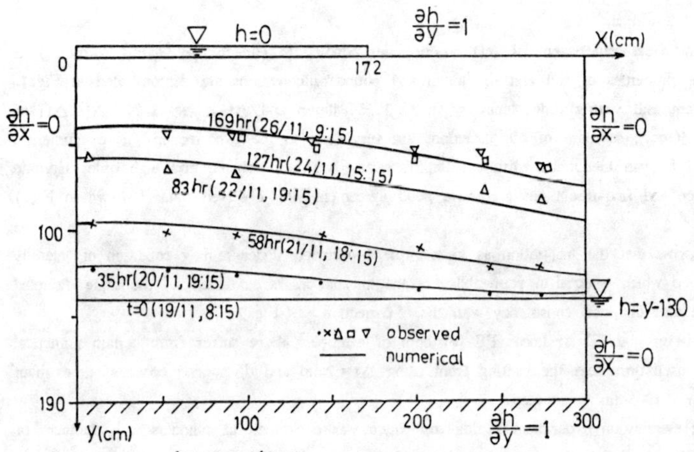

free surfaces

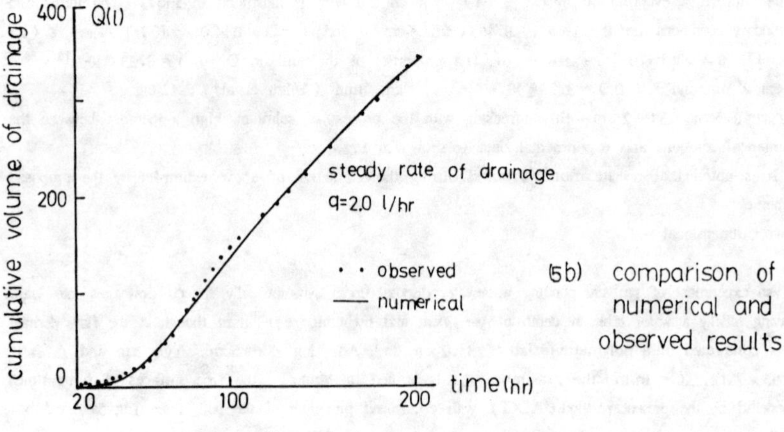

(5b) comparison of numerical and observed results

cumulative drainage process

Fig.5 Numerical simulation of wastewater-ponded disposal

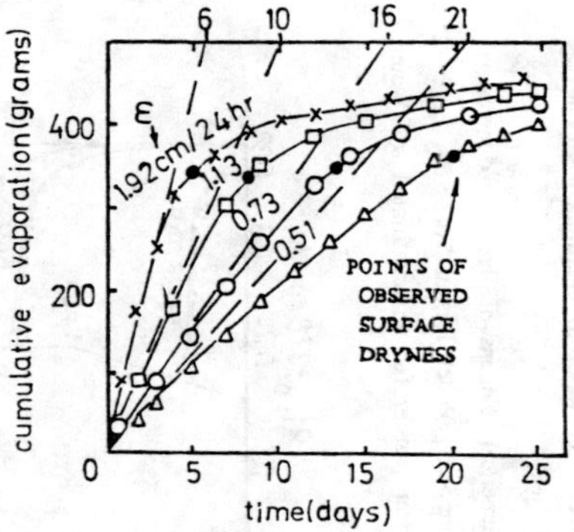

Fig. 3 Cumulative evaporation curves from soil cylinders. ε is evaporation potential (Fritton[1])

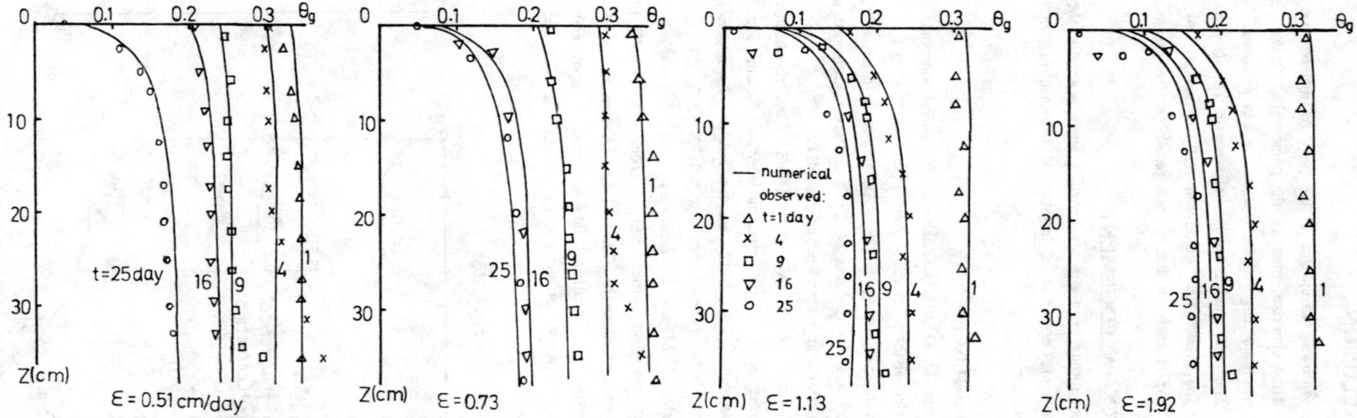

Fig. 4 Moisture profiles under evaporation

CONCLUSION

The following advantages of the proposed methods are demonstrated by examples of application:

1. They eliminate most of the numerical difficulties effectively and ensure high efficiency due to permitting larger discrete sizes and less iterations.
2. They are suitable for general types of boundary conditions.
3. The characteristic method can easily be adopted even for 2-D problems since the characteristic line is always a straight one parallel to the gravitational direction.

ACKNOWLEDGEMENT

The author expresses his heartfelt appreciation for the advice and direction of Wei-Zhen Zhang, professor of Department of Irrigation & Drainage Engineering, Wuhan Univ. of Hydraulic & Electrical Engineering.

REFERENCES

1. Fritton, D. D., and D. Kirkham, (1967), Soil water and chloride redistribution under various evaporation potentials, Soil Sci. Soc. Am. Proc., 31: 599-603.
2. Kang-le, Huang, (1988) An alternating direction Galerkin method combined with characteristic technique for modelling of 2-D saturated-unsaturated transport, in Proceedings of VII Computional Methods in Water Resources., MIT.
3. Luthin, J. N., and A. Orhum, (1975), Coupled saturated-unsaturated transient flow in porous media: exprемental and numerical model, Water Resour. Res., 11 (6).
4. Neuman, S. P., (1984), Adaptive Eulerian-Lagrangian finite element method for advection-dispersion, Int. J. for Num. Meth. Eng., 20, 321-337.
5. Selim, H. M., and D. Kirkham, (1973), Unsteady 2-D flow of water in unsaturated soils above an impervious barriar, Soil Sci. Soc. Am. Proc. 37, 489-495.
6. Van Genuchten, M. Th., (1982), A comparison of numerical solutions of the one-dimensional unsaturated-unsaturated flow and mass transport equations, Adv. Water Resour. 5, 47-55.
7. Warrick, A. W., et al., (1971), Simultaneous solute and water transfer for an unsaturated soil, Water Resour. Res., 7 (5).

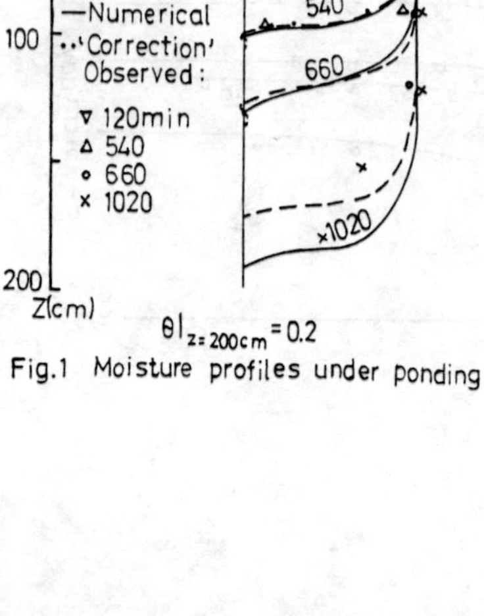

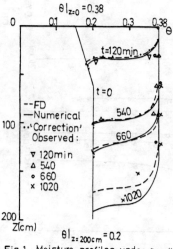

Fig.1 Moisture profiles under ponding

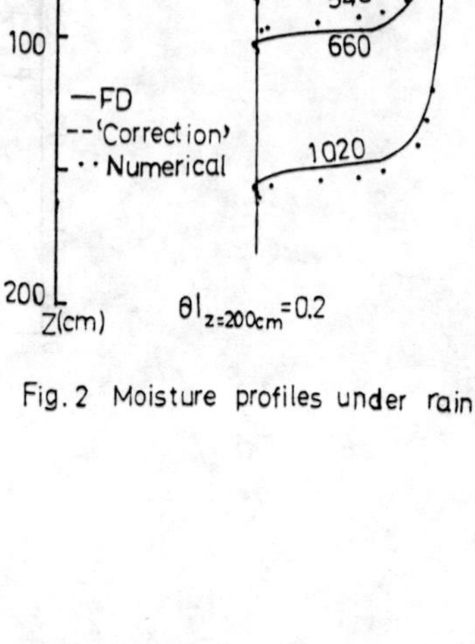

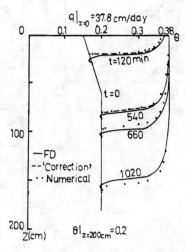

Fig. 2 Moisture profiles under raining

Finite Element Simulation of Nitrogen Transformation and Transport during Hysteretic Flow with Air Entrapment

J.J. Kaluarachchi and J.C. Parker

Virginia Polytechnic Institute and State University, Blacksburg, VA 24061, USA

ABSTRACT

A two-dimensional finite element model was used to predict multispecies transport and transformation of nitrogen accompanying unsaturated flow with hysteresis and air entrapment. Results show that under potential-type boundary conditions, hysteresis substantially influences transport predictions due largely to air entrapment effects. Nonhysteretic simulations using main wetting curve parameters corresponded much more closely to hysteretic results that those using main drainage relations. For contaminant introduction via buried constant strength line sources, hysteresis had little effect on predictions of N transport.

INTRODUCTION

Disposal of secondary wastewater via direct land application or in subsurface drainfields is a potential source of excessive nitrogen loadings into unconfined aquifers. Nitrogen transport is complicated by the occurrence of transformations between various species within the unsaturated zone. These transformation processes are generally mediated by microbial populations. Furthermore, under unsaturated conditions air entrapment and hysteresis in saturation-pressure relationships may substantially affect the accuracy of predictions of flow, which in turn impinges on transport predictions. Previous studies of solute transport with hysteretic flow (Pickens and Gillham, 1980; Jones and Watson, 1987) have been limited to cases without air entrapment and to one dimensional scenarios for single species transport. The objectives of the present study are to evaluate the effects of hysteresis with air entrapment on multispecies nitrogen transport. Since the effect of hysteresis is almost negligible with flux controlled boundary conditions (Kaluarachchi and Parker, 1987), we will focus attention mainly on potential type boundary conditions.

NITROGEN SPECIATION MODEL

The nitrogen speciation model assumes first-order kinetics between solution phase NH_4^+-N and NO_3^--N and organic-N (solid phase) and denitrified-N (gaseous phase) with rate constants K_1 - K_5 as indicated in Fig. 1. Instantaneous and reversible adsorption of NH_4^+-N is assumed here to follow a linear isotherm and NO_3^--N adsorption is regarded as negligible.

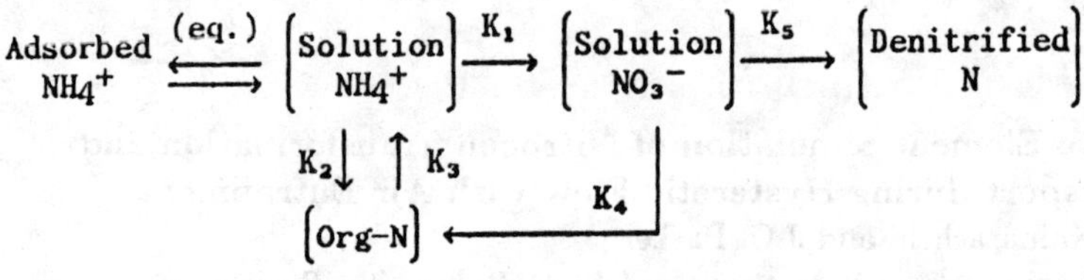

Fig. 1 Nitrogen speciation model

ANALYSIS OF FLOW AND TRANSPORT

Flow of water in an incompressible and variably saturated two-dimensional porous medium under isothermal conditions is described by Richard's equation. The solution of this equation requires numerical solution due to the extremely nonlinear soil hydraulic properties. In this study, we used a Galerkin finite element solution incorporating the method of influence coefficients to avoid costly numerical integration. Complete details of the finite element model are given by Kaluarachchi and Parker (1986).

The convective-dispersive transport equation for species i may be written in the form

$$R_i\theta\frac{\partial C_i}{\partial t} - \frac{\partial}{\partial x}\left[D_x\frac{\partial C_i}{\partial x}\right] - \frac{\partial}{\partial z}\left[D_z\frac{\partial C_i}{\partial z}\right] + q_x\frac{\partial C_i}{\partial x} + q_z\frac{\partial C_i}{\partial z}$$

$$+ Q^*(C_i - C_i^*) + \Phi_i = 0 \quad (1)$$

where C_i is the concentration of the species i in solution (M L^{-3}), θ is the water content, q_x and q_z are flow velocities along the principal x and z directions, D_x and D_z are dispersion coefficients in the x and z directions, R_i is a retardation factor, Q^* and C_i^* are flow rate and concentration of species i in a line source and Φ_i is the net species transformation rate. Defining C_I and C_{II} to be concentrations of NH_4^+-N and NO_3-N, respectively, S_{on} to be the organic-N fraction (MM^{-1}) and ρ to be the soil bulk density, net transformation rates for NH_4^+-N and NO_3^--N, respectively, are

$$\Phi_I = K_1C_I\theta + K_2C_I\theta - \rho K_3S_{on} \quad (2a)$$

$$\Phi_{II} = K_4C_{II}\theta + K_5C_{II}\theta - K_1C_I\theta \quad (2b)$$

The solution of (1) for C_I and C_{II} is obtained by a procedure similar to that for the flow equation except that an upstream weighting technique is utilized to avoid numerical difficulties associated with low dispersion coefficients. Transport of components other than NH_4^+-N and NO_3^--N are not considered. Complete details of the finite element analysis for the N transport model are given by Kaluarachchi and Parker (1987).

HYSTERESIS MODEL

We will employ the hysteretic soil hydraulic property model of Kool and Parker (1987). The hysteretic model is based on van Genuchten's (1980) parametric relations and takes the form

$$\theta^d(h) = \begin{cases} \theta_r + (\theta_s^d - \theta_r)[1+|\alpha^d h|^n]^{-m} & h < 0 \\ \theta_s^d & h \geq 0 \end{cases} \tag{3a}$$

$$\theta^w(h) = \begin{cases} \theta_r + (\theta_s^w - \theta_r)[1+|\alpha^w h|^n]^{-m} & h < 0 \\ \theta_s^w & h \geq 0 \end{cases} \tag{3b}$$

where h is the pressure head, θ_s is the saturated water content, θ_r is the residual water, α and n are shape constants and $m = 1 - 1/n$. The superscript w or d denotes variables pertaining to either the main wetting or main drainage branch of the θ-h relationship, respectively. Hysteresis in hydraulic conductivity, K, versus water content is assumed negligible. Complete details of the hysteresis model are given by Kool and Parker (1987).

SIMULATIONS

Two examples will be considered here for the flow domain illustrated in Fig. 2. The watertable is located 1.5 m below the soil surface and the two end boundaries were assumed to be seepage faces. The assumed soil hydraulic properties are shown in Fig. 3 corresponding to values of parameters θ_s^d, θ_s^w, θ_r, n, α_w and α_d equal to 0.45, 0.38, 0.07, 1.4, 0.045 cm^{-1} and 0.02 cm^{-1}, respectively. The isotropic saturated hydraulic conductivity K_s is 3.0 cm h^{-1}. Rate constants for nitrogen transport K_1, K_2, K_3, K_4 and K_5 were assumed to be 0.02, 0.0063, 0, 0.02 and 0.0063 h^{-1}, respectively. The distribution coefficient for NH_4^+-N adsorption is 1.0 cm^3 g^{-1}. The effective molecular diffusion coefficient (D_O) is taken as 0.12 cm^2 h^{-1} and longitudinal and transverse dispersivities are assumed to be 1.0 and 0.2 cm, respectively.

EXAMPLE I

From an equilibrium initial condition for the flow problem, water was ponded at the soil surface in region AB (Fig. 2) while the remainder of the upper surface and the entire lower boundary received zero flux. Seepage boundaries were stipulated on both sides. Zero initial concentrations were assumed throughout the domain. Influent solution applied at the upper surface was taken to have NH_4^+-N and NO_3^--N concentrations of 50 and 10 ppm, respectively, modeled by third-type boundary conditions. Predicted NO_3^--N concentrations along the center line are shown in Fig. 4. Here and elsewhere, H indicates simulations which employ the hysteretic θ-h relations, ND denotes nonhysteretic analyses with the main drainage θ(h) curve and NW denotes those for the main wetting curve. Rapid movement of NO_3^--N for the ND simulation relative to the H and NW cases is evident. Total NH_4^+-N mass for the ND simulation is also very much higher than that predicted for the H and NW simulations (Fig. 5). These effects mainly reflect reductions in hydraulic conductivity at the soil surface and hence in infiltration rates for H and NW cases due to lower water contents caused by air entrapment. This suggests that if effective rather than actual saturated water contents and conductivities are used to analyze flow, neglecting hysteresis may be justifiable.

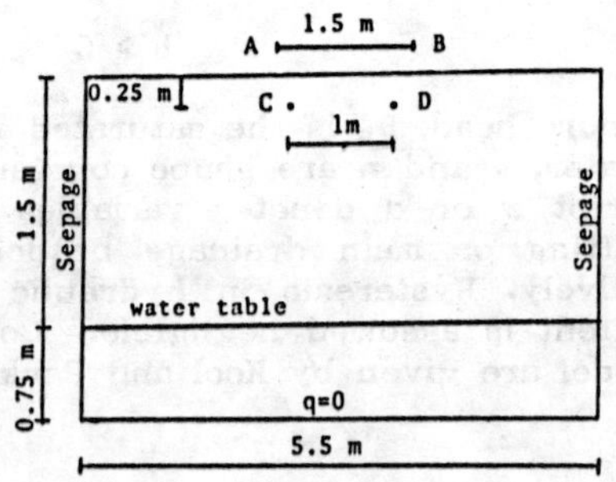

Fig. 2 Flow domain used in simulations.

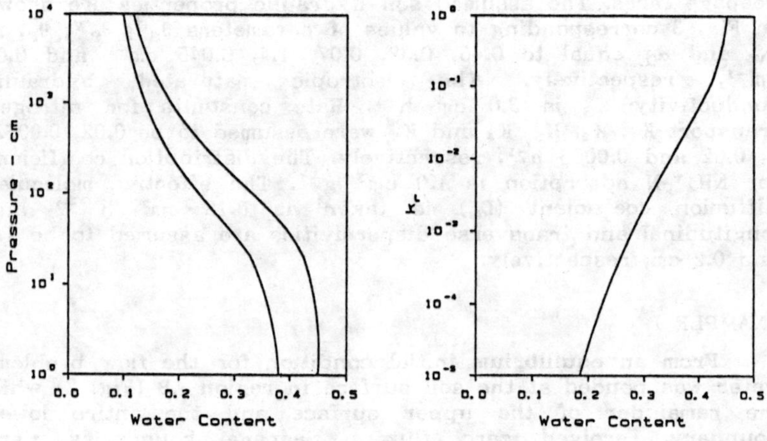

Fig. 3 Soil hydraulic properties used in simulations.

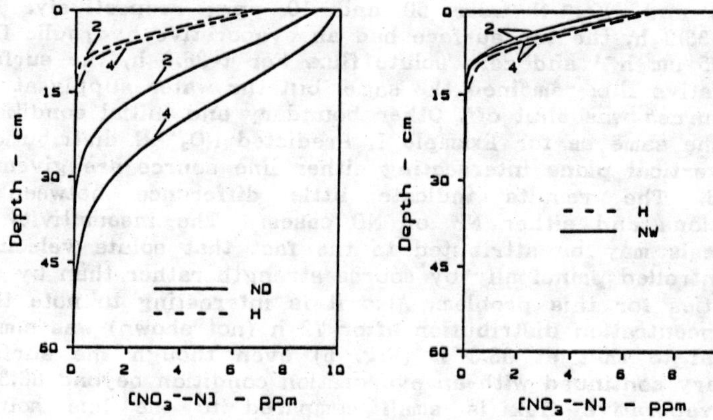

Fig. 4 Distribution of NO_3^--N along the center line for Example I.

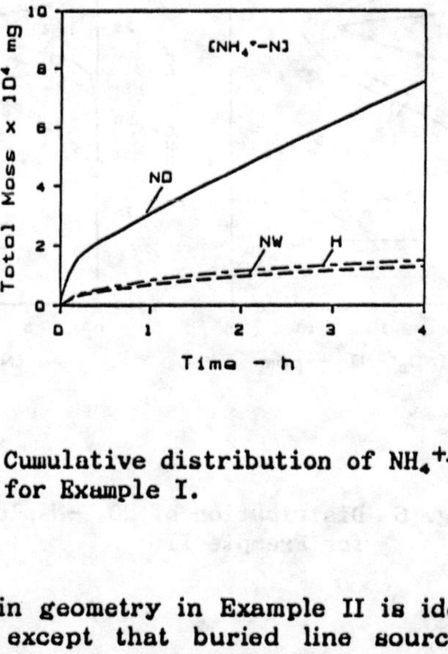

Fig. 5 Cumulative distribution of NH_4^+-N with time for Example I.

EXAMPLE II

The flow domain geometry in Example II is identical to that of the Example I except that buried line sources have been added at locations C and D. For 0<t<1.3 h, the entire top surface had a solute free hydraulic flux of 6.0 cm h^{-1} and flow at each line source of 10 cm^3 cm^{-1} h^{-1}. Influent concentrations of

NH_4^+-N and NO_3^--N were 50 and 10 ppm, respectively. For 1.3h<t<36.3 h, the top surface had an evaporative hydraulic flux of 0.005 cm h^{-1} and zero solute flux. For t>36.3 h, the surface evaporative flux remained the same, but the water supply at the line sources was shut off. Other boundary and initial conditions were the same as for Example I. Predicted NO_3^--N distributions on a vertical plane intersecting either line source are given in Fig. 6. The results indicate little difference between H simulations and either NW or ND cases. The insensitivity to hysteresis may be attributed to the fact that solute velocities are controlled principally by source strength rather than by soil properties for this problem. Also it is interesting to note that the concentration distribution after 72 h (not shown) was almost identical to that at 36.3 h (Fig. 6) even though the surface boundary continued with an evaporation condition beyond 36.3 h. The evaporative flux is small compared to the line source strength and quasi-steady state conditions are maintained.

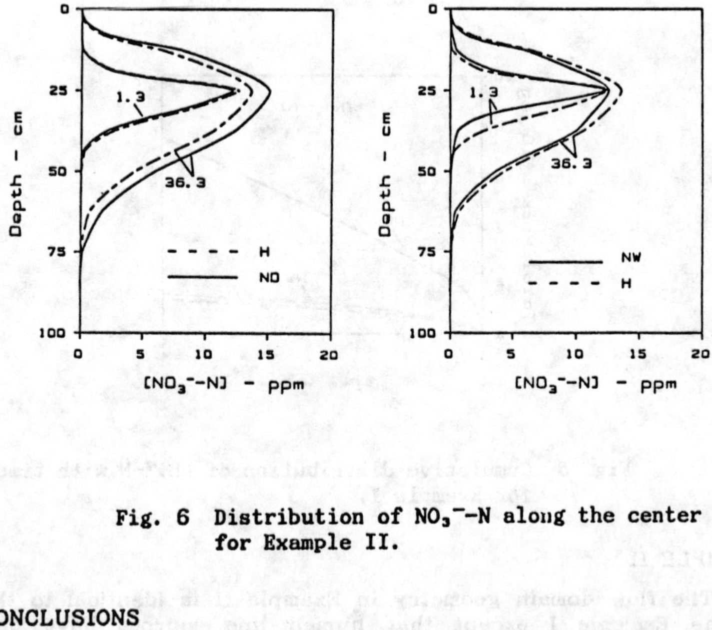

Fig. 6 Distribution of NO_3^--N along the center line for Example II.

CONCLUSIONS

Effects of air entrapment and hysteresis in water retention relations on flow can may have substantial effects on species transport under certain conditions. For potential-type hydraulic boundary conditions, fluid entrapment results in markedly lower solute velocities compared to nonhysteretic simulations based on main drainage θ(h) parameters, while nonhysteretic analyses using the main wetting branch closely approximate hysteretic results. Since these effects are attributable principally to differences in hydraulic conductivities at apparent water saturation for wetting and drying paths, it follows that effects of disregarding hystersis will also be small if data for a primary drainage path beginning from θ_s^w were employed. For solute introduction via controlled flux line sources, effects of hysteresis on transport were observed to be quite small and simulation with either wetting or draining water retention data provided an adequate representation of the hysteretic system.

REFERENCES

Jones, M. J. and R. W. Gillham, Effects of soil water hysteresis on solute movement during intermittent leaching, Water Resour. Res., 23, 1251-1256, 1987.

Kaluarachchi, J. J. and J. C. Parker, Finite element analysis of water flow in variably saturated soil, J. Hydrol., 90,269-291, 1987.

Kaluarachchi, J. J. and J. C. Parker, Effects of hysteresis with air entrapment on water flow in the unsaturated zone, Water Resour. Res., 23, 1967-1976, 1987.

Kaluarachchi, J. J. and J. C. Parker, Finite element model of nitrogen species transformation and transport in the unsaturated zone, J. Hydrol., (in review), 1987.

Kool, J. B. and J. C. Parker, Development and evaluation of closed-form expressions for hysteretic soil hydraulic properties, Water Resour. Res., 23, 105-114, 1987.

Pickens, J. H. and R. W. Gillham, Finite element analysis of solute transport under hysteretic unsaturated flow conditions, Water Resour. Res., 16, 1071-1078, 1980.

van Genuchten, M. Th., A closed-form equation for predicting the hydraulic conductivity of unsaturated soils, Soil Sci. Soc. Am. J., 44,892-898, 1980.

A Characteristic Finite Element Model for Solute Transport in Saturated-Unsaturated Soil

Jin-Zhong Yang and Wei-Zhen Zhang
Department of Irrigation and Drainage, Wuhan Institute of Hydraulic and Electric Engineering, Wuhan, Hubei, People's Republic of China

ABSTRACT

In this paper characteristic finite element method is proposed to solve convection-dispersion equation accounting for mobile and immobile water. The mathematical model using a moving and deforming coordinate system is free of numerical dispersion and oscillation. The suggested method has been used to simulate two-dimensional solute transport experiment. The agreement between calculated and experimental result is good.

GOVERNING EQUATION AND NUMERICAL MODEL

Experiments[1] carried out in laboratory demonstrate that when solute transport in unsaturated and structured soil, immobile water phase does exist in dead end pore, unconnected and semiconnected pore. Due to molecular diffusion the mass exchange occurs between mobile and immobile water phase. The governing equation representing solute transport in unsaturated and saturated soil can be written as:

$$\frac{\partial \theta_{im} C_{im}}{\partial t} + \frac{\partial \theta_m C_m}{\partial t} = \frac{\partial}{\partial x_i} \theta_m D_{ij} \frac{\partial C_m}{\partial x_j} - \frac{\partial}{\partial x_i} q_i C_m \qquad (1)$$

$$\frac{\partial \theta_{im} C_{im}}{\partial t} = \alpha (C_m - C_{im}) \qquad (2)$$

where θ_{im} and θ_m are the immobile and mobile water content ($\theta_{im} + \theta_m = \theta$), respectively, θ is the volumetric water content, C_{im} and C_m the concentration of immobile and mobile water, D_{ij} the component of hydrodynamic dispersion coefficient tensor, q_i the component of water flux vector, α the mass transfer coefficient.

When dispersion dominated in Eq. (1), the equation is parabolic in character, which can be solved successfully by finite difference or finite element methods. When convection dominates, the characterof the equation changes to hyperbolic ; the method of characteristics is effective for solving this kind of equation . The combined use of finite element method and the method of characteristics is effective for both convection-dominated and dispersion-dominated problems.

Expanding the second term of right side in Eq. (1) and assuming the immobile water content to be constant, the following equation results:

$$\theta_m \frac{\partial C_m}{\partial t} + \frac{\partial \theta_{im} C_{im}}{\partial t} = \frac{\partial}{\partial x_i}\left[\theta_m D_{ij} \frac{\partial C_m}{\partial x_j}\right] - q_i \frac{\partial C_m}{\partial x_i} \qquad (3)$$

Let $\vec{y}$ represent a moving point in the interesting domain at time t , $\vec{U}$ represent the velocity of point $\vec{y}$. The new position of point $\vec{y}$ at time t can be calculated from the following equation:

$$\vec{x}\,(\vec{y},t) = \int_{t_0}^{t} \overleftarrow{V}\, dt \qquad (4)$$

The concentration at $\vec{x}$ can be written as:

$$C(\vec{x},t) = C(\vec{x}(\vec{y},t),\ t) \qquad (5)$$

Representing $\partial C_m/\partial t$ in Eq. (3) as concentration change in time of point $\vec{y}$ moving at velocity $\vec{U}$ and rearranging yields:

$$\theta_m \frac{\partial C_m}{\partial t}\bigg|_{\vec{y}} + \frac{\partial \theta_{im} C_{im}}{\partial t} = \frac{\partial}{\partial x}\left[\theta_m D_{ij} \frac{\partial C_m}{\partial x_j}\right] - \left[q_i - \theta_m \frac{dx_i}{dt}\right] \frac{\partial C_m}{\partial x_i} \qquad (6)$$

where dx_i/dt is the velocity of moving point $\vec{y}$. When $\vec{y}$ is taken as a fluid particle, the velocity of the moving point is same as velocity q_i/θ_m. that is:

$$q_i/\theta_m = dx_i/dt$$

substituting above equation in (6) yields:

$$\theta_m \frac{DC_m}{Dt} + \frac{\partial \theta_{im} C_{im}}{\partial t} = \frac{\partial}{\partial x_i}\left[\theta_m D_{ij} \frac{\partial C_m}{\partial x_j}\right] \qquad (7)$$

where DC_m/Dt represents material derivative, here Cm no longer represents concentration at a point in space, but rather concentration of a fluid particle moving at velocity $\vec{U}$.

For two-dimensional problem Eq. (7) can be written as:

$$\theta_m \frac{DC_m}{Dt} + \frac{\partial \theta_{im} C_{im}}{\partial t} = \frac{\partial}{\partial x}\left[\theta_m D_{xx} \frac{\partial C_m}{\partial x} + \theta_m D_{xy} \frac{\partial C_m}{\partial y}\right] + \frac{\partial}{\partial y}\left[\theta_m D_{yx} \frac{\partial C_m}{\partial x} + \theta_m D_{yy} \frac{\partial C_m}{\partial y}\right] \qquad (8)$$

where

$$\theta_m D_{xx} = \theta_m (\alpha_L V_x^2 + \alpha_T V_y^2)/V + D_m$$

$$\theta_m D_{xy} = \theta_m D_{yx} = \theta_m (\alpha_L - \alpha_T) V_x V_y / V$$

$$\theta_m D_{yy} = \theta_m (\alpha_L V_y^2 + \alpha_T V_x^2)/V + D_m$$

Dm is the molecular diffusion coefficient in porous media, α_L , α_T the longitudinal and transverse dispersivity, respectively. The Galerkin finite element method is used to determine the approximate solution of Eq. (8). The nodal points are taken as fluid particles. The trial function has the form:

$$\widetilde{C}_m = \sum_{j=1}^{N} \phi_j C_{mj}(t) \tag{9}$$

where ϕ_j is the basic function, $C_{mj}(t)$ the undetermined time-dependent coefficient at the jth point of the discrete system and N the total number of nodes.

Substituting Eq. (9) into (8) and setting the resulting residual orthogonal to all the function ϕ_i, we obtained:

$$\sum_{\beta}\sum_{j} \frac{1}{4\Delta_\beta} C_{mj}\left[\overline{\theta_m D_{xx}} b_i b_j + \overline{\theta_m D_{xy}}\,(b_i d_j + b_j d_i) + \overline{\theta_m D_{yy}} d_i d_j\right]$$
$$+ \sum_{\beta} \frac{DC_m}{Dt}\Big|_i \cdot \frac{1}{3} \sum_{P} \theta_{m_p} \frac{1+\delta_{ip}}{4} \Delta_\beta + \sum_{\beta} \frac{\partial \theta_{im} C_{im}}{\partial t}\Big|_i \frac{\Delta_\beta}{3}$$
$$= \sum_{e} \int_{\Gamma_2^\beta} \left[\theta_m \left(D_{xx} \frac{\partial C_m}{\partial x} + D_{xy} \frac{\partial C_m}{\partial y} \right) n_x + \theta_m \left(D_{yx} \frac{\partial C_m}{\partial x} + D_{yy} \frac{\partial C_m}{\partial y} \right) n_y \right] N_i \, d\Gamma \tag{10}$$

where:

$$\overline{\theta_m D_{xx}} = \frac{1}{3} \sum_{P=i,j,k} \theta_m(h_p)\,(\alpha_L V_{x_p}^2 + \alpha_T V_{y_p}^2)/V + D_0 \alpha_m e^{\beta_m \theta(h_p)}$$

$$\overline{\theta_m D_{xy}} = \frac{1}{3} \sum_{P=i,j,k} \theta_m(h_p)\,(\alpha_L - \alpha_T)\, V_{x_p} V_{y_p}/V$$

$$\overline{\theta_m D_{yy}} = \frac{1}{3} \sum_{P=i,j,k} \theta_m(h_p)\,(\alpha_L V_{x_p}^2 + \alpha_T V_{y_p}^2)/V + D_0 \alpha_m e^{\beta_a \theta(h_p)}$$

d_p, b_p are the element dimensions, Δ_β the element area.

Replacing the time derivative in Eq. (2) by difference and rearranging yields:

$$\frac{\partial \theta_{im} C_{im}}{\partial t}\Big|_i \approx \frac{\theta_{im}\alpha}{\theta_{im} + Q\alpha\Delta t_k} C_{m_i}^{k+Q} - \frac{\theta_{im}\alpha}{\theta_{im} + Q\alpha\Delta t_k} C^{k+Q} \tag{11}$$

$$C_{im_i}^{k+1} \approx \frac{\alpha + \Delta t_k}{\theta_{im} + Q\alpha \cdot \Delta t_k} C_{m_i}^{k+1} + \frac{\theta_{im} - \alpha\Delta t_k (1-Q)}{\theta_{im} + Q\alpha\Delta t_k} C_{im_i}^{k} \tag{12}$$

where Q is the time weighting factor, $0<Q<1$.

Substituting Eq. (11) into (10) leads to the following system of ordinary differential equation:

$$[D]\{C_m\} + [E]\left\{\frac{DC_m}{Dt}\right\} = \{F\} \tag{13}$$

Replacing material derivative DC_m/Dt by difference yields:

$$\frac{DC_m}{Dt} \approx \frac{C_{m_i}^{k+1} - C_{m_i}^{k}}{\Delta t_k} \tag{14}$$

where $C_{m_i}^{k+1}$, $C_{m_i}^{k}$ are the concentration of moving node i at time t^{k+1}, and t^k, respectively.

Substituting Eq. (14) into (13) and rearranging yields:

$$([D] + [E]/\Delta t_k)\{C_m\}^{k+1} = [E]/\Delta t_k \{C_m\}^k + \{F\} \tag{15}$$

There are various methods to deal with the movements of nodes[2]. To reduce the error caused by projection of concentration in the moving and fixed networks, continuous movement of nodal points is recommended in this paper. The shapes of elements are checked continuously in the process of calculation. When deformation of one of the elements appears, the nodes and elements are rearranged to form a new network. The methods of dividing of elements, automatic generation of elements, adding and eliminating of elements along boundary and the judgement of element shape are proposed to control the deformation of elements and regulate the distribution of elements.

The movement of finite element network can be determined as follows: A nodal point, located at position (x_i^k, y_i^k) at time t^k, will reach a new position (x_i^{k+1}, y_i^{k+1}) at time t^{k+1}, where

$$x_i^{k+1} = x_i^k + \int_{t^k}^{t^{k+1}} V_x dt, \quad y_i^{k+1} = y_i^k + \int_{t^k}^{t^{k+1}} V_y dt \tag{16}$$

Then, adding and eliminating nodal points along boundary andcheckingthe shape of elements, if all the requirements are satisfied, the concentration distribution can be obtained from the solution of Eq. (15). Otherwise, the elements are regulated or generated automatically.

COMPARISON OF NUMERICAL RESULTS WITH ANALYTICAL SOLUTION AND EXPERIMENTAL DATA

The numerical model is tested with analytical solution given by Coast and Smith[3] for one-dimensional solute transport in saturated soil with mobile and immobile water fraction. The comparison is shown in Fig.1 for the case where V=0.05, a=0.00001, al=0.1, θ=0.4, θm=0.34.

Two-dimensional saturated-unsaturated experiment was conducted in laboratory in a soil tank having internal dimension of 300cm in length, 200cm in height and 30cm in thickness (Fig.2). Water flow and solute transport parameters are given in table 1 and 2.

Water flow equation can be written as:

$$\left(\frac{\theta}{u} S_s + \theta(h)\right)\frac{\partial h}{\partial t} = \frac{\partial}{\partial x}\left[K(h)\frac{\partial h}{\partial x}\right] + \frac{\partial}{\partial y}\left[K(h)\frac{\partial h}{\partial y}\right] + \frac{\partial K(h)}{\partial y} \tag{17}$$

$$V_x = -\frac{K(h)}{\theta_m}\frac{\partial h}{\partial x}, \qquad V_y = -\frac{K(h)}{\theta_m}\left(\frac{\partial h}{\partial y} + 1\right) \tag{18}$$

Eq. (17) was solved by finite element method. The solution procedure and comparison with experimental data are given in reference [4] in detail. To obtain more accurate velocity distribution, Eq. (18)is also solved by Galerkin finite element method.

Boundary and initial conditions for Eq. (2) and (8) are:

$$C_m(x,y,t) = 159/L, \ t = 0, \ 0 \leqslant x \leqslant 300, \ 0 \leqslant y \leqslant 190$$

$$C_{im}(x,y,t) = 159/L, \ t = 0, \ 0 \leqslant x \leqslant 300, \ 0 \leqslant y \leqslant 190$$

$$C_m(x,y,t) = 0, \qquad t > 0, \ 0 \leqslant x \leqslant 172, \ y = 190$$

$$\left(\theta_m D_{xx} \frac{\partial C_m}{\partial x} + \theta_m D_{xy} \frac{\partial C_m}{\partial y} \right) n_x + \left(\theta_m D_{yx} \frac{\partial C_m}{\partial x} + \theta_m D_{yy} \frac{\partial C_m}{\partial y} \right) n_y = 0 ,$$

on other boundary

Fig.3 illustrates concentration profile at x=20cm and x=94cm. It can be seen from the figure that the concentration fronts are very sharp in the process of leaching and the agreement between numerical results and the experimental data are satisfactory.

The concentration evolution of some measured points are illustrated in Fig.4. It can be seen from the figure that the experimental concentration distribution lags behind the calculated one and some tailing phenomena appear. Probably, it is due to the delayed response of soil salinity sensor used. The numerical results are in fairly good agreement with the experimental data.

Fig.5 indicates the calculated and experimental results in whole profile. Because the measured points are not enough, the experimental result does not represent the sharp concentration front very well. In general, the agreement of numerical and experimental resultsis good.

REFERENCES

[1] Yang J-Z(1986), Experimental investigation of one-dimensional hydrodynamic dispersion in saturated and unsaturated soils, Journal of Hydraulic Engineering, 3,10-21. In Chinese.

[2] Yang J-Z(1987), Comparison of numerical method for solving convection-dispersion equation, Journal of Wuhan Institute of Hydraulic and Electric Engineering, 5,11-18.

[3] Coast,K.H., Smith,B.D.,(1964), Dead-end pore volume and dispersion in porous media, J.Soc.Petrol.Eng., 4(1),73-84.

[4] Yang J-Z,(1987),Experimental and theoretical study of water flow in saturated-unsaturated soil, Journal of Hydraulic Engineering, In press.

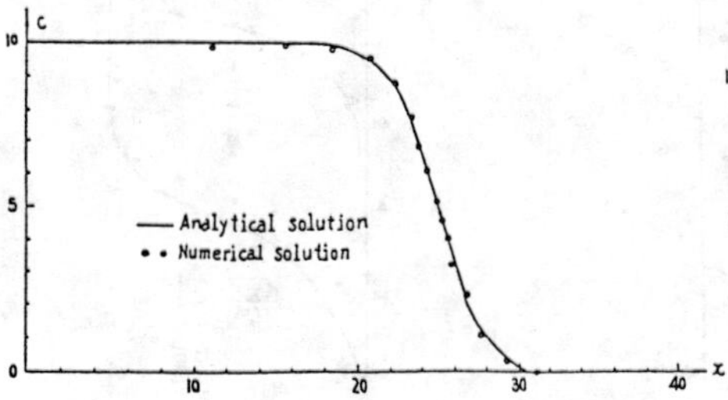

Fig.1. Comparison of numerical model with analytical solution

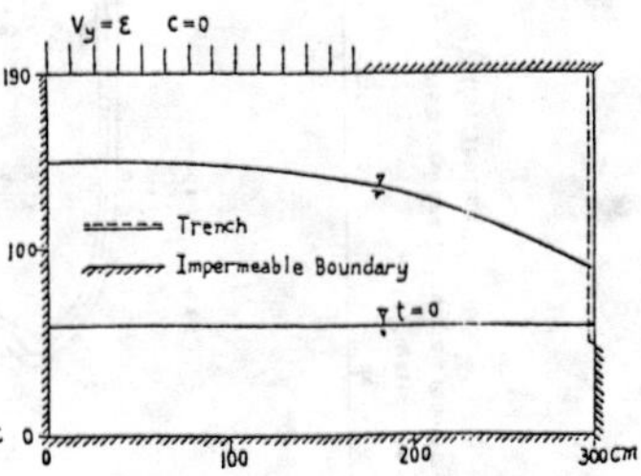

Fig.2 Schematic representation of experimented condition

Table 1. Water Flow Parameters

h cm	θ(h)	C(h) cm^{-1}	K(h) cm·sec^{-1}
-100	0.018		
- 90		7x10^{-4}	
- 80	0.032		1.94x10^{-7}
- 75		1.3x10^{-3}	
- 70	0.045		
- 65		2.0x10^{-3}	
- 60	0.065		6.94x10^{-6}
- 55		3.1x10^{-3}	
- 50	0.094		1.94x10^{-5}
- 45		4.6x10^{-3}	
- 40	0.142		6.94x10^{-5}
- 30		5.9x10^{-3}	8.33x10^{-4}
- 25			1.73x10^{-3}
- 20	0.261		3.72x10^{-3}
- 17.5		4.2x10^{-3}	
- 15	0.282		
- 12.5		2.5x10^{-3}	
- 10	0.294		
- 7.5		1.0x10^{-3}	1.02x10^{-2}
- 5	0.299		
- 4			1.12x10^{-2}
- 2.5		1.6x10^{-4}	
0	0.3	1.07x10^{-5}	1.12x10^{-2}

Table 2. The Transport Parameters

unsatueated longitudinal dispersivity	0.2cm
unsaturated transverse dispersivity	0.04cm
mass transfer coefficient	0.0001 min^{-1}
saturated longitudinal dispersivity	0.12cm
saturated transverse dispersivity	0.03cm
saturated water content	0.412
immobile water content	0.04
infiltration rate	0.003825 cm/min

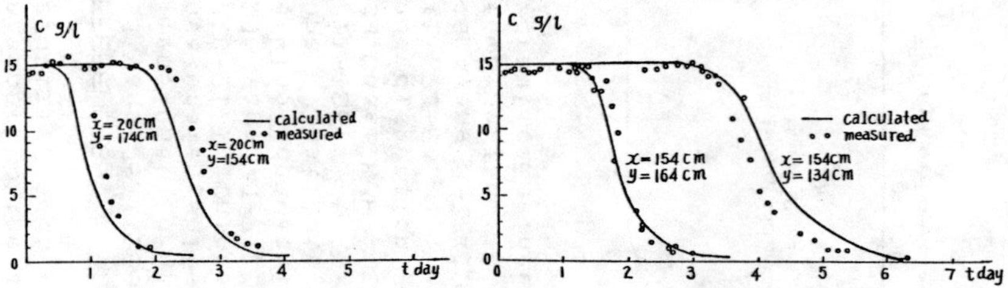

Fig. 4. Measured and calculated time-evolution of concentration at different points

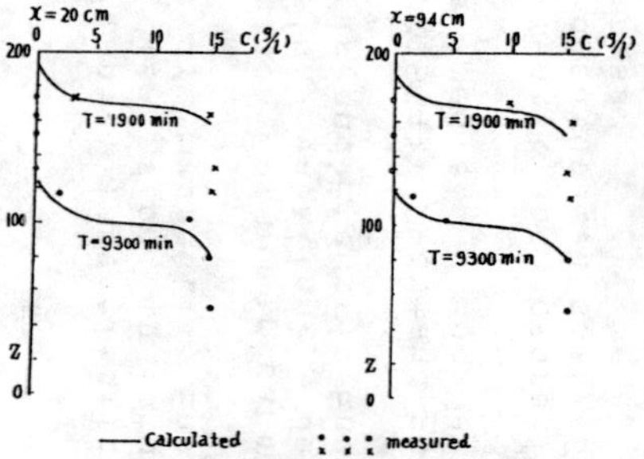

Fig. 3. Comparison between measured and calculated concentration profiles at x=20cm and x=94cm

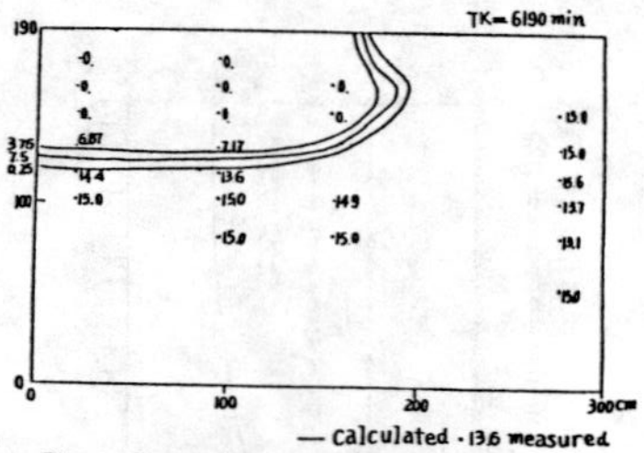

Fig. 5 Measured and calculated concentration distribution

SECTION 2C - CHEMICAL PROCESSES

Solute Transport: Equilibrium vs Non-equilibrium Models

R. Abeliuk
Faculty of Engineering, Catholic University of Chile, Casilla 6177, Santiago, Chile

INTRODUCTION.

In applying transport models to reactive solutes, most authors have assumed that either an equilibrium or a non-equilibrium model can be used to describe the behaviour of the substance under consideration as it infiltrates through the soil. Generally, preference has been shown for equilibrium models mainly because of (1) their mathematical (computational) simplicity as compared with non-equilibrium models and, (2) the equilibrium constants required for their successful application are either readily available or can be determined without difficulties from simple laboratory batch experiments. However, in the past few years, several authors have been preoccupied with the validity of the local equilibrium assumption (LEA) and studies have been carried out to determine the conditions under which this assumption breaks down.

James and Rubin (1979), concluded from their experimental work that the LEA only applies when the coefficient of hydrodynamic dispersion is "nearly the same" as the molecular diffusion coefficient in the soil. This implies that kinetics based models should be used for large flow rates, when the coefficient of hydrodynamic dispersion is significantly larger than the coefficient of molecular diffusion. Jennings and Kirkner (1984), used a numerical approach based on the comparison of dimensionless numbers (the Damkholer number) to identify ranges of parameter values for which equilibrium and non-equilibrium models generate identical concentration profiles. These authors also found that high flow rates lead to large simulation errors if local equilibrium is assumed. Recently, Valocchi (1985) using expressions for the time moments of the solute breakthrough curves, noted that the issue of the validity of the LEA depends upon a complex interplay between transport properties such as flow velocity and coefficient of hydrodyanmic dispersion with mass transfer and distribution coefficients.

The work reviewed above refers only to displacement studies under steady flow conditions and cannot be extended to solute transport through unsaturated porous media. Here, a parameter identification technique applied to numerical experiments was used to test the validity of the different assumptions. The advantage of this approach is that it is applicable to saturated and unsaturated flow conditions.

THEORY.

One-dimensional vertical unsaturated zone transfer processes are generally based on the classical Richards equation for unsaturated water flow -equation (1)-, and the Fickian-based convection dispersion equation for solute transport -equation (2) (Bresler et al., 1982)

$$c(h)\frac{\partial h}{\partial t} = \frac{\partial}{\partial z}\left(K(h)\left(\frac{\partial h}{\partial z} - 1\right)\right) \tag{1}$$

$$\theta\frac{\partial Cs}{\partial t} = \frac{\partial}{\partial z}\left(\theta D(\theta,q)\frac{\partial Cs}{\partial t}\right) - q\frac{\partial Cs}{\partial z} - \rho\frac{\partial S}{\partial t} \tag{2}$$

In these equations, h represents the pressure head in cm of water; c(h) is the soil water capacity or the slope of the water retention curve, $\theta(h)$; θ is the volumetric water content; K(h) is the unsaturated hydraulic conductivity; z is the depth; t is the time; Cs is the solute concentration in the soil solution; $D(\theta,q)$ is the moisture and velocity dependent coefficient of hydrodynamic dispersion; q is the volumetric fluid flux density; ρ is the soil bulk density and S is the solute concentration associated with the solid phase of the soil. Sources and sinks of water and solute in the system have not been considered in expressions (1) and (2).

Analytical or quasi-analytical solutions of equations (1) and (2) only exist for steady-state water flow in homogeneous soils. For unsaturated flow conditions these equations have to be solved by numerical methods (finite differences or finite elements) subject to the appropriate initial and boundary conditions for the problem under consideration.

The parameter estimation technique presented herein was applied to numerical experiments assumed to represent soil columns for which c(h), K(h) and $D(\theta,q)$ were supposed to be described by the following formulae respectively,

$$K(h) = \frac{A}{B + |h|^C} \tag{3}$$

$$c(h) = \frac{\theta s * D * E * |h|^{E-1}}{(D + |h|^E)^2} \tag{4}$$

$$D(\theta,q) = W\left(\frac{q}{n}\right)^n \tag{5}$$

with A=3.18e+07, B=7.33e+06, C=11.7, D=1.05e+08, E=13.0, θs=0.4072, W=0.0452 and n=1.844.

The assumed initial conditions correspond to constant moisture content (0.1000 cm^3/cm^3) and zero solute concentration throughout the column. The upper boundary condition was taken as constant surface water and solute fluxes and the lower as initial moisture and solute concentrations at the bottom end of the column. Mathematically, these conditions are expressed as:

	Water flux	Solute flux	
Initial Condition	$r_1 = 0.1000$	$C_1 = 0.0$	$Z \geq 0$; $t = 0$

Upper boundary	$q_0 = -K(h)(\frac{\partial h}{\partial z} - 1)$	$qCi = qCs - D(\theta,q)\theta\frac{\partial Cs}{\partial z}$	$Z = 0 ; t > 0$
	with $q_0 = 3,41$ cm/h		
Lower boundary	$\theta_L = \theta_i$	$C_L = C_i$	$Z = L ; t > 0$

NUMERICAL EXPERIMENTS.

A parameter identification technique was used to estimate the parameters of the equation governing the transport of solutes that interact with the soil. Rosenbrock's technique (Clarke, 1973) was chosen because it does not require the evaluation of the gradient and because of its computational simplicity.

In this paper, two situations were considered. In the first case, it was assumed that the interaction between the solute and the soil is described by an equilibrium relationship, and in the second case that the interaction between the solute and the soil does not reach equilibrium during the duration of the infiltration event.

Equilibrium situation.

Among the large number of models proposed for describing equilibrium sorption, the linear and Freundlich isotherm models have been the most commonly used (Rao and Jessup, 1983). In this case the linear model given by expression (6) was employed to describe the equilibrium situation.

$$S = K\ Cs \tag{6}$$

In expression (6), K represents the sorption or distribution coefficient and was assigned a value of 1. Cs and s have been previously defined.

A numerical experiment was conducted to generate a data set using the equilibrium model resulting from combining (2) with the differential form of (6). Thus,

$$q\frac{\partial Cs}{\partial t} = \frac{\partial}{\partial z}(\theta D(\theta,q)\ \frac{\partial Cs}{\partial t}) - q\frac{\partial Cs}{\partial z} - \rho K\frac{\partial S}{\partial t} \tag{7}$$

The sequential solution of Richards' equation and the solute transport equation, expressions (1) and (7) respectively, was carried out by the implicit- explicit finite difference technique described by Abeliuk (1987). ρ was assigned a value of 1.0324.

Non-equilibrium situation.

In an analogous form, a data set was generated using the following expression to represent a non-equilibrium kinetics.

$$\frac{\partial S}{\partial t} = k1\ Cs - k2\ S \tag{8}$$

which together with expression (2) constitute the non-equilibrium model. In this case k1 was assigned the value 2.5 and k2, 0.5.

RESULTS

The numerical experiments carried out were designed to test whether equilibrium data could be described by a non-equilibrium model and viceversa. Thus, the parameter estimation technique was used to find the equilibrium and non-equilibrium model parameters that best fitted the numerically generated non-equilibrium and equilibrium data, respectively.

The parameters governing both the equilibrium and non-equilibrium models are those related to the coefficient of hydrodynamic dispersion (namely W and n), and those governing the kinetics of the adsorption process for the non-equilibrium model (k1 and k2) or the distribution coefficient K for the equilibrium model. For the purpose of parameter identification, two distinct situations arise: (1) the coefficient of hydrodynamic dispersion is known, that is W and n have been identified. In this case the parameter identification problem is reduced to the evaluation of one (equilibrium model) or two (non-equilibrium model) parameters and, (2) the coefficient of hydrodynamic dispersion is unknown and therefore three (equilibrium model) or four (non-equilibrium model) parameters have to be identified.

Case 1. Equilibrium model - non-equilibrium data.

Figure 1 shows the results of the application of the parameter identification technique to the estimation of the parameters of the equilibrium model using non-equilibrium data. The data points described in the figure as optimised data equilibrium model (1) correspond to a situation in which the coefficient of hydrodynamic dispersion was assumed to be known. For the purpose of identifying K, two separate optimisation runs were carried out with two initial guesses for K (5.0 and 10.) both converging to a value of 1.61 for K. Figure 1 shows clearly that the identified parameter fails to match the non-equilibrium data in a satisfactory manner. Not only there are large discrepancies between the optimised and the numerically generated data but also the shapes of the curves are quite different. The second set of data points, corresponding to optimised data equilibrium model (2), was obtained by applying Rosenbrock's procedure to the identification of the parameters W, n and K. In this case a good agreement between the numerical data and the optimised data is observed indicating that a 3-parameter equilibrium model can be used to match data generated by an equilibrium model. The implications of this are discussed in a later paragraph.

Case 2. Non-equilibrium model - equilibrium data.

Again two possibilites were considered: a two-parameter model (k1 and k2 from the kinetics expression) and a four-parameter model (k1 and k2 plus W and n corresponding to the coefficient of hydrodynamic dispersion). The results of the application of the parameter identification technique are shown in Figure 2. This figure shows: (a) the two parameter model (optimised non-equilibrium model (1)), gives a reasonably accurate representation of the numerical data, preserving the shape of the curve. (b) the four-parameter model (optimised non-equilibrium model (2)), produces absolute coincidence of the results with a perfect match between the data points at all depths.

The above observations confirm the findings of Valocchi (1985), in that equilibrium and non-equilibrium models, under certain conditions, produce solute profiles that are indistinguishable from one another. Thus, it appears that independently of the velocity of the flow, equilibrium and non-equilibrium models can generate identical solute profiles depending only on the values of the distribution or kinetics coefficients.

If an optimisation technique is to be used to determine the parameters governing equilibrium and non-equilibrium solute transport equations -as is normally done- it is evident then, that more information on the physico-chemical behaviour of the system is needed before the optimisation procedure is applied. More specifically, a knowledge of the kinetics of the adsorption process will certainly be useful to discriminate between the models. It is important to note that if the dispersion coefficient is known (from separate tracer experiments for example), the equilibrium model could not be fitted to non-equilibrium data.

It has been argued (Nielsen et al., 1986) that the form of the dispersion coefficient is uncertain and that the only satisfactory alternative for its evaluation appears to be by means of an optimisation technique (Parker and Van Genuchten, 1984). Under these circumstances, it would seem that all the parameters governing these models, that is those related to the coefficient of hydrodynamic dispersion and to the kinetics of the adsorption process have to be evaluated simultaneously. However, this is not the case because the parameters governing the coefficient of hydrodynamic dispersion can be evaluated from a separate data set.

CONCLUSIONS.

The results of the application of a parameter search technique to the estimation of the values of the parameters governing the flow of a reactive solute, show that information on the physico-chemical aspects of the adsorption process is needed in general in order to be able to discriminate between an equilibrium and a non-equilibrium model. This information can be obtained from simple laboratory adsorption experiments. Once this information is available, it is possible to estimate the kinetics or equilibrium parameters quite readily by Rosenbrock's parameter search technique. The data required are the concentration and moisture profiles for the adsorbed solute, for a simultaneous determination of the dispersive and the kinetics or equilibrium parameters, or moisture and concentration profiles for a tracer and for a sorbed solute for a separate determination of these parameters.

REFERENCES

ABELIUK, R. (1987). Parameter Identification in Unsaturated Flow and Solute Transport Models. Ph.D. Thesis, University of London, 349 pp.

BRESLER, E., McNeal, B.L., and Carter, D.L. (1982). Saline and Sodic Soils: Principles, Dynamics, Modeling. Advanced Series in Agricultural Sciences, 10. Springer-Verlag, Berlin.

CLARKE, R.T. (1973). Mathematical Models in Hydrology. Irrigation and Drainage Paper 19. FAO, Rome.

JAMES, R.V. and Rubin, J. (1971). Applicability of the Local Equilibrium Assumption to Transport Through Soils of Solute Affected by Ion Exchange. In: Chemical Modeling in Aqueous Systems. E.A. Jenne (Ed.) ACS, Washington, DC.

JENNINGS, A.A., and Kirkner, D.J. (1984). Instantaneous Equilibrium Approximation Analysis. Journal of Hydraulic Engineering, ASCE, 110, (12), 1700-1717.

NIELSEN, D.R., Van Genuchten, M.Th., and Biggar, J.W. (1986). Water Flow and Solute Transport Processes in the Unsaturated Zone. Water Resources Research, 22, (9), 89S-108S.

PARKER, J.C., and Van Genuchten, M.Th. (1984). Determining Transport Parameters from Laboratory and Field Tracer Experiments. Virginia Agricultural Experiment Station. Bulletin 84-3, Virginia Polytechnic Institute and State University.

RAO, P.S.C. and Jessup R.E. (1983). Sorption and Movement of Pesticides and other Toxic Organic Substances in Soils. In : Chemical Mobility and Reactivity in Soil Systems. Nelson et al. (Eds.) SSSA Special Publication Number 11.

VALOCCHI, A.J. (1985). Validity of the Local Equilibrium Assumption for Modeling Sorbing Solute Transport Through Homogeneous Soils. Water Resources Research, 21, (6), 808-820.

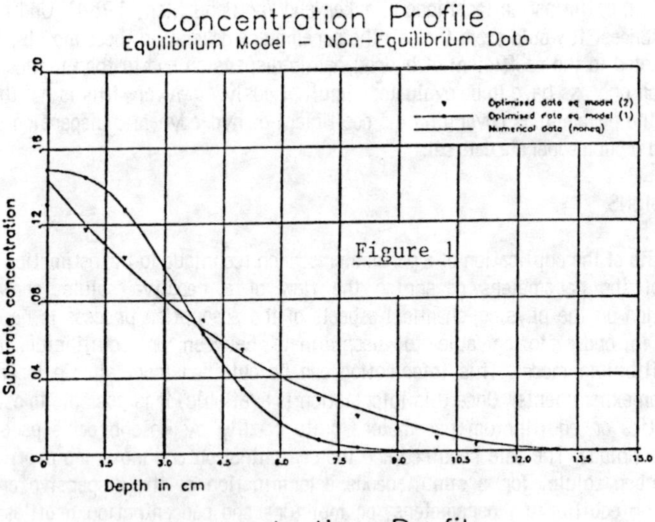

Figure 1

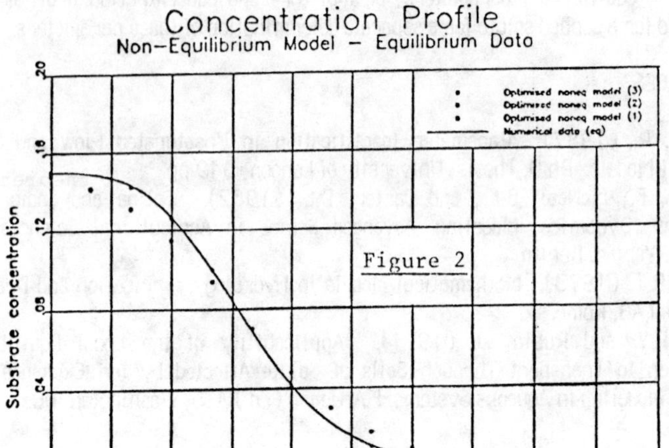

Figure 2

Confrontations Between Computer Simulations and Laboratory Work to Understand Mechanisms Controlling Transport of Mercury

Ph. Behra

Université Louis Pasteur, Institut de Mécanique des Fluides, UA CNRS 854, 2 rue Boussingault, F-67083 Strasbourg Cedex, France

Present Address: Swiss Federal Institute for Water Resources and Water Pollution Control (EAWAG), ETH-Zurich, CH-8600, Dübendorf, Switzerland

ABSTRACT

It is shown how confrontations between computed simulations and discriminating laboratory work allow the development of modelling by validating the assumed concepts, i.e. a step towards a better knowledge of the mechanisms of interactions which act during the exchanges between liquid and solid phases. It appeared that it is necessary to extend the surface complexation concept, that the local equilibrium assumption cannot be assumed, and that the metal uptake is due to several kinds of surface sites with different transfer rates.

INTRODUCTION

The possibilities for metal micropollutants to migrate or to be retained during transport in groundwater are related to their chemical properties and to mechanisms responsible for exchanges between liquid and solid phases. From laboratory work including column experiments, we have observed that the breakthrough of mercury - a trace metal - is differently delayed depending on the chemical composition of the percolating solution (complexing agents, Eh, pH) (e.g., Behra[1]), and on the composition of surface sites (clay minerals, oxides/hydroxides of iron, aluminium and manganese, or silica) (e.g., Behra[1]).

From these observations, a one-dimensional theoretical model is developed for describing the transport of mercury through a natural quartz sand. This model is based on sets of non-linear equations : groundwater mass transport, component speciation in liquid phase, and exchanges between liquid and solid phases. It will be shown how confrontations between simulations and discriminating experiments allow the improvement of modelling.

THEORY

To describe one-dimensional advective-dispersive transport of metal micropollutants, the following multicomponent transport equations are used for each component of the solution (e.g., Cederberg et al.[4]) :

$$\partial[C]/\partial t + \rho/\varepsilon . \partial[C_s]/\partial t = D_L . \partial^2[C]/\partial x^2 - u . \partial[C]/\partial x \qquad (1)$$

where [C] is the concentration of the component C in the mobile phase (mol/l) ; $[C_s]$ is the concentration of the component C on the surface (mol/g soil) ; D_L is the hydrodynamic dispersion coefficient (m^2/s) ; u is the mean pore velocity (m/s) ; ε is the medium porosity (l/l) ; and ρ is the medium bulk density (g soil /l).

$\partial[C_s]/\partial t$ represents the accumulation term of C on the immobile phase while $[C_s]$ is also the sum of the concentrations of all the species sorbed on the surface.

[C] and $[C_s]$ depend on the chemical reactions in the solution and on exchange reactions between liquid and solid phases. In the case of mercury experiments, the chemical reactions in the solution are essentially soluble complexation reactions. The set of equations constitutes the SPECIATION model (e.g., Behra[2]). The general equation and the stability complexation constant are defined as follows for a metal M :

$$M^{2+} + nL^- \rightleftharpoons ML_n^{(2-n)+} \qquad K_{L,n} = [M^{2+}].[L^-]^n \qquad (2)$$

where L^- is a ligand (OH^-, Cl^- or NO_3^-) and n=1 or 2 for NO_3^-, or 1,2,3 or 4 for OH^- and Cl^-.

The other equilibria such as water dissociation, CO_2 dissolution and dissociation in water are also taken into account in this model. Moreover electroneutrality in the solution is also included. Dissolved species are computed from equilibria using mass-law equations ($K_{L,n}$) found in literature (e.g., Behra[2]).

On the other hand, the exchange part is based on surface coordination reactions (e.g., Stumm and Morgan[9]) :

$$M^{2+} + 2(S\text{-}OH) \rightleftharpoons M(S\text{-}O)_2 + 2H^+ \qquad (3)$$

where M^{2+} is a dissolved cationic mercury species ; (S-OH) is a surface site ; and $M(S\text{-}O)_2$ is M_s, the metal species fixed on the surface.

The conditional stability constant of surface complexation reaction (equation (3)) is :

$$\beta_2^* = ([M(S\text{-}O)_2].[H^+]^2)/([M^{2+}].[S\text{-}OH]^2) \qquad (4)$$

During the exchanges, electroneutrality on the solid phase is assumed :

$$N_p = (S\text{-}OH) + M_s \tag{5}$$

where N_p is sites concentration on solid phase (mol/g soil).

The set of equations (1) is numerically computed with finite differences in the case of implicit scheme and SPECIATION model is resolved using the secant method (e.g., Behra[2]).

What about kinetics ? If local equilibrium assumption (LEA) (e.g., Rubin[8] ; Jennings[6]) is not assumed, it is necessary to consider the rate of transfer between liquid and solid phases, i.e. to detail the accumulation term on solid phase. Either chemical kinetics and/or diffusional step can be supposed.

MATERIAL AND METHODS

Experimental apparatus and methods were previously described elsewhere (e.g., Behra[1]). Experimental conditions are indicated in Table I.

Table I : Experimental conditions

Experiment	Mercury salt	C_o (10^{-7}M)	Acid or salt	[Acid] or [salt] (M)	pH	M_s (1) (g)	V_p (l)	u (10^{-5} m/s)	L (2) (m)	V_r/V_p (3)
a	$HgCl_2$	4,6	NaCl	10^{-2}	5,6	690	0,143	8,3	0,20	56
b	$Hg(NO_3)_2$	5,0	HNO_3	10^{-2}	2,0	684	0,144	7,7	0,20	52
c	$HgCl_2$	5,0	NaCl	10^{-2}	5,6	1650	0,350	9,0	0,50	--
d	$HgCl_2$	5,0	NaCl	10^{-2}	5,6	1650	0,350	100	0,50	--

(1) M_s is the mass of sand in a column
(2) L is the length of a column ; columns are 0,05 m in diameter
(3) V_r is the restitution volume determined when C/C_o=1

RESULTS AND DISCUSSION

The proposed exchange model is based on the uptake of free metal ions (Hg^{2+}) by surface complexation. Moreover Davies and Leckie[5] have assumed that the hydroxide metal cation, MOH^+, could be sorbed by surface complexation, i.e. $HgOH^+$ in our case. Figure 1 illustrates two experimental curves from experiments performed with either 10^{-2}M NaCl or 10^{-2}M HNO_3 (e.g., Behra[2]). The simulation experiments are based upon the determination of the product $\beta_2^*.(N_p)^2$ from retardation factors (V_r/V_p) if the processes are not time-dependent. V_r is the restitution volume obtained when C/C_o reaches 1. If $\beta_2^*.(N_p)^2$ is determined for curve a, it is not possible to achieve a correct estimate of curve b : the uptake is too much important. To improve the results, the hypothesis of Davies and Leckie[5] has been completed : the fixation of other cationic mercury species ($HgOH^+$, $HgCl^+$ or $HgNO_3{}^+$) was assumed. The computed breakthrough curves are depicted in figure 1. They are in the same order as are experimental curves. However the slopes of the curves are too steep.

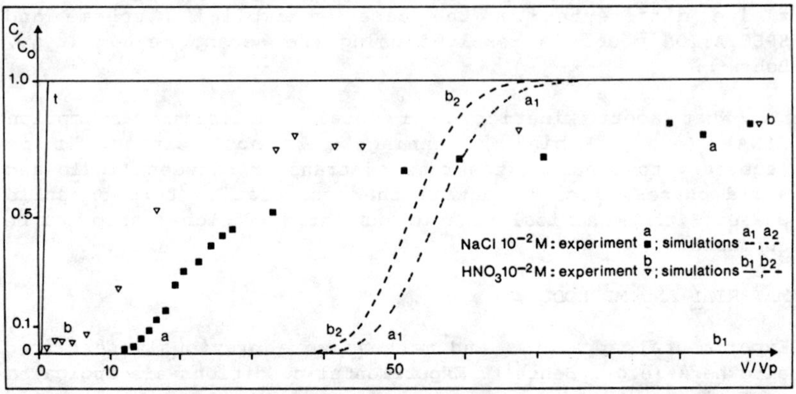

Figure 1 : Breakthrough curves of mercury : comparison of experimental curves and computed curves in the case of LEA

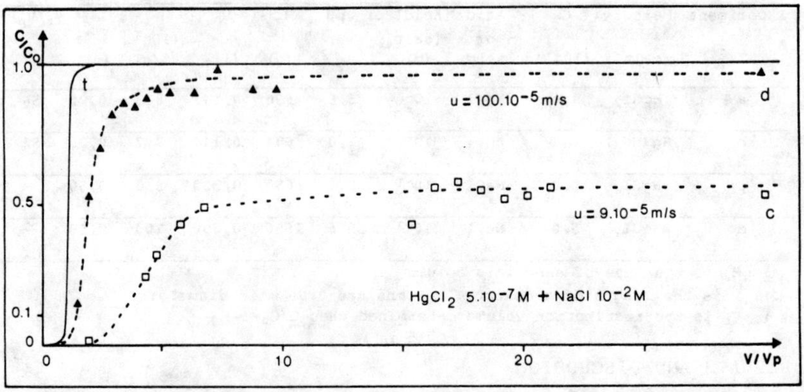

Figure 2 : Breakthrough curves of mercury : effect of mean pore velocity

As the observed discrepancy could be due to LEA, the effect of the contact period was then investigated by modifying the mean pore velocity (Figure 2) : the higher the velocity, i.e. the smaller the contact period between the two phases, the longer the delay of elution. Now it was previously shown that surface complexation processes are instantaneous (e.g., Behra[2], Bourg[3]). So the rate of transfer between the two phases was introduced into the model by taking into account external diffusion equation (e.g., Nicoud[7]) :

$$\rho/\varepsilon.\partial[M_{s,i}^{z+}]/\partial t = k_e.([M_i^{z+}] - [M_i^{*z+}]) \qquad (6)$$

where k_e is a mass-transfer constant depending on the geometric structure of solid particles, on the molecular diffusion, and on hydrodynamics ; $[M_i^{*z+}]$ is the concentration which should be observed at equilibrium with $[M_i^{z+}]$. $[M_i^{*z+}]$ is then calculated from equations (2), (4) and (5).

With this new assumption, it is possible to fit a computed curve (curve b_3, Figure 3) on experimental curve b. However a discrepancy is again observed between experimental curve a and the corresponding computed curve a_3 (Figure 3). Mercury uptake is not important enough. On the other hand, C/C_o does not reaches 1 during the first 100 V_p.

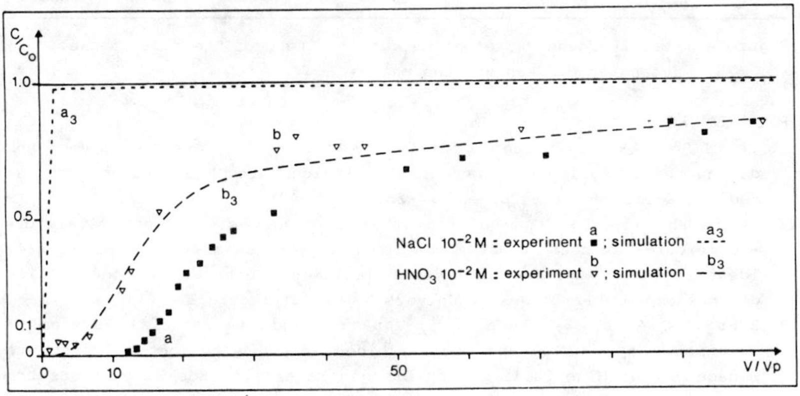

Figure 3 : Breakthrough curves of mercury : comparison of experimental and simulated curves (without LEA)

At this step of the discussion, the question is : how could the model be improved ? We think that two kinds of surface sites could retain mercury cations. This hypothesis is based on previous results where it was shown that mercury is sorbed on the one hand on iron, aluminium and manganese oxides/hydroxides, and on the other hand on clay minerals and silica (e.g., Behra[1]). The diffusion kinetics of one of these surface sitesis fast, while the other is slow : two different k_e values are to be determined. The results obtained for experiment a are reported in figure 4. The beginning of the computed curve exactly fits the experimental curve in the case of the higher k_e value, but C/C_o reaches too quickly 1. On the other hand, the tails of the computed and experimental curves are superimposed with the smaller k_e value.

Work is under way to find out a scheme taking into account the two types of surface sites. So this paper shows how confrontations between computed simulations and laboratory experiments not only allow to build up a model able to describe trace metal transport, but also provide information on mechanisms acting during exchanges between solid and liquid phases.

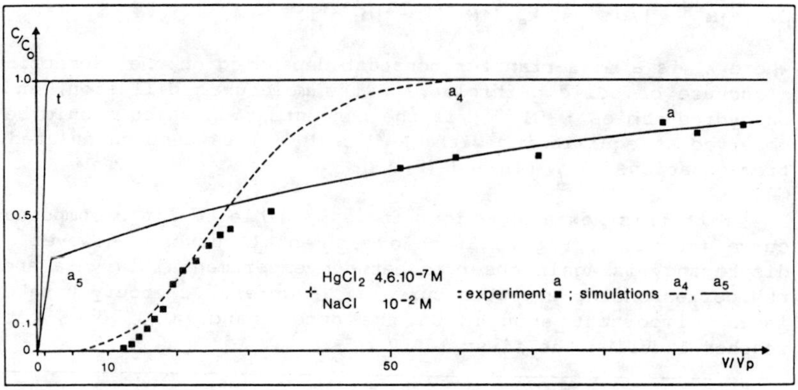

Figure 4 : Breakthrough curves of mercury : effect of the rate of the transfer between the two phases

REFERENCES

1. Behra Ph. (1986). Evidences for the existence of a retention phenomenon during the migration of a mercurial solution through a saturated porous medium, Geoderma, Vol. 38, pp. 20-25.
2. Behra Ph. (1987). Etude du comportement d'un micropolluant métallique -le mercure- au cours de sa migration à travers un milieu poreux saturé: identification expérimentale des mécanismes d'échange et modélisation des phénomènes. Ph-D Thesis, University of Strasbourg I.
3. Bourg A.C.M. (1988). Metals in aquatic and terrestrial systems : sorption, speciation and mobilization. In : Environmental Impact and Management of Mine Tailings and Dredged Materials (Eds W. Salomons and U. Förstner), Springer Verlag. Berlin. In press.
4. Cederberg G.A., Street R.L. and Leckie J.O. (1985). A groundwater mass transport and equilibrium chemistry model for multicomponent systems, Water Resour. Res., Vol. 21, pp. 1095-1104.
5. Davies J.A. and Leckie J.O. (1978). Surface ionization and complexation at the oxide/water interface. II.Surface properties of amorphous iron oxyhydroxide and adsorption of metal ions, J. Coll. Inter. Sci., Vol. 67, pp. 90-107.
6. Jennings A.A. (1987). Critical chemical reaction rates for multi-component groundwater contamination models, Water Resour. Res., Vol. 23, pp. 1775-1784.
7. Nicoud R.M. (1987). Influence respective des facteurs thermodynamiques, hydrodynamiques et diffusionnels sur le fonctionnement des échangeurs d'ions. Application à un procédé d'échange ionique : le Nymphéa. Ph-D Thesis, Institut National Polytechnique de Lorraine.
8. Rubin J. (1983). Transport of reacting solutes in porous media : relation between mathematical nature of problem formulation and chemical nature of reactions, Water Resour. Res., Vol. 19, pp. 1231-1252.
9. Stumm W. and Morgan J.J. (1981). Aquatic Chemistry, Wiley. New York.

ACKNOWLEDGMENTS

This work was supported by the Interdisciplinarian Environment Research Program (PIREN) of CNRS.

A Quick Algorithm for the Dead-End Pore Concept for Modeling Large-Scale Propagation Processes in Groundwater

H.M. Leismann*, B. Herrling, V. Krenn
Institute of Hydromechanics, University of Karlsruhe, West Germany
** Present Address: Paderborner Landstr. 14, D-4770 Soest, Germany*

INTRODUCTION

Calculating the spreading of contaminants in groundwater, a further phenomenon besides the dispersion is noticeable: the well-known tailing. The break-through-curves show a distinct unsymmetry. This effect is physically interpreted by the existence of mobile and immobile water (the so-called dead-ends): The contaminated mobile water drops a part of the pollution by diffusion to the immobile water. There the pollutant is stored temporarily and is given back to the mobile water after the pollution-front has passed.

This phenomenon cannot be described by the convection-dispersion-equation. Already in 1964, Coats and Smith[1] have proposed a mathematical model to describe this phenomenon. This model uses a second unknown variable, the concentration of the immobile water, and a simple linear approach for the diffusive exchange. This model has been successfully used for one-dimensional modeling of break-through-curves found out at soil-columns in the laboratory. Till now, however, it has been seldom applied in large-scale models. The reason is, that two unknowns per node have to be computed, which increases the computational-effort considerably.

In this paper a method is described to reduce the additional computational-effort to almost nil by a computation in two steps. The model of Coats and Smith[1] can now, without disadvantage, be applied to the computation of large-scale propagation.

THE MODEL OF COATS AND SMITH[1]

Coats and Smith[1] add to the convection-dispersion-equation a source-term or a sink-term, which considers the diffusive exchange between the mobile and immobile water (s. Eq. (1)). They assume that the quantity of the exchange is proportional to the concentration-difference. They describe the concentration of the immobile water by a second differential-equation (Eq. (2)), which, according to the definition, contains no moving-terms.

$$n_m \frac{\partial c_m}{\partial t} + v_i \, c_{m,i} - n_m \, D_{ij} \, c_{m,ji} = \beta \, (c_{im} - c_m) \qquad (1)$$

$$n_{im} \frac{\partial c_{im}}{\partial t} = \beta \, (c_m - c_{im}) \qquad (2)$$

where c_m and c_{im} are the concentrations of the mobile and immobile water $[M/L^3]$, n_m and n_{im} are the fractions of soil filled with mobile and immobile water [−], v_i is the Darcy-velocity [L/T], D_{ij} is the dispersion-tensor $[L^2/T]$, β is the exchange-coefficient [1/T], i,j are the indices of spacial co-ordinates and t is the time co-ordinate [T]. With this differential equation set break-through-curves of tracer experiments of soil columns can be described quite well.

THE ALGORITHM

The conventional method for the numerical solution of Eq. (1) and (2) yields a four times larger matrix because of the two unknowns per node. Compared with the solution of the convection-dispersion-equation the computational-effort increases considerably. For this reason, up to now, the differential equation set (Eq. (1) and (2)) is rarely applied to the computation of large-scale propagation. In the following a method is described to split the solution into two steps and to reduce the additional computational-effort to almost nil.

The exchange terms on the right-hand sides in Eq. (1) and (2) and the time-derivative in Eq. (2) are lumped. This is in accordance with the physical facts, because c_{im} at one place is independent of c_{im} in the surroundings (Eq. (2) contains no space-derivative). If the coefficients n_{im} and β vary spacially, they should be averaged around the node of concern.

Because of the lumping of Eq. (2) it can be inserted, in a discretized form, in the discretized Eq. (1). The result is an equation-set, which has only one unknown per node in the new time-level and which therefore requires only the conventional computational-effort. Thereafter, the second unknown is computed nodewise by means of the discretized Eq. (2) without an equation-set. Eq. (2) is discretized:

$$n_{im} \frac{c_{im}^1 - c_{im}^0}{\Delta t} = \beta \, \{\theta \, (c_m^1 - c_{im}^1) + (1 - \theta) \, (c_m^0 - c_{im}^0)\} \qquad (3)$$

where θ is the weighting-factor for the time-levels (θ = 0.5: Crank-Nicholson, θ = 1: backward implicit) and the index 1 stands for the new time-level and 0 for the old time-level.

c_{im}^1 is isolated in Eq. (3):

$$c_{im}^{1} = \frac{n_{im} c_{im}^{0} + \Delta t \, \beta \, \{\theta \, c_{m}^{1} + (1 - \theta) \, (c_{m}^{0} - c_{im}^{0})\}}{n_{im} + \Delta t \, \beta \, \theta} \tag{4}$$

Now the right-hand side of Eq. (1) is discretized:

$$Lc_{m} = \beta \, \{\theta \, (c_{im}^{1} - c_{m}^{1}) + (1 - \theta) \, (c_{im}^{0} - c_{m}^{0})\} \tag{5}$$

where L is the differential operator at the left-hand side of Eq. (1). Eq. (4) is set into Eq. (5):

$$Lc_{m} = \frac{\beta \, n_{im}}{n_{im} + \Delta t \, \beta \, \theta} \, \{c_{im}^{0} - (1 - \theta) \, c_{m}^{0} - \theta \, c_{m}^{1}\} \tag{6}$$

The unknown c_{im}^{1} no longer appears in Eq. (6) (c_{im}^{0} is known from the previous time-level). The term with c_{m}^{1} on the right-hand side is lumped (see above) and results in a diagonal matrix which is added to the matrix, resulting from the conventional discretisation of Lc_{m}. The other terms are known from the previous time-level and remain on the right-hand side. This equation-set for c_{m}^{1} is not bigger than that for the solution of the conventional convectiondispersion-equation, because all terms with c_{im} remain on the right-hand side.

In a second step, after c_{m}^{1} is computed node by node, c_{im}^{1} is computed from Eq. (4). All needed terms are available nodewise, so that Eq. (4) can be evaluated nodewise. No equation-set has to be solved. The required computational-effort is negligibly small. The following time-levels are computed in the same way.

It is stressed that the described method is entirely mathematically identical with the simultaneous computation of the two unknowns in one large equation-set. There are no simplifications.

STABILITY

The numerical solution of Eq. (1) and (2) is, depending on the time-step and the weighting-factor θ, not unconditionally stable or tends to oscillations. A computation of the stability is possible for the special case of $v_i = \underline{0}$ and $D_{ij} = \underline{0}$. The resulting conclusion is transferable to the general case. In the special case of $v_i = \underline{0}$ and $D_{ij} = \underline{0}$, after a very big time-step, the volumetric averaged concentration must be:

$$c_{m}^{1} = c_{im}^{1} = \bar{c} = \frac{n_{m} c_{m}^{0} + n_{im} c_{im}^{0}}{n_{m} + n_{im}} \tag{7}$$

However, depending on the choice of θ, c_{m}^{1} and c_{im}^{1} are oscillating around $\bar{c}$. It can be shown, that the oscillations stop for:

$$\theta \geq 1 - \frac{n_{im} n_{m}}{n_{im} + n_{m}} \cdot \frac{1}{\beta \, \Delta t} \tag{8}$$

The result in the extreme case $\beta \, \Delta t \to \infty$ is $\theta = 1$.

The application of $\theta = 1$ for the exchange terms is practically very successful. But other factors should be used for the temporal weighting of the convective-term and the dispersive-term (e.g. $\theta_{conv.} = \theta_{disp.} = 1/2$ (see Leismann[4], Leismann and Herrling[3]. That simplifies Eq. (4) and (6) to:

$$Lc_m = \frac{\beta\, n_{im}}{n_{im} + \Delta t\, \beta}\left(c^0_{im} - c^1_m\right) \qquad \text{(1st step, matrix)} \qquad (9)$$

$$c^1_{im} = \frac{n_{im}\, c^0_{im} + \Delta t\, \beta\, c^1_m}{n_{im} + \Delta t\, \beta} \qquad \text{(2nd step, nodewise)} \qquad (10)$$

In the extreme case of $\beta \to \infty$, i.e. very quick exchange, the well-known retardation results automatically, where:

$$R = 1 + \frac{n_{im}}{n_m} \qquad (11)$$

which is easy to show. Thus for reaching this, there is only to apply a β, which is sufficiently big. The conventional way of computation with $\beta \to \infty$, however, yields a stagnation, because there are very large numbers on the main diagonal and, in contrast to this, the convective-term and the dispersive-term become numerically insignificant (compare penalty-function in the handling of boundary-conditions).

EXPANSION WITH DECAY

First order decay in the dead-ends can also be dealt with by the described method. The differential equation set is as follows:

$$n_m \frac{\partial c_m}{\partial t} + v_i\, c_{m,i} - n_m\, D_{ij}\, c_{,ji} + n_m\, \lambda\, c_m = \beta(c_{im} - c_m) \qquad (12)$$

$$n_{im} \frac{\partial c_{im}}{\partial t} + n_{im}\, \lambda\, c_{im} = \beta(c_m - c_{im}) \qquad (13)$$

where λ is the decay-parameter $[1/T]$ (half-life period: $t_{half} = -\ln(1/2)/\lambda$). The decay-terms are lumped and, likewise, handled with $\theta = 1$ to avoid oscillations (see above). The discretized equations, which are successively to apply, run as follows ($\theta = 1$ for the exchange and the decay):

$$Lc_m + n_m\, \lambda\, c_m = \frac{\beta\, n_{im}}{n_{im}(1+\Delta t\, \lambda)+\Delta t\, \beta}\left\{c^0_{im} - (1+\Delta t\, \lambda)\, c^1_m\right\} \qquad (14)$$

$$c^1_{im} = \frac{n_{im}\, c^0_{im} + \beta\, \Delta t\, c^1_m}{n_{im}(1 + \Delta t\, \lambda) + \Delta t\, \beta} \qquad (15)$$

EXAMPLE 1

As a first example a one-dimensional computation following the above method is compared with the analytical solution of the differential equation set according to Coats and Smith[1].

Contaminated water with c_0 = 100% flows into a domain which is free of pollution. There is, at the beginning, also no pollution in the dead-ends. The velocity of transport is $5 \cdot 10^{-4}$ m/s, $v_{Darcy} = 10^{-4}$ m/s, $n_m = 0.2$, $n_{im} = 0.1$, $\beta = 10^{-8}$ 1/s, dispersivity $\alpha = 1$ m ($D = 5 \cdot 10^{-4}$ m^2/s).

For the numerical computation a spacial discretisation of 10 m and a temporal discretisation of 10^4 s are chosen. For the computation of Lc_m the method of the finite elements is employed combined with Galerkin-weighting.

Fig. 1 shows the temporal development for three locations, 1000 m, 1500 m and 2000 m. The solid line is the numerical solution, the symbols are the analytical solution according to Coats and Smith[1]; the agreement is perfect. The dashed line is a numerical computation without dead-ends. The tailing in the first solution is, in comparison, obvious.

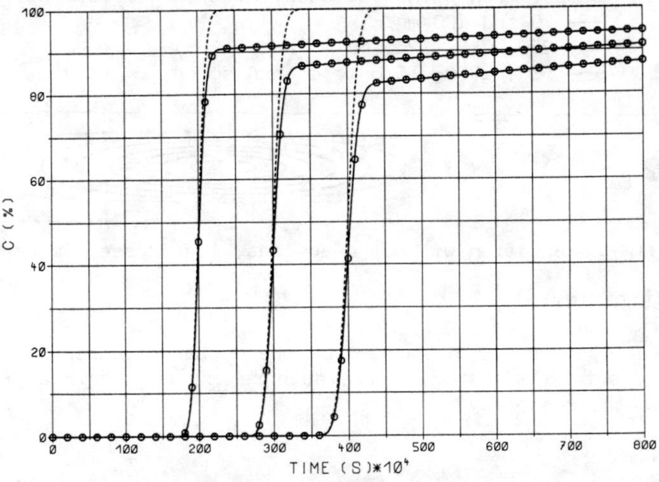

Figure 1. Comparison of numerical and analytical solution
—— numerical solution with dead-ends
ooo analytical solution with dead-ends
--- numerical solution without dead-ends

EXAMPLE 2

This example will demonstrate the effect of tailing in large-scale spreading. Therefore a domain which is 5000 m long and about 1000 m wide is discretized with 3700 equilateral triangles (50 m high). The thickness of the aquifer amounts to 10 m. The time-step is 50 d. Further parameters are: dispersivity α_L = 70 m, α_T = 1 m, velocity of transport v = 0.5 m/d, $n_m = 0.1$, n_{im} = 0.05, $\beta = 10^{-10}$ 1/s. At the beginning the aquifer is (also in the dead-ends) free of contaminants. A tracer with a maximum intensity of 1 kg/d (increasing within 100 d, constant 1 kg/d within 800 d, decreasing to zero within 100 d) seeps into the aquifer on an area of about 35000 m^2 (black hexagon in Figs. 2,3).

Figure 2 shows the result of the numerical computation after 9000 d (about 25 years), Figure 3 shows the same case, but without dead-ends. In the first case the contaminated area is far larger, the maximum concentration, however, is smaller.

This effect is of great importance to the recovery of polluted aquifers (see Herr[2]).

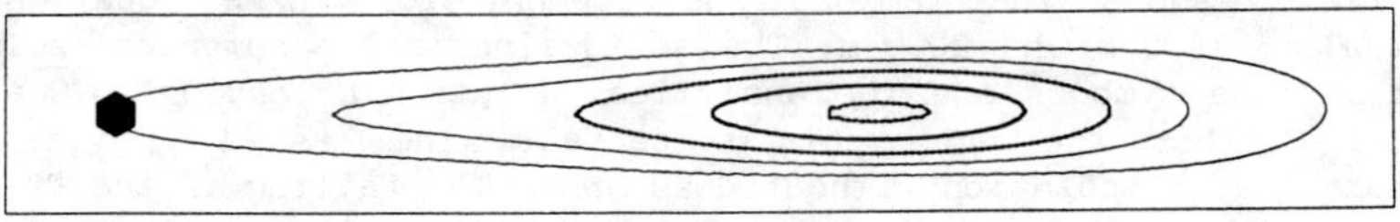

Figure 2. Computation with dead-ends, state after 9000 d
—— 10 and 100 mg/m^3
--- 250, 750, 1250 mg/m^3 and so on
—— 500, 1000, 1500 mg/m^3 and so on

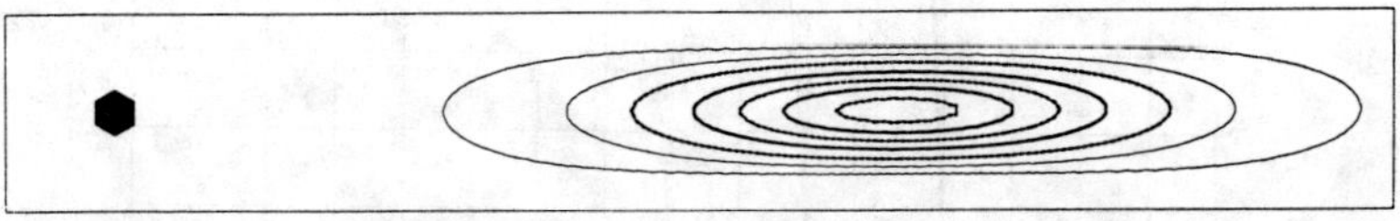

Figure 3. Computation without dead-ends, legend see Figure 2.

ACKNOWLEDGEMENT

The investigation was supported by the German Research Community (DFG), Grant no. Th 159/14. The authors express their appreciation for their subsidy.

REFERENCES

1. Coats, K.H. and Smith, B.D. (1964) Dead-End Pore Volume and Dispersion in Porous Media, Soc. Petr. Eng. J., pp. 73-84.
2. Herr, M. (1986) Grundlagen der hydraulischen Sanierung verunreinigter Porengrundwasserleiter, Mitteilungen Institut für Wasserbau, Universität Stuttgart, Heft 63.
3. Leismann, H.M. and Herrling, B. (1986) Finite Element Schemes for an Improved Computation of Convective Transport in Fluids. In: Finite Approximations in Fluid Mechanics, ed. Hirschel, E.H., Notes on Numerical Fluid Mechanics Vol. 14, pp. 161-174, Vieweg Verlag.
4. Leismann, H.M. (1987) Berechnung von Ausbreitungsvorgängen im Grundwasser mit der Methode der finiten Elemente, Dissertation, University of Karlsruhe.

Simulation of Groundwater Transport Taking into Account Thermodynamical Reactions

B.J. Merkel and J. Grossmann
Department of Hydrogeology and Hydrochemistry, Technical University of Munich, West Germany
A. Faust
PIC-GmbH Software-Inc., Munich, West Germany

ABSTRACT

A groundwater transport simulation concept is given, that can solve two-dimensional transport problems taking into account thermodynamical reactions. The computation of groundwater flow is totally separated from transport simulation. Problems where density differences are not significant can not be solved with this approach. Flow and transport simulation may work on the same spatial grid, but can operate as well on different grids. Convection and dispersion is calculated using a combination of random walk and methods of characteristics. A set of non-linear mass balance equations is solved using Newton-Raphson respectively Gauss-Seidel technique.

INTRODUCTION

From the view of water chemistry and hydrogeology a groundwater aquifer is a large reaction pot affected by

- groundwater flow and groundwater recharge
- different sink and source terms
- dispersivity and diffusivity
- heterogeneous chemical reactions between aquatic species and aquatic species with soil-air and minerals.

Numerical flow and transport models presented by i.e. Prickett et al.[1] or Konikow and Bredehoeft.[2] deal with just one species in solution and interaction of this compound is restricted to non-reversible absorption or a homogeneous reaction. Groundwater problems are mostly much more sophisticated and require computation of several reactions and species simultaneously. Computer programs like WATEQ (Truesdell and Jones[3]) or PHREEQE (Parkhurst et al.[4]) can solve groundwater quality problems, taking into account all relevant thermodynamical data for aquatic species interaction and reactions between aquatic species and minerals. On the contrary these models work only in a zero dimensional spacing and have no book-keeping system for the solid phase. The transport and reaction model presented in this paper solves groundwater problems taking into account groundwater flow, dispersivity and any kind of thermodynamical equilibrium reactions, including exchange and redox reactions. Kinetics of reactions and non-equilibrium reactions are not implemented.

CONCEPT OF THE TRANSPORT AND REACTION MODEL

Groundwater flow has to be computed on a horizontal two-dimensional grid. Finite difference or finite element approaches might be used, but users should remember that transport simulation requires more accurate results from the flow model than commonly is required. The transport model presented here utilizes a rectangular, block-centered finite difference grid and requires a data file with velocity vectors for each node point of the finite difference grid and each time step. If the groundwater flow is simulated with a finite element or a different spaced finite difference grid, the data have to be transformed to the transport grid using a suitable mapping procedure.

The transport equation may be written as

$$\frac{\delta(C_k \cdot b)}{\delta t} = \underbrace{\frac{\delta}{\delta x_i}\left(bD_{ij}\frac{\delta C_k}{\delta x_j}\right)}_{\text{dispersion}} - \underbrace{\frac{\delta}{\delta x_i}(b \cdot C_k \cdot V_i)}_{\text{convection}} - \underbrace{\frac{C_k' W}{\varepsilon}}_{\text{sink/source}} \qquad (1)$$

with

C_k	concentration of dissolved species k (mol/m^3)
C_k'	concentration of dissolved species k (mol/m^3) in a source or sink
D_{ij}	coefficient of hydrodynamic dispersion (m^2/s)
b	saturated thickness of aquifer (m)
x_i, x_j	Cartesian coordinates (m)
V_i	flow velocity in the direction i (m/s)
W	volume flux per unit area (m^3/s)
ε	effective porosity of aquifer

Each chemical species C_k may react either with another aquatic species or with a solid species. These reactions can be discribed by

$$bB + cC = dD + eE \qquad (2)$$

where lowercased letters are the stoichiometric coefficients of species represented by uppercased letters. Equation 2 may be written as mass action equilibrium equation

$$K = \frac{aD^d \cdot aE^e}{aB^b \cdot aC^c} \tag{3}$$

where aB, aC, aD and aE represents the activity of the species B, C, D, E and K is the mass action or equilibrium constant. K is dependent on temperature; this has to be taken into account using i.e. the van't Hoff equation. The activity a_i of a species i is defined by

$$a_i = C_i \cdot g_i \tag{4}$$

where g_i is the activity coefficient of species i. The activity coefficient g_i may be computed from different approaches i.e. Davies equation (5) or extended Debey-Hückel (6)

$$\log (g_i) = \frac{-A \cdot Z_i^2 \sqrt{I}}{1 + \sqrt{I}} - 0.3 \cdot I \tag{5}$$

$$\log (g_i) = \frac{-A \cdot Z_i^2 \sqrt{I}}{1 + B \cdot A_i \sqrt{I}} \tag{6}$$

where A,B constants that depend on the dielectric constant, density and temperature
A_i ion size parameter for the species i
Z_i charge of the species i
I ionic strength (mol/l)

The ionic strength I may be computed from equation 7

$$I = 0.5 \cdot \Sigma(C_i \cdot Z_i^2) \tag{7}$$

The extended Debey-Hückel equation is probably accurate to ionic strengths less than 0.1 while the Davies equation might be used for ionic strengths up to 0.5 (Stumm and Morgan [5]).

NUMERICAL METHODS

To solve the solute-transport equation the model presented uses both a random walk approach and the method of characteristics. At the very beginning a number of traceable particles are placed in each cell of the finite difference grid. Each particle within one cell represents associated concentrations of i species. For each time step every particle is moved according to the time increment, the velocity at the location of the particle and the longitudinal as well as transverse dipersivity. Dispersivity is simulated by a random walk approach which is similar to that of Prickett et al.[1]
The number of particles at each node point is checked. If the program counts less than a minimum number of particles in a area new particles are generated by splitting the existing particles at their position x,y and dividing the species con-

centration information simultaneously. After all particles have been moved, the concentrations for i species of each node are assigned as average concentrations of all particles in that area. At this point the thermodynamic reaction is computed for each node point by solving the given set of equilibrium equations.

The first step in solving the chemical equilibrium is the calculation of the ionic strengths and the activities of i species for all nodes; the second step is solving the set of i equations by Newton-Raphson, Gauss-Seidel or a comparable algorithm. The third step is the computation of new concentrations of i species for each node and according to this updating the concentration information for all particles. If the model deals with problems which are related to solid phase reactions, bookkeeping for the solid phase is essential at this point.

APPLICATION OF THE TRANPORT AND REACTION MODEL

The model presented was used to simulate the effects of salt spreading during winter time by public road service and private activities in a suburb of Munich. The investigated area of Harlaching has a size of about 2.7 km² and is built up by quaternary gravel. The groundwater flow is generally directed from south to north. The saturated thickness of the aquifer varies between 1 and 3 m. Two main streets are served by salt spreading during winter. The thickness of the unsaturated zone varies between 15 and 30 m. Concerning an average seepage velocity of about 4 to 6 m per year (Merkel et al.[6]) there is no direct correlation between the salt application in winter time and groundwater recharge. On the contrary the soil-water of the unsaturated zone beside salted streets has an average salt-concentration of 1000 to 3000 mg/l. Thus results to a rather punctate pollution of the groundwater during periods of groundwater recharge. Though Na and Cl are applicated in the same molar ratio, it is evident that the Na-concentration in groundwater is only half of the Cl-concentration. This is related to cation exchange between Na and Ca as well as Mg (Figure 1).

The groundwater recharge was calculated as a function of the variation of the groundwater level. This function may be determined under certain conditions from the maximum declines of the phreatic line at different levels (Grossmann and Merkel[7]). The groundwater flow was computed using a horizontal two-dimensional finite difference technique (Merkel et al.[6]).

The thermodynamical equations solving the cation exchange reaction between Na, K, Ca, and Mg can be written as

$$\frac{aMe_i^{1/z}}{aMe_j^{1/z}} = -K_{ij}\,\frac{ExMe_i^{z}}{ExMe_j^{z}} \qquad (8)$$

where

aMe_i, aMe_j	activties of the cations i and j in solution
$ExMe_i$, $ExMe_j$	concentration of the cations i and j on the solid exchange phase
z	charge of the cation i or j

Equation 8 yields to three equations. Assuming that the sum of each cation in solution and at the solid phase remains constant, the number of equations to solve the concentrations of the 8 unknown species is reduced to 4. The last equation can be written as

$$\Sigma \; Me(solution) + \Sigma \; (Me(solid) \quad = constant \qquad (9)$$

This set of equations is solved using a Gauss-Seidel algorithm. If more equations and equations that describe solution, precipitation and redox reactions are involved, it is recommended to use a modified Newton-Raphson technique (Parkhurst et al.[4])

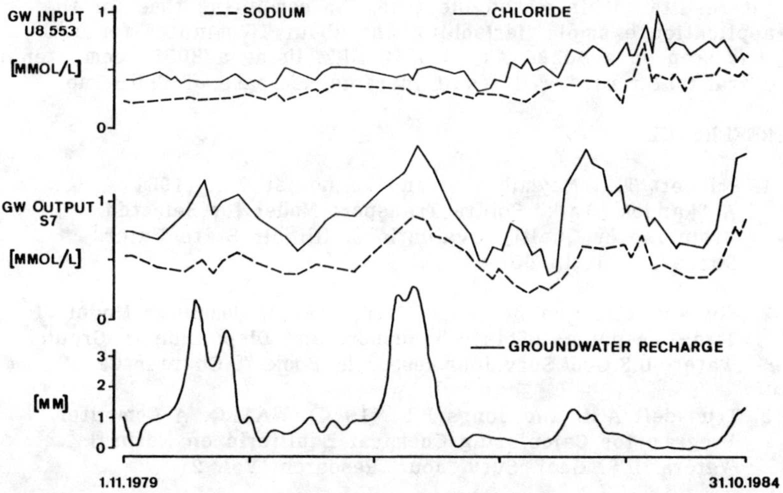

Figure 1. Na and Cl input/output and groundwater recharge

The concentrations of the main ions Ca, Mg, Na, K, HCO_3, Cl, NO_3 and SO_4 and the cation exchange capacity are boundary conditions for the transport model. The groundwater concentration for these ions were kept constant at the input boundary in the south, while the concentration of the soilwater (groundwater recharge) was calculated as a time variant function for Ca, Mg and HCO_3. K and NO_3 were kept constant in the soil water as well while the concentration of Cl and Na depends on the spatial pattern. They were set to natural values of some mg/l at node points without street salting and got concentrations of some hundred mg/l at node points which are affected by salt spreading.

CALIBRATION OF THE MODEL

The flow model can be calibrated totally separate from the transport model. The calibration of the porosity and the dispersion can be done with the help of a conservative tracer. Chloride and Uranin were used in the described application for this purpose. During this calibration procedure the thermodynamic part of the program might be switched off. At the end the thermodynamic part of the program has to be calibrated and verified, which is the most time consuming operation.

PROGRAM FEATURES

The program was first written in FORTRAN 77, but rewritten in MS-Quick-Basic to have interactive screen graphic features. The computer code is segmented in the main program and different subroutines. The program runs on IBM compatible computers. A 80286 or 80386 processor and a math-coprocessor is required to get results within reasonable time. The computing time for the application example Harlaching was about 15 minutes for each time step on a 80286 AT with 10 MHZ. Using a 80386 computer or a transputer card will result in rapid decrease of CPU-time.

REFERENCES

1. Prickett T.A. Naymik T.G. and Lonnquist C.G. (1981). A "Random-Walk" Solute Transport Model for Selected Groundwater Quality Evaluations. Illinois State Water Survey, Bulletin 65

2. Konikow L.F. and Bredehoeft J.D. (1978). Computer Model of Two-Dimensional Solute Transport and Dispersion in Ground water. U.S.Geol.Surv.Jour.Research, Book 7, Chapter C2

3. Truesdell A.H. and Jones B.F. (1974). WATEQ, A Computer Program for Calculating Chemical Equilibria on Natural Waters. U.S. Geol. Surv. Jour. Research, Vol. 2

4. Parkhurst D.L. Thorstenson D.C. and Plummer L.N. (1982). PHREEQE - A Computer Program for Geochemical Calculations. U.S.Geological Survey. Water-Resource Investigations 80-96

5. Stumm W. and Morgan J.J.(1973). Aquatic Chemistry, Wiley

6. Merkel B.J. Freitag G. Grossmann J. Udluft P. and Ullsperger I. (1987). Auswirkungen urbaner Besiedlung auf oberflächennahe Grundwasserleiter. Z.dt.Geol.Ges, 138, Hannover

7. Grossmann J. and Merkel B.J. (1987). Calculating Recharge for Numerical Groundwater Flow Model. GEOMON - Int. Workshop on Geochemistry and Monitoring in Representative Basins, Prague

Multicomponent Solute Transport with Moving Precipitation/Dissolution Boundaries

J.A. Mundell and D.J. Kirkner

Department of Civil Engineering, University of Notre Dame, Notre Dame, IN 46556, USA

Introduction

Multicomponent solute transport problems involving precipitation and dissolution chemical reactions are unique among the wide array of problems encountered in which homogeneous and heterogeneous reactions affect the movement of contaminants in groundwater. The imposition of solubility constraints for various reactive mineral solids may result in the formation or disappearance of moving boundaries which mark the transient location of discontinuities in solid phase concentration, aqueous component flux, and other variables of interest such as porosity. This distinct characteristic of precipitation/dissolution reaction problems poses particular mathematical difficulties to their accurate solution and requires special treatment.

The numerical approaches chosen to solve these types of problems are often suggested directly from the chemical quantities employed as the primary dependent variables (PDVs). While the selection of all species concentrations (Miller and Benson[1]) or the total dissolved concentration of each component (Kirkner et al.[2]) as the PDVs are possible choices, their use results in front-tracking type (FT) schemes in which the moving boundary locations appear explicitly as unknowns in the governing equations. However, choosing the total concentration of each component as the PDV (Walsh et al.[3], Cederburg et al.[4]) removes the explicit presence of the moving boundaries, resulting in formulations that resemble enthalpy-type (ET) methods used for heat flow problems with phase change (Crank[5]). This is an attractive feature which greatly simplifies the numerical approach. However, general finite element ET methods exhibit oscillatory behavior as a result of their inability to accurately simulate the discontinuities across the moving interfaces.

This work will develop and compare fixed-mesh finite element ET and FT formulations for solving precipitation/dissolution chemical transport problems. As will be shown, these schemes result from particular forms of the weak variational statement of the governing differential equations. Throughout the work, it is assumed that all chemical interactions are sufficiently fast that the local equilibrium assumption (LEA) is applicable.

Governing Equations

The following development describes the mass transport of N solute species A_i involved in a precipitation/dissolution reaction with the reactive solid phase C(s) described by

$$C(s) \longleftrightarrow \sum_{i=1}^{N} \upsilon_i A_i \tag{1}$$

where υ_i represents the stoichiometric coefficient indicating the number of moles of the i th species in one mole of the reactive solid. In this presentation, other chemical reactions such as aqueous complexation and sorption are ignored to simplify the treatment, although they may easily be included.

Mass transport equations for the solute species derived from conservation of mass are given by

$$\frac{\partial}{\partial t}(\phi\, C_i) + \nabla \cdot \mathbf{J}_i = \upsilon_i\, I\,, \quad (i=1,\ldots,N) \tag{2a}$$

and for the reactive solid mineral by

$$\frac{\partial}{\partial t}(V_s^{-1}\phi_s) = -\,I \tag{2b}$$

where C_i denotes the concentration of the i th species, $\mathbf{J}_i$ is the mass flux of the i th species, ϕ is the porosity of the porous medium, ϕ_s is the volume fraction occupied by C(s), V_s is the reactive solid molar volume, and I represents the reaction rate expressed in units of moles per unit time per unit bulk volume of porous material. The reaction rate may be eliminated by substituting for I in equation (2a) from equation (2b) resulting in

$$\frac{\partial}{\partial t}\{\phi\, C_i + \upsilon_i V_s^{-1}\phi_s\} + \nabla \cdot \mathbf{J}_i = 0\,, \quad (i=1,\ldots,N) \tag{3}$$

These mass transport equations must also satisfy the chemical solubility inequality constraint posed by equation (1) expressed as

$$K \geq \sum_{i=1}^{N} \{C_i(\mathbf{x},t)\}^{\upsilon_i} \tag{4}$$

where K denotes the equilibrium constant for the reaction and **x** the spacial coordinate vector. It is assumed in this development that unit activity coefficients for the solute species hold, and that the activity of the solid phase is unity. To measure the degree of saturation with respect to the reactive solid, the saturation index function (Lichtner et al.[6]) may be introduced as

$$\xi(\mathbf{x},t) = \sum_{i=1}^{N} \{C_i(\mathbf{x},t)\}^{\upsilon_i} - K \tag{5}$$

where ξ is zero or negative if the aqueous solution is in equilibrium or undersaturated, respectively, with respect to the reactive solid C(s).

If the reactive solid is present throughout the domain, equation (3) represents N partial differential equations and equation (4) one algebraic equation for the N+1 unknowns in the system. However, since all regions of the system may not be saturated with respect to the solid phase during a transport episode, a reaction zone will develop that does not contain the reactive solid. This undersaturated zone will be separated from the saturated region by a sharp reaction front across which the solid phase will appear (precipitate) or disappear (dissolve) depending on the saturation index function in each zone.

Conservation of mass across this front relates the magnitude of the jump discontinuities in the solute species flux and reactive solid volume fraction to the normal velocity v_n of the front according to the generalized Rankine-Hugoniot equations given by (Lichtner et al.[7])

$$v_n = \frac{-\partial s/\partial t}{\mathbf{n} \cdot \nabla s} = \frac{[\,J_i\,] \cdot \mathbf{n}}{\upsilon_i[\phi_s]V_s^{-1}}\,, \quad (i=1,\ldots,N) \tag{6}$$

where v_n denotes the magnitude of the front velocity normal to the surface of discontinuity s with unit outer normal **n**, and the brackets [ ] represent the jump discontinuity in a quantity across the surface s, eg.

$$[J_i] = J_i(s^+,t) - J_i(s^-,t) \tag{7}$$

The moving interface location now represents an additional unknown in the system that may be solved for using equation (6). The choice of the primary dependent variables and the manner in which the requirements of the Rankine-Hugoniot equation are incorporated into the formulation defines the numerical approach to these problem types.

Finite Element Formulation

The transport problem in a single semi-infinite spacial domain will be considered. It is assumed that the solution is initially saturated with respect to the reactive solid phase, and that the reactive solid occupies an initial volume fraction, ϕ_{so}. At time t = 0, the concentrations of all species A_i in solution are lowered on the left boundary and subsequently held fixed at $C_i(o,t) = C_i^o$. This results in a dissolution front moving away from the boundary with its location at time t designated as s(t). From equation (3), it is apparent that the following mass transport equations must be satisfied

$$\frac{\partial}{\partial t}(\phi\, C_i) + \frac{\partial J_i}{\partial x} = 0\,, \quad (i=1,\ldots,N), \quad 0 < x < s(t) \tag{8a}$$

$$\frac{\partial}{\partial t}(\phi\, C_i) + \upsilon_i V_s^{-1}\frac{\partial \phi_s}{\partial t} + \frac{\partial J_i}{\partial x} = 0\,, \quad (i=1,\ldots,N), \; s(t) < x < \infty \tag{8b}$$

The development of a finite element solution requires a weak variational statement of this problem obtained by multiplying equations (8a) and (8b) and the interface boundary condition implied by equation (6) by an arbitrary test function v resulting in

$$\int_0^{s^-}\left[\frac{\partial}{\partial t}(\phi\, C_i) + \frac{\partial J_i}{\partial x}\right] v\, dx + \int_{s^+}^{\infty}\left[\frac{\partial}{\partial t}(\phi\, C_i) + \upsilon_i V_s^{-1}\frac{\partial \phi_s}{\partial t} + \frac{\partial J_i}{\partial x}\right] v\, dx$$

$$+ \left([J_i] - \upsilon_i V_s^{-1}[\phi_s]\frac{ds}{dt}\right) v(s) = 0\,, \quad \forall\, v, \quad (i=1,\ldots,N) \tag{9}$$

Integrating the flux term by parts, equation (9) becomes

$$\int_0^{\infty}\left(\frac{\partial}{\partial t}(\phi\, C_i)\, v - J_i v'\right) dx + \int_{s^+}^{\infty} \upsilon_i V_s^{-1}\frac{\partial \phi_s}{\partial t} v\, dx$$

$$- \upsilon_i V_s^{-1}[\phi_s]\,\frac{ds}{dt}\, v(s) = 0\,, \quad (i=1,\ldots,N) \tag{10}$$

Using generalized function notation, equation (10) may equivalently be written as (Mundell and Kirkner[8])

$$\int_0^{\infty}\left\{\frac{\partial}{\partial t}\left[\phi C_i + \upsilon_i V_s^{-1}\phi_s \theta(x-s(t))\right] v - J_i\, v'\right\} dx = 0\,, \quad (i=1,\ldots,N) \tag{11}$$

where θ(x) denotes the Heaviside step function. The ET formulation arises from the selection of the total component concentration as the PDV. In equation (11), the expression in the square brackets is recognized as the total concentration of the i th species, C_{Ti}, or

$$C_{Ti} = C_i + \upsilon_i \kappa_s \theta(x-s(t))\,, \quad (i=1,\ldots,N) \tag{12}$$

where $\kappa_s = (\phi_s\, V_s^{-1})/\phi$ denotes the solid phase concentration of C(s) per unit volume of aqueous solution. Thus, equation (11) is the weak variational statement corresponding to the local form of the conservation of mass for the total concentration of solute species A_i. Note that the introduction of C_{Ti} into equation (11) as the PDV removes the explicit presence of s(t) in the variational statement. Employing linear basis functions for C_{Ti}, C_i, and v leads to the semi-discrete approximation

$$\mathbf{M}\dot{\mathbf{c}}_{Ti} + \mathbf{K}\mathbf{c}_i = \mathbf{f}_{oi}\,, \quad (i = 1,\ldots,N) \tag{13}$$

where **M** and **K** denote standard mass and coefficient matrices, respectively, for the mass flux conditions specified, and $\mathbf{c}_{Ti}$ and $\mathbf{c}_i$ are the vector of nodal values of $C_{Ti}(x,t)$ and $C_i(x,t)$, and $\mathbf{f}_{oi}$ is the vector containing the boundary condition contribution for the i th solute species. Equation (11) may be fully discretized using the generalized trapezoidal scheme and solved using Newton methods, function iteration, or modifications to these (Mundell and Kirkner[8]).

If equation (10) is rewritten as

$$\int_0^\infty \Big\{ \Big[\frac{\partial}{\partial t}(\phi\, C_i) + \upsilon_i V_s^{-1} \frac{\partial \phi_s}{\partial t}\, \theta(x - s(t)) \Big]\, v - J_i\, v' \Big\}\, dx$$

$$- \upsilon_i V_s^{-1} [\phi_s]\, \frac{ds}{dt}\, v(s) = 0\,, \quad (i = 1,\ldots,N) \tag{14}$$

it is apparent that the unknown value of the interface is explicitly contained in the weak variational statement. Use of all species concentrations as the PDVs suggested by equation (14) leads directly to finite element front-tracking schemes in which s(t) becomes part of the solution. This is also true for the case of the use of the total aqueous concentration when soluble complexation reactions are involved in mass transport.

For the case of dissolution without precipitation, the finite element discretization of equation (14) has been shown[8] to yield the following set of nonlinear ordinary differential equations

$$\mathbf{M}\dot{\mathbf{c}}_i + \mathbf{K}\mathbf{c}_i = \mathbf{f}_{oi} + \upsilon_i V_s^{-1} \phi_{so} \frac{ds}{dt}\, \psi(s)\,, \quad (i = 1,\ldots,N) \tag{15}$$

where $\psi(s)$ is the vector of global linear shape functions evaluated at s. Time discretization using the generalized trapezoidal rule results in the same set of fully discretized equations as for transport without chemical reaction except for the addition of a vector that must be updated by a search algorithm for s[8].

Sample Problem

The one-component diffusion dissolution problem demonstrates the behavior of the ET formulation in comparison to an FT scheme. The problem, originally posed by Lichtner et al.[6], involves a system containing a single aqueous component A in local equilibrium with a reactive solid mineral phase C(s). This requires that the initial concentration of species A everywhere in the domain be constant and equal to the equilibrium constant, K. The reactive solid phase occupies some initial volume fraction ϕ_{so} which is very small in comparison to the system porosity, ϕ. At time t = 0, the concentration of A on the left boundary is reduced and subsequently held fixed at $C_a(0,t) = C_a^{\,o}$ such that the aqueous solution is undersaturated with respect to C(s) and a moving dissolution boundary begins to propagate through the domain.

The problem is completely analogous to the one-phase heat conduction problem with phase change[5] and reduces to solving for the concentration profile in the domain to the left of the moving front. Nondimensionalization of the mass transport equation to the left of the interface results in the parabolic diffusion equation with the leading constant of unity

$$\frac{\partial \hat{c}}{\partial \hat{t}} = \frac{\partial^2 \hat{c}}{\partial \hat{x}^2}, \quad 0 < \hat{x} < \hat{s}(\hat{t}) \tag{16}$$

where $\hat{c} = (C_a - C_a^o)/(K - C_a^o)$, $\hat{x}=x/L$, $\hat{s}=s/L$, $\hat{t}=Dt/L^2$, and D denotes the diffusion coefficient. The interface condition becomes

$$\frac{\partial \hat{c}(\hat{s}^-,\hat{t})}{\partial \hat{x}} = \lambda \frac{d\hat{s}}{d\hat{t}} \tag{17}$$

where the parameter $\lambda = \kappa_s/(K - C_a^o)$ is the dimensionless solid phase concentration and $1/\lambda$ is the so-called Stefan number. It is obvious that λ uniquely controls the velocity of the moving dissolution boundary, and therefore defines the general behavior of the problem.

Figure 1 illustrates the behavior of the concentration solutions for fully implicit ET and FT schemes with $\Delta\hat{x} = 0.025$, $\Delta\hat{t} = 0.0167$, and $\lambda = 100$ at the fixed spacial coordinate $\hat{x} = 0.075$. The greater oscillatory behavior of the ET solution is apparent in comparison to the FT solution, and results from the inability of the ET scheme to accurately describe the discontinuous nature of the total component concentration across the dissolution interface. The L_2 error norm (which measures the root-mean-square of the error over the domain) for the ET method solution of Figure 1 is plotted versus the nondimensional interface location in Figure 2. The magnitude of the L_2 error norm varies considerably as the dissolution front travels through the domain, reaching maximum values when the front moves through the center of an element, and minimum values as it passes through a nodal location.

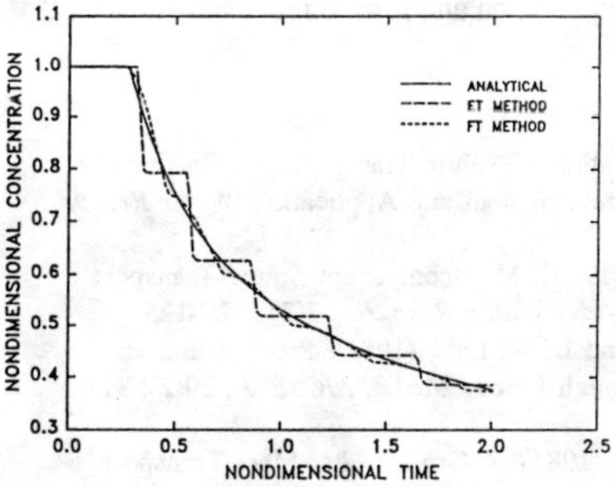

Figure 1. One-component diffusion dissolution ($\lambda = 100$).

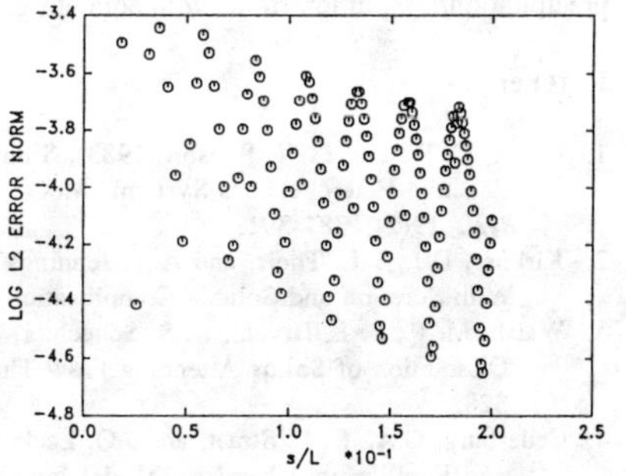

Figure 2. L_2 error norm for the ET scheme in Figure 1.

To determine the effect that the size of the solid discontinuity at the interface has on the magnitude of the error in the concentration solution, the value of λ was varied over a wide range and simulations were performed. The results for fully implicit ET and FT schemes using a fixed domain discretization of $\Delta\hat{x} = 0.025$ are summarized in Figure 3. The mean L_2 error norms between $\hat{s} = 0.0$ and 0.2 were calculated from graphs similar to Figure 2. The results show the superior performance of the FT scheme for values of λ greater than about 10. However, for problems with lower solid concentrations in comparison to solute concentrations, the ET formulation yielded more accurate concentration distributions using less computational effort than the FT method.

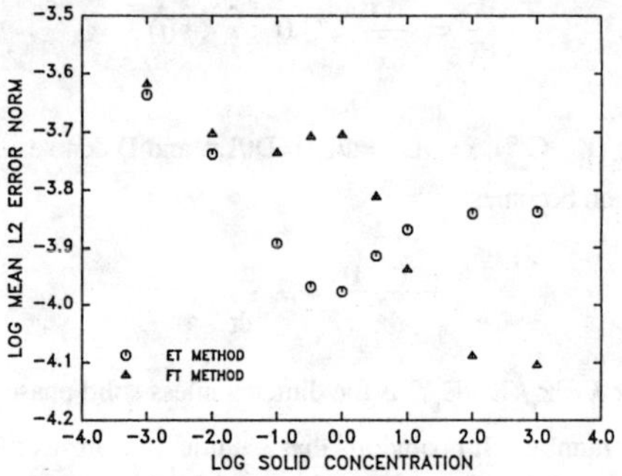

Figure 3. Effect of λ on mean L_2 error norm between s/L = 0.0 and 0.2.

Conclusions

Two fixed-mesh finite element approaches have been presented for solving solute transport problems involving precipitation/dissolution reactions. The results of simulations for one-component diffusion dissolution indicate that despite oscillations inherent in the aqueous concentration profile produced by enthalpy-type (ET) formulations, greater accuracy may be achieved with less computational effort and numerical complexicity than with front-tracking type (FT) schemes for large classes of transport episodes in which the reactive solid concentration does not significantly exceed the aqueous concentration of the solute species. The retention of the simplicity of ET methods is a desirable element in the development of a general multicomponent code for modeling more complex problems involving multiple precipitation/dissolution fronts with aqueous complexation and sorption.

References

1. Miller, C. W. and L. V. Benson (1983), Simulation of Solute Transport in a Chemically Reactive Heterogeneous System: Model Development and Application, *Water Resour. Res.*, 19(2), 381-391.
2. Kirkner, D.J., T.L. Theis, and A.A. Jennings (1984), Multicomponent Solute Transport With Sorption and Soluble Complexation, *Adv. Water Resourc.*, 7(3), 120-125.
3. Walsh, M. P., S. L. Bryant, R. S. Schechter, and L. W. Lake (1983), Precipitation and Dissolution of Solids Attending Flow Through Porous Media, *AIChE J.*, 30(2), 317-328.
4. Cederburg, G.A., R. L. Street, and J.O. Leckie (1985), A Groundwater Mass Transport and Equilibrium Chemistry Model for Multicomponent Systems, *Water Resour. Res.*, 21(8), 1095-1104.
5. Crank, J. (1984), *Free and Moving Boundary Problems,* Clarendon Press, Oxford.
6. Lichtner, P. C., E. H. Oelkers, and H. C. Helgeson (1986), Exact and Numerical Solutions to the Moving Boundary Problem Resulting from Reversible Heterogeneous Reactions and Aqueous Diffusion in a Porous Medium, *J. Geophys. Res.*, 91(B7), 7531-7544.
7. Lichtner, P. C., E. H. Oelkers, and H. C. Helgeson (1986), Interdiffusion with Multiple Precipitation/Dissolution Reactions: Transient Model and the Steady-State Limit, *Geochim. Cosmochim. Acta,* 50(9), 1951-1966.
8. Mundell J. A. and D. J. Kirkner (1988), Numerical Studies of Dissolution Induced Moving Boundary Problems from Aqueous Diffusion in Porous Media, *J. Geophys. Res.* (under review).

The Advantage of High-Order Basis Functions for Modeling Multicomponent Sorption Kinetics

J.A. Pedit and C.T. Miller
Department of Environmental Sciences and Engineering, CB# 7400, 105 Rosenau Hall, University of North Carolina, Chapel Hill, NC 27599-7400, USA

ABSTRACT

The problem of interest is the modeling of sorption of competing solutes on solids in a completely mixed batch reactor. A dual-resistance model is used in which the rate of sorption is described as a series of mass-transfer steps: molecular diffusion through a hydrodynamic boundary layer surrounding a solid particle; and diffusion within the particle itself. The ideal adsorbed solution theory is used to predict multicomponent competitive interactions among the solutes.

The Galerkin finite element method is used to solve the spatial derivative describing diffusion in a spherical geometry. A system of ordinary differential equations result that are integrated using Gear's algorithm. The resulting solutions are compared to an analytical solution in the limiting case where the mass-transfer resistance through the boundary layer is negligible. The main emphasis of this work is to investigate the effect of using high-order Lagrange polynomials as basis functions. Various order basis functions ranging from first to twelfth order are investigated and the clear advantage of high-order schemes is demonstrated.

THEORY

The equation that describes the change in fluid-phase concentration of a solute in response to sorption to a solid phase in a completely mixed batch reactor is

$$\frac{dC_i}{dt} = -\frac{M}{V}\frac{dq_i}{dt} \qquad (1)$$

where C and q are the fluid-phase and the mass-average solid-phase concentrations of solute i at time t; M is the mass of solid phase; and V is the volume of solution in the reactor. The initial conditions are

$$C_i(t=0) = C_{oi} \tag{2}$$

$$q_i(t=0) = 0 \tag{3}$$

where C_o is the initial fluid-phase solute concentration.

In this paper the dual-resistance model is used to describe the rate at which sorption occurs. The dual-resistance model describes sorption as a series of mass-transfer steps: molecular diffusion through a hydrodynamic boundary layer surrounding a solid particle; and diffusion within the particle itself (Miller and Weber[1]). For spherical solid particles, the dual-resistance model can be expressed as

$$\frac{\partial q_{ri}}{\partial t} = \frac{D_{si}}{r^2}\frac{\partial}{\partial r}\left(r^2\frac{\partial q_{ri}}{\partial r}\right) \tag{4}$$

with the boundary and initial conditions

$$\left.\frac{\partial q_{ri}}{\partial r}\right|_{r=R} = \frac{k_{fi}}{D_{si}\rho}(C_i - C_{si}) \tag{5}$$

$$\left.\frac{\partial q_{ri}}{\partial r}\right|_{r=0} = 0 \tag{6}$$

$$q_{ri}(0 \le r \le R, t=0) = 0 \tag{7}$$

where q_r is the solid-phase solute concentration as a function of radial position r; D_s is the intraparticle diffusion coefficient; k_f is the film mass-transfer coefficient; C_s is the fluid-phase solute concentration at equilibrium with the solid-phase solute concentrations at the exterior of the particle; ρ is the density of the solid; and R is the solid particle radius.

If the only source of solute into the solid particle is by diffusion through the hydrodynamic boundary layer then mass-balance considerations allow for the sorption term in Equation 1 to be written

$$\frac{dq_i}{dt} = \frac{3k_{fi}}{R\rho}(C_i - C_{si}) \tag{8}$$

Substituting Equation 8 into Equation 1 gives

$$\frac{dC_i}{dt} = -\frac{3k_{fi}M}{RV\rho}(C_i - C_{si}) \tag{9}$$

Additional expressions are needed to describe the equilibrium distribution of each solute between the fluid and solid phases. For a single solute the Freundlich isotherm equation is often used to describe this distribution

$$q_{ei} = K_{Fi}C_{ei}^{1/n_i} \tag{10}$$

where q_e and C_e are the equilibrium solid- and fluid-phase solute concentrations; and K_F and $1/n$ are the Freundlich isotherm coefficients. When more than one solute is present, solutes may compete for sorption sites within the solid phase, thus altering the final distribution of each solute between the fluid and solid phases. The ideal adsorbed solution theory (IAST) can be used to predict the degree to which competition alters the distribution of each solute (Radke and Prausnitz[2]). If the Freundlich isotherm equation is used to describe single solute behavior in the IAST, then the fluid-phase solute concentration at equilibrium with the exterior solid-phase solute concentrations is (Crittenden, et al.[3])

$$C_{si} = \frac{q_{ri}}{\sum_{j=1}^{n_c} q_{rj}} \left(\frac{\sum_{j=1}^{n_c} n_j q_{rj}}{n_i K_{Fi}} \right)^{n_i} \quad \text{at } r = R \tag{11}$$

where n_c is the number of solutes.

For each solute there is one ordinary differential equation describing the fluid-phase solute concentration as a function of time (Equation 9) and one partial differential equation describing the solid-phase solute concentration as a function of time and radial position (Equation 4). An analytical solution for this system of equations has yet to be derived. A numerical scheme is therefore employed to approximate the solution.

The Galerkin finite element method is used to approximate the spatial derivative in Equation 4, which after integration by parts becomes

$$\int_0^R N_k \frac{\partial \hat{q}_{ri}}{\partial t} r^2 dr = -D_{si} \int_0^R \frac{dN_k}{dr} \frac{\partial \hat{q}_{ri}}{\partial r} r^2 dr + D_{si} N_k r^2 \left. \frac{\partial \hat{q}_r}{\partial r} \right|_{r=0}^{r=R} \tag{12}$$

for $k = 1, ..., n_n$

where $\hat{q}_r$ is the trial solution given by

$$\hat{q}_{ri}(r) = \sum_{l=1}^{n_e} N_l(r) q_{ril} \tag{13}$$

and N_l are the Lagrange polynomial basis functions given by

$$N_l(r) = \prod_{\substack{L=1 \\ L \neq l}}^{n_e} \frac{r - r_L}{r_l - r_L} \tag{14}$$

where r_l and r_L are the location of node l and node L; n_n is the total number of nodes; and n_e is the number of nodes in an element.

Equation 12 was integrated and combined with Equation 9—yielding a system of $n_n + 1$ ordinary differential equations in time for each solute, which was solved using Gear's method for stiff equations (Gear[4]).

VALIDATION

No analytical solution exists to evaluate the accuracy of the numerical solution. However, Crank[5] derived an analytical solution for the single solute case in which the the film mass-transfer resistance is negligible and the Freundlich coefficient $1/n$ is one. Numerical solutions using various order basis functions ranging from first to twelfth order with thirteen evenly spaced nodes are compared to the solution given by Crank[5] in Figure 1. Further simulations were done in which the number of nodes was varied for each order basis function. For each simulation the sum of squares for error was computed for the times indicated by the data points in Figure 1. Figure 2 is a comparison of sums of squares for error as a function of the number of nodes for each order basis function.

BISOLUTE SIMULATIONS

Additional simulations were performed to assess the accuracy of using various order basis functions with a set of more typical film mass-transfer coefficients. The model was used to predict competitive sorption between lindane and nitrobenzene onto an Ann Arbor soil using parameters from Miller and Weber[1]. The results of a simulation using twelfth-order basis functions and 73 nodes (Figure 3) were used as the standard for assessing the accuracy of additional simulations. For each simulation the sum of squares for error was computed for the times indicated by data points in Figure 3. Figure 4 is a comparison of the sums of squares for error as a function of the number of nodes for each order basis function.

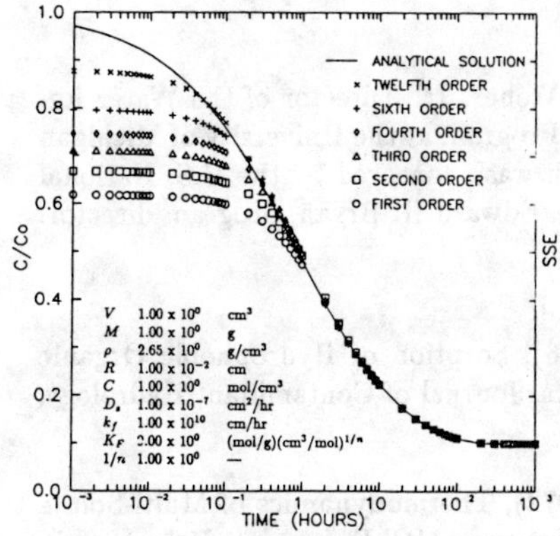

Figure 1. Comparison of Crank's analytical solution with the numerical solutions using various order basis functions with 13 nodes.

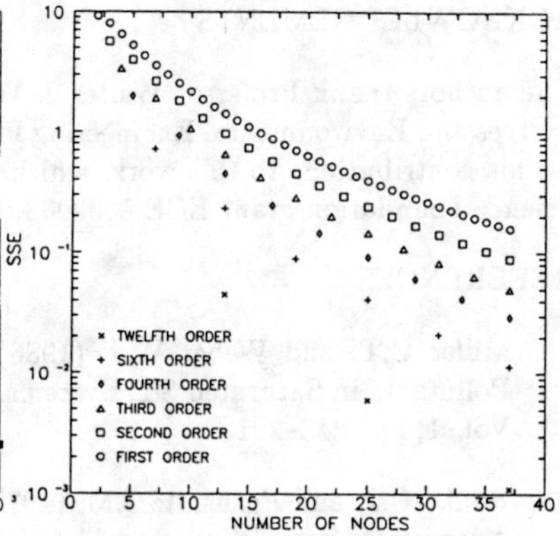

Figure 2. Comparison of sums of squares for error (SSE) as a function of number of nodes for various order basis functions.

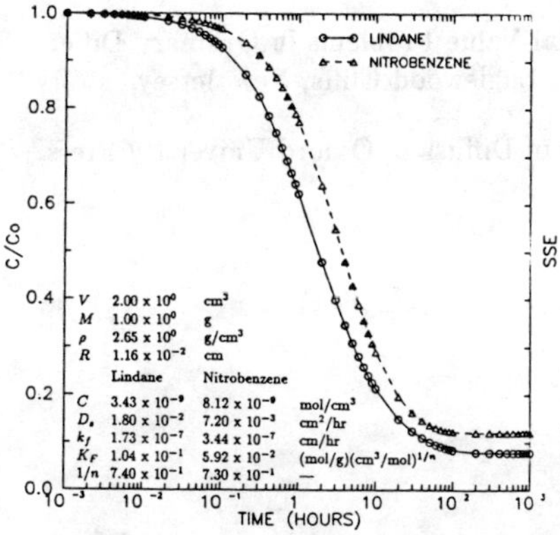

Figure 3. Numerical simulation of competitive sorption between lindane and nitrobenzene on Ann Arbor soil.

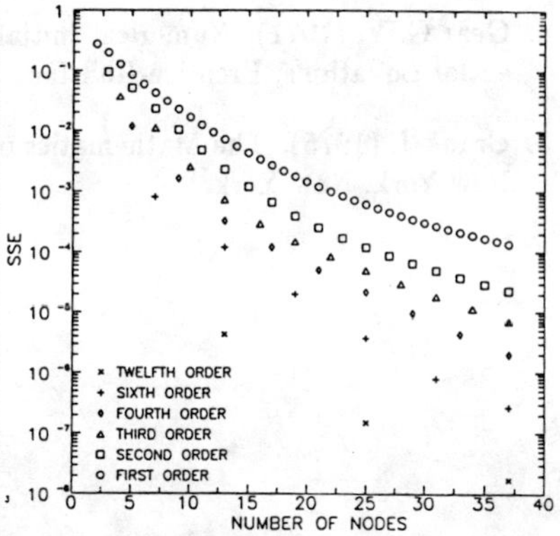

Figure 4. Comparison of sums of squares for error (SSE) as a function of number of nodes for various order basis functions.

CONCLUSIONS

The Galerkin finite element method was used to develop six solutions—using basis functions that varied from first to twelfth order—for a multicomponent, dual-resistance sorption rate model. The results indicate that numerical solutions using high-order basis function are more accurate for situations where the film mass-transfer resistance is negligible and also when more typical parameters are used.

ACKNOWLEDGEMENTS

The authors thank Professor Walter J. Weber, Jr., Director of the Water Resources and Environmental Engineering Program at the University of Michigan for his contributions to this work, which was sponsored by the U.S. National Science Foundation grant ECE 8503903, Edward H. Bryan, program director.

REFERENCES

1. Miller C.T. and Weber W.J. (1986), Sorption of Hydrophobic Organic Pollutants in Saturated Soil Systems, Journal of Contaminant Hydrology, Vol. 1, pp. 243–261.

2. Radke C.J. and Prausnitz J.M. (1972), Thermodynamics of Multi-Solute Adsorption from Dilute Liquid Solutions, AIChE Journal, Vol. 18, pp. 761– 768.

3. Crittenden J.C. Luft P. Hand D.W. Oravitz J.L. Loper S.W. and Ari M. (1985), Prediction of Multicomponent Adsorption Equilibria Using Ideal Adsorbed Solution Theory, Environ. Sci. Technol., Vol. 19, pp. 1037–1043.

4. Gear G.W. (1971). Numerical Initial Value Problems in Ordinary Differential Equations, Prentice-Hall Inc. Englewood Cliffs, New Jersey.

5. Crank J. (1975). The Mathematics of Diffusion, Oxford University Press. New York, New York.

SECTION 2D - HEAT TRANSPORT

A Finite Element Model of Free Convection in Geological Porous Structures

D. Bernard
LEPT-ENSAM, Esplanade des Arts et Métiers, 33405 Talence, France

INTRODUCTION

Free convection is a common phenomenon in geological porous structures. Indeed, due to the complicated geometrical shapes and the spatial variability of the transport properties, in these structures the stability criteria for the saturating fluid are generally not verified [BERNARD[1]].

Convective movements may affect important processes such as mineral diagenesis [RABINOWICZ et al.[2]] or petroleum generation and migration. Precise study of these processes requires estimates of the possible convective flow patterns, of the related thermal perturbation and of the induced fluid velocity field. For this purpose the finite element model DBCONV5 has been developed around the general finite element program package MODULEF (INRIA, France).

After presentation of the numerical aspects of the model, two examples of representative applications are given to illustrate its possibilities; the first one is a real case (ALWYN field -NORTH sea-) for which we only give some preliminary results. The second one is the case of the sloped layer limited by impervious conducting strata which is considered in more details. Various flow patterns have been identified for different values of the filtration RAYLEIGH number and the ratio of the thermal conductivities of the porous and impervious layers [BERNARD [3]].

THE NUMERICAL MODEL DBCONV5

DBCONV5 is a numerical model simulating thermal free convection in two-dimensional complex porous structures. It has been developed around the general-purpose finite element program package MODULEF (INRIA, FRANCE) [HOLLAND[4]]. The computational organization was inspired by MODULEF. It has been chosen to allow future developments toward more complete modelisation of geological structures, while taking into account phenomena such as compaction, diagenesis, etc...

Mathematical model

Natural convection in porous medium can be described, assuming some classical hypothesis [COMBARNOUS, BORIES[5]], by the following set of equations (presented here under dimensionless form):

$$- \nabla.(\underset{\approx}{k}.\nabla \underset{\sim}{p}) = Ra^{*}\ \nabla.(\underset{\approx}{k}.T\ \underset{\sim}{e}) \qquad (1)$$

$$\frac{\partial T}{\partial t} - \nabla.(\underset{\approx}{\lambda}^{*}.\underset{\sim}{\nabla T}) = f - \underset{\sim}{V}.\underset{\sim}{\nabla T} \qquad (2)$$

$$\underset{\sim}{V} = - \underset{\approx}{k}.(\underset{\sim}{\nabla p} + Ra^{*}T\ \underset{\sim}{e}) \qquad (3)$$

This system is composed of the fluid mass balance (eq.1), the energy balance (eq.2) and DARCY's law (eq.3). The unknowns (all dimensionless) are the pressure p and the temperature T. k is the permeability tensor and λ^* the equivalent thermal conductivity tensor. t is the dimensionless time, e the gravity unit vector, f the volumic heat source (radioacticity, chemical reaction,...) and V the filtration velocity. The filtration RAYLEIGH number Ra^* is given by the following expression:

$$Ra^* = \frac{(\rho C)\ \rho\ \beta}{\mu}\ g\ \frac{k\ H\ \Delta T}{\lambda^*} \qquad (4)$$

This dimensionless number explicitly quantifies the relative influence of the driving force for thermal convection, i.e., $\beta\ \rho\ g\ \Delta T$, and the stabilizing effects due to the fluid viscosity μ and to the porous medium thermal conductivity λ^*. The different parameters appearing in eq. 4 are: g the gravity acceleration, β the fluid volumic expansion coefficient, ρ the fluid density, (ρC) the specific heat content of the fluid, ΔT the reference difference of temperature and H the reference length.

Numerical model

Consider Ω the two-dimensional domain to be studied (figure 1). The internal boundary γ divides Ω into two zones: Ω_1 the impervious zone and Ω_2 the porous zone. Within the impervious zone there is no mass transfer and the only equation to be solved is eq.2. In the porous zone both eq.1 and eq.2 have to be considered.

In each triangle of the grid covering Ω $(\Omega_1 \cup \Omega_2)$ the temperature is approximated by a first order polynomial. In each triangle of Ω_2 the pressure is approximated by a second order polynomial.

The differential system (1),(2) is completed by a set of boundary conditions (B.C.) to be specified on the boundaries γ and Γ_2 for the pressure and Γ_1 and Γ_2 for the temperature. Those B.C. can be of two different types; DIRICHLET B.C. or natural B.C.

Time integration

The time integration is performed between the initial time t_0 and the final time t_n by constant time step Δt following the scheme given below:

- *Initialisations ; n = 0*

- *Computation of the pressure* p_n :

$$[K_P].\{p_n\} = \{Ra^* \nabla.(k\ T_n\ e)\}$$

- *Computation of the temperature* T_{n+1}

$$[K_T + M_T/\Delta t].\{T_{n+1}\} = [M_T/\Delta t].\{T_n\} + \{f_n - V_n.\nabla T_n\}$$

where V_n *is given by equation (3)*

K_P, K_T, M_T are respectively the stiffness matrixes of the pressure and the temperature and the mass matrix of the temperature.

The linear systems are stored in a sky-line form and solved using a direct CROUT algorithm.

A REAL CASE : ALWYN FIELD (NORTH SEA)

The importance of mineral diagenesis on the elaboration of the heterogeneities observed in the petroleum field ALWYN has been confirmed by several works of the Compagnie Française des Pétroles (CFP) [JOURDAN[6]]. The Institut Français du Pétrole (IFP) used the model THEMIS [DOLIGEZ et al.[7]] to study at the basin scale (100 km) the creation of this petroleum field. This study gave us the boundary conditions we needed to calculate the temperature and the pressure fields (and then the velocity) induced by natural convection within a porous block (thickness=123m, length=700m, slope=20°) which is well individualized from the hydrodynamic point of view (i.e. it is possible to assume that all the potential convective cells are strictly included in the block limits) and for which precise petrographic informations are available.

Zone number	Filtration velocity (m s^{-1})
1	1.6 10^{-2}
2	2.0 10^{-2}
3	1.9 10^{-2}
4	0.2 10^{-2}
5	0.4 10^{-2}

Table 1. Average velocity in the zones numbered on figure 2.

Around this block and at a distance of about 400m we assume isothermal boundaries (temperatures given by THEMIS). Furthermore the block is supposed to be homogeneous and bounded by impervious surfaces. The computed filtration RAYLEIGH number is equal to 1.29 (H=123m, T_m=115°C, ΔT=4K, λ^*=4.18$Wm^{-1}K^{-1}$). We used a grid composed of 2927 elements (1494 nodes) in temperature and 1040 elements (2187 nodes) in pressure. The velocity field obtained is shown on figure 2. In table 1 averaged values of this velocity are given for the zones numbered on figure 2.

The thermal effect of those fluid movements is negligible in this case. Complementary computations have to be done to take into account the stratification and the anisotropy of the porous block as well as the forced fluid flow related to clay compaction. Next step will be the coupling with a geochemical model in order to investigate the part of free convection in the heterogeneities creation.

THE SLOPED POROUS LAYER

Numerical models like DBCONV5 can be used to study peculiar case as the first example we presented. But it can be also used to produce general results. In this case, the main problem is to define representative general configurations. As an example we propose the configuration described on figure 3a; the homogeneous isotropic porous layer (permeability k) is limited by impervious boundaries. The upper and lower layers are non-porous media (thermal conductivity λ_2). Lateral surfaces are adiabatic, the upper surface is isothermal (T_s) and the lower one is submitted to a constant heat flux (Φ_0).

We present the different flow patterns obtained for various values of the filtration RAYLEIGH number (calculated using H=layer thickness, k, λ_2, and $\Delta T=\Phi_0\ H/\lambda_2$) and of the thermal conductivity ratio $\Lambda=\lambda_1/\lambda_2$. The slope angle is equal to 15°.

On figure 3b three regions can be distinguished corresponding to a multicellular regime (upper part of the graph), to a clockwise one roll regime (lower right) and to a counter-clockwise one roll regime (lower left). The limit Λ=1 between the two one roll regimes has been checked analytically [BERNARD[3]], the other limit is only a numerical approximation.

By analogy to what happens in the case of a porous layer limited by isothermal planes [CALTAGIRONE, BORIES[8]], the stability of the 2D solutions to 3D perturbations has to be studied in the vicinity of the limit between unicellular and multicellular flows.

CONCLUSIONS

Numerical models like DBCONV5 are useful tools that can be used to increase the general knowledge about natural convection in geological porous structures as well as to study simplified real cases. Its modular organization makes easier the necessary extensions of its capabilities (3D flow, time varying transport properties,...) and the coupling with other simulators (geochemical, mecanical, ...).

ACKNOWLEDGEMENTS

The work upon which this publication is based has been partly supported by the IFP. The study of the ALWYN field is part of a project including IFP, CFP, LEPT-ENSAM and several laboratories of the CNRS. Some computations have been done by D. CROIZET during his DEA (Un. of BORDEAUX I).

REFERENCES

1. BERNARD D. (1987), Numerical modeling of free convection in geological porous structures, 24th AIChE/ASME National heat transfer conference, aug. 9-12, Pittsburg, USA.
2. RABINOWICZ M. DANDURAND J.L. JAKUBOWSKI M. SCHOTT J. CASSAN J.P. (1985), Convection in a North Sea oil reservoir: inferences on diagenesis and hydrocarbon migration, Earth and Planet. Sci. Let., 74, 387-404.
3. BERNARD D. (1988), Convection naturelle dans une couche géologique inclinée, Int. conf. on Computer methods and water resources, march 14-18 , Rabat, Maroc.
4. HOLLAND C.J. (1985), MODULEF - A modular finite element program package, Europ. Sci. Notes., 39(10), pp. 464-466.
5. COMBARNOUS M. and BORIES S. (1975), Hydrothermal convection in saturated porous media, Advances in Hydroscience, Vol.10, pp.231-307, Acad. Press.
6. JOURDAN A. (1986), Diagenesis as the control of the Brent sandstone reservoir properties in the Greater ALWYN area (East Shetland Basin), 3rd conf. on petroleum geology of NW Europe, oct. 26-29 , London, U.K.
7. DOLIGEZ B. BESSIS F. BURRUS J. UNGERER P. CHENET P.Y., (1985), Integrated numerical simulation of the sedimentation, heat transfer, hydrocarbon formation and fluid migration in a sedimentary basin: the THEMIS model, Thermal modeling in sedimentary basins (Ed. BURRUS J.), proceedings of the 1st IFP Expl. Res. Conf., june 3-7, Carcans, France, Technip, Paris.
8. CALTAGIRONE J.P. and BORIES S. (1985), Solutions and stability criteria of natural convective flow in an inclined porous layer, J. Fluid Mech., Vol.155, pp. 267-287.

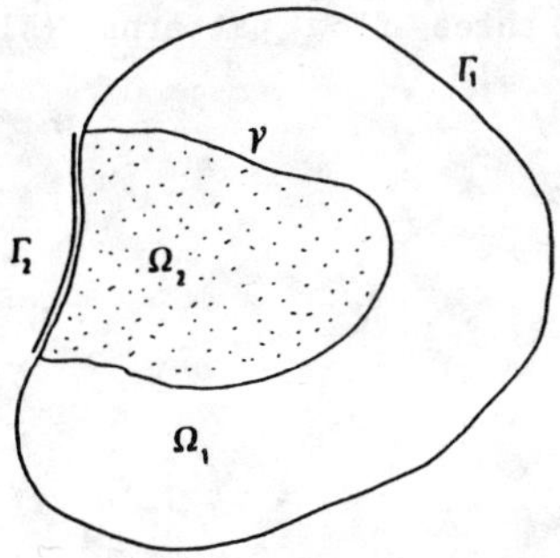

Figure 1. Partition of the 2D studied domain.

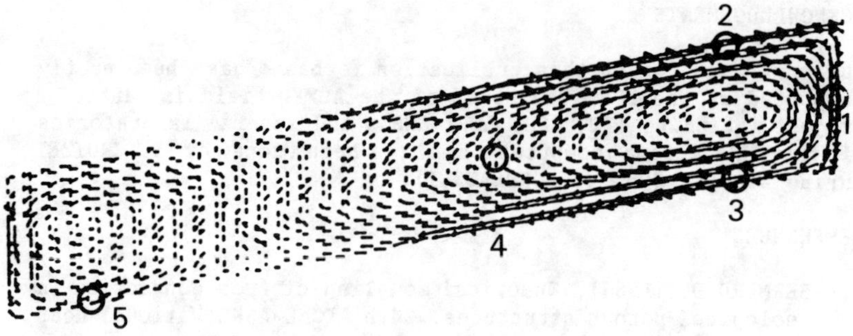

Figure 2. Velocity field in a porous block of ALWYN (NORTH sea)

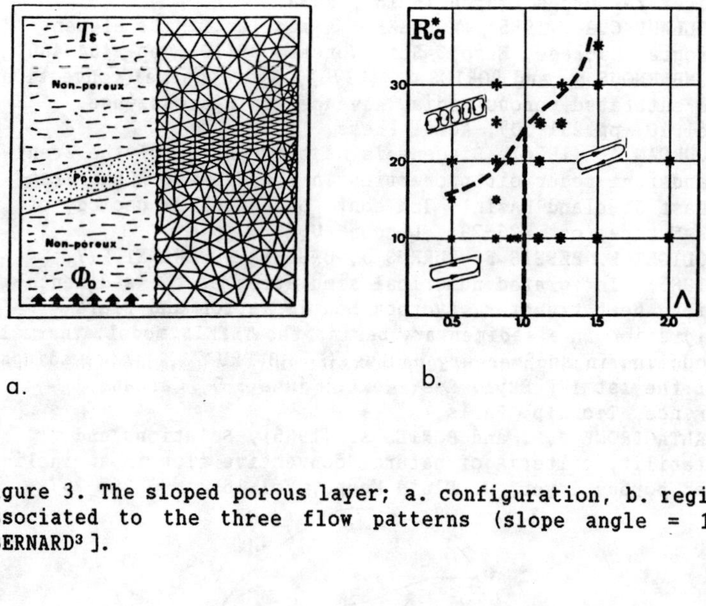

Figure 3. The sloped porous layer; a. configuration, b. regions associated to the three flow patterns (slope angle = 15°) [BERNARD[3]].

Radiative Heat Transfer to Flow in a Porous Pipe with Chemical Reaction and Linear Axial Temperature Variation

A.R. Bestman

International Centre for Theoretical Physics, Trieste, Italy

ABSTRACT

The problem considers the flow in a cylindrical porous pipe in the presence of chemical reaction and radiative heat transfer. If the temperature of the pipe varies linearly with axial distance, this extremely difficult three dimensional problem is tackled for low Reynolds number flows by asymptotic approximation and numerical integration. Consequences of the effect of heat transfer at the wall are discussed quantitatively.

INTRODUCTION

As a result of its technological import in geophysics and reservoir engineering, the problem of flow in porous medium with heat and mass transfer was recently tackled by Raptis and Perdikis[1]. An excellent review of the literature is given in this work.

However in high temperature flows it is difficult to maintain the wall at a constant temperature while the Arrhenius activation energy may not be negligible. Hence this paper is addressed to flow with linear axial wall temperature variation and finite chemical activation energy.

FORMULATION

We consider the flow, of a binary mixture of a dilute chemically reacting gas, in a porous cylinder inclined at an angle Γ to the horizontal. Let (u,v,w) and (q,s,t) be the velocity and radiative flux components in the cylindrical polar coordinates (r,φ,z), while c is the concentration of the depleting species, P is pressure and θ is temperature. If the temperature of the cylinder is of the form $\theta_w + Nz$, where θ_w and N are constant then the non-dimensional equations appropriate for flow in a porous medium are (see Jaluria[4]):

$$\frac{1}{r}\frac{\partial}{\partial r}(ru) + \frac{1}{r}\frac{\partial v}{\partial \varphi} + \frac{\partial w}{\partial z} = 0 \qquad (1)$$

$$(\nabla^2-\sigma^2)\left[\frac{1}{r}\frac{\partial}{\partial r}(rv)-\frac{1}{r}\frac{\partial u}{\partial \varphi}\right]+Gr\cos\Gamma\left(\frac{\partial\theta}{\partial r}\sin\varphi+\frac{1}{r}\frac{\partial\theta}{\partial\varphi}\cos\varphi\right)+$$

$$Gc\cos\Gamma\left(\frac{\partial c}{\partial r}\sin\varphi+\frac{1}{r}\frac{\partial c}{\partial\varphi}\cos\varphi\right)=0 \qquad (2)$$

$$(\nabla^2-\sigma^2)w+Gr\sin\Gamma(\theta-1)+Gc\sin\Gamma(c-1)=\frac{\partial p}{\partial z} \qquad (3)$$

$$RePr\left(u\frac{\partial\theta}{\partial r}+\frac{v}{r}\frac{\partial\theta}{\partial\varphi}+w\frac{\partial\theta}{\partial z}\right)=\nabla^2\theta-\frac{1}{r}\frac{\partial}{\partial r}(rq)-$$

$$\frac{1}{r}\frac{\partial s}{\partial\varphi}-\frac{\partial t}{\partial z} \qquad (4)$$

$$ReSc\left(u\frac{\partial c}{\partial r}+\frac{v}{r}\frac{\partial c}{\partial\varphi}+w\frac{\partial c}{\partial z}\right)=\left[\nabla^2-k_r^2 f(\theta)\right]c, f(\theta)=$$

$$\theta^{\eta}\exp)(-E/\theta) \qquad (5)$$

$$\frac{\partial\Delta}{\partial r}\cdot\frac{3}{N_0^2}q-\frac{3}{B_0}\theta^3\frac{\partial\theta}{\partial r}=0=\frac{1}{r}\frac{\partial\Lambda}{\partial\theta}-\frac{3}{N_0^2}s-\theta^3\frac{1}{r}\frac{\partial\theta}{\partial\varphi}=$$

$$\frac{\partial\Delta}{\partial z}-\frac{3}{N^2}t-\frac{3}{B_0}\theta^3\frac{\partial\theta}{\partial z} \qquad (6)$$

where

$$\Delta=\frac{1}{r}\frac{\partial}{\partial r}(rq)+\frac{1}{r}\frac{\partial s}{\partial\varphi}+\frac{\partial t}{\partial z},\quad \nabla^2=\frac{\partial^2}{\partial r^2}+\frac{1}{r}\frac{\partial}{\partial r}+\frac{1}{r^2}\frac{\partial^2}{\partial\varphi^2}$$

Equations (1)-(6) are to be solved subject to the boundary conditions

$$(u,v,w)=0,\quad \theta=\theta_w+N z,\quad c=c_w,\ \left(\frac{1}{\varepsilon_w}-\frac{1}{2}\right)q-\frac{N_0}{4}\frac{\partial q}{\partial r}=0 \quad \text{on } r=1$$

$$(u,v,w),\theta,c,(q.s.t)<\infty \qquad \text{on } r=0 \qquad (7)$$

Equation (2) is obtained by eliminating the pressure gradients in the r and φ momentum equations. The inertia terms are neglected in equations (2) and (3) and this is valid except for very highly porous medium. In equation (5) E is the activation energy. $k^2{}_r$ is the chemical rate constant and η is a constant exponent while in equation (7) ε_w is the emittence of the wall. The problem depends on the following parameters: Gr and Gc are the buoyancy parameter due to heat and mass transfer, Re is the Reynolds number, Pr is the Prandtl number, Sc is the Schmidt number, N_o and B_o are the radiation parameters while σ is the porosity parameter. All the various non-dimensional quantities have been discussed in Bestman[2,3].

The mathematical statement of the problem is to solve equations (1)-(6) subject to (7). This is a very difficult problem and to make any headway with the solution we shall make the realistic assumption that Re is small. Further only linear terms in z will be retained in the analysis and it is in anticipation of this that all derivatives of z higher than the first are dropped in equations (1)-(6).

PERTURBATION AND BASIC SOLUTION

We seek the asymptotic series expansion
$u = u^{(0)} + Re\, u^{(1)} + \ldots$ etc.,
such that $s^{(0)} \equiv 0$, while $\theta^{(0)}$, $q^{(0)}$, $c^{(0)}$ and $t^{(0)}$ are expressed as
$\theta^{(0)} = \Theta_0^{(0)}(r) + z\,\Theta_1^{(0)}(r)$ etc.
and

$$\frac{\partial p^{(0)}}{\partial z} = \pi_0^{(0)} + z\pi_1^{(0)}(r),$$

$$u^{(0)} = (U_0^{(0)}(r) + zU_1^{(0)}(r))\mathrm{Cos}\varphi, v^{(0)} = (V_0^{(0)}(r) + V_1^{(0)}(r))\sin\varphi$$

However we shall assume that the flow is fully developed so that w is independent of z (i.e. $w^{(0)} = W^{(0)}(r)$) and $\pi_0^{(0)}$ is an externally impressed constant pressure gradient.

In the light of these the basic equations for the temperature and the radiative flux become

$$\left(\frac{d^2}{dr^2} + \frac{1}{r}\frac{d}{dr}\right)\Theta_0^{(0)} = \frac{1}{r}\frac{d}{dr}(rQ_0^{(0)}) + T_1^{(0)}$$

$$\left(\frac{d^2}{dr^2} + \frac{1}{r}\frac{d}{dr}\right)\Theta_0^{(0)} = \frac{1}{r}\frac{d}{dr}(rQ_1^{(0)})$$

$$\frac{d}{dr}\left[\frac{1}{r}\frac{d}{dr}(rQ_1^{(0)}) + T_1^{(0)}\right] - \frac{3}{N^2}\Theta^{(1)} - \frac{3}{B_0}\Theta_0^{(0)3}\frac{d\Theta_0^{(0)}}{dr} = 0$$

$$\frac{d}{dr}\left[\frac{1}{r}\frac{d}{dr}(rQ_0^{(1)})\right] - \frac{3}{N^2}Q_0^{(1)} - \frac{3}{B_0}\left(\Theta_0^{(0)3}\frac{dQ_1^{(0)}}{dr} + \right.$$

$$\left. 3\Theta_0^{(0)2}\frac{d\Theta_0^{(0)}}{dr}\Theta_1^{(0)}\right) = 0 \tag{8}$$

$$\frac{3}{N_0^2}T_1^{(0)} = -\frac{9}{B_0}\Theta_0^{(0)2}\Theta_1^{(0)2} \tag{9}$$

$$\Theta_0^{(0)} = \theta_w, \quad \Theta_1^{(0)} = \Gamma, \quad \left(\frac{1}{\varepsilon_w} - \frac{1}{2}\right) Q_{0,1}^{(0)} - \frac{N_0}{4}\frac{dQ_{0,1}^{(0)}}{dr} = 0 \quad \text{on } r=1$$

$$\Theta_0^{(0)} < \infty, \; \Theta_0^{(0)} < \infty, \; \Theta_1^{(0)} = 0 = Q_1^{(0)} \qquad \text{on } r=0 \qquad (10)$$

Thereafter

$$\frac{3}{N_0^2} T_0^{(0)} = \frac{1}{r}\frac{d}{dr}(rQ_1^{(0)}) - \frac{3}{B_0^2}\Theta_0^{(0)3}\Theta_1^{(0)}.$$

Equations (8)-(10) are a system of coupled nonlinear equations and numerical algorithms exist for such problems. The point r=0 in the present problem presents difficulties and as a result we evolve a simple algorithm for this problem. We set $T_1^{(0)} = SO_j$ = constant in Equation (8). Then following Bestman[2,3] the equations governing $\Theta_0^{(0)}$ and $Q_0^{(0)}$ reduce to

$$Q_0^{(0)2} = K_0 + K_1\Theta_0^{(0)} + \frac{3}{N_0^2}\Theta_0^{(0)2} + \frac{3}{10B_0}\Theta_0^{(0)5} - f(\Theta_0^{(0)})$$

in which the constants K_0 and K_1 are given by

$$K_0 + K_1 + \frac{3}{N_0^2} + \frac{3}{10B_0} = 0$$

$$\left[\left(\frac{1}{\varepsilon_w} - \frac{1}{2}\right) + \frac{N_0}{4}\right]^2 (K_0 + K_1\theta_w + \frac{3}{N^2}\theta_w^2 + \frac{3}{10B_0}\theta_w^5) -$$

$$- \frac{N_0^2}{16}\left(\frac{1}{2}K_1 + \frac{3}{N^2}\theta_w^2 + \frac{3}{4B_0}\theta_w^4 - \frac{3}{4}SO_i\right)^2 = 0$$

such that $\Theta_0^{(0)}$ and f are governed by the initial value problem

$$\frac{d\Theta_0^{(0)}}{dr} = Q_0^{(0)} + \frac{1}{2}SO_i\, r$$

$$\frac{df}{dr} = \frac{2}{r^2}Q_0^{(0)2} + SO_i.r\left(\frac{1}{2}K_1 + \frac{3}{N^2} + \frac{3}{N^2}\Theta^{(0)} + \frac{3}{4B_0}\Theta^{(0)4}\right)$$

$$\Theta_0^{(0)} = \theta_w, \quad f=0 \qquad \text{on } r=1,$$

which is easily tackled by the Runge-Kutta method. In similar vein, for $\Theta_1^{(0)}$ and $Q_1^{(0)}$ the following expressions may be deduced:

$$Q_1^{(0)2} = K_3\Theta_1^{(0)} + \frac{3}{N_0^2}\Theta_1^{(0)2} + g(\Theta_1^{(0)})$$

$$\left[\left(\frac{1}{\varepsilon_w}-\frac{1}{2}\right)+\frac{N_0}{4}\right]^2\left(\frac{3}{N_0^2}\Gamma^2+K_3\Gamma\right)=\frac{N_0^2}{16}\left(\frac{1}{2}K_3+\frac{3}{N^2}\Gamma+\frac{3}{B_0}\theta_w^3\Gamma\right)^2$$

$$\frac{d\Theta_1^{(0)}}{dr}=Q_1^{(0)},\ \frac{dg}{dr}=2\left(\frac{3}{B_0}\Theta_0^2\Theta_1-\frac{1}{r}Q_1^{(0)}\right)Q_1^{(0)}$$
$$Q_1^{(0)}=\Gamma,\quad g=0 \qquad \text{on} \qquad r=1$$

We start the computation with $SO_i|i=0=-3\theta^2{}_w\Gamma^2N^2{}_0/B_0$, then having computed $\Theta^{(0)}{}_{0,1}$ and $Q^{(0)}{}_{0,1}$ value of $T_1{}^{(0)}$ now follows from (9) the mean value of which constitutes SO_1. The procedure is then repeated and so on,. The iteration converges for i=2.

Next for the concentration the governing equations are

$$\left(\frac{d^2}{dr^2}+\frac{1}{r}\frac{d}{dr}-k_r^2Q_0^{(0)\eta}e^{-E/Q_0^{(0)}}\right)C_0^{(0)}=0$$

$$\left(\frac{d^2}{dr^2}+\frac{1}{r}\frac{d}{dr}-k_r^2\Theta_0^{(0)\eta}e^{-E/\Theta_0^{(0)}}\right)C_1^{(0)}=k_r^2\Theta_0^{(0)\eta}e^{-E/\Theta_0^{(0)}}\left(\eta+\frac{E}{\Theta_0^{(0)}}\right)\frac{\Theta_1^{(0)}}{\Theta_0^{(0)}}C_0^{(0)} \qquad (12)$$

$$C_0^{(0)}=c_w,\ C_1^{(0)}=0 \quad \text{on } r=1,\ C_0^{(0)}<\infty,\ C_1^{(0)}=0 \quad \text{on } r=0.$$

If we take $dC_0{}^{(0)}/dr=0$ on r=0, by virtue of symmetry, the Equations (12) are linear and may be tackled by Runge-Kutta method.

Finally the velocity field is obtained from the system of equations

$$\left(\frac{d^2}{dr^2}+\frac{1d}{r\,dr}-\sigma^2\right)w_0=\pi_0^{(0)}-GrSin\Gamma\left(\Theta_0^{(0)}-1\right)-Ge\ Sin\Gamma\left(C_0^{(0)}-1\right)$$

$$\pi_1^{(0)}=GrSin\Gamma\Theta_1^{(0)}+GcSin\ \Gamma C_1^{(0)}$$
$$\frac{1}{r}\frac{d}{dr}(rU_0^{(0)})+\frac{1}{r}V_0^{(0)}=\frac{1}{r}\frac{d}{dr}(rU_1^{(0)})+\frac{1}{r}V_1^{(0)}$$
$$\left(\frac{d^2}{dr^2}+\frac{1}{r}\frac{d}{dr}-\frac{1}{r^2}-r^2\right)\left[\frac{1}{r}\frac{d}{dr}(rV_{0,1}^{(0)})+\frac{1}{r}U_{0,1}^{(0)}\right]=-GrCos\Gamma\frac{d\Theta_{0,1}^{(0)}}{dr}-$$
$$GcCos\Gamma\frac{dC_{0,1}^{(0)}}{dr} \qquad (13)$$

subject to homogeneous boundary conditions at the wall. We find it expedient to solve these equations by expanding $\Theta_0{}^{(0)}$, $C_0{}^{(0)}$ and $d\Theta^{(0)}{}_{0,1}/dr$, $dC^{(0)}{}_{0,1}/dp$ in terms of Fourier Bessel series with Bessels function of order zero and unity respectively (see Bestman[2,3]).

HIGHER APPROXIMATION

For $\theta^{(1)}$, $q^{(1)}$, $t^{(1)}$, $w^{(1)}$, $c^{(1)}$ we seek the separation of variables

$$\theta^{(1)} = [\Theta_0^{(1)}(r) + z\Theta_1^{(1)}(r)]\text{Cos}\ \varphi$$

while for $s^{(1)}$ we set

$$s^{(1)} = [S_o^{(1)}(r) + zS_1^{(1)}(r)]\text{Sin}\ \varphi$$

and for $u^{(1)}$ and $v^{(1)}$ we put

$$u^{(1)}=[U_0^{(1)}(r)+zU_1^{(1)}(r)]\text{Cos}2\varphi,\ v^{(1)}=[V_0^{(1)}(r)+zV_1^{(1)}(r)]\text{Sin}2\varphi$$

It is then possible to reduce the velocity and radiative flux equations to essentially two sets of coupled linear equations for $\Theta^{(1)}{}_{0,1}$ and $Q^{(1)}{}_{0,1}$. These are then tackled by the finite difference scheme.

The concentration and velocity fields are also reduced to the forms similar to Equations (12) and (13). They are then tackled by methods used in analysing Equations (12) and (13).

HEAT FLUX AT THE WALL

The non-dimensional heat flux is defined as

$Q = \partial\theta/\partial r|r=1$.

Its computations are entered in Table 1 for z=1=N, Pr=0.71, Sc=5,Gr=5=Gc, Re=0.5, k_r=5, E=η=1, Γ=45°, Θ_w=10, ε_w=1/2. We are primarily interested on the effect of porosity and radiation on the heat flux.

Table 1: HEAT FLUX

N_o	B_o	σ	Q
0.05	0.5	0.5	0.1236
0.07	1.0	0.5	0.0687
0.05	0.5	1.5	0.0835

We find that increase in the radiation parameters or porosity parameter causes a decrease in the heat flux at the wall.

ACKNOWLEDGEMENT

The author is indebted to Prof. Abdus Salam, IAEA and UNESCO for hospitality at the International Centre for Theoretical Physics.

REFERENCES

1. Raptis A. and Perdikis C.(1987), Mass Transfer and Free Convection Flow Through a Porous Medium, Energy Research, Vol. ii, pp.423-428.
2. Bestman A.R., Nwabuzar S.S. and Metibaiye, J.W.E. (1987), Radiative Heat Transfer to Flow in a Porous Medium, EPMESC, Guangzhou (China).
3. Bestman A.R., Nwabuzar S.S. and Metibaiye, J.W.E. (1987), Radiative Heat Transfer to Flow of a Binary Reacting Gas Mixture Through a Porous Medium , Computational Techniques and Applications Sydney (Australia).
4. Jaluria Y. (1980), Natural Convection Heat and Mass Transfer, Pergamon Press, New York.

Assessment of Thermal Impacts of Discharge Locations using Finite Element Analysis

Y.C. Chang

Stone and Webster Engineering Corporation, Boston, MA, USA

D.P. Galya*

ERT, A Resource Engineering Company, Concord, MA, USA

**Formerly with Stone and Webster Engineering Corporation*

INTRODUCTION

Stone & Webster Engineering Corporation (Stone & Webster) conducted an evaluation of the thermal plume from Florida Power Corporation's (FPC) Crystal River Power Station in Citrus County, Florida. Finite-element modeling was used to assess the impact of various discharge locations on the extent and magnitude of the plume.

Crystal River Power Station (total average heat load - 2.65×10^{11} Btu/day) is located on Crystal Bay, on the eastern shore of the Gulf of Mexico (see Figure 1). The bay tends to be very shallow with depths of less than 10 feet up to three miles offshore and less than 20 feet as far offshore as 10 miles. The nearshore region of the bay consists of shallow basins separated by long strings of oyster bars or reefs. The reefs act to restrict advective flux and dispersive mixing between the basins. Flows in the study region are further influenced by river inflows from the Withlacoochee River and Cross Florida Barge Canal to the north and Crystal River to the south.

This paper describes the modifications made to the finite-element models to reflect the restricted flow and their applications in simulating thermal plumes for different discharge locations in Crystal River Bay.

MODEL DESCRIPTION

A pair of two-dimensional, finite element models developed by the Massachusetts Institute of Technology (MIT) was used in this study: CAFE, a hydrodynamic model; and DISPER, a dispersion model. Both models have been verified extensively by their originators and others in the engineering community (Wang and Connor[1], Wang[2], Leimkuhler et al[3], Galya and Colangelo[4], MIT[5]). Brief descriptions of the models and the modifications made by Stone & Webster for this study are presented below. Details of the mathematical formulation of the models are provided elsewhere (Wang and Connor[1], Wang[2], Leimkuhler et al[3]).

CAFE

CAFE uses input parameters, such as study area geometry, bottom topography and roughness, and boundary conditions, in solving the vertically integrated equations of mass and momentum. The output from CAFE consists of temporally varying tidal heights and vertically integrated velocities within the study region.

Two basic types of boundary conditions are employed in CAFE: specified water levels and specified flux. In regions subject to tidal variations, the water-level specification is generally used to simulate tidal height fluctuations. The specified flux conditions are generally used to simulate no-flux boundaries as well as fluxes into or out of the study region. For this study, modifications were made to CAFE to add a boundary condition that would allow the simulation of the semipermeable or restricted flow characteristics of the oyster reefs. The restricted flow simulation is represented by adding a line of nodes on each side of the semipermeable barrier causing the restricted flow. Flux through the barrier is specified as a boundary condition.

Flows between basins through the semipermeable boundaries are assumed to be composed of weir-like flows over the tops of the oyster reefs and orifice-like flows through spaces in the reefs. When the water level is below the top of the oyster reefs, there may be orifice-like flow but no weir-like flow. As the water level increases during flood tide, an increase in the cross-sectional area of the spaces results in increased orifice-like flow. Weir-like flow begins as the water level reaches and exceeds the level of the reefs. As the water level increases further, weir-like flow tends to predominate. During ebb tide, the scenario occurs in reverse order. The specification of the flux coefficients assumes the form of a step function with the value of the coefficients based on the height of the water level.

DISPER

Input parameters of DISPER, a far-field model, include study area geometry, dispersion coefficients, plant intake and discharge flow rates, initial and boundary conditions of temperature or concentration, and the temporally varying values of water velocity and depth from CAFE output. The model employs these parameters in solving the vertically integrated form of the advective dispersion equation. Output from DISPER consists of temporally varying values of temperature or concentration within the study region.

A change was made to DISPER to complement the change to CAFE incorporating the ability to simulate semipermeable boundaries. In DISPER, the semipermeable boundary formulation used grid geometry, water levels and velocities from CAFE, and either concentration or temperature data to determine the flux of either material or heat out of the nodes comprising the upstream portion of the semipermeable boundary. An equivalent flux of material was added at the nodes along the downstream portion of the boundary to represent the flux of material through the element sides.

MODEL APPLICATION

Application of CAFE and DISPER to a particular study region involves the development of a finite element grid and specification of appropriate boundary conditions. The finite element grid is developed by subdividing the study region into an array of discrete elements to provide sufficient resolution to reflect the pertinent physical features of the region. Figure 2 shows the finite element grid for the Crystal River study region; in this grid, oyster reefs are represented as strings of node pairs, shown as parallel line segments between subregions of the study area.

The application of the models was complicated by the presence of open boundaries on three sides of the study region. Specification of tidal height fluctuations at all three boundaries might have resulted in large errors in model predictions. This situation was precluded by specifying tidal height fluctuations along the western (or offshore) boundary to provide the major forcing mechanism, and by specifying tidal fluxes throughout the tidal cycle along the northern and southern boundaries to simulate alongshore flow. Additional boundary conditions consisted of a no-flux condition for the eastern land boundaries; flux perpendicular to the land boundaries for the river inflows and for the intake and discharge flow; and flux across semipermeable boundaries representing oyster reefs.

The models were calibrated to Crystal Bay using current velocities, water temperatures, conductivities, and tidal heights collected over two 1-month periods (Galya and McDougall[6]). The bottom friction factor and semipermeable boundary coefficients in CAFE and the dispersion coefficient in DISPER were varied until a good comparison with field data was obtained.

NEAR-FIELD MODEL

As actual field data indicated shallow water depths and small vertical temperature differences near the existing point of discharge (POD), a vertically mixed near-field model was developed based on the conservation of mass, momentum, and heat. A solution was obtained by assigning a normal distribution to both velocity and temperature rise and by assuming the entrainment velocity is proportional to the local centerline velocity (FPC[7]).

The near-field model formulation assumes the entrainment of ambient, unheated water. However, the far-field model, DISPER, predicts elevated temperature conditions near the discharge point. The near-field model provides specific detail by predicting the temperature rise above the far-field value. The actual temperature rise near the discharge point is the result of adding the near- and far-field predictions. The near- and far-field results are merged to provide the complete temperature field.

THERMAL EVALUATION OF DISCHARGE LOCATIONS

The discharge locations simulated were: a) existing POD; b) 3.5 miles offshore on the north side of the existing intake jetty; c) 1 mile offshore from the existing POD with a north jetty added to the existing discharge jetty; and d) 2 miles offshore from the end of the existing POD by extending both jetties. Various heat loads reflecting different helper tower cooling modes were also simulated. Table 1 lists the conditions of the simulations.

Table 1. CRYSTAL RIVER SIMULATION CONDITIONS

Location	Case	Temperature Rise (°C)
a	a	9.3
b	b	9.3
c	c.1	7.6
	c.2	6.5
	c.3	5.4
d	d.1	7.6
	d.2	6.5
	d.3	5.4

Note: Discharge flow rate was 83 m^3/sec in all cases.

RESULTS

The thermal isotherms of each case were plotted for the high-water slack (HWS) and low-water slack (LWS). The HWS condition results in a shoreward plume reflective of the currents on the flood tide, while the LWS condition results in a plume extending farther offshore, reflective of the currents on ebb tide. Because of shallow depths, the thermal plumes in the LWS are larger than those of HWS. The area identified by the model for each isotherm was measured for biological impact assessment. Figure 3 shows the isotherms for the LWS condition for four of the cases investigated. The results indicate that the thermal plume size could be reduced by discharging offshore at deeper water locations where the influence from the shoreline and oyster reef is less significant. Other factors such as environmental impacts during construction and economics would have to be considered in the evaluation of discharge schemes.

SUMMARY

Stone & Webster used the finite element computer models CAFE and DISPER to simulate thermal plumes for effluent discharges into Crystal Bay. The models were modified to reflect the restrictions in flow due to oyster reefs. The models, in conjunction with a vertically mixed near-field model, were successfully applied to simulate thermal plumes for offshore discharges. The results clearly show the influence of discharge locations and identify areas to be included in a biological impact assessment.

REFERENCES

1. Wang J.D. and Connor J.J. (1975). Mathematical Modeling of Near Coastal Circulation. Techn. Rept. No. 200, R.M. Parsons Lab. for Water Resource and Hydrology, Massachusetts Institute of Technology.

2. Wang J.D. (1978). Verification of Finite Element Hydrodynamic Model CAFE. Verification of Mathematical and Physical Models in Hydraulic Engineering. ASCE, New York, New York.

3. Leimkuhler W.F. Connor J.J. Wang J.D. Christodolou G. and Sundgren S. (1975). Two-Dimensional Finite Element Dispersion Model. Symposium on Modeling Techniques, "Modeling 75", San Francisco, California.

4. Galya D.P. and Colangelo P.M. (1981). Finite Element Modeling of Complex Embayment Systems. Proc. Third Waste Heat Management and Utilization Conference, Miami Beach, Florida.

5. Massachusetts Institute of Technology (MIT). (1977). Computer Models for Environmental Engineering and Research in Near-Coastal Environments. MIT Marine Industry Colloquium Workshop.

6. Galya D.P. and McDougall D.W. (1985). Finite Element Modeling of an Offshore Region with Restricted Flow. Proc. Oceans 85, Ocean Engineering and the Environment, San Diego, California.

7. Florida Power Corporation. (1985). Crystal River Units 1, 2, and 3, 316 Demonstration.

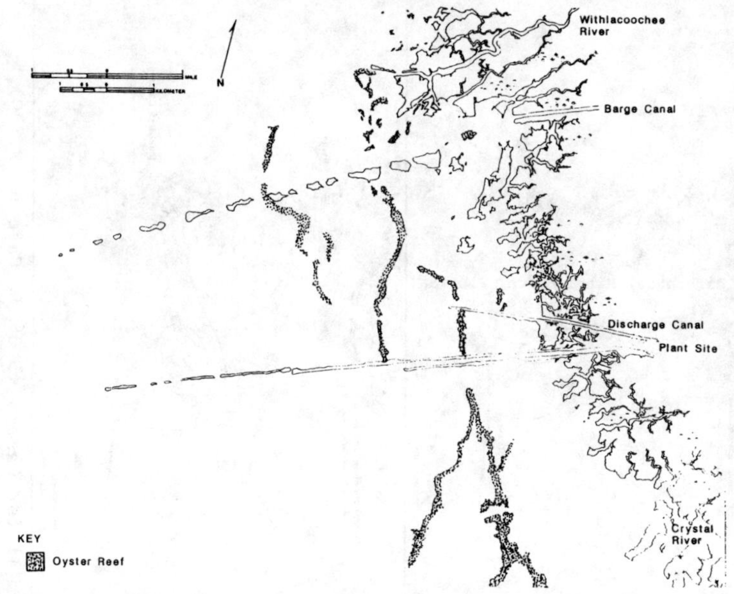

Figure 1. Crystal River study region

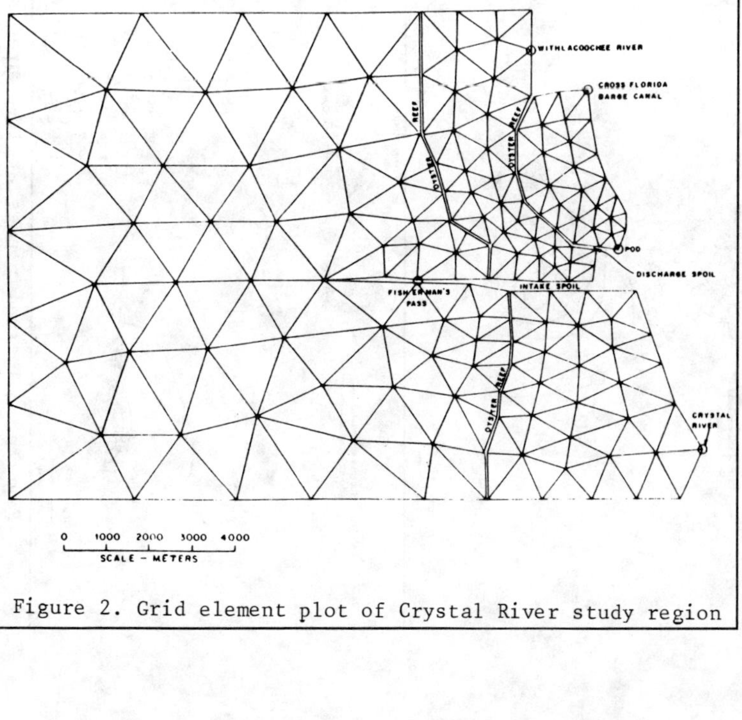

Figure 2. Grid element plot of Crystal River study region

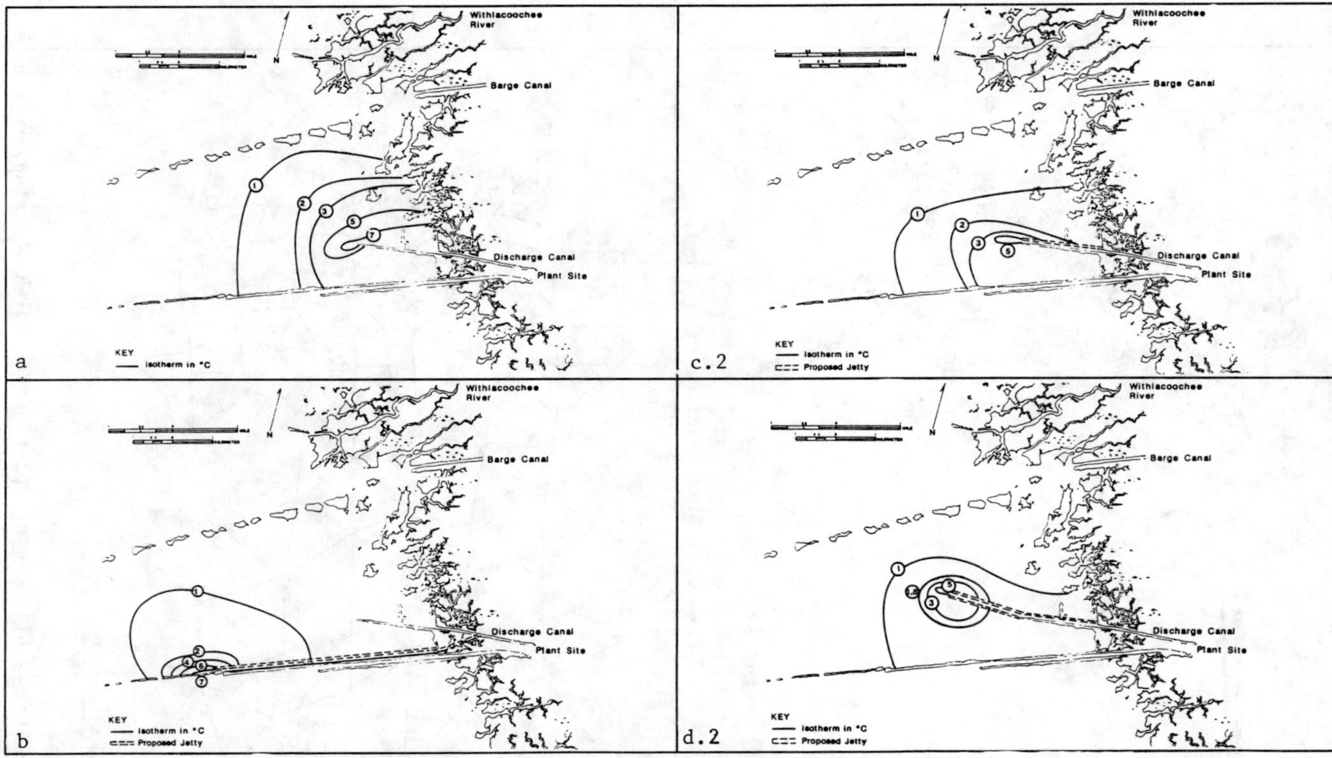

Figure 3. Thermal plumes at low water slack (LWS) for Case a (9.3°C), Case b (9.3°C), Case c.2 (6.5°C), and Case d.2 (6.5°C)

Validation of Finite Element Simulation of the Hydrothermal Behavior of an Artificial Aquifer Against Field Performance

H. Daniels

Institut fuer Wasserbau und Wasserwirtschaft, Aachen University of Technology, Federal Republic of Germany

ABSTRACT

The effect of natural convection and density driven flow is of major importance to the flow field and the temperature field in the artificial aquifer for thermal energy storage that has been built at the Stuttgart University. This paper describes the Finite Element Approximation that has been used to calculate the transient temperature field in the aquifer and shows some validation results against measurement data from field experiments in Stuttgart.

THE TEST SITE

At the Institute for Thermodynamics of the Stuttgart University an artificial aquifer has been constructed within the German underground thermal energy storage program. Fig. 1 shows a vertical cross section of the axisymmetric aquifer. Cold water is exchanged through holes near the bottom of the pressure chamber in the central tube while hot water is fed into or released from the aquifer by a drainage ring conduit at a radius of about 9,70 m at the top of the aquifer. A storage layer of gravel 8-16 mm is bounded vertically by two charging layers of gravel with 16-32 mm grain size. This vertical layering as well as the location of the water inlet facilites was emplaced as a result of design studies for different geometries of the aquifer that were done in Aachen. The Stuttgart aquifer contains several hundred temperature measuring probes. The thermal performance of the test aquifer can be observed exactly. Finite Element Analysis of the aquifer's hydrothermal behavior was fairly complicated because of the huge permeability of the gravel filling and temperature differences up to 45°C in the water-saturated aquifer.

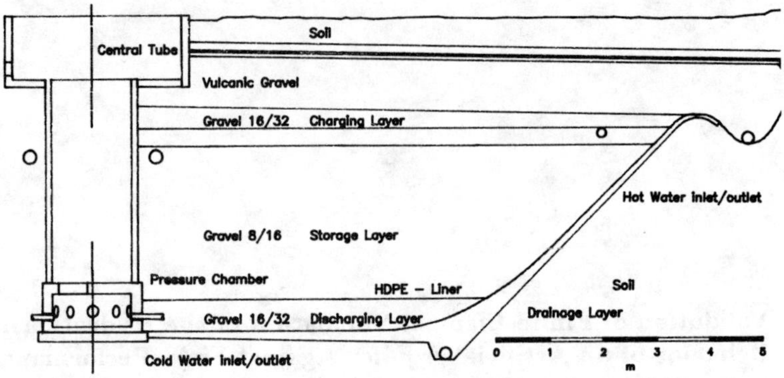

Fig. 1: Vertical cross section of the artificial aquifer

MATHEMATICS OF ENERGY TRANSPORT IN POROUS MEDIA FLOW

The description of transient energy transport in porous media is well understood since more than one decade. Suppose the velocity field v_i is given, the convection-diffusion equation can be written:

$$\rho_b c_b \frac{\partial T}{\partial t} + \frac{\partial}{\partial x_i} (c_f \rho_f v_i T) - \frac{\partial}{\partial x_i} \left[(n_{eff} \rho_f c_f D_{m,ij} + \lambda_{b,ij}) \frac{\partial T}{\partial x_j} \right] = 0. \qquad (1)$$

In eq. (1) local equilibrium between fluid and matrix temperature is assumed. As the velocity field is not known a priori, it has to be calculated before the convective transport can be evaluated. Major difficulties arose and efforts had to be made when calculating the strongly density dependent flow field in the artificial aquifer. Different mathematical formulations were used in order to find an appropriate approximation for the hydraulic behavior of the artificial aquifer.

Hybrid Formulation

Assuming DARCYs law to be valid and substituting it into the continuity equation ends up with a hybrid formulation for the pressure distribution p

$$S_{o,p} \frac{\partial p}{\partial t} - \frac{\partial}{\partial x_i} \left[\frac{\rho_f K_{ij}}{\eta_f} \left(\frac{\partial p}{\partial x_j} + \rho_f g \frac{\partial z}{\partial x_j} \right) \right] = 0. \qquad (2)$$

Multi Array Formulation

If DARCYs law

$$v_i = - \frac{K_{ij}}{\eta_f} \left(\frac{\partial p}{\partial x_j} + \rho_f g \frac{\partial z}{\partial x_j} \right) \qquad (3)$$

is not inserted into the continuity equation

$$\frac{\partial v_i}{\partial x_i} = 0, \tag{4}$$

more than one independent variable, i.e. p and v_i, are to be calculated simultaneously.

Quasi NAVIER-STOKES Formulation

Analogous to the laminar viscous flow problem, porous media flow can be written in a NAVIER-STOKES Formulation for isotropic permeability (GARTLING 1987). In this case eq. (3) can be replaced by

$$\frac{\rho_o}{n} \frac{\partial v_i}{\partial t} + \left(\frac{\rho_o \hat{a}}{\sqrt{K}} |v| + \frac{\eta_f}{K}\right) v_i =$$

$$\frac{\partial}{\partial x_j} \left[- p + \eta_{eff} \left(\frac{\partial v_i}{\partial x_j} + \frac{\partial v_j}{\partial x_i}\right)\right] - \rho_f \, g \, \frac{\partial z}{\partial x_j} \tag{5}$$

Other than eq. (2) and (3), eq. (5) takes viscous forces into account and does not necessarily require a zero-order potential field as standard DARCYs law does.

Discussion on Buoyancy Flow

A zero-order potential field $\underline{F}$ in the way

$$v_i = - \frac{\rho_f K_{ij}}{\eta_f} \left(\frac{1}{\rho_f} \frac{\partial p}{\partial x_j} + g \frac{\partial z}{\partial x_j}\right) = - \underline{k}_f \underline{F} \tag{6}$$

does not exist if the fluid density ρ_f is variable in space, because we obtain

$$\text{rot } \underline{F} = \text{grad } p \times \text{grad } \frac{1}{\rho_f} \neq 0 \tag{7}$$

in this case (STÖSSINGER 1979). Only if the spacial variability of ρ is small, the flow field can approximately be handled as in eq. (2) and (3). In terms of Finite Element Analysis the element spacing necessary is strongly dependent on grad $1/\rho$ and at the same time on the value of grad $1/\rho$ compared to the value of grad p or the matrix permeability K, respectively.

NUMERICAL SOLUTION

The Finite Element Formulations based on the BUBNOV-GALERKIN method for eqs. (1) to (4), are assumed to be familiar. Here the Finite Element form of the quasi NAVIER-STOKES flow eq. (5) will be considered. Using the GAUSS theorem to reduce the second order diffusion terms in eq. (5) the GALERKIN formulation finally leads to

$$\left[\int_V \frac{\rho_o}{n} \phi_m \phi_n \, dV \right] \frac{\partial v_{i,n}}{\partial t} + \left[\int_V \frac{\rho_o \hat{a}}{\sqrt{K}} \phi_m |v| \phi_n \, dV \right] v_{i,n} +$$

$$\left[\int_V \frac{\eta_f}{K} \phi_m \phi_n \, dV \right] v_{i,n} + \left[\int_V \eta_{eff} \frac{\partial \phi_m}{\partial x_j} \frac{\partial \phi_n}{\partial x_j} dV \right] v_{i,n} +$$

$$\left[\int_V \eta_{eff} \frac{\partial \phi_m}{\partial x_j} \frac{\partial \phi_n}{\partial x_i} dV \right] v_{j,n} - \left[\int_V \frac{\partial \phi_m}{\partial x_i} \phi_n \, dV \right] p_n =$$

$$\left[\int_V \phi_m \rho_f \, g \frac{\partial \phi_n z_n}{\partial x_i} dV \right] + \left[\int_{\partial V} \phi_m \tau_{ij} n_j \, \partial V \right] \qquad (8)$$

FINITE ELEMENT GRID

Because of the domination of buoyancy flow in the artificial aquifer, a Finite Element grid that consists of equilateral triangles proved to be the most appropriate discretization. Fig. 2 shows two ideas for grids that performed about the same accuracy during model validation. Grid a), developed according to KINZELBACH and FRIND (1986) did not only have twice as much nodes than grid b), but also required a much smaller time step to obtain a numerically stable solution.

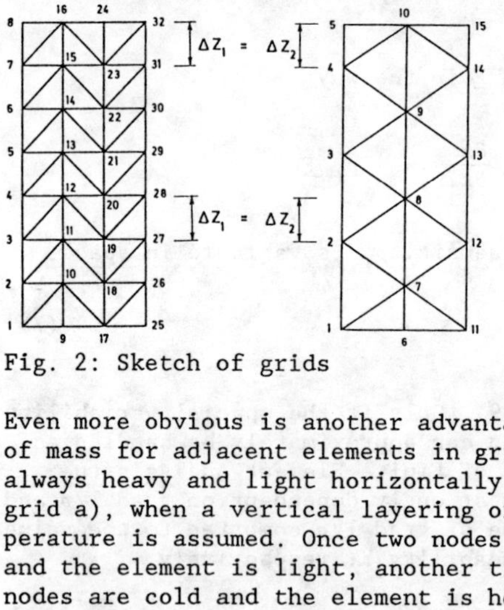

Fig. 2: Sketch of grids

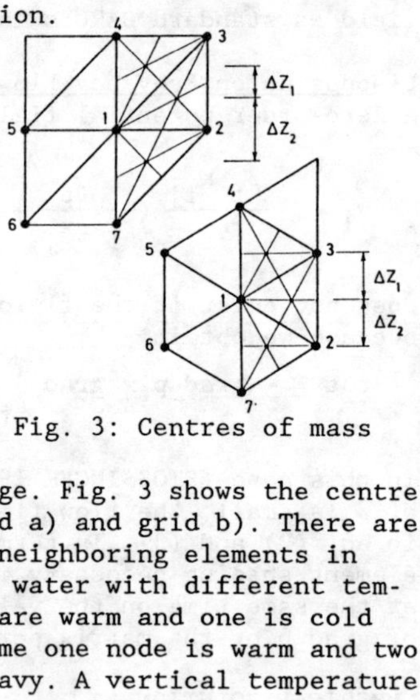

Fig. 3: Centres of mass

Even more obvious is another advantage. Fig. 3 shows the centre of mass for adjacent elements in grid a) and grid b). There are always heavy and light horizontally neighboring elements in grid a), when a vertical layering of water with different temperature is assumed. Once two nodes are warm and one is cold and the element is light, another time one node is warm and two nodes are cold and the element is heavy. A vertical temperature layering and a stable flow field are very hard to obtain with grid a). There are no such problems with grid b).

RESULTS OF MODEL VALIDATION

Fig. 4 shows the parameters used during validation of the numerical model for the aquifer in Fig. 1. Aquifer permeability is very high. The relation of longitudinal to transversal dispersion coefficient aI/aII = 2/1 is reasonable due to the very homogeneous and isotropic gravel filling.

	K $[m^2]$	n $[-]$	n_{eff} $[-]$	a_I $[m]$	a_{II} $[m]$	λ_b $[\frac{W}{mK}]$	ρ_b $[\frac{kg}{m^3}]$	c_b $[\frac{J}{kgK}]$
charging layers	$2 \cdot 10^7$	0,37	0,34	0,25	0,125	2,2	1.910	1.994
storage layer	$4 \cdot 10^8$	0,37	0,33	0,25	0,125	2,2	1.910	1.994

Fig. 4: Hydrothermal parameters in the aquifer

The aquifer itself was discretized into axisymmetric equilateral triangles of 0,20 m spacing for the simulation. Outside the aquifer, where only heat conduction is assumed, the flow field equation was not solved at all and the elements were streched. Fig. 5a gives an impression of the two-dimensional temperature field inside the aquifer calculated during the charging cycle. A stable vertical layering of water of different temperatures can be observed. Charging temperatures were about 50 C. Due to natural convection, temperatures in the vicinity of the hot water inlet stayed below 50 C until the end of the charging cycle. Fig. 5b shows temperture over time for three points in the aquifer. Measured and calculated temperatures fit very well in the whole aquifer all the time, though buoyancy flow is much stronger than forced convection and though there is not a single DIRICHLET boundary condition for the temperature field.

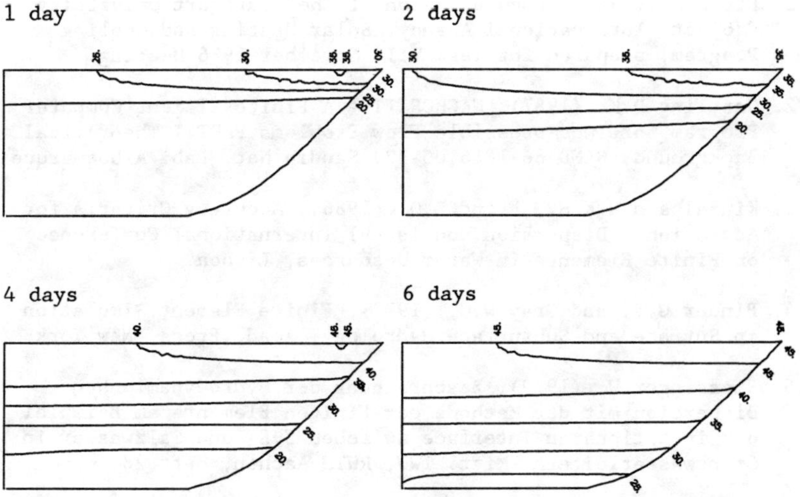

Fig. 5a: Two-dimensional temperature field

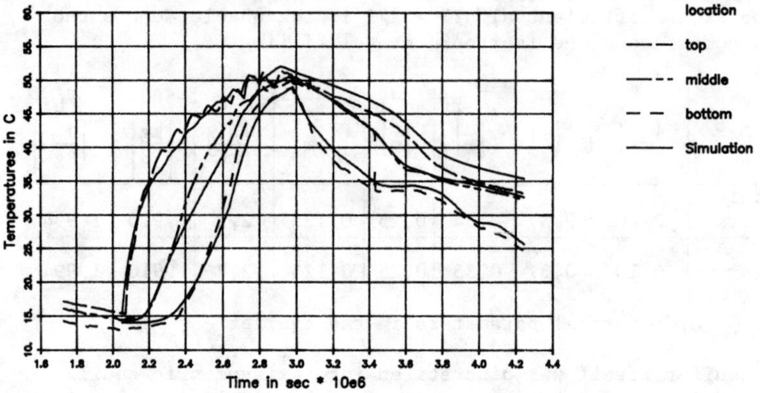

Fig. 5b: Temperature over time measured and calculated

CONCLUSIONS

Permeability and temperature differences are very high in the Stuttgart aquifer test site. The temperature field can be calculated accurately using equilateral triangles according to Fig. 2. A quasi NAVIER-STOKES formulation for the flow field equation may be advantageous compared to standard DARCY as far as natural convection in coarse gravel is concerned. A comparison of accuracy and computational effort between different flow field formulations will be given elsewhere.

REFERENCES

1. Fisch N. (1986), Documentation of the Stuttgart University Project, International Agency, Solar Heating and Cooling Program, prepared for Task VII, November 1986 Meeting.

2. Gartling D.K. (1987), NACHOS II - A Finite Element Computer Program for Incompressible Flow Problems PART I-Theoretical Background, SAND 86-1816.UC-32, Sandia Nat. Lab. Albuquerque

3. Kinzelbach W.K.H., Frind E.O. (1986), Accuracy Criteria for Advection - Dispersion Models, VI International Conference on Finite Elements in Water Resources, Lisboa.

4. Pinder G.F. and Gray W.G. (1977), Finite Element Simulation in Surface and Subsurface Hydrology, Acad. Press, New York

5. Stössinger W. (1979), Beschreibung der Hydrodynamischen Dispersion mit der Methode der Finiten Elemente am Beispiel der instationären Interface zwischen Süß- und Salzwasser in Grundwasserleitern, Mitt. IWW, RWTH Aachen, Heft 28.

6. Zienkiewicz D.C. and Morgan K. (1983), Finite Elements and Approximations, A Wiley Interscience Publication, New York.

Numerical Modeling of Hot Water Storage in Aquifer by Finite Element Method

B. Goyeau and J. Gounot

LEPT-ENSAM, Esplanade des Arts et Métiers, 33405 Talence Cedex, France

P. Fabrie

U.E.R. Mathématique, 351 Cours de la Libération, 33405 Talence Cedex, France

INTRODUCTION

A three dimensional finite element model is developed for the simulation of hot water storage in a stratified aquifer saturated by cold water. The hydrodynamic instabilities of non-isothermal flow through porous media are investigated when both density and viscosity vary with temperature. The instabilities are due to a density gradient directed upward (Rayleigh-Bénard instabilities), and to the displacement of a more viscous fluid by a less viscous one inducing fingering (Saffman-Taylor instabilities). Both instabilities may drastically affect the efficiency of hot water storage, and the purpose of this simulation is to identify the conditions for which the effects of hydrodynamic instabilities are minimized.

The simulations are performed for cylindrical three-dimensional aquifers. The equations used to model the system are discretized using Q_1 finite elements in space and an Euler's semi-implicit scheme in time. The matrix system is solved by the preconditioned conjugate gradient method with Choleski's incomplete factorization.

In first part, natural convection problems were used to test the validity of the model. The coupling of Rayleigh-Bénard and Saffman-Taylor instabilities is investigated during the storage in homogeneous aquifer.

PHYSICAL MODEL

We consider a horizontal cylindrical aquifer limited by two impervious adjacent beds. Hot water is injected and pumped via a thoroughly penetrating well (figure 1). The temperature and filtration velocity obey the equation of mass, momentum and energy. In their most general form, the equations which describe the flow in porous media can be written as follows:

$$\epsilon\frac{\partial\rho}{\partial t} + \vec{\nabla}.(\rho\vec{V}) = 0 \qquad (1)$$

$$\frac{\rho}{\epsilon}\frac{\partial\vec{V}}{\partial t} + \vec{\nabla}P - \rho\vec{g} + \mu(T)\tilde{\tilde{K}}^{-1}.\vec{V} = 0 \qquad (2)$$

$$(\rho c)^{*}\frac{\partial T}{\partial t} - \vec{\nabla}.(\tilde{\tilde{\lambda}}.\vec{\nabla}T) + (\rho c)_{f}\vec{V}.\vec{\nabla}T = 0 \qquad (3)$$

where ϵ is the porosity, ρ is the density, g the acceleration due to the gravity, μ is the dynamic viscosity, K and λ are respectively the effective tensor of permeability and thermal conductivity. V is the filtration velocity, P the pressure and T the temperature. The heat capacity of fluid and solid is respectively $(\rho c)_f$ and $(\rho c)_s$. Thus $(\rho c)^* = \epsilon(\rho c)_f + (1-\epsilon)(\rho c)_s$ represents the equivalent heat capacity of the medium. The density and viscosity variations with temperature are given by:

$$\rho = \rho_0(1 - \beta(T - T_0)) \quad (4); \qquad \mu = a\,e^{-bT} \quad (5)$$

The following assumptions are made:

- The Boussinesq approximation which consists of taking into account the density variations only in the buoyancy term in the Darcy equation is valid.
- The density variations are linear with temperature.
- The geothermal gradient is neglected.

The boundary conditions are (see figure 1):

$$\text{on } \sigma_1 \quad (-\tilde{\tilde{\lambda}}.\vec{\nabla}T).\vec{n}_1 = (\rho c)_f U_0 (T-T_1) \qquad (6)$$

$$\vec{V}.\vec{n}_1 = -U_0 \qquad (7)$$

$$\text{on } \sigma_{2,4} \quad (-\tilde{\tilde{\lambda}}.\vec{\nabla}T).\vec{n}_{2,4} = 0 \quad (8); \qquad \vec{V}.\vec{n}_{2,4} = 0 \qquad (9)$$

$$\text{on } \sigma_3 \quad (-\tilde{\tilde{\lambda}}.\vec{\nabla}T).\vec{n}_3 = 0 \quad (10); \qquad \vec{V}.\vec{n}_3 = U_0\frac{Ri}{Re} \qquad (11)$$

where n_i is the outward normal vector to the boundary σ_i, T_1 and V_0 are respectively the temperature and the filtration velocity of the injected water. In order to resolve this system, it is tractable to adopt a pressure formulation, and therefore, the dimensionless equations are:

$$\vec{\nabla}.(\frac{-\tilde{\tilde{K}}}{\mu(\theta)}.(\vec{\nabla}P - Ray^{*}\theta\,\vec{k})) = 0 \qquad (12)$$

$$\frac{\partial\theta}{\partial t} = \vec{\nabla}.(\tilde{\tilde{\lambda}}.\vec{\nabla}\theta) + (\frac{\tilde{\tilde{K}}}{\mu(\theta)}.(\vec{\nabla}P - Ray^{*}\theta\,\vec{k})).\vec{\nabla}\theta \qquad (13)$$

with the boundary conditions:

$$\text{on } \sigma_1 \quad (-\tilde{\tilde{\lambda}}.\vec{\nabla}\Theta).\vec{n}_1 = Pe^* \frac{\Delta R}{Ri} (\Theta - 1) \tag{14}$$

$$(\frac{-\tilde{\tilde{K}}}{\mu(\Theta)}.\vec{\nabla}P).\vec{n}_1 = -Pe^* \frac{H}{Ri} \tag{15}$$

$$\text{on } \sigma_{2,4} \quad (-\tilde{\tilde{\lambda}}.\vec{\nabla}\Theta).\vec{n}_{2,4} = 0 \tag{16}$$

$$(\frac{-\tilde{\tilde{K}}}{\mu(\Theta)}.\vec{\nabla}P).\vec{n}_{2,4} = Ray^* (\frac{-\tilde{\tilde{K}}}{\mu(\Theta)} \Theta.\vec{k}).\vec{n}_{2,4} \tag{17}$$

$$\text{on } \sigma_3 \quad (-\tilde{\tilde{\lambda}}.\vec{\nabla}\Theta).\vec{n}_3 = 0 \quad (18); \quad (\frac{-\tilde{\tilde{K}}}{\mu(\Theta)}.\vec{\nabla}P).\vec{n}_3 = Pe^* \frac{H}{Re} \quad (19)$$

The main dimensionless parameters of the problem are:

$$\text{The Rayleigh number} \quad Ray^* = \frac{\rho_0\, g\, (\rho c)_f K_0\, H\, (T_1 - T_0)}{\lambda_0^*\, \mu_0} \tag{20}$$

$$\text{The Peclet number} \quad Pe^* = \frac{U_0\, (\rho c)_f Ri}{\lambda_0^*} \tag{21}$$

which are respectively characteristic of the Rayleigh-Bénard and Saffman-Taylor instabilities.

NUMERICAL MODEL

The simulation is performed for cylindrical three dimensional aquifers by finite element method. To formulate the matricial equation system, we adopt a weighted residual method with a Galerkin's approach.

The system of equations is discretized by using Q_1 finite elements. The evaluation of integrals over each element which are needed to calculate the element matrix is made by a Gaussian quadrature technique.

The time integration is achieved by using Euler's semi-implicit scheme. At each time step, the nonlinear term $(V(\Theta).\nabla\Theta))$ in Eq.(13) is calculated by sucessive iterations on $\nabla\Theta$. Then the discretized problem leads to solve several sparse linear systems like Ax=b, where A is an N*N symmetric positive definite matrix.

Each system is solved by a Preconditioned Conjugate Gradient Method with Choleski's incomplete factorization. This is an interesting descent method insuring convergence in few iterations.

Validation of the program

To validate this simulation, we tested the model for natural convection with constant viscosity in various configurations:

→ bi-dimensional cartesian X,Z
→ bi-dimensional radial R,Z
→ Tri-dimensional cylindrical R,Φ,Z

For instance in the simple case of a square layer of homogeneous and isotropic porous medium heated from below, cooled from above and with adiabatic lateral walls, we were able to verify the good agreement between our results and those formed in [6], [7]. An example of the calculated temperature field is shown in fig.2(a).

All the results presented here, have been obtained on two dimensional grid 33*33 containing 1089 nodes and 1024 elements. The computation is made on IBM MVS 3090 and the program is vectorized.

RESULTS AND DISCUSSION

Natural convection

Few works in natural convection take into account viscosity variations with temperature. We present fig.2(b), the new temperature field with the same Rayleigh number as in the case shown fig.2(a), but with the viscosity depending on temperature. We observe that the flow symmetry disappears because the filtration velocities are more important in the hot region. Furthermore, the global heat transfer, characterized by the Nusselt number, is greater in this case (see figure captions).

Storage in aquifer

We consider here the storage of hot water in homogeneous and isotropic aquifers saturated by cold water.

When the Rayleigh number is nearly equal to zero, the flow is one dimensional and the isotherms are vertical, fig.3(a). This is no longer the case when Rayleigh number increases as it is illustrated on fig.3(b).

The effect of viscosity variations is illustrated in figure 4(a)-(b). The hot fluid penetration in the cold fluid at the upper part of the layer is increased with diminuting viscosity ratio and with the increasing of Rayleigh and Peclet number.

One can notice at this point that all these results can be applied to isothermal miscible flow because of the analogy between the heat transfer equation and the dispersion equation.

CONCLUSION

All these phenomena may increase the heat transfer in the aquifer and drastically affect the efficiency of the storage. Their study must permit to identify the conditions for which the effects of hydrodynamics instabilities are minimized.

The main numerical difficulty in this problem is to take into account the viscosity variations. Indeed, it is necessary to do assembling again at each time step and the global computational time is long. For this reason, we are studying a new mathematical formulation which would permit an improvement of the efficiency of the algorithm without loss of precision.

REFERENCES

1. ALEXANDRE M. (1982), Approche numérique des phénomènes de diffusion convection lors d'un stockage d'eau chaude en aquifère. Thèse, Institut polytechnique de toulouse.
2. AXELSSON D. and BARKER V.A. (1984), Finite Element Solution of Boundary Value Problems. Academic Press.
3. BACHU S. and DAGAN (1979), Stability of Displacement of a Cold Fluid by a Hot Fluid in Porous medium, Phys.Fluids, Vol.22, N°1, pp.54-59.
4. BERNARD D. (1987), Numerical Modeling of free Convection in Geological Porous Structures, 24th AICHE / ASME National Heat Transfer Conference Pittsburg.
5. BERTIN H. and QUINTARD M. (1984), Stabilité d'un écoulement miscible radial en milieu poreux: étude théorique et expérimentale, I.F.P., Vol.39,pp.573-587.
6. COMBARNOUS M. and BORIES S. (1975), Hydrothermal Convection in Saturated Porous Media, Advances in Hydroscience, Vol.10, pp.231-307 Academic Press.
7. CALTAGIRONE J.P. (1975), Thermo-Convective Instabilities in a Horizontal Porous Layer, J. Fluid. Mech. 72
8. FABRIE P. (1987), Contribution à l'étude de la convection naturelle en milieux poreux, Thèse, Université de Bordeaux I.
9. GOYEAU B. and GOUNOT J. (1986), Approche experimentale de la mise en place de structures convectives en milieu poreux, C.R.Acad.Sc. Paris, t.303,Serie II,pp.779-784.
10. KORSON DROST-HANSEN W. and MILLERO F. (1969), Viscosity of Hot Water at Various Temperature, The Journal of Physical Chemistry, Vol.73.
11. QUINTARD M. (1983), Stabilité des écoulements miscibles en milieux poreux homogènes: injection d'un fluide chaud dans un massif poreux saturé par le meme fluide froid, Thèse, Université de Bordeaux.
12. ZIENKIEWICZ O.C. (1977), The Finite Element Method, 3rd ed. Mc Graw-Hill, London.

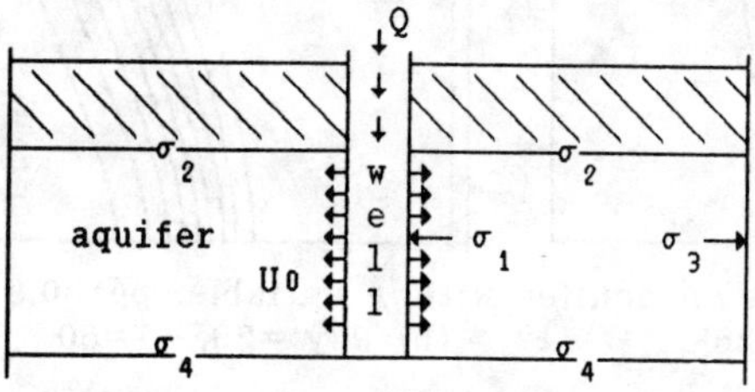

figure 1 : aquifer

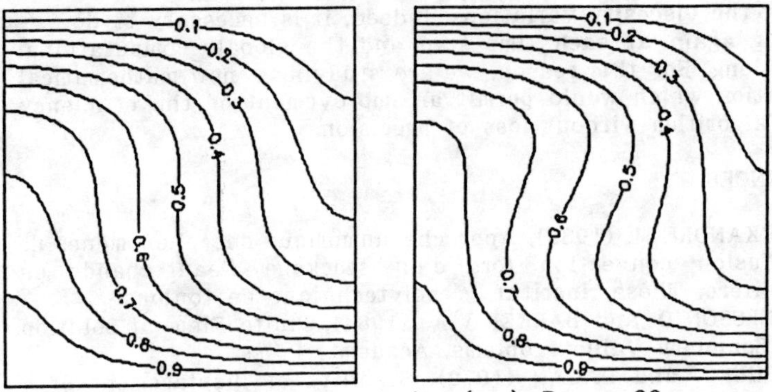

fig.2: Natural Convection (x,z), Ray = 80

a) ν constante, Nu*=2.28 b) ν variable, Nu*= 4,1

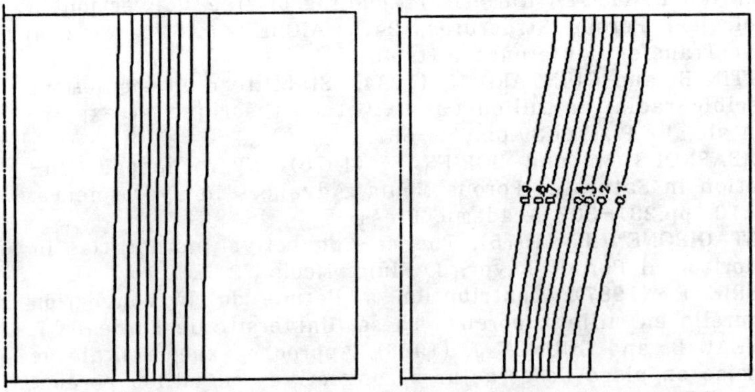

fig.3: Storage in aquifer with ν constante, pe*=0.8

a) Ray*=2 b) Ray*=20

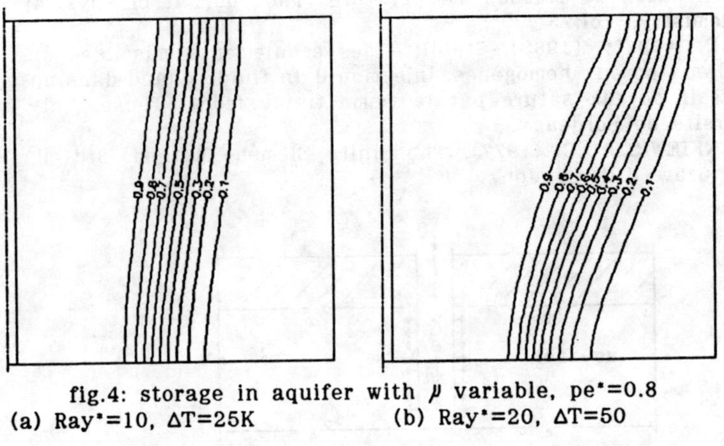

fig.4: storage in aquifer with ν variable, pe*=0.8

(a) Ray*=10, ΔT=25K (b) Ray*=20, ΔT=50

Modelling the Regional Heat Budget in Aquifers

J. Trösch
Laboratory of Hydraulics, Hydrology and Glaciology at the Federal Institute of Technology, CH-8092 Zurich, Switzerland
H. Müller
Suiselectra, Malzgasse 32, CH-4010 Basel, Switzerland

INTRODUCTION

During the last decade an increasing number of heat pumps have been installed in aquifers. They use the relatively warm groundwater as a natural energy resource. Through a well the water is extracted to a heat exchanger. After cooling it, the cold water has to be reinjected into the aquifer. The extracted energy is renewed mainly by atmospheric heat input, infiltrations of river water and precipitation during the warmer seasons. As the same aquifers are used as drinking water resources, it is important to prevent overuse of the groundwater in terms of energy, water quality and quantity.

For the management of the groundwater an adequate model is needed. This model should allow the simulation of the prevailing physical processes that determine the flow of water and energy in the aquifer. In shallow aquifers the water flows horizontally. The energy is advected and dispersed by the flow. On the other hand the energy exchange between atmosphere and groundwater occurs through the confining unsaturated layer, and the predominant process is in the vertical direction. Thus the whole physical system is fully three-dimensional with a nonlinear coupling of hydrodynamic and thermodynamic equations.

To circumvent a three-dimensional treatment of regional aquifer problems, an interlaced numerical model based on the finite element technique was developed. The vertical and the horizontal directions are separated and discretized with one-dimensional and two-dimensional submodels. For the heat flux from the atmosphere through the confining layer and through the aquifer one-dimensional vertical models are used. The flow and the advection of the energy in the aquifer are calculated with the well-established concept of two-dimensional horizontal models where the unknowns are vertically integrated with the help of some simplifying assumption, as the Dupuit assumption in flow calculations.

MATHEMATICAL FORMULATION

Flow equation

The flow in an unconfined aquifer is described by the differential equation

$$(T_{ij}h,_i),_j = S\,h,_t \qquad (i,j = 1,2) \tag{1}$$

where T_{ij} is the transmissivity tensor, h the water level, S the specific yield and t the time. The equation includes the Dupuit assumption and is valid for horizontal two-dimensional aquifers where vertical velocities can be neglected. As the density differences due to the heat pumps are small, flows induced by density differences are neglected. The boundary conditions are given values of water level and flux.

Two-dimensional, horizontal heat transport

The equation for the heat transport in a two-dimensional aquifer of variable thickness is

$$mT,_t + m\frac{n_e c_{v,w}}{c_v} u_i T,_i - m(D_{ij}T,_i),_j + \frac{R}{c_v} = 0 \tag{2}$$

where m is the thickness of the aquifer, T the temperature, n_e the fraction of specific yield, $c_{v,w}$ the volumetric heat capacity of the water, c_v the volumetric heat capacity of saturated soil, R the radiation and D_{ij} the thermal dispersion. The velocities u_i are calculated from the solution of the flow equation (1) with the Darcy law.

The boundary conditions are given values of temperature and for the heat flux.

The thermal dispersion D_{ij} is a sum of the mechanical dispersion through the porous matrix and the heat conductivity of the soil-water mixture.

One-dimensional vertical heat transport

In the layer above the aquifer only vertical heat transport is important, for taking into account the atmospheric heat input.

$$T,_t - \frac{\lambda}{c_B} T,_{zz} = 0 \tag{3}$$

where T is the temperature, t the time, λ the heat conductivity of the soil, c_B the volumetric heat capacity and z the vertical coordinate.

The boundary condition is a given value of the temperature at the surface, in the top layer of the soil. As a lot of measured values of this temperature are available, the calculation of heat flux at the surface depending on meteorological data is not necessary.

FINITE ELEMENT FORMULATION

The flow equation (1) is integrated with the well established Galerkin formulation (Zienkiewicz [1]). Isoparametric quadrilateral elements are used with mixed linear, quadratic or cubic sides. The program used here is well documented in (Trösch [2]). It uses a problem oriented language easy to use for hydrogeologists.

The two-dimensional horizontal energy equation (2) is solved with the same framework of shapefunctions, numerical integration and linear equation solver as the flow problem. The convective terms are calculated based on the velocities obtained from the flow model.

The integration of the one-dimensional vertical energy equation (3) uses also a Galerkin formulation, the shapefunctions are only quadratic. The element matrices as a function of the length of the element are once calculated by hand. Then for each element the length and the material properties have to be introduced and the element matrix can be added to the global matrix. The band width of the global matrix is in this case only 3 so the system is solved with a band solver.

MODELLING CONCEPT

The three models have to be connected with a modelling concept. As density driven flow is not considered, the flow model can be regarded as independent. With that in a first step the flow model is calculated, stationary or non-stationary, and the head results are kept in storage.

The two heat models are not independent, as the one-dimensional vertical model gives the atmospheric heat input for the two-dimensional horizontal model. For this reason the models are interlaced. For each timestep the vertical model is run first, giving a temperature distribution from the ground surface down to some distance below the impervious layer. For every point of the horizontal model such a distribution is calculated. The mean value of the temperature on the thickness of the aquifer is then used as starting value for the horizontal heat transport model. For the convection of the heat the velocities are calculated from the head results of the flow model. The new mean values of temperature are then used to update the temperature profiles of the vertical models, keeping the form but shifting them proportionally (Fig. 1).

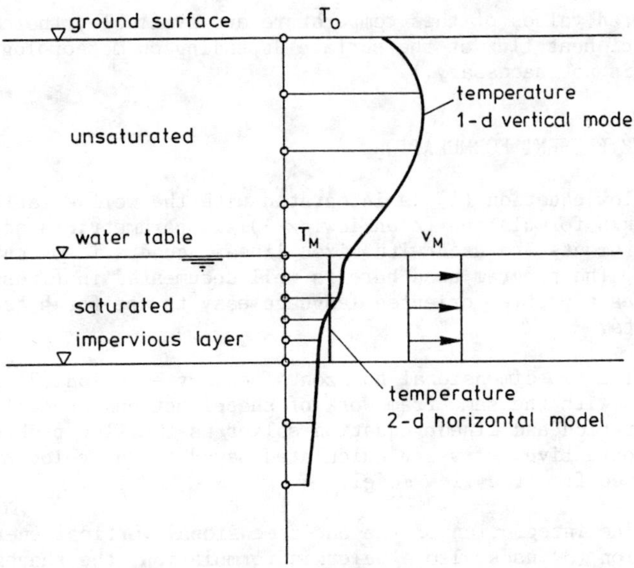

Figure 1: Cross section with one-dimensional vertical and two-dimensional horizontal models

RESULTS

The heat transport models were first tested separately and compared with analytical solutions. The vertical model is normally subdivided in 8 elements for the upper "dry" layer, 5 elements for the aquifer and 3 elements in the impervious layer below the aquifer. As each quadratic element has one midside and two corner nodes one vertical model is built up by a total of 33 nodes. This model was in good agreement with an analytical solution for the seasonal cycle during one year.

The two-dimensional model was compared to the analytical solution of a point source in a homogeneous flow field. If the condition Pe ≤ 2 is not violated, where

$$Pe = \left| \frac{u_i \ \Delta x}{D_{ii}} \right|$$

is the grid Peclet number, u_i a typical velocity, Δx a grid length and D_{ii} the diffusion, then the numerical results compare well with the analytical solution. In real world problems this condition implies often a finer discretisation as would be needed for flow models only.

The simulation of a real domain was started. The horizontal mesh contains 562 elements with 619 nodal points, the timestep is one day. As first calculations show, it may take several years until a new equilibrium is reached. The most important problem is a thorough determination of the soil parameters. Even with a lot of measured data over one year and in different locations this task remains the crucial point.

CONCLUSIONS

The decoupling of the two-dimensional transport in the aquifer from the vertical heat conduction from the atmosphere to the aquifer gives good results for heat budget calculations. It is a possibility to circumvent an expensive three-dimensional treatment of the problem. The execution time on a VAX 8600 for a simulation of a whole year is around one hour. But with the simplifications made it is not possible to represent for instance the near field of a coldwater plume.

REFERENCES

1. Zienkiewicz O.C. (1977). The Finite Element Method, Third Edition, McGraw-Hill, London.

2. Trösch, J. (1975). Numerische Simulation Dupuit'scher Grundwasserströmungen. Mitteilung der Versuchsanstalt für Wasserbau, Hydrologie und Glaziologie, Nr. 14/15.

A Thermal Energy Storage Model for a Confined Aquifer

Yuqun Xue and Chunhong Xie
Department of Geology and Department of Mathematics, Nanjing University, Nanjing, 210008, People's Republic of China
Qingfen Li
Ministry of Geology and Minerals Resources, People's Republic of China

Introduction

Here we propose a three-dimensional convection-heat dispersion model, which is used for describing a series of seasonal aquifer thermal energy storage experiments in China. The simulated results are compatible with the field data.

Mathematical model and numerical method.

The governing equation used in this sort of problem generally had convection and conduction items without having heat dispersion item. But in practice we have discovered that like the mechanical dispersion phenomena in mass transfer problems, heat mechanical dispersion can not be neglected. So the mathematical description of heat transfer in a confined aquifer will be obtained as follows:

$$\frac{\partial T}{\partial t} = \frac{1}{c}\,\mathrm{div}(\lambda\,\mathrm{grad}\;T) - \frac{c_w}{c}\,\mathrm{div}(T\underline{V}) \qquad (1)$$

where T is temperature; c and c_w are the heat capacity of the porous medium and water, respectively; $\underline{V}$ is the filtration velocity vector of groundwater; t is time; λ is the coefficient of heat dispersion, $\lambda=\lambda_e+\lambda_v$; λ_e is the thermal conductivity of porous medium; λ_v is the coefficient of heat mechanical dispersion. If we define $d\underline{r}/dt=c_w\underline{V}/c$ and use the Lagrangian coordinates system, where $\underline{r}=x\underline{i}+y\underline{j}+z\underline{k}$, $\underline{i}$, $\underline{j}$ and $\underline{k}$ are the unit vector along x,y and z axes, respectively, the equation (1) could be written as:

$$\frac{dT}{dt} = \frac{1}{c}\,\mathrm{div}(\lambda\;\mathrm{grad}\;T) - \frac{c_w}{c}\,T\,\mathrm{div}\;\underline{V} \qquad (2)$$

In steady flow without source and sink, we yield

$$\frac{dT}{dt} = \frac{1}{c}\,\mathrm{div}(\lambda\;\mathrm{grad}\;T) \qquad (3)$$

The initial and boundary conditions are simplified and given by

$$T(x,y,z,0) = T_o(x,y,z) \qquad (x,y) \in F,\ z_1 \leq z \leq z_2 \qquad (4)$$

$$T(x,y,z,t)|_\Gamma = \varphi(x,y,z,t) \qquad (x,y) \in F,\ z_1 < z < z_2 \qquad (5)$$

$$T(x,y,z_1,t) = T_1(x,y,t) \qquad (x,y) \in F \qquad (6)$$

$$T(x,y,z_2,t) = T_2(x,y,t) \qquad (x,y) \in F \qquad (7)$$

$$T(x,y,z,t)|_{w_i} = \psi_i(z,t) \qquad w_i \in F,\ z_1 < z < z_2 \qquad (8)$$

where Γ is the boundary of the computational domain F ; z_1 and z_2 are the distance from the datum plane to the upper and lower boundary of domain F, respectively; φ, T_1 and T_2 are prescribed functions; T_o is the temperature under natural regime; w_i is the surface of injection well i; ψ_i is its temperature.
The characteristic alternating direction implicit scheme is proposed for solving the above problem.

Example
We applied this model to the experimental data extracted from a test site in China (there were 4 injection-production wells and 34 observation wells). The testing aquifer is a confined aquifer consisted of fine and medium sand to coarse sand with a few pebbles. The aquifer's top is 67-74m and bottom is 91-95m below the ground surface; above and below the aquifer there are mild clay and clay. In the site, the hydraulic gradient was 3.4×10^{-4} to 4.1×10^{-4}, the yearly variation of groundwater temperature is very slight. Because the physical features of the top and bottom surfaces of the aquifer are gentle except in individual places and the natural hydraulic gradient is very slight, the aquifer can be regarded as horizontal and having a uniform thickness of M with no natural groundwater flow. Since the duration of the transient flow period was short in comparison with the length of time required to reach thermal equilibrium, flow could be assumed to be steady without other flow interference. The flow field caused by injection or production could be described as

$$V = \frac{Q}{2\pi rM}$$

where V is the filtration velocity at a distance r from well; Q is the injection or pumping rate.
According to the testing data, the zone of thermal influence being affected by injecting or pumping water is only limited to 40-60 meters horizontally and 3M/2 vertically. Outside this scope, groundwater temperature is all but natural. Thus the computational domain was easily ascertained accordingly with the first type boundary circled out by the observation data under natural regime, the temperature in z direction conforms to the local geothermal gradient. If the coordinate system is placed such

that the x-y plane coincides with the midplane of the aquifer in the test site, the region where $|z| \leqslant M/2$ is the aquifer (energy storage layer) and where $|z| > M/2$ is non-storage layers. So we have

$$c = \begin{cases} c_a \\ c_c \end{cases} \quad \lambda = \begin{cases} \lambda_a = \lambda_{ea} + \lambda_{va} & |z| \leqslant M/2 \\ \lambda_c = \lambda_{ec} & |z| > M/2 \end{cases}$$

$$T = \begin{cases} T_a \\ T_c \end{cases} \quad \underline{V} = \begin{cases} \underline{V}_a & |z| \leqslant M/2 \\ 0 & |z| > M/2 \end{cases}$$

where subscript a represents the aquifer and subscript c represents non-storage layers above or below the aquifer. Thus the mathemetical model of the test site is constructed by (3) and (4)-(8), there $z_1 = 3M/2$, $z_2 = -3M/2$. Equation (3) could be used to describe both aquifer and non-storage layers, but with different parameters.

Three experiments have been accomplished. They were (1) single well experiment with cold water injected into well w_1; (2) double well experiment in which warm water was injected into well w_2 and cold water formerly injected was produced from well w_1; (3) multiwell experiment in which warm water was injected into wells w_2 and w_4 and cold water injected by another experiment was produced from wells w_1 and w_3 (Fig.1). We performed numerical simulation for all the experiments. Here we only describe the multiwell simulation result.

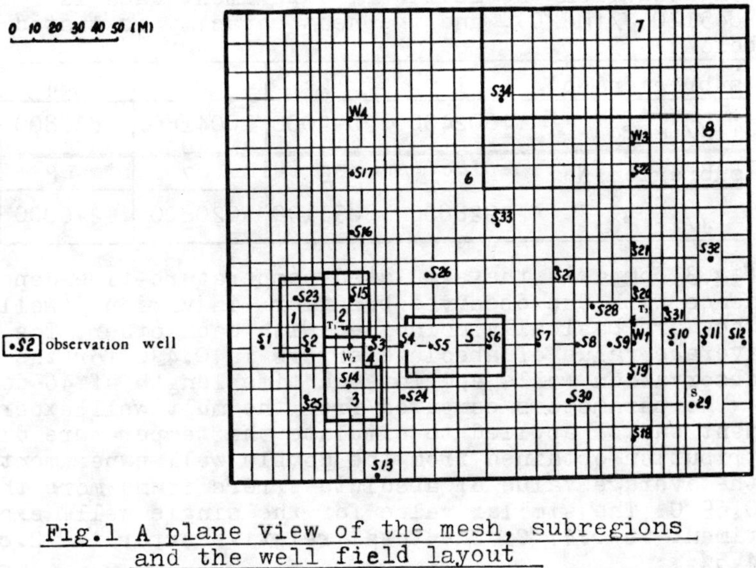

Fig.1 A plane view of the mesh, subregions and the well field layout

In the test site, the aquifer is inhomogeneous and isotropic, so the computational domain was divided into eight subregions in the plane and three subregions in the vertical direction represented the upper and lower non-storage layers and aquifer itself, respectively. The mesh used is shown in Fig.1 and 2. The mesh element within the region of thermal influence should decrease in size with increasing distance from the injection and production wells.

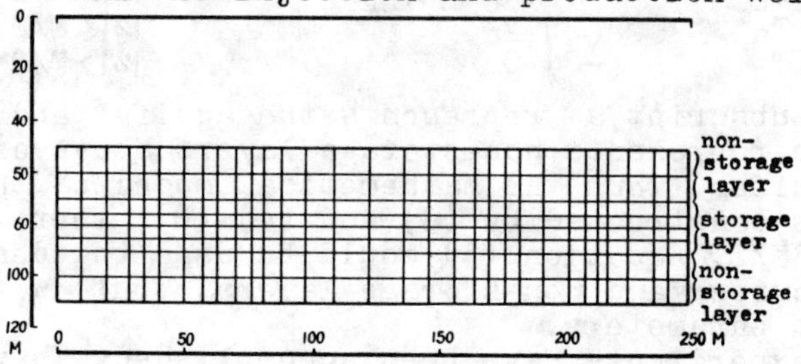

Fig.2 A profile of the mesh used in simulation.

The parameters for upper and lower non-storage layers were taken directly from laboratory testing data, the thermal conductivity λ_{ec} is 141112.6 and 180430.5 J/(md °C), the heat capacity c_c is 2510400 and 2092000 J/(m^3 °C), respectively. Considering the difficulties to take the undisturbed sample for measuring, we did not take parameters from laboratory values. Rather, we determined them through observational data in experiments. The thermal conductivity computed from the multiwell experiment data is 2845120 J/(md °C), and the heat capacity is listed below:

subregion No.	1	2	3	4
c_a J/(m^3 °C)	4602400	2092000	1004160	836800
subregion No.	5	6	7	8
c_a J/(m^3 °C)	2928800	3891120	5020800	6276000

Fig.3 compares the simulated temperature-time dependence with the observed behavior in various wells. They are basically coincided with each other. The average value of absolute errors is 0.43°C for the 33 observation wells and for the time length of 46 days.

The parameters computed from the multiwell experiment can be applied to simulate the temperature distribution obtained from the double well experiment, the average value of absolute errors is no more than 0.69 °C. The similar value for the single well experiment is 0.47 °C. The total relative error is 2.8-4.5%.

Among the parameters obtained earlier, the λ_a/c_a (0.453-3.4m^2/d) of aquifer is far bigger the λ_c/c_c (0.055-0.088m^2/d) of nearby non-storage layers. This is because $\lambda=\lambda_e+\lambda_v$ and $\lambda_v=0$ for non-storage layer, so λ_a is much bigger than λ_c. Also, the λ_a/c_a(1.36-3.4 m²/d) near the injection well is larger than that (0.68-0.97 m²/d) far from it, and this reflects that the λ_v near the injection well is large. That is because the velocity at that place is fast. Note that c_a obtained is larger than that from laboratory measurements (178489.4-2178608.8 J/(m³°C). It also conforms to reality because the effect of flow rate may not be considered in the laboratory measurement.

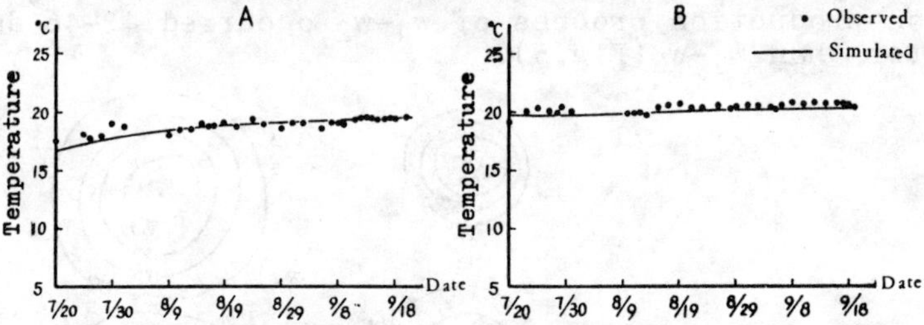

Fig.3 The simulated and observed temperatures for A(well S30 at 75m depth) and B(well S18 at 85m depth)

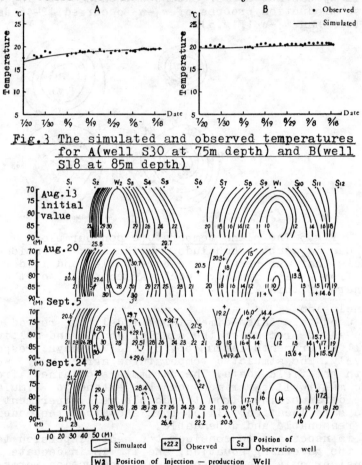

Fig.4 The simulated temperature contours and the observed temperature

The simulated and observed temperatures in the vertical profile are also conformed to each other, without relevant numerical dispersion. This distribution shows how heat is transported in the aquifer during injection or production period(Fig.4). It seems clearly that there is no effect of convection because the temperature difference between the cold and warm water is small. The plane temperature contour plots also reflect the above mentioned characteristics. The warm water body around injection well w_2 has been effected by production well w_1 and thus the thermal front has extended towards w_1. But the tongue formed by the thermal front near injection well w_4 was not so far extended because the injection-production process of w_4-w_2 occurred 42-46 days later than w_1-w_2(Fig.5).

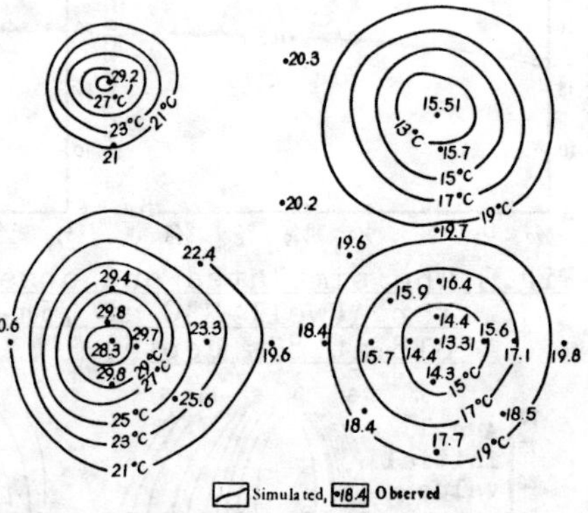

Fig.5 The plane temperature contour plots

We can no doubt conclude from the above mentioned facts that our results are dependable, and it derived from a high quality model, as compared to the others of this kind.

Conclusion

We used this three-dimensional model to reproduce the whole process of the single, double and multiple well experiments lasting 88, 63 and 46 days, respectively. The simulated temperatures agree very well with the field data. The parameters obtained from the multiwell experiment can be extend to simulate the data obtained from the double well experiment at the same location. All of this showsthat the model is reasonable and dependable.

It is necessary to consider the heat dispersion term in the governing equation, and it is inadequate to consider only the convection and conduction terms.

References

Papadopulos, S.S. and S.P.Larson, Ground Water, 16(4), 242-248, 1978.

Sykes, J.F., R.B.Lantz, et al., Ground Water, 20(5), 1982.

Tsang, C.F. et al., Water Resour. Res. 17(3), 1981.

SECTION 3 - HYDROLOGY

SECTION 3A - GENERAL HYDROLOGY

Numerical Analysis of Transients in Complex Hidropower Scheme

S.A. Furlani and G.J. Corrêa

Centrais Elétricas do sul do Brasil S/A - ELETROSUL, Department of Hydropower Engineering, Rua Deputado Antonio Edu Vieira, 353 88048, Florianópolis, SC, Brazil

ABSTRACT

The designing of a penstock (or powertunnel) requires from an Engineer a long term responsibility, beginning at the preliminary studies with the selection of a convenient diameter and accessories, in order to provide the required turbine regulation stability during transient and permanent operation.
This report describes the procedure that was used for the analysis of hydraulic transients during studies of the Forçacava Project that will be part of the Fontes Hydrogenerating System in the Rio de Janeiro State.

1. INTRODUCTION

A numerical method was used to determine the conduit required diameter, the generating unit inertia and its overspeed after a load rejection, the resulting overpressure (or underpressure) and the influence of this occurrence at adjacent powerhouses and units including the torque increase, under several load transient conditions.

2. MATHEMATICAL MODEL

2.1 Theoretical Basis

Considering the different theories for hydraulic transient analysis, we made the option for the "Method of Characteristics" to obtain the fundamental informations about the water flow at diferent points

of the circuit at each interval.
The method consists in the solution of the equations:

$$g \cdot A \cdot H_x + Q_t + \frac{f}{2DA} \cdot Q \cdot |Q| = 0$$

$$g \cdot A \cdot H_t + a^2 \cdot Q_x = 0$$

solving the two pairs of characteristics equations:

$$\frac{dH}{dt} \pm \frac{a}{gA} \cdot \frac{dQ}{dt} + \frac{fa}{2gDA2} \cdot Q \cdot |Q| = 0$$

$$\frac{dx}{dt} = \pm a$$

For the sake of accuracy and to save computation time, we used the "MOC Method" (1) that avoids the two independent hydraulic transient solutions that occur on normal transient solution methods.

2.2. Hydrogenerating Unit Inertia

For evaluation of required inertia of the rotating parts of the generating unit (speed regulation purposes), we used the results of the hydraulic analysis of the transient at each time interval, computed in form of energy delivered to the turbine runner, that is spent for accelerating the unit.
This unit inertia may be expressed by the equation:

$$GD^2 = \frac{730 \cdot P \cdot HE}{((1 - \Delta n)^2 - 1)\, n^2} \qquad (2)$$

Where

$$HE = \frac{1}{Q_I \cdot H_I} \int_{ti}^{tf} Q(t) \cdot H(t)\, dt$$

The analysis of the required GD^2 is made by comparing the above result, with the natural inertia that the generator would have if its design was economically oriented. In long conduits the normally required GD^2 is bigger than the natural, and the customer has to pay for the difference. When the rotational speed is high the diameter of the generator rotor is small and it may be very expensive to increase the generator inertia.

2.3. Overspeed

To obtain overspeed against time graph we used the same equation mentioned in the last paragraph. After a sensitivity test we fixed a value for the GD^2 of the unit and the graph was computed.

2.4 Shaft Torque

For a complex system it is not sufficient to solve the regulation stability of the unit itself, it is important to know what happens during transients to the adjacent units and to the adjacent powerhouses, that have a section of the conduit in common use.
For this analysis it is necessary to introduce into the computer memory a convenient data base containing the characteristic turbine hill charts (performance charts). In the Forçacava Preliminary Studies we only knew the hill charts of the existing units, because at this stage the turbine supplier was unknown. We used the charts of a unit with similar head range, and specific speed (ns). In this case the main objective was not to have a precise answer but to have a feeling about the technical feasibility of the project.

3. VALIDATION OF THE MATHEMATICAL MODEL

The aforementioned mathematical model was validated by comparing the results of measured conditions of the Salto Santiago Hydropower Plant with the results of the numerical simulation.
Salto Santiago Hydropower Plant has a total projected output of 2000 MW (six 333 MW units) 106 m nominal gross head. Four of these units are in operation since 1980. The water supply is done by six independent penstocks with 7.60m inside diameter, 168m long and $345 m^3/s$ maximum flow.
As shown in fig. 1 the measured and the calculated pressures are very similar, and the imprecision during the first second may be attributed to the dead time between the command and the effective start of distributor closing. The diference at the fourth second may be attributed to draft tube influence. For this analysis other differences may be neglected.

4. MODEL APLICATION IN FORÇACAVA HYDROPOWER PROJECT

4.1 General Description

Forçacava Hydropower Project is located in the Rio de Janeiro city neighborhood (43 km far) and aims to introduce 432 MW of peak load availability to the Fontes Hydrogenerating System, with the construction of an underground powerhouse with 3 x 147 MW Francis type hydraulic turbines for 306m head.
The water power system comprises a single power tunnel with $57 m^2$ cross section and 638m long from the intake to the valve chamber where the water is diverted through three derivations (one duly plugged) to the existing Fontes Nova powerhouse (3 x 48 MW) and Nilo

Peçanha Powerhouse 4 x 78 MW and 2 x 36 MW and to the future Forçacava Powerhouse, see fig. 2. The pressure tunnel after the valve chamber is 690,0 m long up to the diversion for the first unit and for the selected layout the inner diameter is 6,0 m, each unit has a 2,20 inner diameter butterfly valve.

4.2 Analysis of the Results

From the simulation of various different conditions we concluded that 6 meter inner diameter is convenient for the power tunnel. For the worst condition of simultaneous load rejection of three Forçacava units , considering the inertia (GD^2) of 4.200 tm^2, the maximum transient overpressure was 36 percent at the unit 1 turbine entrance (farthest), and 11,3 percent maximum for a single Forçacava unit load rejection. The inertia of the unit was defined in order to admit a maximum overspeed of 45% at the worst condition.
The analysis at the adjacent units showed that the torque increase was 17 percent of full load with a single unit load rejection and 35 percent with two units full load rejection. The analysis showed that at the adjacent powerhouses there is no significant disturbance.
The investigation also showed that the introduction of a surge chamber would not bring benefits for the transient conditions, because it would not be physically placed in a convenient position.

5. CONCLUSION

The developed software was very important for the evaluation of feasibility of such a complex hydropower System as Forçacava. The analysis flexibility of different alternatives and the accuracy attained proved in the comparison to the Salto Santiago Powerplant made us believe in the usefulness of this method as a tool for hydraulic transient analysis.

6. NOMENCLATURE

a	- Waterhammer wave velocity	(m/s)	GD^2	- Inertia	($t\ m^2$)
A	- Cross section area	(m^2)	H	- Head	(m)
Δn	- Overspeed	(%)	H_I	- Initial head	(m)
D	- Penstock inner diameter	(m)	P	- Output	(KW)
f	- Friction factor		n	- Rotation speed	(rpm)
g	- Gravity	(m/s^2)	Q	- Water flow	(m^3/s)
			Q_I	- Initial water flow	(m^3/s)
			t	- Time	(s)

FIG. 1 — HYDRAULIC TRANSIENT

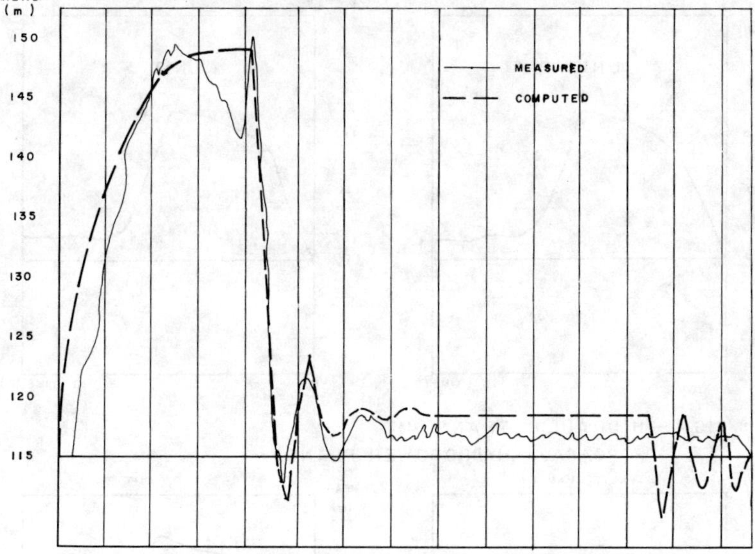

SALTO SANTIAGO HYDROPOWER PLANT

FIG. 2

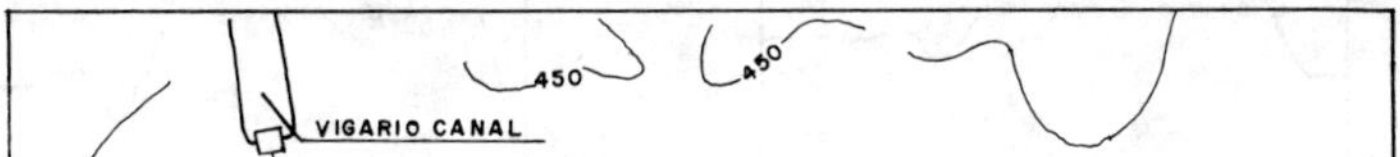

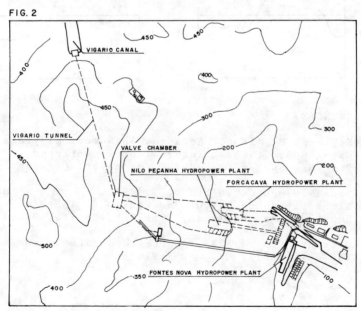

FONTES HYDROGERATING SYSTEM

FIG. 3 — TORQUE VARIATION AFTER UNIT 1 FULL LOAD REJECTION

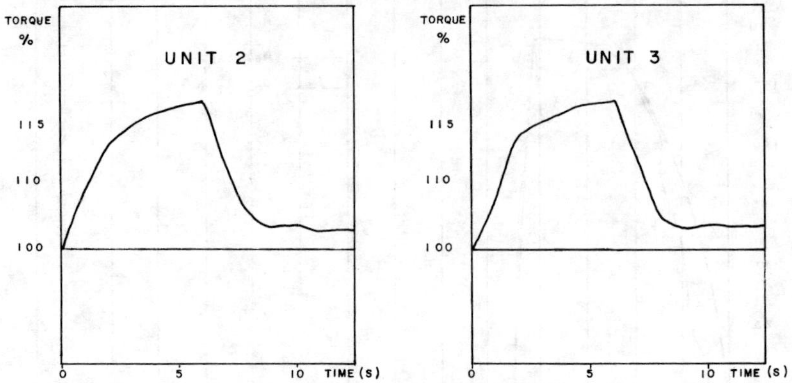

FIG. 4 — HIDRAULIC TRANSIENT
FORÇACAVA HYDROPOWER PLANT

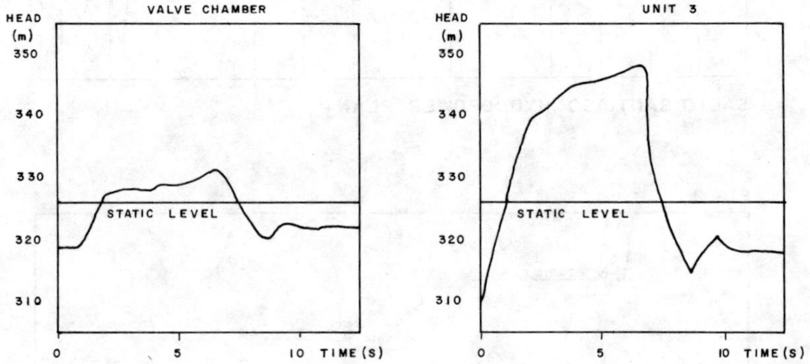

REFERENCES

1. Schimada M. (1986), Advances in Numerical Analysis Using MOC, 5th BHRA International Conference on Pressure Surges, Paper TN2, Hannover, Germany.

2. Paes N.N and Castelato D. (1979). Método Simplificado de Cálculo do Efeito de Inercia (GD^2) Necessário à Regulação, V SNPTEE, Recife, BRASIL.

3. Chaudhry M.H. (1979). Applied Hydraulic Transients, Van Nostrand R. Co. New York.

Some Aspects of Kalman Filtering Application in Hydrologic Time Series Processing

M. Markuš

Jaroslav Černi Institute for the Development of Water Resources, 11000 Belgrade, Yugoslavia

D. Radojević

Mihailo Pupin Institute, 11000 Belgrade, Yugoslavia

ABSTRACT

A procedure of hydrological forecasting using linear ARX model and Kalman filtering is presented. Model parameters have been determined adaptively. The application of Kalman filter techniques leads to autoregression (AR) and crossregression (X) coefficients estimates after each time increment. An example is given and conclusions proposed.

INTRODUCTION

Accurate determination is a prerequisite of efficient forecasting models of efficient water resources operation. A forecasting model is developed by making use of a linear Kalman filter. The main principle of the model is based on recursive algorithm of parameter estimation which gives an updated state estimation after each computation step. The initial ideas of the paper were derived from conventional filter scheme (Gelb,1974; Anderson & Moore,1979) and its application to hydrological forecasting (O'Connell, 1978; Wood & Szöllösi-Nagy,1978).During the research work some new aspects were examined.

THE ARX MODEL

The ARX model represents the forecasting dependence used for forecasting the output of hydrological systems on the basis of known values of the output and of the input.

$$Q(t+\tau)=A_1(t)Q(t)+A_2(t)Q(t-1)+...+A_p(t)Q(t-p+1)+$$
$$+B_1(t)INP(t)+B_2(t)INP(t-1)+........+B_p(t)INP(t-q+1)+$$
$$+ C_o+v(t)$$

where

$Q(t)$ - known value of the outflow at the time t

$Q(t+\tau)$ - forecasted value of the outflow at the time t for τ steps ahead ($\tau=1,2,3,..$)

$A_i(t)$ - autoregression coefficients, $i=1,2,3,..$)

$B_j(t)$ - crossregression coefficients, $j=1,2,..,q$

$INP(t)$ - known value of the inflow at the time t

$C_o(t)$ - independent term

$v(t)$ - model error

KALMAN FILTER

The numerical procedure of the model is based on the simplified state-space equations.

The state equation:

$$x(t+1) = x(t) \quad (1)$$

The measurement equation:

$$q(t) = C(t)x(t)+v(t) \quad (2)$$

where

$x(t)$ - the state vector of the system in the time t

$q(t)$ - output of the system in the measurement eq.

The noise term $v(t)$ includes both of the effects of model error and measurement noise.

$$x(t) = [A_1(t),\ A_2(t),\ ..,A_p(t),B_1(t),\ B_2(t),....B_q(t)]^T \quad (3)$$

$$C(t) = [Q(t-1),Q(t-2),...,Q(t-p),\ INP(t-1),INP(t-2),...INP(t-q)] \quad (4)$$

The appropriate Kalman filter recursive matrix equations are:

$$v(t) = Q(t)-C(t)x(t-1|t-1) \quad (5)$$
$$\Sigma(t) = C(t)P(t|t-1)c^T(t)+R \quad (6)$$
$$K(t) = k\ P(t|t-1)C^T(t)\Sigma^{-1}(t) \quad (7)$$
$$x(t|t) = x(t|t) = x(t|t-1)+K(t)v(t) \quad (8)$$
$$P(t+1|t) = P(t|t-1)-K(t)C(t)P(t|t-1) \quad (9)$$

The meaning of matrices (5)throughout (9) is given in Fig.1.The Kalman gain equation (7) extended by the variable product - k. It was found that changes in the term k considerably influence the accuracy of the forecasting.Based on this conclusion an optimimisation model of k is created, which gives the best forecasting model with the smallest forecasting error. A similar model is created for the covariance matrix of the measurement noise, since the first attempt to use a recursive equation for R was not successful.

The most convenient pair of k and R from the identification period of forecasting is assumed to be valid for the verification of the model.

APPLICATION OF THE MODEL

The model is applied to two gauging stations at the Danube river in Yugoslavia (Bezdan-input, Slankamen-output). The input parameters were:

$$p = 2;\ q = 2;\ A_1 = 0.956;\ A_2 = -0.117;$$

$$B_1 = 0.567;\ B_2 = 0.297;\ C_o = -12.53;$$

$$k = 0.50;\ R = 1.0 \times 10^3$$

The efficiency of the forecasting can be evaluated by means of the relationship

$$S = (S_1 - S_2)/S_1$$

where

S_1 - standard forecasting error of the ARX model without Kalman filter

S_2 - standard forecasting error of the ARX model with Kalman filter

S - effect of Kalman filtering

In the described example the following values were obtained:

$$S_1 = 97\ m^3/s$$

$$S_2 = 79\ m^3/s$$

$$S = 19\%$$

The effect of the Kalman filtering is obvious and consisted in 19% error reduction.

CONCLUSIONS

The recursive adaptive procedure is usually superior to the conventional non-recursive methods. By getting new information, the state of the hydrological system is updated and the time varying parameter estimates are obtained.These parameters, as a rule, after several steps of computation tend to converge about some constant values (Fig.3).A good hydrological forecast can be obtained by using the best input parameters of the model.

REFERENCES

1. Anderson,B.D.O.and Moore J.B.(1979). Optimal filtering,Prenticehall,Englewood Cliffs,N.J.
2. Gelb,A.(1974).Applied Optimal Estimation,The Analytic Science Corporation,Reading, Massachusetts.
3. O'Connell P.E.and Clarke R.T.(1981),Adaptive Hydrological Forecasting, Hydrological Sciences Bulletin,Vol.26.

4. Wood E.F and Szöllösi-Nagy A. (1978), An Adaptive Algorithm for Analyzing Short-Term Structural and Parameter Changes in Hydrologic Prediction Models, Water Resources Research, Vol.14, No.4.
5. Salas J.D., Delleur J.W., Yevjevich V.and Lane W.D.(1984), Modelling of Hydrologic Time Series, Water Resources Publications.
6. O'Connell P.E.(1980), Real-Time Hydrological Forecasting and Control, Proceedings of the First International Workshop, Wallingford,England,198o.

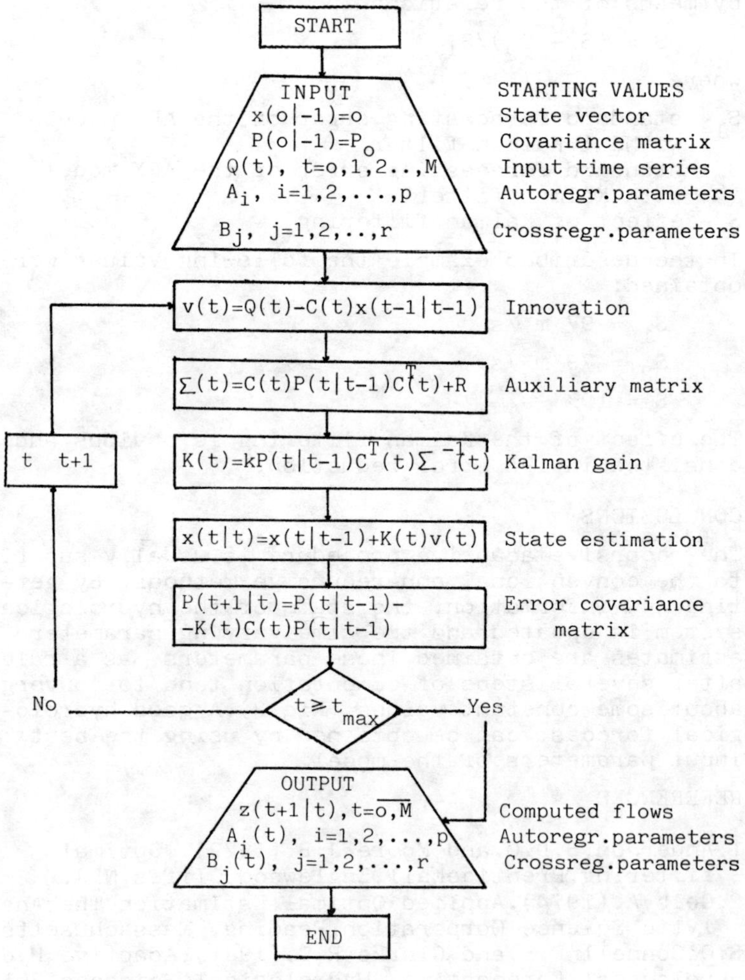

Fig.1 Computer program flow-chart

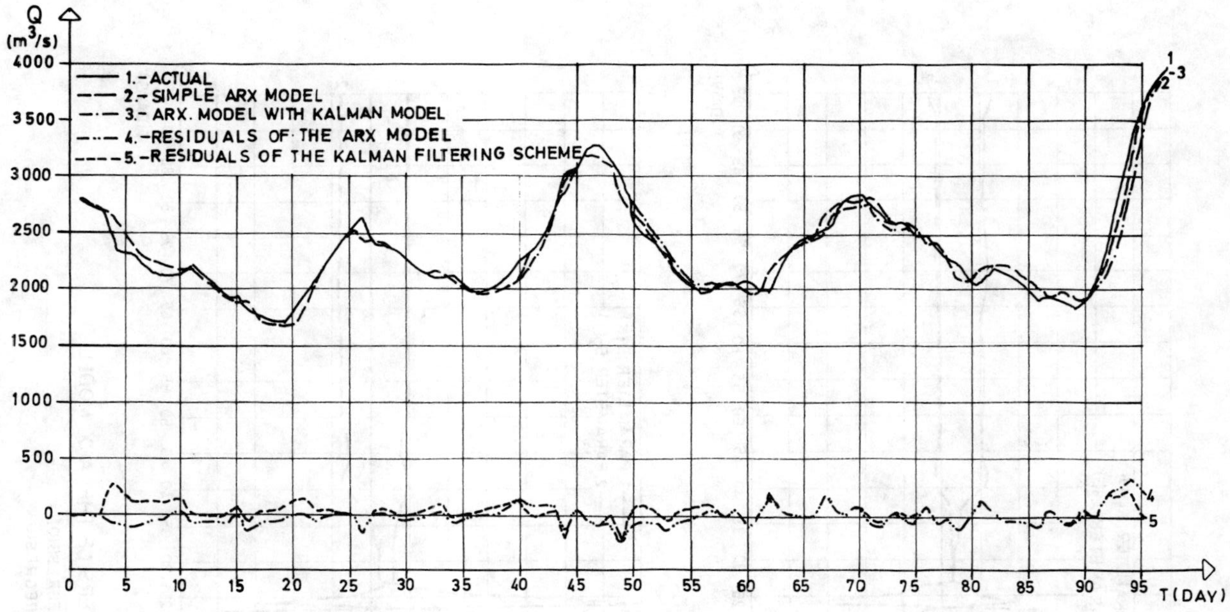

FIG. 2 - THE COMPARISON BEETWEN OBSERVED ANDCALCULATED DISCHARGES AND MODELS ERRORS.

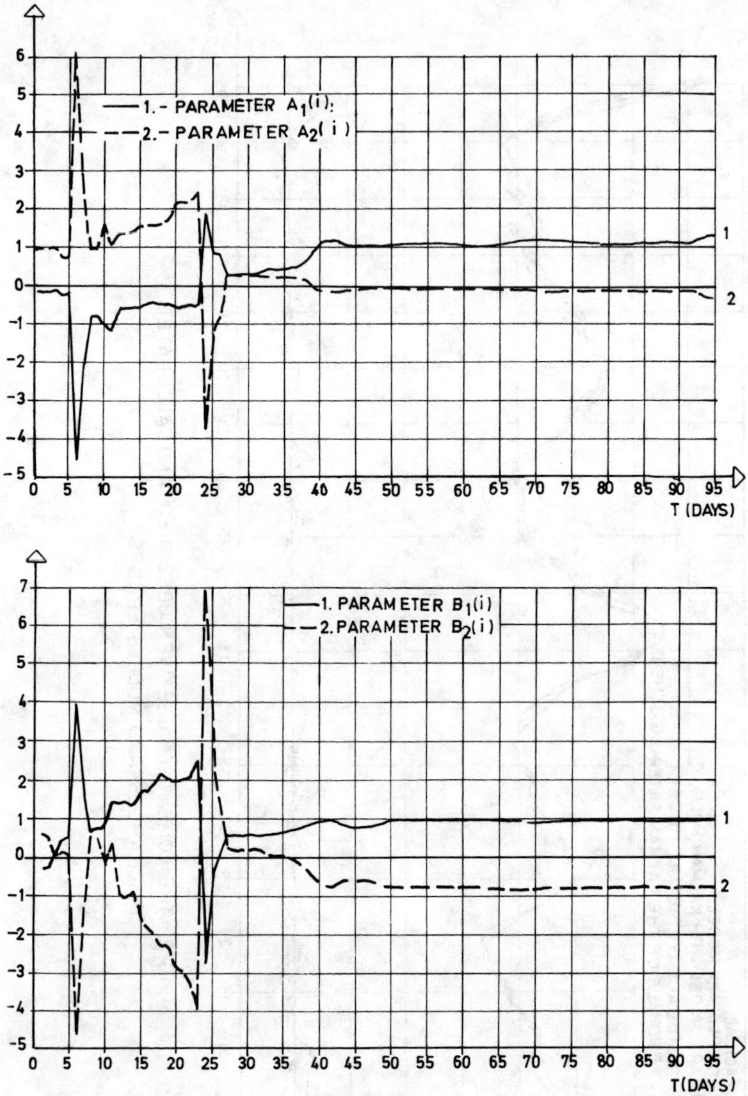

FIG. 3.-PARAMETERS OF THE ARX MODEL
A.- AUTOREGRESSION
B. - CROSSREGRESION

A Computer Model for the Estimation of Effluent Standards for Priority Pollutants From a Wastewater Discharge Based Upon Aquatic Life Criterion of the Receiving Stream

J.R. Nuckols
Certified Professional Hydrologist, Lexington, Kentucky, USA
S.F. Thomson
University of Kentucky Computing Center, Lexington, Kentucky, USA
A.G. Westerman
Kentucky Department for Environmental Protection, Frankfort, Kentucky, USA

ABSTRACT

A computer model, the Parameter Estimation Model (PEM), has been developed to provide the user with a numerical calculation of risk associated with the discharge of priority pollutants, as defined by the United States Environmental Protection Agency (U.S. E.P.A.), on the aquatic ecosystem of the receiving stream. The model uses an application of the SAS™ statistical computing system in conjunction with standard mass balance analytical techniques to determine the risk associated with the discharge of any pollutant for which ambient water quality criterion has been established. The model was developed primarily to assist in the determination of specific numerical values to be used as limits for priority pollutants in permits required by the Federal Clean Water Act. This paper discusses the development of the model, data requirements and procedure for its application, and provides examples of use of the model as a policy-support tool in prescribing specific effluent limits for two Publicly Owned Treatment Works (POTW) in Kentucky.

INTRODUCTION

Historically, the estimation of waste assimilation capacity for a stream receiving discharge from a municipal and/or industrial wastewater treatment works has been based on dissolved oxygen depression from organic loading of the stream. Currently, it has been determined that there are design factors other than maintaining desired oxygen levels that have a significant influence on the waste load assimilation capacity of a stream. The U.S. Environmental Protection Agency[5] has identified at least 129 pollutants which may adversely impact aquatic ecosystems and the health, safety, and welfare of humans who might use water resources that are downstream to wastewater discharge points. There are an estimated 50,000 additional substances which have yet to be tested for their potential impact on the aquatic environment (U.S.E.P.A.[5]; Birge, et al.[1]). The need for establishing rational standards for these pollutants as they occur in wastewater effluents is an important topic to environmentalists, public health advocates, and the industrial and municipal sectors of our society (Garber[3]; Chalmers[2]; Stephan[4]).

The purpose of this paper is to introduce a computer model that was developed for the purpose of assisting the design engineer, community planner, regulatory authorities, and other citizens that might be concerned about the potential impact of a wastewater discharge on the aquatic life of a receiving stream. The Parameter Estimation Model (PEM) can be used to calculate the risk associated with effluent limits for any pollutant that has been evaluated by the U.S.E.P.A. in its Ambient Water Quality Limits series (U.S.E.P.A.[6]). The model was designed to be used in addition to, not as a replacement for, waste load assimilation models for organic loading such as Biological Oxygen Demand (BOD), etc.

THE PARAMETER ESTIMATION MODEL (PEM)

The Parameter Estimation Model (PEM) is based upon essentially four criteria: (1) streamflow frequency of flow duration data for the proposed receiving stream; (2) ambient water quality data for the receiving stream at the proposed discharge point; (3) a profile of expected pollutants from the discharge; and (4) the recommended ambient water quality limit for each pollutant to be analyzed as prescribed by the U.S.E.P.A.

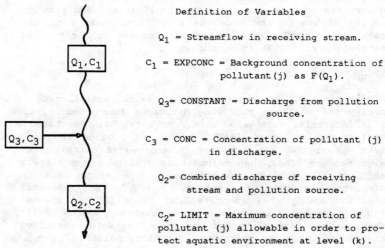

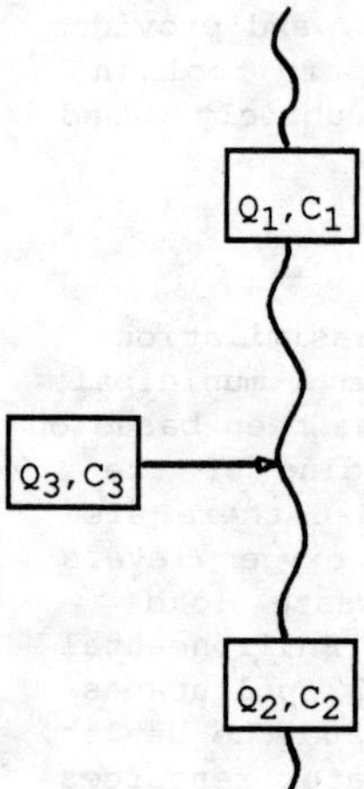

Definition of Variables

Q_1 = Streamflow in receiving stream.

C_1 = EXPCONC = Background concentration of pollutant(j) as F(Q_1).

Q_3= CONSTANT = Discharge from pollution source.

C_3 = CONC = Concentration of pollutant (j) in discharge.

Q_2= Combined discharge of receiving stream and pollution source.

C_2= LIMIT = Maximum concentration of pollutant (j) allowable in order to protect aquatic environment at level (k).

Figure 1. Schematic Diagram of PEM Model

Basically, the model uses the following flow dilution formula to evaluate the risk potential for a range of values for each pollutant to be analyzed:

$$CONC = \frac{(Q2 * LIMIT) - (Q1 * EXPCONC)}{CONSTANT} \quad (1)$$

This mass balance relationship is shown schematically in Figure 1. along with a definition of the variables. The model has been designed so that it can simultaneously evaluate any number of pollutants for different levels of protection to the aquatic ecosystem of the receiving stream. These levels of protection currently include acute aquatic life impact, chronic aquatic life impact, and impact on use of the stream as a water supply and fishery by humans.

Frequency of flow duration data is required as input data for the model in order to calculate the RISK factor in the output. The United States Geological Survey (U.S.G.S.) publishes flow duration tables for most active and discontinued stream gauging stations operated by that agency. Tables reflecting current data for most U.S.G.S. gauging stations can be assembled from the output of computer program A969 in the U.S.G.S.'s National Water Data Storage and Retrieval System, (acronym WATSTOR; U.S.G.S., [7]). In the PEM model, the RISK factor of assigning specific numerical effluent standards to the wastewater stream at the discharge point being analyzed is the inverse of the probability that a flow value will be exceeded. These probabilities are available from the flow duration tables published by the U.S.G.S.

The PEM model requires the input of ambient water quality data for each of the pollutants that are desired to be analyzed. The data are inputted as measured average concentration values of the pollutant and must be entered with the corresponding flow value for variable Q1 at the upstream station for each measurement taken. It should be attempted by the model user to provide a measured value for each pollutant over the range of design streamflow values desired to be analyzed. It is especially important to input such data points at the endpoints of the flow duration curve to be analyzed.

For each pollutant parameter selected to be analyzed by the PEM model, a value for the desired final stream concentration of the pollutant must be entered into the model by the user. These values can be obtained from published data such as the Water Quality Criterion documents published by the U.S.E.P.A.[6], which are continually being updated. Data is available for approximately 177 different pollutants in these documents.

OPERATIONAL INFORMATION AND MODEL OUTPUT

The modeling process is shown in Figure 1. Basically, PEM

computes an effluent concentration limit, variable CONC, which corresponds to each value of the design streamflow variable, Q1, specified by the user. The Q2, or downstream discharge values, are calculated in the model by adding the average discharge value of the point source being analyzed, variable CONSTANT, to the upstream streamflow variable, Q1. The expected background instream concentrations of the pollutant being analyzed, variable EXPCONC, are computed by linear interpolation of the user specified Q1 flow values with the observed concentration and flow values. Finally the program associates the RISK values input when specifying the values of Q1, to the calculated CONC values. The actual analysis was performed using an integrated reporting and statistical computing language, SAS™ on the IBM 3081 at the University of Kentucky. SAS has sorting, data manipulation, and statistical capabilities that made it particularly easy to implement the PEM model.

Recommended Effluent Limit

Level of Protection: No acute Impact on Aquatic Life

	Pollutant			
Risk	Dis. Cadmium Concentration Limit	T. Chromium Concentration Limit	T. Copper Concentration Limit	T.Cyanide Concentration Limit
0	1.79	16.00	9.22	22.00
2	1.34	16.80	9.74	23.59
5	0.00	19.83	11.73	29.65
10	0.00	24.29	14.66	38.58
15	0.00	29.39	18.01	48.78
20	0.00	36.72	22.82	63.45
25	0.00	45.65	28.68	81.30
30	0.00	56.81	36.01	103.62
40	0.00	84.26	58.61	171.22
50	0.00	83.51	102.70	293.01

North Elkhorn Creek below Georgetown No.2 POTW

Scott County, Kentucky

Figure 2. Tabular Output Format of the PEM Model.

The PEM model provides final output in two formats, a tabular form and a graphical form. An example of the tabular format is presented in Figure 2. In this example, the model was used to evaluate the potential for adverse impact of a proposed publicly owned wastewater treatment discharge in terms of acute toxicity for a small stream in Central Kentucky. An example of the graphical output format from an application of the PEM model is presented in Figure 3. In this application, the model was used to assist a client in determining which pollutants originating from a leather tanning operation should be subject to regulation in order to protect a receiving stream in Southeastern Kentucky. In the graphical output format of the model, the risk associated with discharging each pollutant as a function of its maximum allowable concentration is presented for each level of protection addressed by the model.

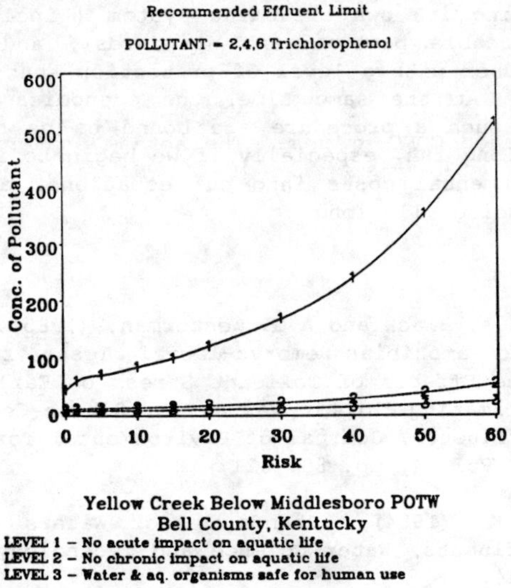

Figure 3. Graphical Output Format of the PEM Model.

DISCUSSION

The PEM model was developed with a number of "end-product" applications in mind; including implementation of policy set forth by the federal Clean Water Act, scientific research in regards to protection of aquatic ecosystems, and development of a scientific method for establishing reasonable and attainable effluent limitations for priority pollutants on a site specific basis . In its present form, the model is useful as a policy support tool. For example, in the applications cited above, the primary motivation for implementation of the model was to provide a screening mechanism for pollutants which should be of concern to the regulatory agency. In both cases, the regulatory agency involved had made the policy decision not to require effluent limits for any of the non-conventional priority pollutants in the Pollution Discharge Elimination System (NPDES) permits for the two Publicly Owned Treatment Works (POTW's) being constructed. This was in spite of the fact that each of the POTWs involved were known to be recipients of significant industrial waste loads; and that the receiving streams had, prior to receiving these discharges, been used extensively for contact recreation and sports fisheries, and in one case, as a secondary water supply. The PEM model was developed to provide a basis for requesting specific limits for priority pollutants that could be discharged by these POTWs. The PEM model provides a basis for implementation of a reasonable first definition of pollution control. The procedure prescribed by the model provides the regulatory agency with a basis for establishing specific limits, the discharger with a basis for optimizing its own treatment system (including establishing reasonable pretreatment standards), and the downstream water user with a level of protection that has a scientific basis. At the same time, the economics for implementation of such a procedure are bound to be cost-effective in the long run, especially if we begin to include social and environmental costs into our equations, as we should have been doing all along.

REFERENCES

1. Birge, W.J., J.A. Black and A.G. Westerman. (1985), Short-term fish and amphibian embryo-larval Tests for Determining the Effects of Toxicant Stress on Early Life Stages and Estimating Chronic Values for Single Compounds and Complex Effluents, Journal of Environmental Toxicology and Chemistry, Vol. 4, pp. 807-821.

2. Chalmers, R. K. (1984), Standards for Waters and Industrial Effluents, Water Science and Technology, Vol. 16, pp. 219-244.

3. Garber, W.F. (1977), Effluent Standards - Effect Upon Design, Journal of the Environmental Engineering Division, Vol. 103, pp.1115-1127, American Society of Civil Engineers. New York.

4. Stephan, C.E. (1985), Are the Guidelines for Deriving Numerical National Water Quality Criterion for the Protection of Aquatic Life and Its Uses Based on Sound Judgements, Aquatic Toxicology and Hazard Assessment, Proceedings of the 7th Symposium of the American Society for Testing and Materials. ASTM STP 854.

5. U.S. Environmental Protection Agency. (1986). Quality Criteria for Water 1986. Office of Water Regulations and Standards. EPA 440/5-86-001. Washington, D.C.

6. U.S. Environmental Protection Agency. (1980, Revised 1986). Water Quality Criteria Documents. Federal Register, Vol. 45 No. 231, Friday, Nov. 28, 1980. pp. 79318-79379. Washington, D.C.

7. U.S. Geologic Survey. (1981). WATSTOR: A Water Data Storage and Retrieval System. Branch of Distribution. U.S.G.S. Alexandria, VA.

Network Model Assessment to Leakage of Fill Dam

T. Sato and T. Uno
Department of Civil Engineering, Gifu University, Japan

INTRODUCTION

Finite difference or finite element procedure has frequently been used in assessments for predicting a change in the groundwater condition (Bachmat et al. (1980)). However, some obstacles arise when it is applied to a real problem. One of the troubles is to construct a geological model within a flow region. Geologists sometimes give us the information about soil profiles, but the insufficiency of the information and/or the scattering of boring test results make actual application difficult.

Another useful approach to get rid of obstacles has been developed by Wu et al. (1973). They recommend using a probabilistic model instead of a deterministic one. The model, which is called the network model, is made of some node points and branches connecting with them. The node point shows an imaginary spatial coordinate of soil deposit. The branch plays the role of a groundwater conduit in the model.

Groundwater flow analysis using the network model does not require a deterministic geological structure. All parameters in the model, such as spatial coordinate of the node point, thickness, length and permeability of the branch, are approximated by probabilistic variables. To get a solution, the Monte Carlo simulation is used in this analysis.

This paper deals with a case history for an application of the network model for assessment work of leakage through a fill dam and subsoils. Widely scattered boring and in-situ test results showed us that the deterministic approach is not suited to this case. The network model in this study was constructed as simply as possible to save on computation time. Three-dimensional finite element analysis was also performed to check the results.

MODELING OF GEOLOGICAL STRUCTURE AND FUNDAMENTAL EQUATIONS

The fill dam is located in Hokkaido, Japan. The plane view of the vicinity of the dam is shown in Fig. 1. Boring investigation was carried out at 39 different places. Based on these investigations, the cross-sectional view of soil profile is roughly drawn at section B–B' as shown in Fig. 2.

The first thing we had to do was to assign node points at each zone of the network. All soil layers appearing in the boring log are nominated in this step . This nomination should be limited to the layers related to the groundwater flow. Md6 and Ss7 are eliminated from the model since these two are deposited above the water table of the reservoir. This also allowed us to save on computation time.

The previous step is followed by making a decision for the groundwater flow path. The flow path has some branches which play the role of the groundwater flow conduit.

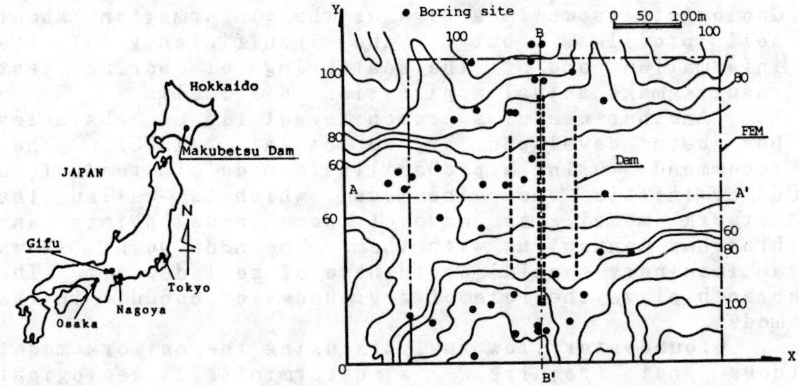

Figure 1. Location and plane view of the vicinity of the dam.

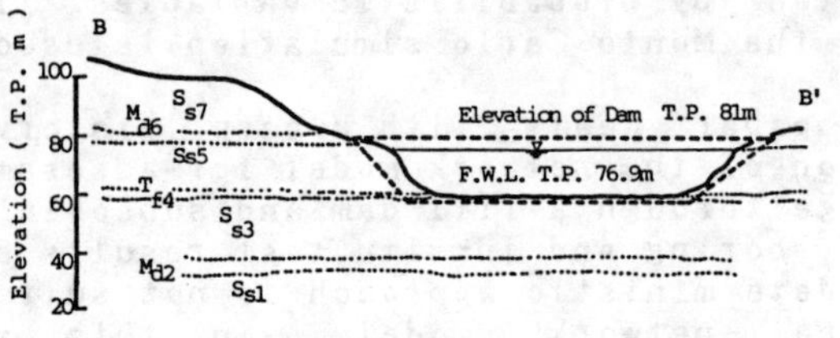

Figure 2. Subsoil profile along B–B' section.

Figure 3 shows a network model in the case study. The model has 11 node points and 22 branches. Groundwater flow in the network originates from the reservoir and ends at the exits 1,2,3,4,5 and 6. The way of leakage is described by the 19 different flow paths shown in Fig. 3.

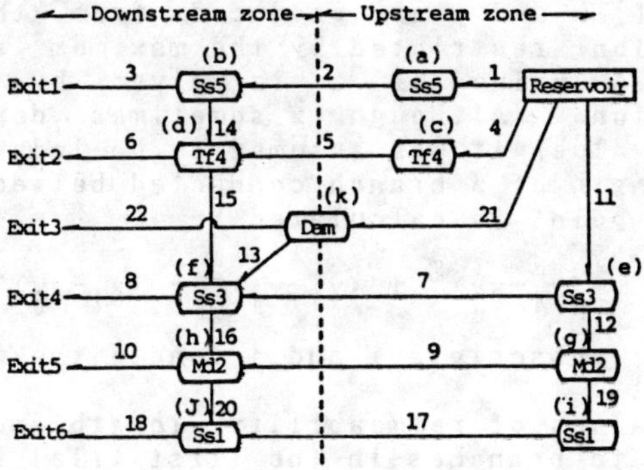

Figure 3. Network model set up for the case study.

Fundamental equations to simulate groundwater flow are as follows:

$$\sum_{j=1}^{n} D_{ij}X_{ij} - \Delta H_i = 0 \qquad (i=1,2,\text{-------}\ N) \qquad (1)$$

$$\sum_{j=1}^{m} K_{pj}t_{pj}W_{pj}X_{pj} = 0 \qquad (p=1,2,\text{-------}\ M) \qquad (2)$$

D_{ij} : length of the j-th branch in the i-th flow path
X_{ij} : hydraulic gradient of the j-th branch in the i-th flow path
ΔH_{ij} : head loss through the i-th flow path
n : number of branch in the i-th flow path
N : total number of the flow paths in network
k_{pj} : permeability of the j-th branch connected with the p-th node point
t_{pj} : thickness of the j-th branch connected with the p-th node point
W_{pj} : width or length of the j-th branch connected with the p-th node point
X_{pj} : hydraulic gradient of the j-th branch connected with the p-th node point
m : number of branches connected with the p-th node point
M : total number of the node points in network

MONTE CARLO SIMULATION

Boring and in-situ permeability test results give

us values for the network. They were found to be widely scattered. The Monte Carlo simulation helped us compute considering those scattered values. The procedures carried out in this study are summarized as follows:

1) Spatial coordinates of the node points are given to the model in the first trial computation. Values of x and y are produced from the constant distribution restricted by the maximum and minimum values. A value of z is given by the normal distribution. Although z sometimes depends on x and/or y value, it was assumed to be independent.

2) Length of a branch connected between the i and j-th node point is calculated by

$$D_{ij} = \sqrt{(x_i - x_j)^2 + (y_i - y_j)^2 + (z_i - z_j)^2} \qquad (3)$$

in which subscripts i and j denote the i and j-th node points.

3) Values of permeability, length and thickness are given to branches in the first trial computation. The value of permeability is approximated by the log normal distribution. The others are produced from the normal distribution.

4) Eqs. (1) and (2) are solved by ΔH_i.

5) Leakage from the reservoir, Q is calculated by

$$Q = \sum_{r=1}^{R} K_r t_r W_r X_r \qquad (4)$$

in which suffix r denotes the r-th branch facing the reservoir or the terminals and R denotes total number of branches connecting with the reservoir or the terminals.

6) Calculations from 1) to 5) are repeated in total trial numbers.

RESULTS AND DISCUSSIONS

Computation results are shown in Fig. 4. Total trial numbers amount to one thousand in this case. The vertical axis of the figure shows the probability and the horizontal axis the quantity of leakage. The real line shows the results of the computation using a blanket below the reservoir to decrease the leakage. The dashed line shows the one not using it. Expected value of leakage in the use of the blanket is estimated as 240 m^3/d. The one in the bare case is estimated as 310 m^3/d. These computations followed that both cases do not exceed the restriction of leakage regulated by the Japanese Ministry of Agriculture, Forest and Fishery, 0.005 percent of the total amount of dam pondage.

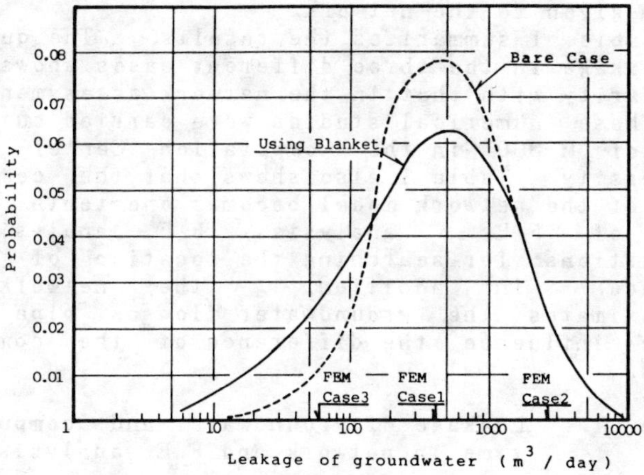

Figure 4. Leakage of groundwater calculated by the network and F.E. models.

Finite element analysis was also performed to check the results. Plane view of F.E. analysis area is shown in Fig.1. The vertical sectional view at B-B' section is shown in Fig. 5. A precise deterministic geological model is required for F.E. analysis. Insufficiency of test sites and scattering of boring test results makes it difficult. Then the average model was used in the F.E. analysis. Tf4 which deposits like a lens is neglected in this analysis.

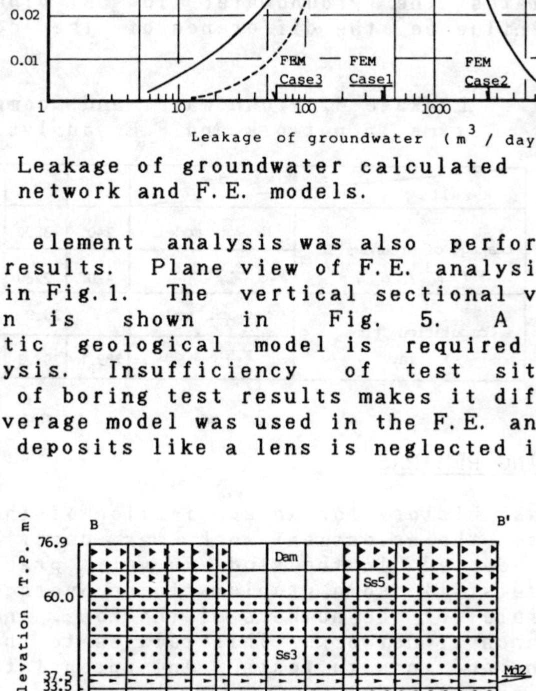

Figure 5. Geological model for F.E. analysis.

F.E. model consists of 6114 tetrahedron elements and 1428 nodal points. Quantity of leakage computed by F.E. are also shown in Fig. 4. F.E. computations were done according to three different cases. The values of permeability in these three cases respectively coincide with that of μ(case1), $\mu+2\sigma$(case2) and $\mu-2\sigma$(case3) of the permeability model given to the network.

Table 1 summarizes the results. The quantity of leakage in the three different cases shows a good similarity with that in the network assessment.

These numerical studies were carried out by the use of M-360 in the Computation Center of Gifu University. Table 1 also shows that the computation time of the network model becomes one-tenth of each case of F.E. analysis. F.E. analysis needs repetitions for searching the location of phreatic surface. In addition, the network model approximates the groundwater flow as pipe flow. These influence the difference of the computation times.

Table 1. Leakage of groundwater and computation time in network and F.E. analysis.

Results \ Model case		1	2	3
Leakage of groundwater (m^3 / d)	Network model	240	3800	14
	F.E. model	420	2260	60
Computation time (sec)	Network model	25.9		
	F.E. model	289.1	264.3	339.8

CONCLUDING REMARKS

A case history for an application of the network model to the assessment work was described. The method adopted in the study is very practical but there are still some problems to be overcome. There is no rule for the numbers of the zone, node point and branch of network. This relates to the numbers of boring and in-situ tests. An appropriate way for optimizing the model must be researched reflecting on the quantity and quality of investigations. The model is only available for the steady-state condition of groundwater flow. It should be expanded so as to compute the transient flow.

REFERENCES

Bachmat, Y. et al. (1980) : Groundwater Management-The use of numerical models, American Geophysical Union.

Wu, T.H., Vyas, S.K. and Chang, N.Y. (1973) : Probabilistic Analysis of Seepage, ASCE, Vol.99, No.SM4, pp.323-340.

SECTION 3B - PARAMETER ESTIMATION

Groundwater Monitoring Network Design

H.A. Loaiciga
Department of Geological Sciences, Wright State University, Dayton, Ohio, 45435, USA

BACKGROUND AND PROBLEM OVERVIEW

The design of a groundwater quality monitoring network is an important aspect in aquifer restoration and prevention of groundwater pollution. By network design it is understood in this study the selection of the number and locations of sampling wells at a site where there exists contamination. The problem of network design is described in simple terms by classical sampling theory (Cochran, 1977). Suppose a set of independent, identically distributed data are sampled with the purpose of estimating the sample mean. Under further assumptions the optimal sample size, n*, is given by the expression

$$n^* = (\Psi \sigma^2 / C)^{1/2} \quad (1)$$

in which Ψ = a factor related to the expected loss incurred in estimating the population mean; σ^2 = the (assumed known) population variance and C = the additional cost per sampling well. Eq. (1) contains the three factors of overriding importance in network design: (1) a measure of risk aversion or loss related to the fact that a practical decision will be made based on the results from the sample and that a quantifiable loss is expected due to errors in estimates derived from a data set; (2) the variance, a measure of spread, of the population or phenomenon under study; (3) the cost of sampling. Equation (1) indicates that, other things being constant, the sample size increases as the coefficient associated with expected loss becomes larger. The same increasing trend applies as the variance of the process or population increases. Intuitively, the larger the cost of adding one more sampling well, the more pronounced the trade-off between expected loss (due to errors in results) and the cost of sampling will be, resulting in a smaller sample size that best balances these competing objectives. Within the context of a temporal/spatial process such as subsurface contamination, knowledge of the number of sampling points is not sufficient. The locations of wells are required as well. The purpose of this paper is to develop a network design procedure that considers the spatial nature of contaminant plumes.

NETWORK DESIGN CONSIDERATIONS

The problem of network design involves some background information. Foremost is hydrogeologic data such as: (1) geologic setting (lithology and stratigraphy); (2) groundwater flow patterns and volumes; (3) recharge areas and rates; (4) aquifer characteristics (e.g., hydraulic conductivity, dispersion coefficients); (5) existing monitoring wells and their locations. A description of general geomorphoclimatic characteristics of a site is also important. This might help to understand the subsurface environment, and could also influence site accessibility, monitoring costs and potential threats of contamination to the surrounding environment. Data on precipitation, temperature, evapotranspiration, topography, accessibility, site size, proximity to ecosystems/population centers would be most valuable. In other words, network design is not a purely statistical optimization exercise. Rather, it has a broader context, and fortunately, such context may be incorporated, to some extent, in the network design approach given herein.

PLUME DETECTION

Prior to undertaking the network design problem, one must have a predetermined set of N locations at which new wells could be developed for monitoring purposes. In addition, the grid layout containing the sampling points and the geometry of such layouts (i.e., shape and distance between sampling points) must be known. Evidently, the sampling grid must be such that, with probability one, one or more wells hit or intersect the contaminant plume. However, there may be instances in which a grid layout that would hit the "hot spot" is not readily obvious unless some information is available about the contaminant plume geometry. This section explains how to use some basic results of hydrodynamic dispersion to approximately assess the probability of failing to detect a plume with some regular arrays of sampling wells.

For large Peclet numbers, contaminant plumes developed by hydrodynamic dispersion (under suitable hydrogeological heterogeneity and anisotropic conditions) are approximately ellipsoidal in shape. The semi-axes of the ellipsoidal plume may be obtained from the square roots of the variances of the plume particles displacements, σ^2_{ij}. The variances are related to macrodispersivity parameters, A_{ij}, by the expression (see Bear, 1972, pp. 610).

$$\sigma^2_{ij} = 2LA_{ij} \tag{2}$$

in which L = expected distance of the plume's center of mass to the source of contamination along the direction of average groundwater velocity. A_{ij} may be inferred from hydrogeological data. For example, along the major principle axis (Neuman et al., 1987) of macrodispersivity,

$$A_{11} = (3/8)\sigma_y^2 \ell \quad (3)$$

where σ_y^2 = variance of log-hydraulic conductivity; and ℓ = correlation length of the log-hydraulic conductivity. The major principal axis is assumed to be oriented in the direction of average groundwater velocity. For the other principal axes $A_{22} = A_{33} \cong f \bullet A_{11}$, where f is a fraction between 0 and 1. There is some controversy in the literature as to what f should be. Neuman et al. (1987) suggest that f = 0 (for large Peclet numbers), while Gelhar and Axness (1983) proposed a nonzero f. For practical purposes and $1/10 \leq f \leq 3/10$ seems reasonable based on empirical evidence reported in the literature. In order to illustrate how to use Eq. (2) for plume geometry delineation assume a normal plume distribution

$$C(x,y,z) = [C_0/(2\pi)^{3/2} \; \sigma_{11}\sigma_{22}\sigma_{33}] \bullet$$
$$\exp\,[-(1/2)(m_1^2/\sigma^2{}_{11} + m_2^2/\sigma^2{}_{22} + m^2{}_3/\sigma^2{}_{33})] \quad (4)$$

in which $m_1 = x - (x_0 + L)$; $m_2 = y - y_0$, and $m_3 = z - z_0$ are the coordinates relative to the contaminant plume center of mass and (x_0, y_0, z_0) represents the location of the contaminant source; C_0 is the concentration at the source (in theory the source should be a point source). The axes x and m_1 are aligned with the direction of the mean groundwater flow velocity, which in turn is taken as the direction of the major principal axis of the dispersivity tensor. By taking the logarithm on both sides of Eq. (4), solving for the quadratic expression under the exponent on its right-hand side, and normalizing coordinates to obtain the standard ellipse equation (4) yields

$$m_1'^2/\sigma^2{}_{11} + m_2'^2/\sigma^2{}_{22} + m_3'^2/\sigma^2{}_{33} = 1 \quad (5)$$

where $m_i' = m_i/\sqrt{k}$ (i = 1,2,3) and $k = -2\ell n(C/C_0)\,(2\pi)^{3/2}\,(\sigma_{11}\sigma_{22}\sigma_{33})$. Letting the concentration level C(x,y,z) be a fraction p of C_0, Eq. (5) represents the ellipsoid at a concentration pC_0 with semiaxes equal to σ_{11}, σ_{22}, and σ_{33} (see Eq. 2). By selecting p sufficiently small so that pC_0 will fall below detection limits Eq. (5) would represent the plume configuration of interest. Certainly, the ellipsoid semiaxes must be sufficiently large relative to the distance between sampling points so that various wells will intersect the contaminant plume. Otherwise, i.e., if the dimensions of the ellipsoidal plume are relatively small compared to the distance between grid points, then the probability of the plume not being hit by any sampling well becomes an issue.

NETWORK DESIGN

Expected Loss

One approach to determining the number and location of sampling wells is to minimize the total monetary cost involved in the development of the sampling network plus the expected loss incurred in a decision through an estimation error υ on the variable of interest. It is being assumed that groundwater quality data serves as a basis for practical decisions and that losses from such decisions are quantifiable in monetary terms. Sampling theory enables us to find the frequency distribution $g(\upsilon,n)$ of υ which, for a specified sampling method will depend on the sample size n. Under classical assumptions (i.e., independent, identically distributed observations), the expected loss for a sample of size n is

$$R(n) = \int s(\upsilon)g(\upsilon,n)d\upsilon \qquad (6)$$

where $s(\upsilon)$ is the loss incurred through an error of estimate υ. When the space dimension becomes an issue, such as in groundwater networks, Eq. (6) is further complicated by the fact that there are multiple possible arrangements of n wells and each arrangement may produce a different error of estimation υ. Matters are simplified considerably if one assumes a quadratic loss, $s(\upsilon) = \Psi\upsilon^2$ where Ψ is a loss coefficient. If the estimation error has zero mean, then the expected loss is $R(\upsilon) = E(s(\upsilon)) = \Psi V(\upsilon)$ where $V(\upsilon)$ denotes the variance of estimation error. The coefficient Ψ may be estimated by utility analysis within a mean-variance framework.

Geostatistical Estimation

Suppose that groundwater quality data is used in calculating an average concentration, C_0, over a domain. Such domain could be the vicinity of a well, or any other spatial section of aquifer suitable defined by the nature of groundwater investigations. The geostatistical approach is quite suitable for the purpose of concentration estimation based on data collected at nearby sampling wells (control points). Let us assume that n wells provide concentration data ($n \leq N$, where N is the total number of possible wells, see section on PLUME DETECTION). The concentration (it may be a spatial average) estimate is $C^* = \Sigma \lambda_i C_i$ $(i = 1,2,\ldots,n)$, where the λ's are suitable weights imposed on the measured concentrations C_i. It is well known that the geostatistical approach (see, e.g., Journel and Huijbregts, 1978) yields optimal weights λ_i that minimize the variance of estimation error $\upsilon = C_0 - C^*$ and produce an unbiased estimate of C_0 (i.e., the expected value of C^* equals the mean value of C_0). The application of either simple or universal kriging (or any other related variation of these) presumes that the covariance of the concentration field is known. In addition, for spatially variable fields, such as concentration, the mean concentration is variable, i.e., there is a trend, introducing additional parameters. Therefore, we see that network design has imbedded into it the estimation of correlation and trend parameters. There is an ample body of literature addressing this estimation problem and due to space limitations this subject may not be pursued further herein. It will be seen, though, that in the process of network design one must solve a parameter estimation problem for each network configuration considered.

<u>The Criterion: Network Configuration</u>

Let x_i denote a binary variable that can take the values 1 or 0 depending on whether a sampling well is developed or not at the ith location. The ith location is any of the possible well sites. Then, the estimator of C_0 is $C^* = \Sigma \lambda_i X_i C_i$ $(i = 1,2,\ldots,N)$. The expected loss is proportional to the variance of estimation error as proposed previously. Consequently, the network design criterion is to

$$\underset{\lambda_i, X_i}{\text{Minimize}} \sum_{i=1}^{N} K_i X_i + \Psi \left[\sigma^2 + \sum_{i=1}^{N} \sum_{j=1}^{N} \lambda_i \lambda_j X_i X_j \sigma_{ij}^2 - 2 \sum_{i=1}^{N} \lambda_i X_i \sigma_{i0}^2 \right\} \tag{7}$$

in which K_i is the cost of a sampling well at the ith location; σ^2 is the variance of concentration C_0; σ_{ij}^2 and σ_{i0}^2 are the covariances between the ith and jth and the ith and 0 location respectively. There is a budget (B) constraint appended to Eq. (7), i.e.,

$$\sum_{i=1}^{N} K_i X_i \leq B \tag{8}$$

If the estimator C* is unbiased, then $EC^* = m_0$ where m_0 is the expected value of the estimated concentration C_0, that is $\Sigma \lambda_i x_i m_i = m_0$ $(m_i = E(C_i))$. Assume that the trend at the ith location is

$$m_i = \sum_{k=1}^{Q} b_k^i \beta_k$$

in which β_k are (usually unknown) parameters and b_k^i are functions of the location coordinates of the ith sampling well. Then, the unbiasedness condition is easily shown to be

$$\sum_{i=1}^{N} \lambda_i X_i b_k^i = b_k^0, \quad k = 1,2,\ldots,Q \tag{9}$$

Equations (7) - (9), along with the binary condition X_i = [1 or 0] depending on whether a site is developed or not, respectively, constitute the mathematical network design problem.

On the Solution of the Network Design Problem

The design problem as outlined above is rather complex to solve for the following reasons: (i) the decision variables X_i are binary; (ij) the problem is nonlinear in the objective function as well as in the constraints (see Eq. (9)); (iii) the covariance structure (determined by the hydrogeologic setting) of the concentration field (see σ_2, σ_{ij}^2 and σ_{i0}^2 in Eq. (7)) are most likely to be unknown implying that there must be a parameter estimation module imbedded in the optimization program. The author believes that with the availability of high-speed "super" computers the solution of the optimization problem is more efficiently approached with a random, combinatorial, search for the optimal number and location of sampling wells.

CONCLUSIONS

The design of groundwater monitoring networks is a function of (1) the statistical heterogeneity and geologic anisotropy of the aquifer; (2) the hydrodynamics of plume migration; (3) the practical decisions that result from a data acquisition program; (4) budgetary constraints imposed on available resources. The joint treatment of such factors and their incorporation on a mathematical (combinatorial) formulation of the problem to determine the best well locations was given in this paper. The next step should be the application to an actual design. Efforts are in progress in this direction. However, the mathematical conceptualization and commanding factors in groundwater network design has been set in this work.

REFERENCES

Bear, J. (1972). Dynamics of Fluids in Porous Media, American Elseiver, New York, N.Y., pp. 764.

Cochran, W.G. (1977). Sampling Techniques, 3rd ed., John Wiley, New York, N.Y., pp. 428.

Gelhar, L.W., and Axness, C.L. (1983). "Three-dimensional stochastic analysis of macrodispersion in aquifers," Water Resour. Res., 19(1), 161-180.

Journel, A.G., and Huijbregts, Ch. J. (1978). Mining Geostatistics, Academic Press, p. 600.

Neuman, S.O., Winter, C.L., and Newman, C.M. (1987). "Stochastic theory of field-scale Fickian dispersion in anisotropic porous media," Water Resour. Res., 23(3), 453-466.

Adjoint-State and Sensitivity Coefficient Calculation in Multilayer Aquifer System

A.H. Lu
Basalt Water Isolation Project, Westinghouse Hanford Company, Richland, Washington 99352, USA
C. Wang and W. W-G. Yeh
Civil Engineering Department, UCLA, Los Angeles, California 90024-1593, USA

ABSTRACT

In this paper, a computational code using the variational method has been developed to calculate the sensitivity coefficients. The sensitivity coefficients are defined as the derivative of the hydraulic head with respect to the hydraulic conductivities. Both the three-dimensional flow equations and the adjoint sensitivity equations are solved using a method which combines the local mass balance and finite element approximation. The code is designed for use as a module for parameter estimation by incorporating a generalized least-squares algorithm. The developed methodology is especially suitable for a large number of groundwater parameter estimations for which a limited number of measurements may exist. The computational advantage and applications of the developed code will be discussed.

INTRODUCTION

The Basalt Waste Isolation Project is investigating deep basalt formations for the disposal of high-level nuclear wastes. The ongoing hydrogeologic characterization plan is aimed at a thorough understanding of the hydrogeology of this site. Along with several parallel efforts to achieve the goal, numerical programs are being developed in conjunction with field testing plans to estimate three dimensional representative values of the hydraulic parameters. The objectives of the numerical program development are: 1) to provide computationally efficient simulations of multilayer groundwater flows, and 2) to estimate parameters using the inverse or parameter estimation technique.

To overcome burdensome computational problems in three-dimensional simulations of groundwater flow, various numerical algorithms have been proposed by many authors (e.g. Babu and Pinder, 1984[1]; Huyakorn, 1986[2]). In this paper a multiple-cell mass balance method (MCB) (Sun and Yeh, 1983[3]; Wang et al, 1986[4]) is employed. The method uses local mass balance and finite element approximation.

The discretized system of the flow equation is derived from the integral form that expresses the balance of mass in each element. A modified successive over relaxation (SOR) method (Wang, 1986[5]) was used for solving the discretized system.

In this paper, the adjoint-state equation is developed and solved in the same fashion to provide sensitivity analysis. Sensitivity analysis is a useful tool for examining model sensitivity, guiding an adjustment of parameters in the parameter identification procedure, and estimating variance and confidence intervals for the hydraulic heads and their gradients. There are three different methods to calculate the sensitivity coefficient: 1) the influence coefficient method, 2) the sensitivity equation method, and 3) the variational method. A comparison of the accuracy of the three calculation methods was made by Li et al (1986[6]). Wang and Yeh (1986[7]) compared the three methods in terms of accuracy, computational effort and computer storage requirement. We employed the variational method in this paper because it is computationally more efficient for problems involving a large number of parameters and a small number of observation wells. The developed code will be used as a module in the three-dimensional inverse code which is an extension of the 2-D inverse code (Lu et al, 1988[8]).

The adjoint-state theories based on the variational method were introduced by Chavent et al (1975[9]), Carter et al (1974[10]), and Chen et al (1974[11]). In this paper, algorithms for calculation of adjoint-state and sensitivity coefficients are developed for an analysis of a groundwater flow system in the Columbia basalt beneath the Hanford reservation in south central Washington State.

The geology of Columbia basalts is composed of intact, thick basalt flow interiors of very low permeability interspaced by thin flow top layers of higher permeability. The flow system is conceptualized as a three dimensional flow with uniform-thickness multilayer stratigraphic unit layers.

FLOW EQUATIONS AND NOTATION

The governing equation for a three-dimensional solute transport process in a saturated porous medium is generally written as (Bear, 1972[12] and Bear, 1979[13]).

$$\nabla\cdot(K\nabla h) = S_s\frac{\partial h}{\partial t} + Q \qquad (1)$$

$$h(x,y,z,0) = f(x,y,z) \qquad (x,y,z) \in D \qquad (2)$$

$$-K\frac{\partial h}{\partial n} = \alpha(h - H_0) + Q \qquad (x,y,z) \in \Gamma \qquad (3)$$

where ∇ is the gradient operator (i.e., $\nabla(\cdot) = (\frac{\partial}{\partial x}, \frac{\partial}{\partial y}, \frac{\partial}{\partial z})(\cdot)$; K is the hydraulic conductivity tensor; h is the hydraulic head; S_s is the specific storage; Q is source or sink; t is time; x, y, z are the spatial variables; D is the flow region; Γ is the boundary of D; $\frac{\partial}{\partial n}$ is the normal derivative; α is the parameter controlling type of boundary conditions: if $\alpha = 0$, Equation 3 represents a Neumann condition; if $\alpha = \infty$, Equation 3 is a Dirichlet condition; otherwise, it is a mixed condition. H_0 is the prescribed boundary head. Only the Neumann and Dirichlet conditions are discussed in this paper.

ADJOINT EQUATIONS

Taking the first variation of equations 1 through 3 yields

$$\nabla \cdot (\delta K \nabla h) + \nabla \cdot (K \nabla \delta h) - S_s \frac{\partial \delta h}{\partial t} \tag{4}$$

$$\delta h(x,y,z,0) = 0 \tag{5}$$

$$- \delta K \frac{\partial h}{\partial n} - K \frac{\partial \delta h}{\partial \delta n} + \alpha \delta h \tag{6}$$

Let q(x, y, z, t) be an arbitrary function which will be chosen later. Multiplying Equation (4) by q, integrating over D and (0, t), the following is obtained:

$$\int_0^t \int_D q S_s \frac{\partial \delta h}{\partial t} dxdydzd\tau = \int_0^t \int_D q[\nabla \cdot (K \nabla \delta h) + \nabla \cdot (\delta K \nabla h)]dxdydzd\tau \tag{7}$$

Integrating by parts, applying Green's first identity, and substituting (6) into the result expression yields

$$\int_D q(t) S_s \delta h dxdydz = \int_0^t \int_D \delta h [S_s \frac{\partial q}{\partial t} + \nabla \cdot (K \nabla q)]dxdydzd\tau -$$

$$- \int_0^t \int \Gamma \delta h K \nabla q \cdot nd\Gamma d\tau - \int_0 \int_D \nabla q \cdot \delta K \nabla h dxdydzd\tau$$

$$- \int_0^t \int \Gamma \, q \alpha \delta h d\Gamma d\tau \tag{8}$$

With q satisfies

$$S_s \frac{\partial q}{\partial t} + \nabla \cdot (K \nabla q) = \delta(x-x_0) \, \delta \, (y-y_0) \, \delta \, (z-z_0) \, \theta(\tau) \tag{9}$$

$$q(x,y,x,t) = 0 \tag{10}$$

$$k \nabla q \cdot n + \alpha q = 0 \tag{11}$$

where $\delta(x-x_0) \, \delta \, (y-y_0) \, \delta \, (z-z_0)$ is the Dirac delta function, and $\theta(\tau) = 0$ for $\tau \le 0$; $\theta(\tau) = 1$ for $\tau > 0$,

Equation (8) can be written as

$$\int_0^t \delta h \, (x_0,y_0,z_0,\tau)d\tau = \int_0^t \int_D \, q(x,y,z,t-\tau) \cdot \delta K \, h(x,y,z,\tau)dxdydzd\tau \tag{12}$$

Differentiating equation (12) with respect to t leads to the following expression:

$$\delta h(x_0,y_0,z_0, t) = \int_0^t \int_D \nabla (\frac{dq}{dt}) \cdot \delta K \nabla h dxdydzd\tau. \tag{13}$$

When the MCB method is used to solve Equations (9) through (11), the same coefficient matrices as in solving Equations (1) through (3) obtained. Therefore, the same solution algorithm can be used to solve both problems.

FINITE ELEMENT APPROXIMATION AND BASIC FUNCTIONS

In order to apply the multiple cell balance method (MCB), one must divide the flow region into elements, select the location of nodes and define the basis functions. Dirichlet and Neumann conditions are prescribed on the boundaries.

The flow region is divided into triangular prism elements. The tops and bottoms of these triangular prisms are parallel to the xy-plane. The other sides are normal to the xy-plane. The six vertices of each element are taken as nodes. Within each element, the head h and solution for adjoint equation q are expressed.

$$h = \Sigma\phi_i h_i = z^-[\phi_i h_1+\phi_j h_2+\phi_k h_3] + z^+[\phi_i h_4+\phi_j h_5+\phi_k h_6] \tag{13}$$

$$q = \Sigma\phi_i q_i = z^-[\phi_i q_1+\phi_j q_2+\phi_k q_3] + z^+[\phi_i q_4+\phi_j q_5+\phi_k q_6] \tag{14}$$

where $z^- = \dfrac{z_1-z}{z}$; $z^+ = \dfrac{z-z_0}{z}$; $z = z_1-z_0$; and ϕ_i, ϕ_j, ϕ_j, and ϕ_k are two-dimensional linear basis functions for the triangular element Δ_{ijk} which is the top or bottom of e (Figure 1), i.e.,

$$\phi_p(x,y) = \frac{1}{2\Delta}(a_p + b_p x + c_p y)$$

$$(x,y) \in \Delta_{ijk},(p=i,j,k) \tag{15}$$

where $a_i = x_j y_k - x_k y_j$ $\quad b_i = y_j - y_k$ $\quad c_i = x_k - k_3$

and the remaining coefficients are obtained by cyclic permutation of subscripts: Δ is the area of ∇_{ijk}.

FIGURE 1. Configuration of the aquifer of the example

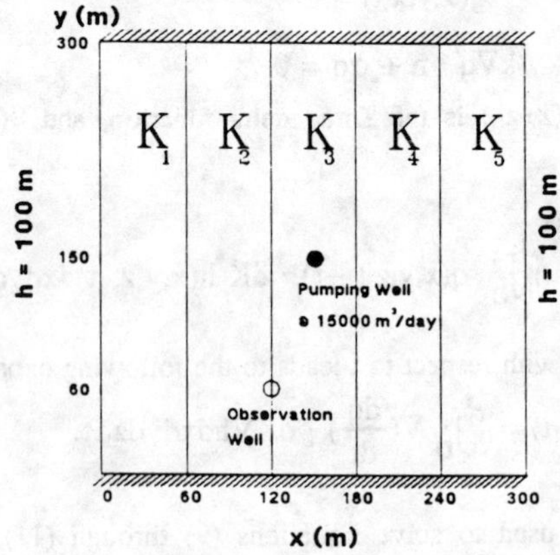

SENSITIVITY COEFFICIENTS

Assuming the hydraulic conductivity K_n is constant in each element, equation (13) can be written in the following form:

$$\frac{dh_o}{dK_n} = \int_0^t \iiint_{D^e}^{s} \nabla\dot{q}(x,y,z,t-\tau)\ \nabla h\ (x,y,z,\tau)dxdydzd\tau \qquad (16)$$

where o = a node number at which the observation well is located, and D^e = a flow region of element e.

Carrying out the integration of eq. (16) yields:

$$\frac{dh}{dK_n} = \int_0^t I\ d$$

$$\begin{aligned}
I = \{ & \frac{\Delta z}{12\Delta_e} [b_i(h_1\dot{q}_1+h_4\dot{q}_4) + b_ib_j\ (h_1\dot{q}_2+h_2\dot{q}_1+h_4\dot{q}_5+h_5\dot{q}_4) \\
& + b_ib_k(b_1\dot{q}_3+h_3\dot{q}_1+h_4\dot{q}_6+h_6\dot{q}_4) + b_j(h_2\dot{q}_2+h_5\dot{q}_5) \\
& + b_jb_k(h_2\dot{q}_3+h_3\dot{q}_2+h_5\dot{q}_6+h_6\dot{q}_5) + b_k(h_3\dot{q}_3+h_6\dot{q}_6)] \\
& + \frac{\Delta z}{24\Delta_e} [b_i(h_1\dot{q}_4+h_4\dot{q}_1) + b_ib_j\ (h_1\dot{q}_5+h_5\dot{q}_1+h_2\dot{q}_4+h_4\dot{q}_2) \\
& + b_ib_k(b_1\dot{q}_6+h_6\dot{q}_1+h_3\dot{q}_4+h_4\dot{q}_3)+ b_j(h_2\dot{q}_5+h_5\dot{q}_2) \\
& + b_jb_k(h_2\dot{q}_6+h_6\dot{q}_2+h_3\dot{q}_5+h_5\dot{q}_3) + b_k(h_3\dot{q}_6+h_6\dot{q}_3)] \\
& + \frac{\Delta z}{12\Delta_e} [c_i(h_1\dot{q}_1+h_4\dot{q}_4) + c_ic_j(h_1\dot{q}_2+h_2\dot{q}_1+h_4\dot{q}_5+h_5\dot{q}_4) \\
& + c_ic_k(b_1\dot{q}_3+h_3\dot{q}_1+h_4\dot{q}_6+h_6\dot{q}_4) + c_j(h_2\dot{q}_2+h_5\dot{q}_5) \\
& + c_jc_k(h_2\dot{q}_3+h_3\dot{q}_2+h_5\dot{q}_6+h_6\dot{q}_5) \\
& + c_jc_k(h_2\dot{q}_3+h_3\dot{q}_2+h_5\dot{q}_6+h_6\dot{q}_5)+ c_k(h_3\dot{q}_3+h_6\dot{q}_6)] \\
& + \frac{\Delta z}{24\Delta_e} [c_i(h_1\dot{q}_4+h_4\dot{q}_1) + c_ic_j(h_1\dot{q}_5+h_5\dot{q}_1+h_2\dot{q}_4+h_4\dot{q}_2) \\
& + c_ic_k(b_1\dot{q}_6+h_6\dot{q}_1+h_3\dot{q}_4+h_4\dot{q}_3) + c_j(h_2\dot{q}_5+h_5\dot{q}_2) \\
& + c_jc_k(h_2\dot{q}_6+h_6\dot{q}_2+h_3\dot{q}_5+h_5\dot{q}_3) + c_k(h_3\dot{q}_6+h_6\dot{q}_3)] \\
& + \frac{\Delta_e}{6\Delta z} [(h_4-h_1)\ (\dot{q}_4-\dot{q}_1) + (h_5-h_2)\ (\dot{q}_5-\dot{q}_2) + (h_6-h_3)\ (\dot{q}_6-\dot{q}_3)] \\
& + \frac{\Delta_e}{12\Delta z} [(h_4-h_1)\ (\dot{q}_5-\dot{q}_2) + (h_5-h_2)\ (\dot{q}_4-\dot{q}_1) + (h_4-h_1)\ (\dot{q}_6-\dot{q}-3) \\
& + (h_6-h_3)\ (\dot{q}_4-\dot{q}_1) + (h_5-h_2)\ (\dot{q}_6-\dot{q}_3) + (h_6-h_3)\ (\dot{q}_5-\dot{q}_2)]\}
\end{aligned} \qquad (17)$$

If several elements can be characterized by the same conductivity K_n (i.e. number of zones differs from number of elements), Equation (17) can be applied by integrating over the zone.

EXAMPLE

The following example is used to demonstrate the accuracy and the computational efficiency of the proposed scheme. A hypothetical aquifer (Figure 1) is 300 x 300 m in the x-y plane with a thickness of 30 m. The top and lower boundaries are impervious. No flow conditions are imposed on boundaries at y=0 and y=300 m, while the boundaries at x=0 and x=300 m are constant head of 100 m. There is a complete penetration well located at the center of the aquifer, (x, y) = (150m, 150m), with a pumping rate of 15000 m^3/day. There is a monitoring well located at (x, y) = (120m, 60m) and taking samples only at (x, y, z) = (120m, 60m, 30m). The aquifer is at a steady state with an initial head of 100 m before the start of pumping.

Assume the hydraulic conductivity of the aquifer can be grouped into five different zones. Within each zone the hydraulic conductivity is homogeneous and isotropic. The question is what the proper value of the parameter of each zone is. A common practice is to formulate the parameter identification problem as a nonlinear least-squares optimization problem and then to apply a gradient search technique to solve the optimization problem. When a gradient method is applied, it is required to evaluate the sensitivity coefficient matrix repeatedly. The success of such an inverse solution technique depends on the accuracy and efficiency with which the sensitivity coefficient matrix is calculated.

In this example, assume the initial estimate of parameters are K_1 = 20 m/day, K_2 = 16 m/day, K_3 = 22 m/day, K_4 = 22 m/day, and K_5 = 10 m/day. When the influence coefficient or the sensitivity equation methods are applied to calculate the sensitivity coefficient matrix, the required number of simulation runs is the number of parameters + 1, which is 6 in this case. While the number of simulation runs required by the proposed method is the number of observation locations + 1, which is 2. Therefore, in each iteration step, the proposed method will save four simulation runs when compared to the other two methods.

To numerically demonstrate the accuracy of the proposed method, the calculated results are compared to those obtained by the influence coefficient method with carefully chosen perturbation factors. Table 1 indicates that almost identical results were obtained by both methods.

TABLE 1. The Sensitivity Coefficient Matrix, t = 0.2 day

Methods	Parameter Zones				
	1	2	3	4	5
Variational	0.0021	0.0051	-0.021	0.0024	0.0017
Influence Coefficient	0.0021	0.0051	-0.022	0.0024	0.0017

CONCLUDING REMARKS

An efficient and effective method has been proposed to calculate the sensitivity coefficients. The advantages of the method are: 1) The sensitivity coefficent matrix can be accurately calculated by the method; and 2) When the number of observation locations is less than the number of parameters, which is always the case, the proposed method will achieve computational efficiency. However, since numerical integration over the time domain is required by the method, in order to obtain accurate results, the head and adjoint variable at every time step have to be stored. This may limit the application of the method to a large scale system over a long simulation period if the computer storage is a concern.

Although the developed code can be used to perform 3-D groundwater simulation and calculate the uncertainty of performance measures (Sykes et al, 1985[14]), the main purpose of the development in this paper is to be used as a module for parameter estimation by incorporating an inverse code (Lu et al, 1988). The inverse method can be roughly described as a means for using lognormal a priori information in a standard least-squares algorithm. The underlying theorem is Bayes' theorem. The approach emphasizes the sequential nature of the problem. The methodology is especially suitable for a large number of groundwater parameter estimations for which a limited number of measurements may exist.

REFERENCES

1. Babu, D.K. and Pinder, G.C.F. (1984). A Finite Element-Finite Differences Alternating Direction Algorithm for Three Dimensional Groundwater Transport, pp. 165-174, Proceedings of the 5th Int. Conf. on Finite Elements in Water Resources, New York, 1984.

2. Huyakorn, P.S., Jones, B.G. and Andersen, P.F. (1986), Finite Element Algorithms for Simulating Three-Dimensional Groundwater Flow and Solute Transport in Multilayer Systems, Water Resources Research, Vol. 22(3), pp 361-374.

3. Sun, N.Z. and Yeh, W.W-G. (1983), A Proposed Upstream Weight Numerical Method for Simulating Pollutant Transport in Groundwater, Water Resources Research, Vol. 19(6), pp. 1489-1500.

4. Wang, C., Sun, N.Z. and Yeh, W.W-G. (1986), An Upstream Weight Multiple Cell Balance Finite Element for Solving 3-D Convection-Dispersion Equations, Water Resources Research, Vol. 22(11), pp. 1575-1589.

5. Wang, C. (1986). A Three-Dimensional Finite Element Model Coupled with Parameter Identification for Aquifer Solute Transport, Ph.D. Dissertation. University of California, Los Angeles.

6. Li, J., Lu, A.H., Sun, N.Z. and Yeh, W.W-G.(1986), A Comparative Study of Sensitivity Coefficient Calculation Methods in Groundwater Flow, pp. 347-358, Proceedings of the 6th Int. Conf. on Finite Elements in Water Resources, Lisbon, Portugal, 1986.

7. Wang, C. and Yeh, W.W-G. (1986). Sensitivity Analysis of an Aquifer Solute Transport Model, paper presented at the American Geophysical Union Fall Meeting, San Francisco.

8. Lu, A.H., Schmittroth, F. and Yeh, W.W-G. (1988), Sequential Estimation of Aquifer Parameters, submitted to Water Resources Research.

9. Chavent, G.(1975), History Matching by Use of Optimal Control Theory, Soc. Petroleum Engr. J., Vol. 15(1), pp. 74-86.

10. Carter, R.D., Kemp, L.F. Jr., Pierce, A.C. and Williams, D.L. (1974), Performance Matching With Constraints, Soc. Petroleum Eng. J., Vol. 14(2), pp. 187-196.

11. Chen, W.H., Gavalas, G.R., Seifeld, J.H. and Wasserman, M.L.(1974), A New Algorithm Automatic History Matching, Soc. Petroleum Engr. J., Vol. 14, pp. 593-608.

12. Bear, J. (1972). Dynamics of Fluids in Porous Media, Elsevier. New York.

13. Bear, J. (1979). Hydraulics of Groundwater, McGraw-Hill. New York.

14. Sykes, J.F., Wilson, J.L. and Andrews, R.W. (1985), Sensitivity Analysis for Steady State Groundwater Flow Using Adjoint Operators, Water Resources Research, Vol. 21(3), pp. 359-371.

Identification of IUH Ordinates Through Non-Linear Optimization

J.A. Raynal Villasenor and D.F. Campos Aranda
Water Resources Program, Universidad Nacional Autonoma de Mexico, 04510 Mexico, D.F., Mexico

ABSTRACT

The process of identification of ordinates of the instantaneous unit hydrograph is an important step in its application to forecast river flows. The process itself has several interesting features, like the appearence of the overestimation condition, that is when more equations than variables exist, undesirable outcomes like negative ordinates and oscillations. To avoid these problems, several approaches based in optimization techniques have been proposed in the literature using linear, quadratic and non-linear optimization. The latter approach is used to construct the procedure of identification of IUH ordinates. A comparison of the proposed approach is made with some options available in the literature.

INTRODUCTION

The spread use of the instantaneous unit hydrograph as an essential component of real time river flows forecasting models and due to the generalized use of computers to perform the required operations to produce forecast hydrographs, the phase of identification of ordinates of the instantaneous unit hydrograph (IUH) becomes a very important step of the whole process. Often this identification stage produces undesirable results like negative values and oscillations on the ordinates of the IUH, which are difficult to explain and to back up from the physical basis of the phenomenon.

Some efforts have been made to avoid such problems, Deininger[1] proposed two approaches based in linear programming, namely the MINISAD (minimize the sum of absolute deviation) and MINIMAD (minimize the maximum absolute deviation) methods for identification of IUH ordinates. More recently, Mays and Taur[3] have applied a non-linear optimization code to perform the cited identification stage.

THE MINIMSE APPROACH TO IDENTIFY IUH ORDINATES

The discrete form of the convolution between effective rainfall and direct runoff is, Morel-Seytoux[5]:

$$q(n) = \sum_{\nu=1}^{n} \delta(n - \nu + 1)\, r(\nu) \;;\; n = 1,2,... \quad (1)$$

where q(n) is the direct runoff hydrograph ordinate at time n, $\delta(.)$ is the IUH ordinate and r(.) is the mean effective rainfall rate. In Eq. (1), q(.) and r(.) must have consistent units.

From Eq. (1), the error of the river flow forecast, e(n) is:

$$e(n) = q^{\circ}(n) - \sum_{\nu=1}^{n} \delta(n - \nu + 1)\, r(\nu) \;;\; n = 1,2,... \quad (2)$$

and the mean square error of the river flows forecast is:

$$MSE(e) = \left\{\sum_{n=1}^{N} \frac{e^2(n)}{N}\right\}^{1/2} \quad (3)$$

where MSE(e) is the mean square error of the river flows forecast and N is the number of ordinates of the direct runoff hydrograph.

Now, setting the problem in an adequate form to be handled by an optimization technique, the following objective function is used:

$$\min_{\underline{\delta}}(MSE(e)) = \min_{\underline{\delta}} \left\{\sum_{n=1}^{N} \frac{e^2(n)}{N}\right\}^{1/2} \quad (4)$$

subject to the following constraints, Deininger[1]:

a) Non-negativity constraints:

$$\delta_i \geq 0 \;;\; i = 1,..., M \quad (5)$$

where M is the memory time

b) Ordering constraints, Deininger[1]:

$$\delta(1) \leq (2) \leq ... \leq \delta(p) \quad (6)$$

$$\delta(p) \geq \delta(p+1) \geq ...$$

where $\delta(p)$ is the ordinate of the peak of the IUH.

The optimization problem has been tackled by using the well-known Rosenbrock method for constrained multivariables,

Küester and Mize[4].

COMPARISON WITH OTHER APPROACHES

The proposed MINIMSE method was tested with other procedures, namely MINISAD and MINIMAD.

The proposed approach was applied to the data contained in Deininger[1] and the resulting IUH with those obtained with the MINISAD and MINIMAD methods are contained in Table 1, the corresponding direct runoff hydrographs are depicted in Table 2. Both are displayed in Fig. 1.

TABLE 1

Instantaneous Unit Hydrograph Ordinates

Time (min)	Method		
	MINISAD	MINIMAD	MINIMSE
1	0.100	0.050	0.051
2	0.170	0.118	0.114
3	0.190	0.208	0.206
4	0.160	0.208	0.202
5	0.050	0.014	0.071
6	0.010	0.014	0.011
7	0.010	0.014	0.009
8	0.010	0.014	0.006
9	0.010	0.014	0.003
10	0.010	0.014	0.003
11	0.010	0.014	0.003

The corresponding final values of the objective functions considered for each case and the computation of the values of the other schemes are contained in table 3. From this data and for this example, it is clear that the MINIMSE method achieves the best overall results giving the least Mean Square Error (MSE), a very near value to the minimum Maximum Absolute Deviation (MAD) and the second value of the Sum of Absolute Deviations (SAD).

TABLE 2

Real and Forecast Direct Runoff Hydrographs

Time (min)	Mean Effective Rainfall (cm/hr)	Direct Runoff Hydrographs (cm/hr)			
		Real	MINIMSE	MINISAD	MINIMAD
1	9.65	0.13	0.50	0.96	0.43
2	11.18	0.88	1.66	2.75	1.64
3	14.48	2.52	3.98	5.14	3.95
4	13.21	5.16	6.55	7.38	6.60
5	19.56	9.32	8.39	9.14	7.88
6	15.24	10.45	9.50	10.25	9.02
7	13.72	10.83	10.33	10.63	9.70
8	14.48	10.96	10.60	10.73	10.15
9	13.97	10.71	10.00	10.28	9.30
10	7.37	9.19	9.24	9.40	8.99
11	2.03	6.80	8.21	7.83	8.18
12	1.02	4.66	6.12	5.59	6.10
13	1.52	3.27	3.58	3.43	3.58
14		2.14	1.84	2.01	2.19
15		1.51	1.13	1.39	1.76
16		1.13	0.71	0.96	1.28
17		0.88	0.33	0.60	0.76
18		0.63	0.15	0.40	0.58
19		0.50	0.10	0.25	0.38
20		0.38	0.05	0.12	0.18
21		0.25	0.03	0.05	0.08
22		0.13	0.00	0.03	0.025
23		0.08	0.00	0.03	0.025

TABLE 3

Comparison of Objective Functions Value

CRITERION	Method		
	MINISAD	MINIMAD	MINIMSE
SAD	12.90	14.83	14.00
MAD	2.62	1.44	1.46
MSE	0.90	0.87	0.75

SAD: Sum of absolute derivations

MAD: Maximum absolute deviation

MSE: Mean square error

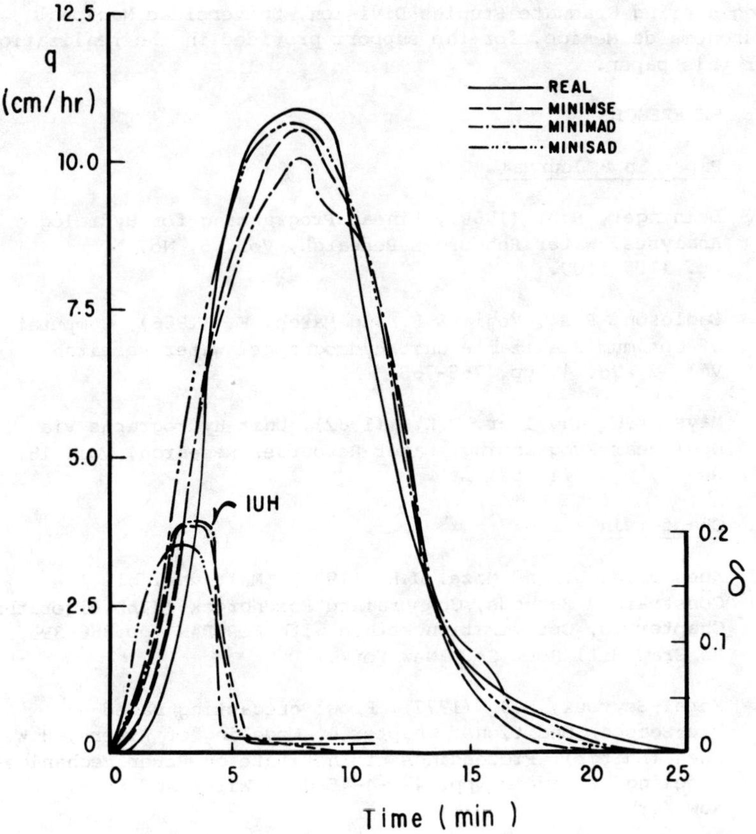

Figure 1. Direct Runoff and Instantaneous Unit Hydrographs

CONCLUSIONS

A procedure to identify the ordinates of the instantaneous unit hydrograph, based in non-linear optimization, has been presented. The proposed methodology has some attractive features, as: simplicity on problem formulation and computer code design, which eases the application of the methodology to solve real problems and consistency of results when it is compared with other options. The authors strongly support their application.

ACKNOWLEDGEMENTS

The authors wish to express their deepest gratitude to the

Engineering Graduate Studies Division, Universidad Nacional Autonoma de Mexico, for the support provided in the realization of this paper.

REFERENCES

Paper in a Journal

1. Deininger, R.A. (1969), Linear Programming for Hydrologic Analyses, Water Resources Research, Vol. 5, No. 5, pp. 1105-1109.

2. Eagleson, P.S., Mejia-R,R. and March, F. (1966), Computation of Optimum Realizable Unit Hydrographs, Water Research, Vol. 2, No. 4, pp. 755-765

3. Mays, L.W. and Taur, C.K., (1982), Unit Hydrographs via Nonlinear Programming, Water Resources Research, Vol. 18, No. 4, pp. 744-752.

Chapter in a Book

4. Kuester, J.L. and Mize, J.H. (1973), Multivariable Constrained Methods, Constrained Rosenbrock (HILL Algorithm), Chapter 10, Optimization Method with FORTRAN, pp.386-398, Mc Graw Hill Book Co., New York.

5. Morel-Seytoux, H.J. (1977), Flow Forecasting Based on Pre-Season Conditions, Chapter 4, Modeling of Rivers, H.W. Shen (Editor), Proceedings of Institute on River Mechanics-Modeling of Rivers, pp. 4-1-4-45 John Wiley and Sons, New York.

SECTION 3C - OPTIMIZATION

Numerical Aspects of Simulation and Optimization Models for a Complex Water Resources System Control

M. Baošić

Jaroslav Černi Institute, Belgrade, POB 530, Yugoslavia

B. Djordjević

Faculty of Civil Engineering, University of Belgrade, Yugoslavia

ABSTRACT

A very complex (both in a hydraulic and a numerical sense) simulation model aiming at the effective control of a water resources system (WRS) is designed.The non -standard problems of an unsteady flow with increased number of discharge changes within a WRS (tributaries inflows,pumping stations for irrigation and drainage) are solved using this flow in open channels modelling theory (Preissman scheme,Cholesky scheme etc).

INTRODUCTION

A chain of mathematical models (MM) for the simulati= on, estimation and optimization, using decomposition of WRS according to its function,space,time and nume= rical aspects has been developed to control a complex and large WRS.The control is realized by the weirs constructed on the rivers and channels.The flood control is the most delicate purpose of this multi= purpose WRS.It is performed by the synchronous operation of the weirs which direct water according to a complex hydraulic scheme towards various recipi= ents depending on the hydraulic situation.The aim is to evacuate most effectively flood water and the water pumped out of the drainage systems.

The highest operational demands within the chain of mathematical models (MM) are set for the unsteady flow model.In order to determine the optimal control of the weirs used to minimize the levels on the cri= tical sectors,it is necessary to make quick iterative calculations of propagation for various weirs maneuv=

res.The numerical contributions made during the deve= lopment of the MM considerably improved the effectiv= eness of the MM which is the subject of this paper.

NUMERICAL ASPECTS

The unsteady flow models are based on the solution of complete St.Venant's differential equations which could be written as follows:

$$\Delta(Q, C_{Q1}) + \Delta(Z, C_{Z1}) = 0$$

$$\Delta(Q, C_{Q2}) + \Delta(Z, C_{Z2}) = 0$$

where:

$$C_{Q1} = \{1, 0, 0, 0\}$$

$$C_{Z1} = \{0, B, 0, q\}$$

$$C_{Q2} = \left\{\frac{2Q}{A^2}, \frac{1}{A}, W, 0\right\}$$

$$C_{Z2} = \{g, 0, 0, 0\}$$

when:

$$\Delta(\varsigma, C) = C_1 \frac{\partial \varsigma}{\partial x} + C_2 \frac{\partial \varsigma}{\partial t} + C_3 \cdot \varsigma + C_4$$

$$C = \{C_1, C_2, C_3, C_4\}$$

$$W = -\frac{Q}{A^3} \cdot \frac{\partial A}{\partial x} + g \frac{|Q| \cdot n^2}{A^2 . R^{4/3}}$$

Solving this system of partial differential equations actually means the determination of the dependent variables Z and Q for the points defined in advance in space (the chosen cross-sections in longitudinal direction x) and time (the chosen constant interval of time),providing that the system of equations be satisfied.The required values Z (level) and Q (disch= arge) represent the control components within the MM.

This system of partial equations was solved using the Preissmann s finite increments implicit method with discretization,having the following shape:

$$\varsigma = \Psi [\theta \cdot \varsigma_{i+1}^{j+1} + (1-\theta) \cdot \varsigma_{i+1}^{j}] + (1-\Psi) [\theta \cdot \varsigma_i^{j+1} + (1-\theta) \varsigma_i^j]$$

$$\frac{\partial \varsigma}{\partial x} = \frac{\theta}{\Delta x} (\varsigma_{i+1}^{j+1} - \varsigma_i^{j+1}) + \frac{1-\theta}{\Delta x} (\varsigma_{i+1}^{j} - \varsigma_i^{j})$$

$$\frac{\partial \varsigma}{\partial t} = \frac{\Psi}{\Delta t} (\varsigma_{i+1}^{j+1} - \varsigma_{i+1}^{j}) + \frac{1-\Psi}{\Delta t} (\varsigma_i^{j+1} - \varsigma_i^{j})$$

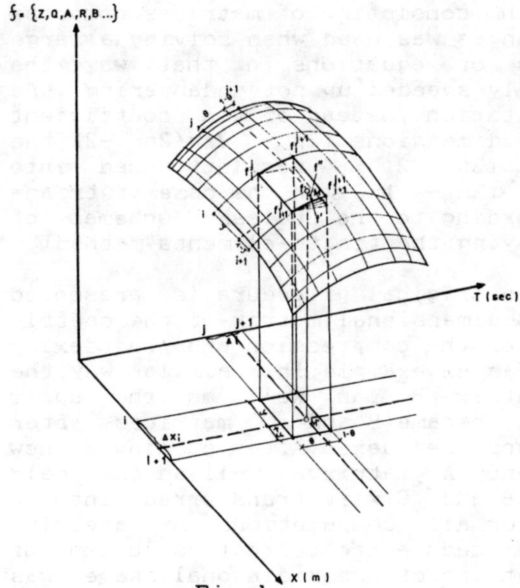

Fig.1

Fig.1 represents three-dimensional quantization of "hydraulic areas" in space (x,t,f) at which nonlinear coefficients in the system of equations are determined. It is represented by a set of inclined iterative areas. The iterative algorithm applied here is an original generalization of Verwey s variant of Preissmann s scheme with weighting coefficients both in time and in space.

The specific feature of this model is a frequent change of the discharge along the flow at the places where tributaries are joining it (Q > o),as well as at water intakes (Q < O) or water inflows (Q > o) within the irrigation system (Fig.2).

Taking into account that the distance between the sections (i) and (i+1) is very small it could be assumed, similarly to the internal boundary conditions scheme, that:

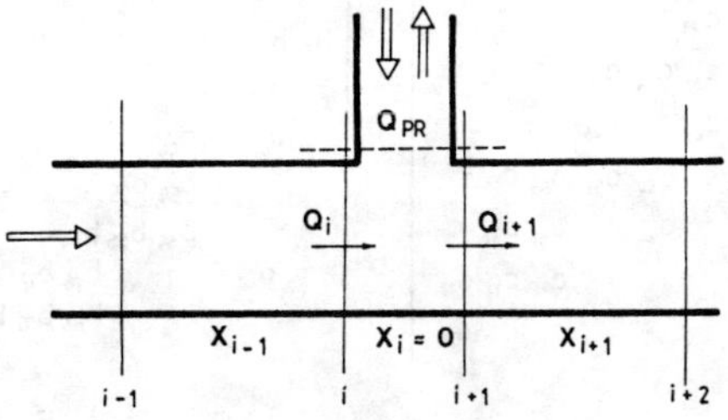

Fig.2

$$Q_{i+1}^{j+1} = Q_{i}^{j+1} + Q_{PR}^{j+1} \quad ; \quad Z_{i+1}^{j+1} = Z_{i}^{j+1}$$

The values of Qpr(t) could be defined by the dependences Qpr = F(t),discretely given,namely:

a) Qpr > o for the case of tributaries and drainage,
b) Qpr < o for the case of irrigation.

The unfavourable circumstance lies in a fact that the control purpose of this model requires,in a certain moment of flood approaching,calculations to be repeated several times.This is why a special attention was paid to the effectiveness of the model.

An original algorithm.consisting of matrices compres= sions and index changes was used when solving a large number of systems of equations.In that way the process was extremely speeded up not endangering the correctnessof computation.Instead of the coefficient matrix having the dimensions (2n - 2)x(2n -2),the coefficient matrix (2n - 2) x 4 was introduced into the computation,and (2n - 2)x3 (in the case of trian= gle matrices according to the Cholesky schemes of equations when applying the finite-elements method).

The essence of this original procedure is presented in Fig.3,where an 8-dimensional matrix of the coeffi= cient A ,which after the compression and preindexing became A .was used as an example.In a similar way,the lower triangle matrix B ,as well as the upper triangle matrix C became B and C matrices after being compressed and preindexed.After gaining a new shape,the coefficient A matrix as well as the cal= culated matrices B and C were transferred into a computer and used for all computations. By applying a simple program procedure,the correct solution of the system of equations of quadridiagonal shape was easily obtained.The procedure was separately tested which gave very good results.

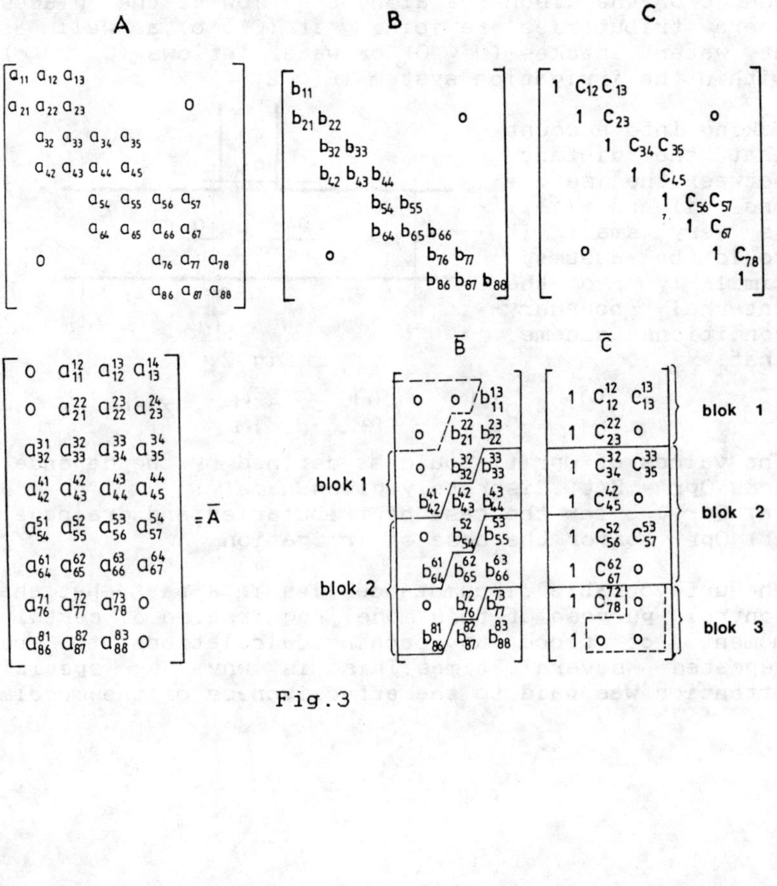

Fig.3

The model output results are presented in Fig.4 - 7, as a threedimensional graphic interpretation. This makes the decision making control much easier. The first case represents a flood control period when water is intensively released from the WRS even through the inlet weir (weir Botoš) of this subsystem (Qinp < o) (Fig.5). The second case is the one when the inlet weir is closed (this weir separates two subsystems) and the whole discharge from all the cut waterflows and irrigation channels directed towards the outlet (Qinp = o) (Fig.6). The third case is when the flood control in a recipient is alleviated by directing water from one river through the inlet weir towards another river, i.e. outlet recipient (Qinp > O) (Fig.7).

This mathematical model (MM) is very efficacious(ope= rative) so that it processes one variant for about 2o seconds on the Digital - VAX computer. It is one of the most operative models of the kind.

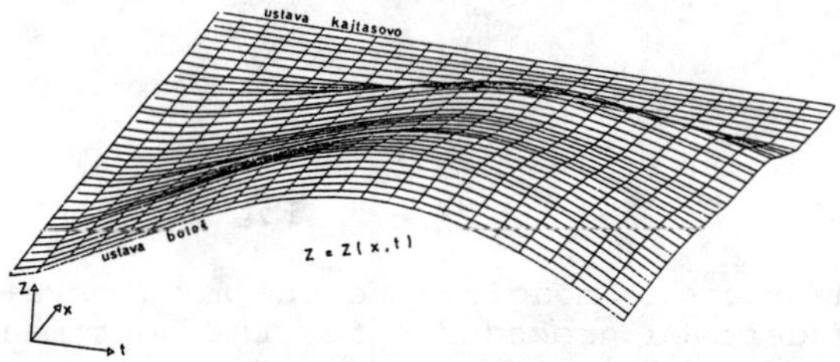

Fig.4

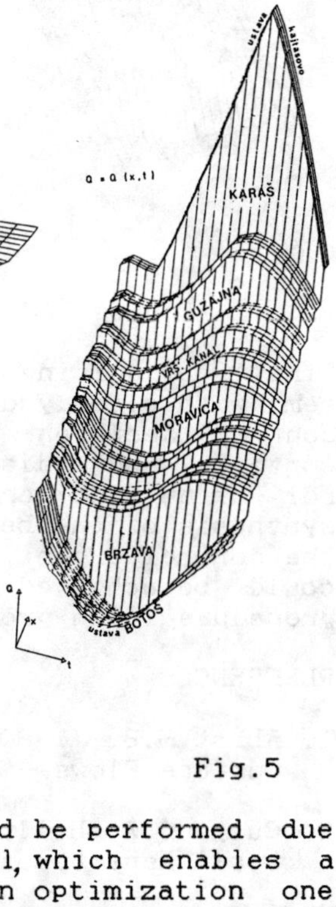

Fig.5

A large number of iterations could be performed due to the effectiveness of this model, which enables a simulation model to develop into an optimization one if the criterion for the control evaluation be intro= duced, (e.g. level minimization on the critical sec=

tors, damage minimization, etc.), the most effective control of the weirs could be achieved in accordance with the chosen criterion for the evaluation of the control actions.

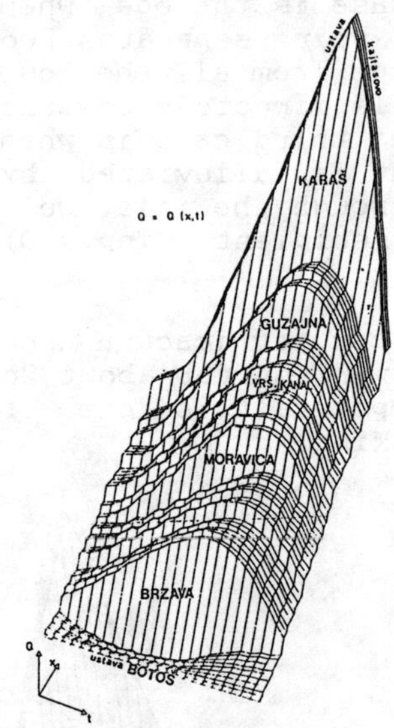

Fig.6

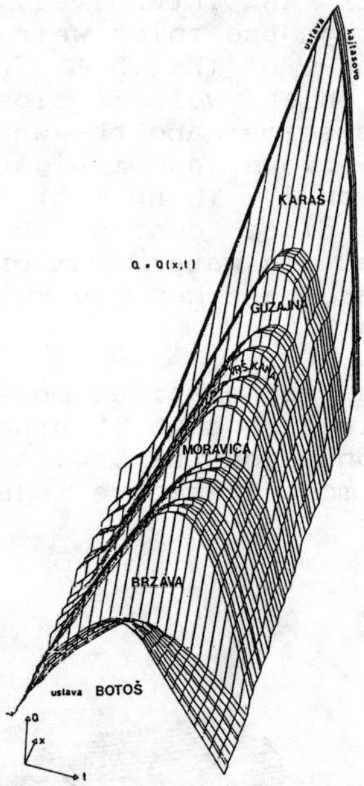

Fig.7

Finally, by uniting all the models into one model-sys= tem with clearly defined procedures for the optimal control decision - making and for establishing a continual and undisturbed dialogue between the cont= rol organ and the model (the model represents a synthesis of the best solutions and proposes them to the control organ), the highest level of control model could be achieved by creating an EXPERT SYSTEM. The procedure is in progress.

REFERENCES

1. Abbot M.B. (1980). Elements of the Theory of Free Surface Flows - Computational Hydraulics. London.

2. Cunge J.A., Holly F.M. and Verwey A. (1980). Prac= tical Aspects of Computational Hydraulics. London.

3. George A. and Liu J. (1981). Computer Solution of Large Sparce Positive Definite Systems. New Jersey.

Optimal Operation of a Reservoir System with Network Flow Algorithm

P.B. Correia and M.G. Andrade Filho
Universidade Estadual de Campinas - UNICAMP, Faculdade de Engenharia Elétrica, Departamento de Engenharia de Sistemas, Cx. Postal 6101 - CEP 13081, Campinas, SP, Brazil

Abstract

This paper treats the problem of optimal operation of a reservoir system with multiple use of the water in the state of São Paulo. The operation of this system needs to consider conflicting objectives. These are: water supply, electrical energy production, irrigation and minimum and maximum out flow constraints due to water quality and flood control problem respectively.

The problem is formulated as a network flow optimization model with a parametrical objective function. The solution consists of reservoir optimal daily operation rules and the curve representing the trade-off between water supply and electrical energy production. We have also analysed the water quality due to system optimal operation.

INTRODUCTION

The problem of multiple use of water has been widely discussed in the past few years. Vedula and Rogers [7] study the problem of irrigation planning in the development of hydrographic basin, formulating a multiobjective problem as a model of linear programming. Nakamura and Rilley [6] explore the structure of the multiobjective problem of water treatment,adapting the single objective Branch and Bound method to the multiobjective situation. Yeh and Becker [8] use a modified algorithm of linear and dynamic programming to determine the optimal operation of multiple reservoir system with five objective fuctions.

In this paper we will discuss the problem of daily operational optimization of a reservoir system whose objectives are water supply and electric power production. We will also examine minimum and maximum constraints imposed by water quality problems and flood control, respectively. The problem was formulated as a model of network flow optimization, Kennington [4], with a parametric objective function.

This model was applied to a reservoir system responsible for part of the metropolitan São Paulo supply (Cantareira System). The solution obtained consists of the daily optimal reservoir operation and that of each of the hydroelectric plants belonging to the system, in three different hydrological circumstances: dry, intermediate and rainy

We will also present the effect of this operation on the level of the Piracicaba River, emphasizing the water quality problem and will construct the trade-off curve between the objective: water supply and production of electrical power.

MATHEMATICAL MODEL

The problem of multiple use of water under discussion in this article has two objective functions. The first represents the quantity of water destined for supply during the planning time span, Equation (1), and the second represents hydroelectric production, Equation (2).

$$Z_1(a(t)) = \sum_{t=t_0}^{T} a(t) \tag{1}$$

$$Z_2(u_j(t)) = \sum_{t=t_0}^{T} [\sum_{j \in M} \alpha_j u_j(t)] \tag{2}$$

where $a(t)$ is the daily flow of the reservoirs to supply the region, α_j is the average productivity of plant j, $u_j(t)$ is the flow during period t, and M is the index set of hydroelectric plants in the system. adapting the weight factor ω between the objective fuction (1) and (2), we can express the multiple use of water problem as an optimization problem with one sole objective, Equations (3) to (10)

$$\max \quad \sum_{t=t_0}^{T} [\sum_{j \in M} \alpha_j u_j(t) + \omega a(t)] \tag{3}$$

$$\text{s.a} \quad x_j(t+1) = x_j(t) + y_j(t) + d_{j-1}(t) - v_j(t) - a(t) - d_j(t) \tag{4}$$

$$x_j(t+1) = x_j(t) + y_j(t) - u_j(t) - v_j(t) + \sum_{i \in \Omega_j} [u_i(t) + v_i(t)] \tag{5}$$

$$\underline{x}_j \leq x_j(t) \leq \overline{x}_j \tag{6}$$

$$\underline{u}_j \leq u_j(t) \leq \overline{u}_j \tag{7}$$

$$\underline{v}_j \leq v_j(t) \leq \overline{v}_j \tag{8}$$

$$\underline{d}_j \leq d_j(t) \leq \overline{d}_j \tag{9}$$

$$\underline{a} \leq a(t) \leq \overline{a} \tag{10}$$

$$x_j(t_0) \text{ dado}, \ t = t_0, \ t_0 + 1, \ ..., \ T$$

where $x_j(t)$ is the volume cl water stored in the reservoir j during period t, Ω_j is the index set of plants and reservoir, immediately above the plant j, $y_j(t)$ is the incremental inflow to reservoir j during period t, $v_j(t)$ is the outflow of reservoir j during period t, $d_{j-1}(t)$ is the transfer of water from reservoir $j-1$ to reservoir j during period t and $\omega \in [0, \infty)$ is the dimensional factor that serves to weight as well as to integrate the units among the objectives.

The optimal operation of a hydroelectric system can be obtained by network algorithms, Carvalho e Soares [2]. In this way Equations (4) and (5) are represented by a network flow to a plant or reservoir where each node (j, t) corresponds to a plant or reservoir j at t point in time. The problem (3) - (10) can be resolved for various values of ω. The algorithm used can be summarized in the following steps:

begin
initiate make $\omega_k = 0$;
repeat
resolve the problem of network defined by (3) – (10), B_k being the optimal base found;

determine the maximum increment ϵ_k, so that for $\omega_k < \omega \leq \omega_k + \epsilon_k$ the B_k base remains optimal;

if $\epsilon_k < \infty$ **then** make $\omega_{k+1} = \omega_k + \epsilon_k$;
until $\epsilon_k = \infty$
end

A CASE STUDY

The region under consideration in the article is considered, in terms of water quality, one of the most critical in the State of São Paulo. A detailed description of the subject subzone can be found in DAEE-[3] and MME/DNAEE-[5]

The complex of the hydroelectric plants have a 47.8 MW maximum production and the reservoir corresponds to 1000 hm^3 stored capacity. The minimum water supply demand was established in this paper at 23 m^3/s during the rainy and intermediate seasons and 15 m^3/s during the dry season.The maximum demand considered was 60 m^3/s, imposed by the limitations of the capitation canals.

We have also considered the flow constraints of 40 m^3/s for the Piracicaba River, which corresponds to the minimum necessary to maintain the dissolved oxygen level at its lowest limit of 3 mg/l, Canejo e Carneseca [1], and the maximum flow of 600 m^3/s to prevent floods.

As a result of the optimal system operation, we observed the behavior of the reservoir illustrated in Figure 1 (a - b - c), where the change in the operation of Jaguari reservoir draws special attention during the intermediate season, when we adopted water supply as a priority. The fluxes of the rivers are shown in Figure 1 (d - e - f), where we can observe that all values are above the critical flow imposed by the water quality problem. The trade-off curve, Figure 2, between

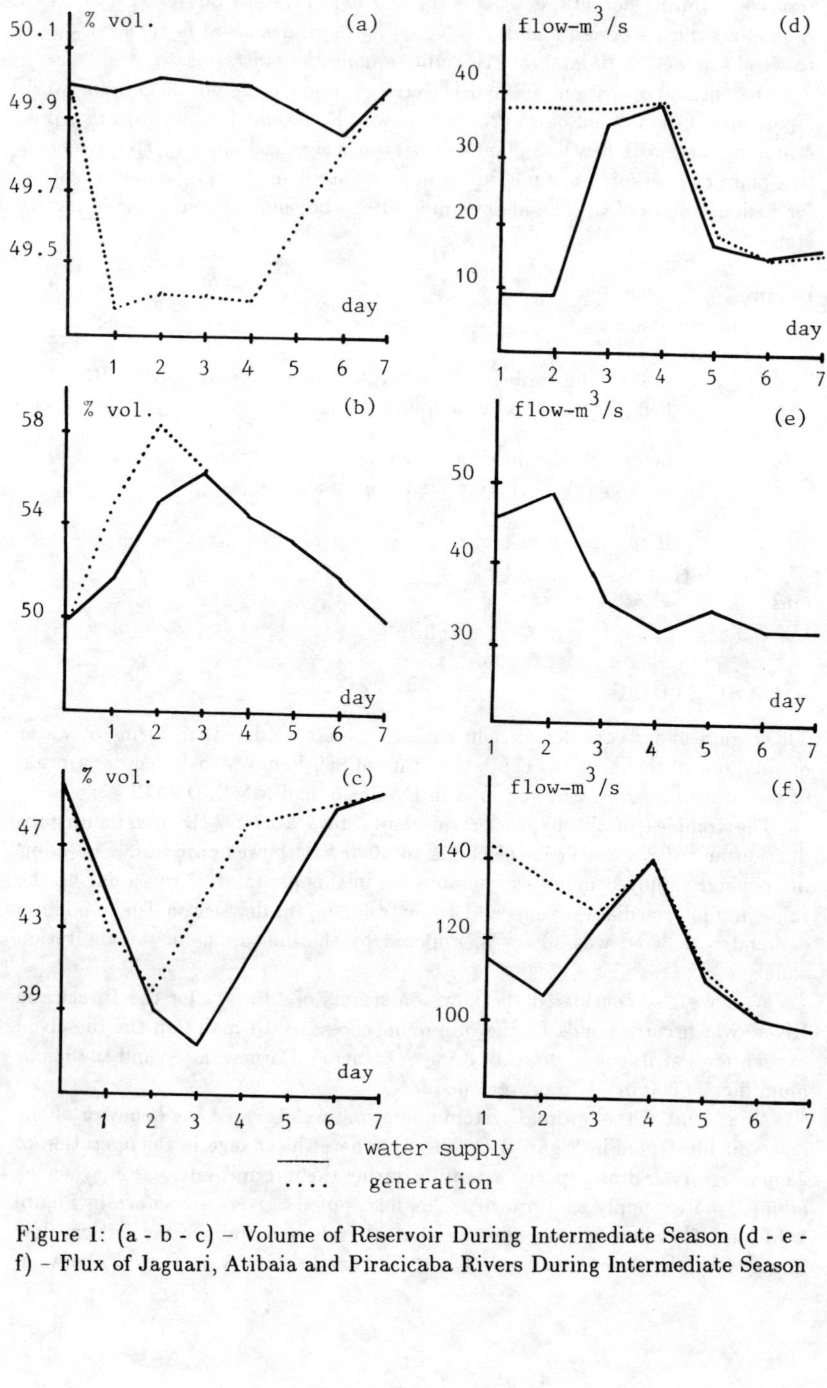

Figure 1: (a - b - c) – Volume of Reservoir During Intermediate Season (d - e - f) – Flux of Jaguari, Atibaia and Piracicaba Rivers During Intermediate Season

water supply and electric power production during the intermediate season shows a variation of 28% in supply versus 17% in production; a decrease of 2.5 m^3/s in supply for every 1 MWh produced.

During the dry and rainy seasons, there is practically no conflict between the objectives. The variations noted were 20% in supply versus 5% in electric power production during the dry season and, during the rainy season, the objectives were totally achieved.

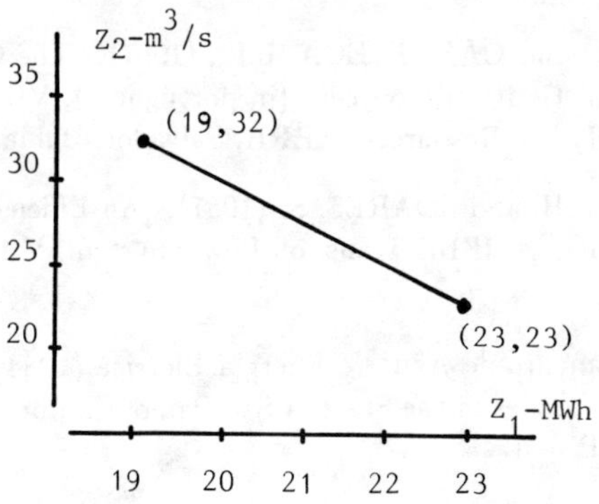

CONCLUSION

In this article we have examined the problem of optimal operation of a reservoir system and hydroelectric plants with multiple use of water. The objectives under consideration are: water supply and electric power production. The model takes into consideration the maximum and minimum flow necessary for flood control, and water quality problems, respectively.

The system was modeled as a problem of optimization of a multiobjective and linear network flow. The original problem is transformed into a monocriterial problem by the weighting method and is resolved as a network optimization problem with an parametrical objectives function.

The model was applied to the Cantareira System, located in the State of São Paulo. This system is made up of three reservoirs that supply water and three hydroelectric plants.

The study was made for a time span of one week. The initial and final volumes of the reservoir were fixed at pre-established values, according to medium term planning objectives. The parametric analysis of the objective function showed the variation in the operation of the reservoir as the priority for supply grew. We also evaluated the trade-off between the supply of water and the production of electric power. The study was conducted throughout three distinct hydyological periods (dry, intermediate and rainy) and we observed conflict only during intermediate season. During the rainy season, the objectives are totally attained and, during the dry season, the maximum water supply demand that the system is capable of, without infringing the minimum outflow restrictions, is 15 m^3/s. Another important result obtained was that, in all cases studied, we noted that the fluxes of the rivers were above the critical values necessary to guarantee the quality of the water.

References

[1] CANEJO, J.G.L., and CARNESECA, L.F., (1987), "The Quality of Water in the Planning of Hydric Resources", (in portuguese), VII Brazilian Symp. on Hydrology & Hydric Resources, ABRH, Salvador, Bahia, Brazil.

[2] CARVALHO, M.F.H. and SOARES, S. (1987), "An Efficient Hydrothermal Scheduling Algorithm", IEEE Trans. on Power System, Vol. PWRS-2, No.3, pp. 537-542.

[3] DAEE - Departamento de Águas e Energia Elétrica (1984), "Characterization of Hydric Resources in the State of São Paulo, (in portuguese), RT-095-A04, São Paulo, Brazil.

[4] KENNINGTON, J.L. and HELGASON, R.V. (1980), "Algorithms for Network Programming, John Wiley & Sons, Inc., Texas, USA.

[5] MME/DNAEE/Divisao de Controle de Recursos Hídricos (1985), "Report on Synthesis - Diagnostic and Planning of the Utilization of the Hydric Resources of the Jaguari/Piracicaba Basin", (in portuguese) Brasilia-DF, Brazil

[6] NAKAMURA, M. and RILY, J.N. (1981), "A Multiobjective Branch and Bound Method for Network-Structured Water Resources Planning Problems", Water Resources Research, Vol. 18, No. 5, pp. 1349-1359.

[7] VEDULA , S. and ROGERS, P.P. (1981), "Multiobjective Analysis of Irrigation Planning in River Basin Development", Water Resources Research, Vol. 17, No. 5, pp. 1304-1310.

[8] YEH, W. W-G. and BECKER, L. (1982), "Multiobjective Analysis of Multireservoir Operation", Water Resources Research, Vol. 18, No. 5, pp. 1326-1336.

Optimization of Water Quality in River Basin

I. Dimitrova and J. Kosturkov
Institute of Water Problems, Bulgarian Academy of Sciences, Sofia 1113, Bulgaria

SUMMARY

An alternative management scheme for water quality maintenance in river basin is presented. The twofold basic planning problems usually are to determine the desired level of water quality and to develop a waste management program. In this paper the optimization problem is formulated. The objective is to minimize the sum of the waste removal costs for some dischargers. This is carried out in terms of determining an optimal design procedure for the individual plants and then optimizing the overall treatment costs with constraint of stream quality preservation. Because of water deficit in the experimental river basin in low water period, the model gives the possibility to control the river discharge. The separable programming for solving the optimization problem is used. The results from the optimization are given.

INTRODUCTION

The pollution of rivers has sharply increased during the last several decades under the influence of anthropogenous and technogenous factors. The maintenance of their water quality within the standard is carried out by regulating waste water quality and quantity discharged in. This may be achieved in different ways, and in the first place by waste water treatment, improvement of treatment methods, implementation of water recycling in water supply and wasteless technologies, use of waste water, etc. In addition, maintenance of water quality within the standard may be carried out by increasing river selfpurification capacity, namely by increasing water quantity in low water periods, dam building, artificial aeration, water transfer, etc. River pollution is mostly reduced by waste water treatment in waste water treatment plants. Thus the problem related to finding out a combination of stages in treatment of waste water by different dischargers. Such a combination is to ensure a water quality required by the standard in the river at minimum total costs in part of or the whole river basin.

FORMULATION OF THE PROBLEM

According to figure 1 the part of the river basin considered is the following: q_1=0.53m^3/s are discharged for flowing in the river from a dam built for industrial water supply, irrigation and recreation purposes. The river considered is small, flowing in a densely populated industrial area and subject to intensive pollution. During the low water-level period river flow is considered totally formed by waste water and a run-off from the dam provided for flowing in the river. There are five waste discharges downstream's of the dam: q_2=4.64m^3/s, q_3=3.50m^3/s q_4=2.31m^3/s, q_5=0.70m^3/s and q_6=0.48m^3/s. Waste water quality of each waste discharger is shown in table 1. The six parameters i.e. solubles, chemical oxygen demand (COD), biological oxygen demand (BOD_5), nitrogen, phosphorus and oil products are exceeding the standard specified for the river under consideration. The following technological processes are selected for treating the waste water of dischargers: filtration, precipitation, flotation, biochemical treatment, coagulation, active carbon adsorption, chlorination. The residual concentrations obtained for the different parameters of pollution (i=1..6) after using different technological combinations of treatment as per the treatment methods mentioned above are shown in table 1 for each waste discharger (j=1..6). This table also shows the cost data (total costs S_j) of each technological combination of treatment.

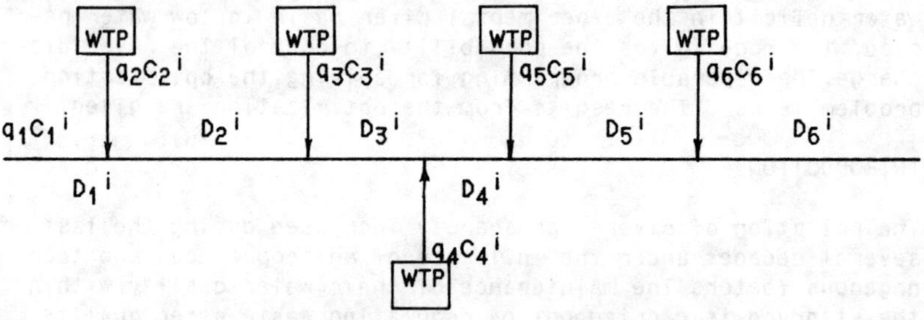

Figure 1. Waste loading situation

The mathematical formulation of the problem related to looking for an optimum combination of stages for waste water treatment of dischargers in a river basin is as follows. Water quality in a river may be characterized by the values of parameters in six control stations, the place of which in a section between two waste dischargers is selected in such a way that concentration ($D_K{}^i$, k=1..6) per parameter should be the maximum one. River water quality meets the sanitary standard ($D_s{}^i$) per parameter (or group of parameters) of pollution in case their values ($D_K{}^i$) do not exceed the standard for each parameter.

$$D_K{}^i \leq D_s{}^i \qquad (1)$$

This concentration ($D_K{}^i$) is the result of a superposition of mixed quantities of i-parameter of pollution of all waste dischargers "j" located upstream.
The transfer coefficient is equal to the ratio between concentration ($D_{Kj}{}^i$ in K-station of i-parameter as discharged by j-waste dischargers and its concentration in the discharger ($C_j{}^i$)

$$T_{Kj}{}^i = \frac{D_{Kj}{}^i}{C_j{}^i} \qquad (2)$$

Ignoring diffusion particularities of each parameter of water quality we accept that the coefficients of transfer are identical for each of them, i.e. $T_{Kj}{}^i = T_{Kj}$. Each coefficient's values T_{Kj} is inversely proportional to the waste discharge dilution.

Table 1. Waste loading, version of treatment and cost data

Version of treatment	Total cost mill. leva/ year	Parameters of waste loading, ppm					
		Insolubles	COD	BOD_5	Nitrogen	Phosphorus	Oils
		j = 1					
-	-	15.5	14.8	2.0	0.4	0.1	0.0
		j = 2					
1	0.0	1073	294.4	211.5	34.8	5.0	52.4
2	15.0	13.7	24.2	10.6	17.4	3.5	10.5
3	16.1	8.8	19.4	9.2	17.4	3.5	8.1
4	17.4	5.1	15.9	5.3	17.4	3.5	3.2
5	25.4	4.0	5,0	3.1	8.4	1.5	1.1
6	33.7	1.8	2.9	1.1	3.0	0.3	0.3
		j = 3					
1	0.0	7.7	92.5	17.4	10.2	0.5	15.9
2	0.8	3.4	60.1	10.5	10.2	0.5	6.5
3	9.7	0.4	9.6	3.7	7.7	0.2	0.8
4	10.3	0.3	4.6	1.7	3.0	0.1	0.2
		j = 4					
1	0.0	40.0	56.4	14.3	12.5	0.9	25.0
2	0.6	3.6	36.7	3.6	12.5	0.9	7.3
3	10.9	1.6	2.8	1.4	8.1	0.3	0.3
4	14.5	0.8	1.8	1.4	3.5	0.1	0.1
		j = 5					
1	0.0	68.0	164.0	73.8	14.6	2.1	3.0
2	5.4	13.4	14.1	3.7	7.3	1.5	1.6
3	8.6	6.4	9.4	2.8	7.3	1.5	0.9
4	15.8	3.4	4.9	2.0	7.3	0.8	0.5
5	18.7	2.2	2.9	1.8	3.1	0.2	0.2
		j = 6					
1	0.0	41.0	37.6	16.7	0.8	1.4	0.0
2	2.4	12.1	28.6	12.8	0.5	1.0	0.0
3	8.6	8.1	10.8	5.0	0.4	1.0	0.0
4	10.7	3.3	6.5	2.5	0.4	0.5	0.0
5	12.3	2.1	3.8	1.0	0.3	0.3	0.0

Determination of these coefficients in practice is made by the theory of mixing.When the information needed is missing,the control stations accept waste into waste outlet,where their full mixing is, or:

$$T_{kj}=\frac{q_j}{\sum_{j=1}^{k} q_j}$$

Thus,these coefficients depend on river water quantity and waste water outlet diffusors design.Therefore,river water quality in the different control stations may be expressed by means of the following matrix equation:

$$\begin{vmatrix} T_{11} & 0 & 0 & 0 & 0 & 0 \\ T_{21} & T_{22} & 0 & 0 & 0 & 0 \\ T_{31} & T_{32} & T_{33} & 0 & 0 & 0 \\ T_{41} & T_{42} & T_{43} & T_{44} & 0 & 0 \\ T_{51} & T_{52} & T_{53} & T_{54} & T_{55} & 0 \\ T_{61} & T_{62} & T_{63} & T_{64} & T_{65} & T_{66} \end{vmatrix} \cdot \begin{vmatrix} C_1{}^i \\ C_2{}^i \\ C_3{}^i \\ C_4{}^i \\ C_5{}^i \\ C_6{}^i \end{vmatrix} = \begin{vmatrix} D_1{}^i \\ D_2{}^i \\ D_3{}^i \\ D_4{}^i \\ D_5{}^i \\ D_6{}^i \end{vmatrix}$$

or $$TxC=D \qquad (3)$$

T_{kj} for k=1..6 and j=1..6 are the transfer coefficients ac cepted as identical for each quality parameter "C"; $C_k{}^i$ for k=1 ..6 are the concentrations which are different for the different parameters i=1..6; $D_k{}^i$ for k=1..6 are concentrations in the control stations which are different for the different parameters i=1..6.

In this way the left part of all limitations (1) characterizing river water qualities will be of kind (3) where the values of $C_k{}^i$ represent a function of total costs (S_j) for waste water treatment of all waste dischargers.

$$C_j=C_j(S_j) \qquad (4)$$

These functions are shown in table 1 where S_j are the costs for waste water treatment of j-discharger. The optimum is looked for by selecting such a vector C at which the system of limitation (1) is fulfilled and the function of the costs is minimized:

$$(S_j)\text{---> min}$$

Such a solution of the problem related to looking for an optimum combination of stages for waste water treatment of different waste dischargers in a river basin is presented by Gordin (1987)/1/.

The problem formulated in this way is a non-linear programmed one. It is solved by means of separable functions as per the method established by Hadley(1964)/2/.The problem mentioned above may be solved in two stages:the first one is to look for an optimum combination of stages of waste water treatment with a view to achieving one parameter of water quality in the river within the standard,and the second one is,to select from the degrees of treatment obtained in the first stage,a combination which will meet all parameters. The problem allows to obtain optimum solution in case of increasing water quantity discharged from the dam and flowing in the river during low water-level periods. In this specific case the value of the water let flow in the river is not included in the total costs because of lack of information, although the optimization model allows their estimation. The problem is solved by means of a computer programme, the solution being:

1-2,2-1,3-1,4-1,5-1
15.0

This recording means that the optimum combination of treatment stages for this area of the river basin is the following: the second version for the first waste discharger, the first version for the second discharger, the first version for the third discharger, the first version for the fourth discharger and the first version for the fifth waste discharger. This being the distribution and the standard of river water quality will be achieved at minimum costs of 15,000,000 leva/year. The results are shown in table 2.

Table 2. Computational Results by Present Study

Parameters	Standard ppm D_s^i	q_1 m^3/s	Optimal version million leva/year
Insolubles	80.0	0.53	1-2,2-1,3-1,4-1,5-1 15.0
COD	70.0	0.53	1-2,2-1,3-1,4-1,5-1 15.0
BOD_5	25.0	0.53	1-2,2-1,3-1,4-1,5-1 15.0
Nitrogen /ammonia/	3.0	0.53	1-6,2-4,3-4,4-5,5-1 77.2
Phosphorus /soluble/	0.5	0.53	1-6,2-4,3-1,4-1,5-1 44.0
Oil products	0.3	0.53	1-6,2-4,3-3,4-3,5-1 63.5
Insolubles, COD, BOD_5, nitrogen, phosphorus, oils	-	0.53	1-6,2-4,3-4,4-5,5-1 77.2
Insolubles, COD, BOD_5, phosphorus, oils	-	13.48	1-6,2-3,3-3,4-1,5-1 54.3
Insolubles, COD, BOD_5, phosphorus	-	4.65	1-6,2-1,3-1,4-1,5-1 33.7

CONCLUSION

The method proposed for establishing an optimum combination of stages of waste water treatment of different waste dischargers in a river basin enables to solve the basic problems related to maintenance of water standard quality required in a river basin at minimum sum of the waste removal costs. An optimum management programme for waste water treatment may be obtained for different river water standard qualities required.

REFERENCES

Chapter in a book

1. Gordin, I.V. (1987), Tehnologicheskie sistemi vodoobrabotki, Dinamicheskaja optimizaczija, Leningrad, Himia.
2. Hadley, G., (1964), Nonlinear and dynamic programming, Addison Wesley Publishing company, Mass., London.

Coupling of Unsteady and Nonlinear Groundwater Flow Computations and Optimization Methods

A. Heckele and B. Herrling

Institute of Hydromechanics, University of Karlsruhe, West Germany

INTRODUCTION

Optimization procedures which couple the numerical calculation of groundwater flow and optimization have been known for some time and applied with success. Gorelick[2] reviews the development and application of such methods. In addition to Gorelick other investigations show that for the optimization of complex groundwater systems with unsteady flow, whose approximation requires a fine discretization, the use of influence functions is superior to other solution formulations. The influence (or answer) functions describe the dependence of the state variables - primarily the groundwater levels - on the decision variables, which in general are composed of unknown inflow and outflow distributions. From the viewpoint of optimization, the influence functions, which completely describe the properties of the groundwater system, can be regarded as physical constraints. These physical constraints together with additional ones (economical or ecological constraints) and the objective function define the groundwater optimization problem. If linear differential equations are used to describe the groundwater flow, then by applying the superposition principle, linear influence functions can be derived. Using the linearity assumption, quite complex groundwater systems have been studied. Examples include those of Maddock[8], Willis[12], Heidari[6] and Lindner and Marotz[7].

Various solution methods are known to incorporate the nonlinearity due to the dependence of the aquifer parameters on the groundwater level. Using the perturbation theory Maddock[9] developed influence functions in form of power series, where in the actual optimization the series were truncated after only a few terms. The groundwater level constraints will not be linear using this method, so that complex optimization algorithms are necessary. Danskin and Gorelick[1] capture the nonlinearity through iterative correction of the aquifer parameters, as is known from conventional flow calculations. The iteration is done for each time step, so that using this simple method, optimization is only possible within a time step. Wanakule[11] et al., in contrast to the previous references, did not explicitly calculate the influence functions. In their nonlinear optimization algorithm they require values of the objective and constrain functions and their derivatives with respect to each decision variable (Jacobian matrix). This method is independent of the type of the state equations and is therefore also used in the optimization of groundwater quality problems (Gorelick[3] et al.) whose basis is the transport equation. The disadvantage of this technique, in addition to the time-consuming evaluation of the Jacobian matrices, is the fact that nonlinear optimization algorithms are required even for problems which are basicly linear in character, except for the dependence of the parameters on the groundwater levels.

Willis and Finney[13] calculate the influence functions recursively. They divide the computational time domain into several optimization intervals with constant values of the decision variables. The influence functions, calculated quasilinearily, are functions which describe the groundwater levels at the end of the optimization interval as a function of the decision variables, active during the time interval, and of the groundwater levels at the beginning of the interval. The nonlinearity is handled by iterative correction of the influence functions after each optimization.

The optimization algorithm explained here is based on the method of Willis and Finney[13]. However, for the influence functions used by the present authors, the groundwater levels are exclusively functions of the decision variables, and the optimization problem can be simply reduced to an equation system which only contains decision variables. The calculation of the influence functions are accomplished almost completely with algorithms known from the numerical calculation of groundwater levels.

BASIC EQUATIONS

The optimization problem for plane horizontal groundwater flows can be defined by the following equations and inequalities:

objective function: $$Z(e,h) \rightarrow \min|\max \tag{1}$$

constraints: $$g_T(e,h,x_i,t) \begin{smallmatrix}\le\\=\\\ge\end{smallmatrix} 0 \tag{2}$$

differential equation describing the groundwater flow:

$$e + \frac{\partial}{\partial x_i}\left(T_{ij}\frac{\partial h}{\partial x_j}\right) + q^* = S\frac{\partial h}{\partial t} \tag{3}$$

boundary conditions: $$h - \bar{h} = 0 \tag{4}$$

$$q_i n_i + \bar{q} = 0 \tag{5}$$

initial conditions: $$h^o = h(t_o) \tag{6}$$

for i,j = 1,2 and T = 1,2,...,NR (NR: number of constraints). The Einstein summation convention involving a summation over identical indices is used in the above equations. The two unknowns of the problem are the state variable h, which describes the time and space dependent distribution of the groundwater levels throughout the flow domain, as well as the decision variable e, which characterize the optimal source and sink distribution. Using linear optimization methods, Eqs. (1) and (2) must be linear functions of the two unknowns. The groundwater flow system is described by Eq. (3) together with Eqs. (4) - (6). Eq. (3) is, in the general form for flows in a phreatic aquifer, nonlinear due to the dependence of the transmissivity tensor T_{ij} or of the specific yield S on the groundwater level h. The term q*, in contrast with e, represents the fixed distribution of sources and sinks. In the boundary condition Eqs. (4) and (5), $\bar{h}$ is the given groundwater level, $\bar{q}$ the given inflow perpendicular to the domain boundary, q_i the flow vector and n_i is the unit vector perpendicular to the boundary. h^o in Eq. (6) describes the groundwater level distribution throughout the domain at the beginning of the calculation.

PRINCIPLES OF THE NUMERICAL SOLUTION

The numerical approximation of Eqs. (1) - (6) was handled in space by the finite element method (Galerkin method) using triangular elements with linear basic functions, and in time applying finite differences (e.g. Crank-Nicholson method). For clarity, only the nodal point inflows Q'_{KS} are used out of all possible inflows which can be calculated with the decision variable e. The index K describes the location and S the time of an unknown nodal point inflow. Formally, there is an unknown inflow at every node and at every time. For most problems, however, the unknown inflows, which should be optimized, are limited to a few discretization nodes or can be grouped in spacially and temporally uniform blocks, so that separate indexing of nodal variables and actual decision variables of the problem turns out to be appropriate. Through the matrix D_{KSV} the nodal inflows Q'_{KS} are related to the decision variables e_V:

$$Q'_{KS} = D_{KSV}\, e_V \tag{7}$$

with K = 1,2,...,N (N: number of nodes), S = 1,2,...,NZ (NZ: number of time steps) and V = 1,2,...,NQ (NQ: number of decision variables). Using Eq. (7), the objective function and the constraints are discretized to:

$$Z = c^1_V\, e_V + c^2_{KS}\, h_{KS} \rightarrow \min|\max \tag{8}$$

$$g^1_{TV}\, e_V + g^2_{TKS}\, h_{KS} \lesseqqgtr R_T \tag{9}$$

The values and dimensions of the objective function coefficients c^1_V and c^2_{KS}, of the constraint matrices g^1_{TV} and g^2_{TKS} and the right-hand side R_T depend on the actual problem. Practically, the formulation of the objective function and the constraints is made directly in the discrete quantities h_{KS} and e_V.

Lastly, the approximation of the continuum problem Eqs. (3) - (6) leads to a system of linear influence functions:

$$h_{KS} = \hat{h}_{KS} + D^*_{KSV}\, e_V \tag{10}$$

that describe the dependence of the nodal values h_{KS} on the decision variables e_V. The vector $\hat{h}_{KS}$ contains the groundwater levels which result from the boundary conditions and the known inflows, while the second term on the right-hand side of Eq. (10) with the influence matrix D^*_{KSV} produces the groundwater changes due to the decision variables e_V.

From the viewpoint of optimization, the influence functions Eq. (10) constitute constraints which define a linear optimization problem together with the objective function Eq. (8) and the other constraints Eq. (9). In this form, the problem can be solved using any linear optimization method, however a good approximation of the true groundwater flow requires such a fine discretization in space and time that the optimization problem becomes much too large for a direct solution. For this reason, before the actual optimization, the scope of the problem is reduced by combining Eq. (10) with Eqs. (8) and (9). The substitution leads to the reduced problem

$$Z = c_V\, e_V + c \rightarrow \min|\max \tag{11}$$

$$g_{TV}\, e_V \begin{matrix} \le \\ = \\ \ge \end{matrix} R^*_T \tag{12}$$

which contains only the decision variables e_V as unknowns. The use of the Simplex algorithm (3-Phase-Simplex algorithm (Neumann [11])) gives the desired vector of decision variables $e_{V,opt}$, which minimizes (or maximizes) the objective function Eq. (11) and satisfies the constraints Eq. (12). The groundwater levels corresponding to the solution are produced by substituting the vector $e_{V,opt}$ into the influence function Eq. (10).

CALCULATION OF THE INFLUENCE FUNCTIONS

The influence functions Eq. (10) are calculated one after another for each time step r = 1,2,...,NZ. Fig. 1 shows schematically the calculation procedure at each time step (indices defined above). The first line contains the initial conditions. In the first time step the influence matrix $D_{KV} = 0$ because the unknowns e_V do not yet play a role. Line 2 in Fig. 1 contains the individual terms of the equation system in their approximate form (Eqs. (3) - (6)). K_{LK} is, as usual, the symmetrical banded coefficient matrix that appears in all transient groundwater flow calculations. R_L contains known sinks and sources. H_{LK} is the matrix which results from the second and last term in Eq. (3), which when multiplied by the initial conditions forms a portion of the right-hand side in conventional computations. The term $D_{LV}e_V$ identifies the effect of all decision variables that are active in the current time step.

The third line of Fig. 1 contains some of the terms from line 2

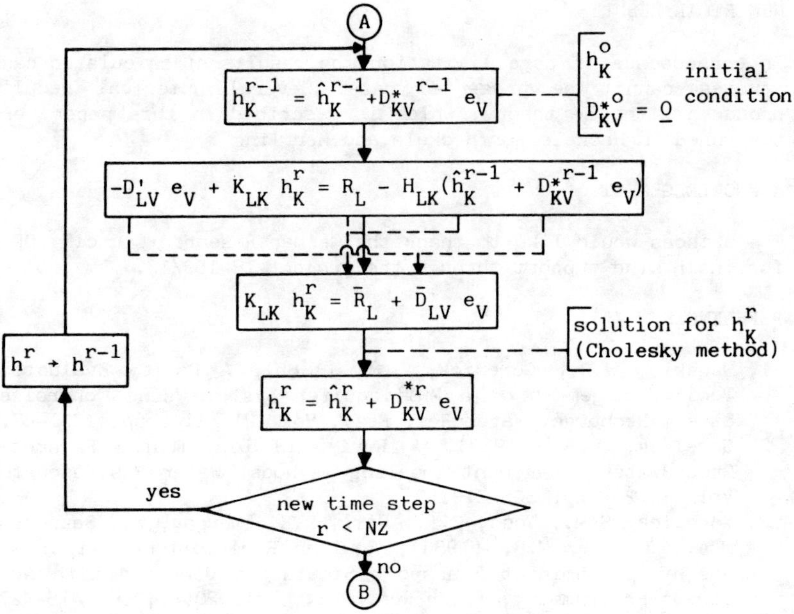

Figure 1. Calculation of the influence functions in a time step

in combined form. In the first time step (r=1) $\bar{R}_L = R_L - H_{LK} h_K^o$ and $\bar{D}_{LV} = D'_{LV}$. The third line is solved using the method of Cholesky for the groundwater levels h_K^r for upto (1 + NQ) right-hand sides. For example, for the case that at r = 1, there are 3 decision variables active, 4 right-hand sides must be calculated. For r = NZ there are always (1 + NQ) right-hand sides.

The fourth line in Fig. 1 shows the influence function for a time step r and corresponds to the influence function in Eq. (10) for a specific r = S. For the next time step ($r \to r - 1$) the entire fourth line is used as initial condition. In this manner for S = 1,2,...,NZ the influence functions are calculated.

NONLINEAR GROUNDWATER SYSTEMS

The nonlinearities due to the dependence of the aquifer parameters on the groundwater level were handled iteratively. In comparison with a conventional flow calculation, in which the iterative correction of the aquifer parameters can occur within a time step, for an optimization the influence functions for the entire time period of the calculation and the optimal solution must be computed before the parameters in the following iteration can be adjusted. For several case studies (see Heckele[4]) converged solutions were achieved after only a few iterations.

NUMERICAL RESULTS

In consequence of page limitation the results of calculated case studies cannot be presented here. Several numerical results produced with the theory, which is described in this paper, are published in Heckele[4] or Heckele and Herrling[5].

ACKNOWLEDGEMENT

The authors would like to thank the German Research Council (DFG) for their kind support through the project TH 159/13.

REFERENCES

1. Danskin, W.R., Gorelick, S.M. (1985), A Policy Evaluation Tool: Management of a Multiaquifer System Using Controlled Stream Recharge. Water Res. Res., Vol. 21 (11), pp. 1731-47.
2. Gorelick, S.M. (1983), A Review of Distributed Parameter Groundwater Management Modeling Methods. Water Res.Research, Vol. 19(2), pp. 305-319.
3. Gorelick, S.M., Voss, C.I., Gill, P.E., Murray, W., Saunders, M.A., Wright, M.H. (1984), Aquifer Reclamation Design: The Use of Contaminant Transport Simulation Combined with Non-linear Programming, Water Res. Res., Vol. 20(4), pp. 415-427.
4. Heckele, A. (1987), Numerische Optimierungsverfahren zur Bewirtschaftung flächig ausgedehnter Grundwassersysteme. Fortschr. Ber. VDI R. 4, Nr. 81, 171 S., VDI-V., Düsseldorf.
5. Heckele, A., Herrling, B. (1987), Numerische Optimierungsverfahren zur Bewirtschaftung nichtlinearer Grundwassersysteme mit instationären Strömungsverhältnissen. Z.dt.geol.Ges., Band 138, Teil 2, Seite 1-14.
6. Heidary, M. (1982), Application of Linear System's Theory and Linear Programming to Groundwater Management in Kansas. Water Res. Bull., 18(6), pp. 1003-1012.
7. Lindner, W., Marotz, G. (1985), Steuerung von Grundwasserentnahmen in Mittelgebirgstälern unter Einhaltung land- und forstwirtschaftlich erforderlicher Grundwasserstände. Zeitschr. f. Kulturt. u. Flurber., 26, S. 149-160.
8. Maddock, T. (1972), Algebraic Technological Functions from a Simulation Model. Water Res. Res., Vol. 8(1), pp. 129-134.
9. Maddock, T. (1974), Nonlinear Technological Functions for Aquifer whose Transmissivities Vary with Drawdown. Water Res. Research, Vol. 10(4), pp. 877-881.
10. Neumann, K. (1975), Operations Research Verfahren, Band I. Carl Hanser Verlag, München.
11. Wanakule, N., Mays, L.W., Lasdon, L.S. (1986), Optimal Management of Large-Scale Aquifers: Methodology and Applications. Water Res. Research, Vol. 22(4), pp. 447-465.
12. Willis, R. (1977), Optimal Groundwater Resources Mangement Using the Response Equation Method. In: Finite Elements in Water Resources. Ed. by W.G.Gray et al., Pentech Press.
13. Willis, R., Finney, B.A. (1985), Optimal Control of Nonlinear Groundwater Hydraulics: Theoretical Development and Numerical Experiments. Water Res. Research, Vol. 21(10), pp. 1476-1482.

Reliability Constrained Markov Decision Programming and its Practical Application to the Optimization of Multipurpose Reservoir Regulation

Liang Qingfu
Department of Hydraulic Engineering, Tsinghua University, Beijing, China

ABSTRACT

This paper considers the optimization of regulating a reservoir with multiobjective. The system is modeled as a periodic Markov Decision Programming (MDP). A procedure is proposed in which the Dynamic Programming (DP) is combined with use of probability computation in conjunction with penalty costs to derive long-term operating policies that will maximize one of the objectives and yet not violate constraints on the others. The application of this procedure to a reservoir with four objectives demonstrates its power.

INTRODUCTION

This paper considers the optimization of regulating a reservoir with the objectives of flood control, hydroelectric generation, reliability of generation, navigation and irrigation. Although flood control is a very important objective, it cannot be quantified so that it may be treated as the constraint of reservoir storage. Because of the serious lack of electricity in China, hydroelectric generation is usually maximized subject to the set quota limits on the other objectives.

The inflows to the reservoir are treated as stochastic random variables and are correlated with the inflow of the preceding period.

This system can be modeled as a periodic MDP. The method used for solving it is the combined use of DP and Successive Approximation Method (SAM) originally developed by White (1963) and then extended by Su and Deininger (1972) to find optimal operating policies that maximize the average expected annual electric energy. However, in most cases, these policies cannot satisfy the reliability requirements of the generation and other objectives of the system. Though such reliability values can be estimated by simulation, they are not probability values but frequency ones. Based on the limiting property of Markov process, this paper derives the probability formulas from their frequency ones so that the reliability values can get rid of influences of the initial state and the number of operation years. A procedure is proposed that DP and SAM is combined use of probability computation in conjunction with penalty costs to derive an operating policy that will maximize the expected annual electricity and yet not violate the constraints on the other objectives. Simulation is also carried out. Both results are almost the same.

MATHEMATIC MODEL

Consider a given reservoir with the objectives which must be calculated:- the hydroelectric generation, reliability of generation, navigation and irrigation. The objective was to find operating policies for the system that maximize the average expected annual hydroelectric production subject to the set quota limits on the other objectives so the model is

$$\text{maximize} \quad g(d)$$

subject to

$$\begin{aligned} A(d) &\geqq Ag \\ B(d) &\geqq Bg \\ I(d) &\geqq Ig \end{aligned} \qquad (1)$$

where

- d — operating policy of the system;
- g(d) — hydroelectric generation associated with the policy d;
- A(d) — reliability of generation associated with the policy d;
- B(d) — reliability of navigation associated with the policy d;
- I(d) — reliability of irrigation associated with the policy d;
- Ag — reliability quota of the generation;
- Bg — reliability quota of the navigation;
- Ig — reliability quota of the irrigation.

OPTIMAL OPERATING POLICIES FOR GENERATION

A year is divided into T periods. If the inflows to the reservoir in each period are serially correlated and the information of the inflow in preceding period is available, the variable of the system can consist of the storage at the beginning of the period and the inflow of the preceding period. Decision variable is the outflow from the reservoir. Conditional probabilities between the inflows in consecutive periods define the state transition probability matrices. Reward function is the hydroelectric generation in the period and objective is the average expected annual electric energy. So the model is a MDP with finite states and decisions, limited reward, discrete time and average objective. Because the inflow is a periodic stochastic process with an annual cycle, the state transition probability matrices, the reward function and the steady operating policy are periodic process with an annual cycle. Therefore, the model is a periodic MDP with an annual cycle.

Suppose that the system is currently at the beginning of a period that in nT+t ($n \geq 0$). At this time the system is observed to be state i. If a decision e is taken from finite set D of feasible decisions, the system moves to a state j at the end of the period according to the probability $P_{ij}(t, e)$, and an immediate return in this transition is the hydroelectric generation $R_{ij}(t, e)$.

Let $V_n(t, i)$ be the maximal total expected electric energy from the nT+t periods. By the principle of optimality, the recursive equation can be written as

$$V_n(t,i) = \max_{e \in D} \sum_{j=1}^{N} P_{ij}(t,e)[R_{ij}(t,e) + V_n(t+1,j)] \tag{2}$$

where N is the total discrete number of the system state.

This equation is used to find the optimal operating policy d which maximizes the average expected annual electric energy. Therefore, backward recursive computation is carried out according to DP within the cycle and SAM is used between the consecutive cycles until the policies between two successive cycles are identical or the values of the average expected annual electric energy converge to a constant.

GENERATION RELIABILITY CONTROL

Frequency value of generation reliability
Let firm power of the hydropower be Ng kw and the amount of electricity be $Rg = Ng \times \Delta t$ (where Δt is the length of a period). According to policy d, the reservoir runs successively by simulation for n years from an initial state i* on, the amount of average annual electric generation is

$$g_n(i^*,d) = \frac{1}{n} \sum_{y=1}^{n} \sum_{t=1}^{T} \sum_{i=1}^{N} \sum_{j=1}^{N} R_{ij}(t,e) \tag{3}$$

If the actual electric generation $R_{ij}(t,e) \geq Rg$ in the period t, this period is regarded as reliable one.

Definition

$$A_{ij}(t,e) = \begin{cases} 1 & \text{if } R_{ij}(t,e) \geq Rg \\ 0 & \text{if } R_{ij}(t,e) < Rg \end{cases} \tag{4}$$

The generation reliability of the system which runs successively for n years can be written as

$$A_n(i^*,d) = \frac{1}{nT} \sum_{y=1}^{n} \sum_{t=1}^{T} \sum_{i=1}^{N} \sum_{j=1}^{N} A_{ij}(t,e) \tag{5}$$

Obviously, both $g_n(i^*, d)$ and $A_n(i^*,d)$ depend not only on the operating policy d but also on the initial state i* and the operation years of n if the system runs successively according to the inflow series, either historical records or synthetic inflows. That is why they are referred to as frequency values.

Probability formula of generation reliability

Howard (1960) studied the finite homogeneous Markov processes with rewards. If let P(d) be one-step transition matrix, which is the product of the transition matrix in each period within a cycle, and assume that the operation process of the system, which runs along a track and P(d) for a long time, is a completely ergodic Markov process, the model here is the same as Howard's. If it is extended to the periodic Markov process (transition between the periods within a cycle) and decision e at state i is specified by policy d in period t, the average annual electric energy is

$$g(d) = \sum_{t=1}^{T} \sum_{i=1}^{N} \Pi_i(t,d) \sum_{j=1}^{N} P_{ij}(t,d)\, R_{ij}(t,e) \tag{6}$$

where $\Pi_i(t,d)$ is steady storage probability of state i at the beginning of period t. If the energy shortages are not penalized, g(d) should be equal to what is calculated by DP and SAM.

Let $Y_{ij}(t,d,n)$ represent the number of transitions which satisfy formula (4), formula (5) can be rewritten as

$$A_n(i^*,d) = \frac{1}{nT} \sum_{t=1}^{T} \sum_{i=1}^{N} \sum_{j=1}^{N} Y_{ij}(t,d,n)\, A_{ij}(t,e) \tag{7}$$

Let $X_i(t,d,n)$ be the number of the transitions which are observed to be state i at the beginning of period t during n years, formula (7) can be rewritten as

$$A_n(i^*,d) = \frac{1}{T} \sum_{t=1}^{T} \sum_{i=1}^{N} \frac{X_i(t,d,n)}{n} \sum_{j=1}^{N} \frac{Y_{ij}(t,d,n)}{X_i(t,d,n)} A_{ij}(t,e) \tag{8}$$

If n is large enough, it can be proved (Liu Wen, 1978)

$$P\left[\lim_{n\to\infty} \frac{X_i(t,d,n)}{n} = \Pi_i(t,d)\right] = 1 \tag{9}$$

$$P\left[\lim_{n\to\infty} \frac{Y_{ij}(t,d,n)}{X_i(t,d,n)} = P_{ij}(t,d)\right] = 1 \tag{10}$$

Let A(d) denote the probability value of generation reliability,

$$P\left[\lim_{n\to\infty} A_n(i^*,d) = A(d)\right] = 1 \tag{11}$$

or

$$A(d) = \frac{1}{T} \sum_{t=1}^{T} \sum_{i=1}^{N} \Pi_i(t,d) \sum_{j=1}^{N} P_{ij}(t,d)\, A_{ij}(t,e) \tag{12}$$

Because this is the probability formula of generation reliability, the value calculated by it can get rid of influences of the initial state and the number of the operation years and depends only on the nature of the system itself.

Computational experiences show that the completely ergodic assumption still holdswater although Howard's completely ergodic condition cannot be surveyed directly.

Method of adjusting generation reliability.
If an optimal operating policy that maximizes the average expected annual electric generation does not satisfy the requirement of generation reliability, a penalty cost is used to force optimization algorithm into selecting a new operating policy. Therefore, the optimization of regulating the reservoir is a reliability constrained MDP.

Definition of new reward function

$$W_{ij}(t,e) = R_{ij}(t,e) + F\left[A_{ij}(t,e) - 1\right] \qquad (13)$$

where F ($F \geq 0$) is penalty cost. The reward in formula (2) is substituted by the new one to select a new operating policy dp. Let G(dp) denote the value calculated by DP and SAM, g(dp) and A(dp) represent the values from formula (6) and (12) respectively, it is not difficult to find the relationship

$$g(dp) = G(dp) + FT\left[1 - A(dp)\right] \qquad (14)$$

RELIABILITY COMPUTATION OF NAVIGATION AND IRRIGATION

Improving the navigation conditions of the inland river waterway and raising its reliability is an important objective of regulating the reservoir. In order to guarantee the shipping, the discharge from reservoir should be as constant as possible and not less than the minimum flow dg that maintains enough depth of the waterway. Therefore, dg can be treated as constraint of the decision variable for the generation.

Definition

$$B_{ij}(t,e) = \begin{cases} 1 & \text{if } e \geq dg \\ 0 & \text{if } e < dg \end{cases} \qquad (15)$$

The frequency formula of the navigation reliability, similar to that of the generation, can be written as

$$B_n(i^*,d) = \frac{1}{nT} \sum_{y=1}^{n} \sum_{t=1}^{T} \sum_{i=1}^{N} \sum_{j=1}^{N} B_{ij}(t,e) \qquad (16)$$

The probability formula is

$$B(d) = \frac{1}{T} \sum_{t=1}^{T} \sum_{i=1}^{N} \Pi_i(t,d) \sum_{j=1}^{N} P_{ij}(t,d) B_{ij}(t,e) \qquad (17)$$

The water allocated to irrigation includes diversion of water from the reservoir and withdrawal of water from the river system down the reservoir. If the reservoir level is too low to channel the water into irrigation canal, the water required is pumped into it. One of the study purposes is to demonstrate whether a pumping station should be built or not. Let Id(t) and Iw(t) denote the demands for the diversion and the withdrawal of water respectively and $I_p(t)$ be the actual diversion of water in period t.

Definition

$$I_{ij}(t,e) = \begin{cases} 1 & \text{if} \quad e \geq d_g + I_w(t) \text{ and } I_p(t) \geq I_d(t) \\ 0 & \text{others} \end{cases} \qquad (18)$$

The reliability formulas of the frequency and probability of the irrigation can be respectively written as

$$I_n(i^*,d) = \frac{1}{nT} \sum_{y=1}^{n} \sum_{t=1}^{T} \sum_{i=1}^{N} \sum_{j=1}^{N} I_{ij}(t,e) \qquad (19)$$

$$I(d) = \frac{1}{T} \sum_{t=1}^{T} \sum_{i=1}^{N} \pi_i(t,d) \sum_{j=1}^{N} P_{ij}(t,d) I_{ij}(t,e) \qquad (20)$$

Formulas (6), (12), (17) and (20) indicate that the computations of the four objective functions are mathematically the same.

COMPUTATION RESULTS AND CONCLUSIONS

To show the validity of the model, the procedure proposed in this paper has been applied to a real-time operation of Danjiangkou Reservoir in Hubei Province, China. The model involved 36 decision periods, 64 storage and 10 inflow states and used the historical inflow records as data to evaluate the derived policy by simulation. The results show that the average yearly amount of electric generation, compared with that from conventional rules, can be increased by at least 4.1% under the conditions of not reducing the requirements of the other objectives.

Optimization of regulating a reservoir with multiobjectve is a very complex and nonlinear problem. The approach used for solving it should be determined by the requirements of the local economic development. Constrained Optimization algorithm here, which finds an optimal operating policy that maximizes hydroelectric power production subject to the set quota limits on the other objectives, is suitable for the recent conditions of China, and the computational experiences demonstrate its significant power.

Optimal Multiobjective Operational Planning of a Water Resources System

S. Soares and M.G. Andrade Filho

Department of Systems Engineering, Faculty of Electrical Engineering, State University of Campinas - UNICAMP, PO Box 6101-13081, Campinas, SP, Brazil

Abstract

This paper considers the optimal multiobjective operational problem of a water resources system, with two conflicting objectives, the generation of electric energy and the use of water for irrigation. The work analyses the effect of the use of water for irrigation on the reliability of system load supply. The hydroelectric system is represented by an energy composite reservoir model and the randomness of inflows is considered. The optimal solution is obtained by Stochastic Dynamic Programming. The model is applied to the hydroelectric system of the São Francisco River in the Northeast region of Brazil. As a result the curves of risk of load shedding × irrigation and average load shedding × irrigation are presented. These results show the trade-off between the two conflicting objectives for the considered system and represent important information for the decision-making process.

INTRODUCTION

One of the most important applications of multi-objective programming is the problem of multiple use of water resources. This subject has been considered in many publications such as Duckstein and Oprocovic–1980, Vedula and Rogers–1981, Yeh and Becker–1982, Hall at al.–1968, Gershon at al.–1982, Oven-Tompson at al.–1982.

In this paper a particular approach to the multi-objective operational planning of a multi-reservoir system is suggested, where two conflicting objectives are considered: the use of water for hydroelectric generation and for irrigation. The trade-off between these objectives is obtained through the weight method, Cohon [3], which transforms the multi-objective problem into a single-objective one. This single-objective problem consists of the optimal operation of a stochastic multi-reservoir system which represents a complex and not well solved problem. The approach followed was to aggregate the multi-reservoir system into a composite reservoir, Arvanitidis [2], and to carry out its optimization by Stochastic Dynamic Programming. This approach was applied to the hydroelectric system of the São Francisco River in the Northeast of Brazil, an arid region where the water resource is scarce and where the conflict in its use has become crucial for the region, Andarde [1]. The results show the trade-off between the use of water for hydroelectric generation and for irrigation purposes.

The paper is written in a comprehensive way, from simple to more complex formulations. First, a general static two-objective problem is analysed. The specific problem of the water use between hydroelectric generation and irrigation is considered. Then the dynamic two-objective problem is formulated for a single reservoir system and a static-dynamic decomposition is suggested in order to permit the solution to be carried out by Stochastic Dynamic Programming. Finally, the approach is generalized for multi-reservoir systems through the composite reservoir model and applied to the hydroelectric system of the São Francisco River.

THE STATIC TWO-OBJECTIVE PROBLEM

Suppose there are two utilities for a given resource, each one associated with a specific benefit denoted here by the functions $B_1(x_1)$ and $B_2(x_2)$, where x_1 and x_2 represent the amount of the resource allocated to utilities 1 and 2 respectively. If the benefit functions are concave, as illustrated in figure 1, the best way for allocating a certain amount of resources between the two utilities is given by the solution of the following problem

$$\max \quad B_1(x_1) + B_2(x_2) \tag{1}$$

$$\text{s.t.} \quad x_1 + x_2 = r \tag{2}$$

$$x_1 \geq 0, \; x_2 \geq 0 \tag{3}$$

If the non-negativity constraints (3) are not binding, the solution of (1)-(3) will equalize the marginal benefits of the two utilities, that is, the solution will be given by

$$\hat{x}_1 + \hat{x}_2 = r \tag{4}$$

$$\frac{dB_1(\hat{x}_1)}{dx_1} = \frac{dB_2(\hat{x}_2)}{dx_2} \tag{5}$$

as illustrated in figure 1

Suppose now that the resource considered is water and that the utilities are hydroelectric generation and irrigation. The benefit of the use of water for hydroelectric generation is associated with the cost reduction due to replacement of thermal generation and eventual load sheddings in the hydrothermal system. The economic use of the non-hydroelectric sources (coal, oil, gas, nuclear, load shedding) by its incremental costs assure that the total cost in a hydrothermal system will be a convex increasing function of the non-hydroelectric generation. Thus, the benefit from the use of water for hydroelectric generation is a concave function that depends on the particular thermal sub-system, the total load demand and the social-economic cost of load shedding, as illustrated in figure 2.

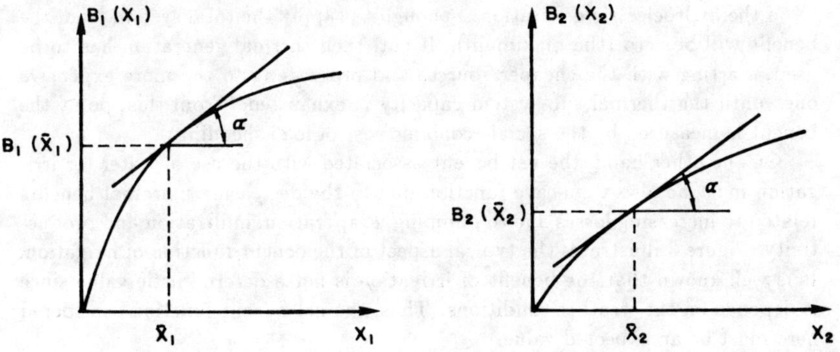

Figure 1: Benefit functions for two-objectives

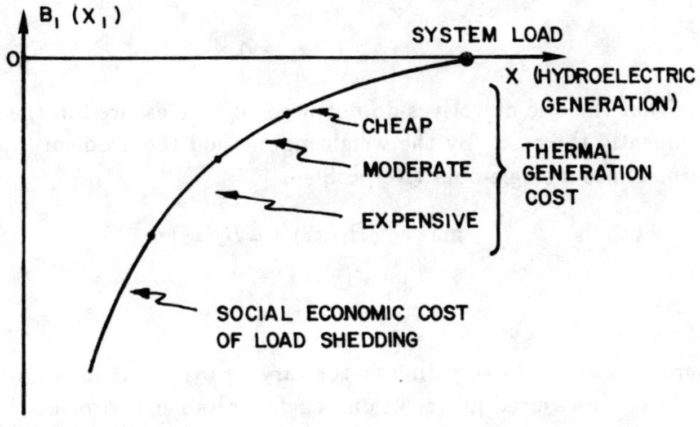

Figure 2: Benefit function for hydroelectric use of water

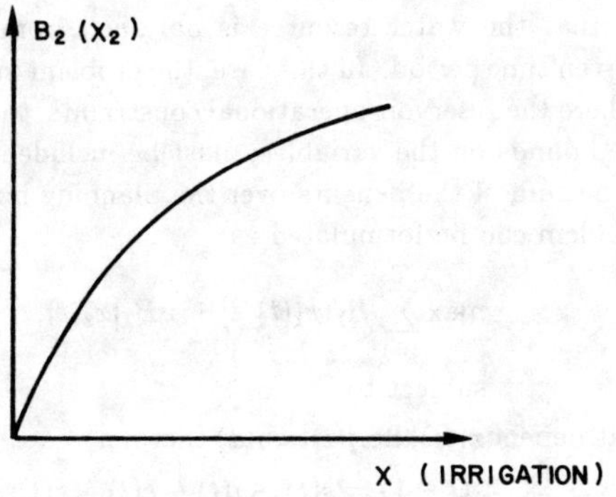

Figure 3: Benefit function of irrigation

If the hydroelectric generation is enough to supply the total system load, the benefit will be zero (the maximum). If not, then thermal generation has to be used, starting with the cheaper sources and progressing to the more expensive ones until the thermal sub-system capacity is exhausted. From this point the benefit is measured by the social-economic cost of load shedding.

On the other hand, the net benefit associated with the use of water for irrigation must be also a concave function due to the decreasing marginal benefits related to increasing losses due to pumping, evaporation, infiltration and productivity. Figure 3 illustrates the typical aspect of the benefit function of irrigation. It is well known that the benefit of irrigation is not a deterministic value since it depends on the weather conditions. Thus, the net benefit function considered here must be an expected value.

For the purpose of this paper the static two-objective problem can be stated as

$$\max \quad \boldsymbol{B}(\boldsymbol{x}) = [B_1(x_1), B_2(x_2)] \tag{6}$$

$$\text{s.t.} \quad x_1 + x_2 = r \tag{7}$$

$$x_1 \geq 0, \; x_2 \geq 0 \tag{8}$$

Note that the two objectives do not need to be measured in the same units, as is frequently the case. By the weighting method the problem (5)–(8) can be transformed into a single-objective problem

$$\max \quad B_1(x_1) + \omega B_2(x_2) \tag{9}$$

$$\text{s.t.} \quad x_1 + x_2 = r \tag{10}$$

$$x_1 \geq 0, \; x_2 \geq 0 \tag{11}$$

where ω is as much weighting factor varying over $[0, \infty)$, as a unit transformation factor, measured in irrigation benefit units per hydroelectric generation benefit units.

THE DYNAMIC TWO-OBJECTIVE PROBLEM

Suppose now that the water resource is obtained from the operation of a reservoir over a given time period. In this case, the problem must be formulated as a dynamic one where the reservoir operational constraints, the water conservation equation and the bounds on the variables, must be included. The objective now turns out to be the sum of the benefits over the planning horizon. The dynamic two-objective problem can be formulated as

$$\max \sum_{t=1}^{T} B_1[x_1(t), t] + \omega B_2[x_2(t), t] \tag{12}$$

$$\text{subject to} \tag{13}$$

$$x_1(t) + x_2(t) = r(t) \tag{14}$$

$$s(t+1) = s(t) + i(t) - e(t) - r(t) \tag{15}$$

$$\underline{s}(t) \leq s(t) \leq \overline{s}(t) \tag{16}$$

$$\underline{r}(t) \leq r(t) \leq \overline{r}(t) \tag{17}$$

$$s(1) \text{ given}, \; x_1(t) \geq 0, \; x_2(t) \geq 0 \tag{18}$$

where $s(t)$ is the water storage at the begining of period t; $\underline{s}(t)$ and $\overline{s}(t)$ are the bounds on storage at the begining of period t; $r(t)$ is the water release at period t; $\underline{r}(t)$ and $\overline{r}(t)$ are the bounds on release at period t; $i(t)$ is the incremental water inflow at period t; $e(t)$ is the water evaporated from the reservoir at period t (a function of its storage at each period t); $x_1(t)$ is the amount of the water used for hydroelectric generation at period t; $x_2(t)$ is the amount of water used for irrigation at period t; and T is the horizon of planning. The dynamic problem (12)–(17) has two distinct decisions. The first decision concerns the total release from the reservoir at each period, the dynamic decision. The second decision concerns the partitioning of the total release between the two conflicting uses, the hydroelectric generation use and the irrigation use, the static decision. This suggests the decomposition of problem (12)–(17) by projecting it onto the release decision sub-space. The projected dynamic problem becomes

$$\max \sum_{t=1}^{T} B[r(t)] \tag{19}$$

$$\text{subject to} \tag{20}$$

$$s(t+1) = s(t) + i(t) - e(t) - r(t) \tag{21}$$

$$\underline{s}(t) \leq s(t) \leq \overline{s}(t) \tag{22}$$

$$\underline{r}(t) \leq r(t) \leq \overline{r}(t) \tag{23}$$

$$s(1) \text{ given} \tag{24}$$

where B[r(t)] is the optimal value of the static problem

$$\max \quad B_1[x_1(t), t] + \omega B_2[x_2(t), t] \tag{25}$$

$$\text{s.t.} \quad x_1(t) + x_2(t) = r(t) \tag{26}$$

$$x_1(t) \geq 0, \; x_2(t) \geq 0 \tag{27}$$

Finally, as the inflow $i(t)$ is a random variable, the dynamic problem (18)–(22) must be replaced by its stochastic version where the objective function (18) is replaced by the maximization of the expected benefit with respect to the inflow $i(t)$. The solution of the projected stochastic dynamic problem can be carried out by Stochastic Dynamic Programming, Yakowitz [9].

MULTIPLE-RESERVOIR SYSTEM

In order to apply the described approach on a multiple-reservoir system, the composite reservoir model has been adopted, as in Arvanitidis [2]. By this approach, a multi-reservoir system is represented by a single energy reservoir where the reservoir variables, measured in amounts of water, are transformed into energy variables through the average productivity of each reservoir (the average power generated by the discharge of one square foot per second). Thus, the water storage at each reservoir is transformed into the energy storage in the composite reservoir, the incremental water inflow at each reservoir is transformed into the energy inflow into the composite reservoir, the water release at each reservoir is transformed into the energy release from the composite reservoir, composed by energy discharge through the turbines and energy displaced for irrigation, and so on. Figure 4 illustrates the individual system and its equivalent aggregated system.

Although the composite model does not allow us to obtain optimal individual decisions but only an aggregate decision for the whole system, it is very useful for long term planning purposes where the objective is to evaluate the expected operational cost and the system supply reliability. This is the case of the study presented as follows.

A CASE STUDY

The approach presented in this paper was applied to the hydroelectric system of the São Francisco River in the Northeast region of Brazil. This system is composed of a hydroelectric plant with a large reservoir, called Sobradinho, with 1050 MW of installed power, and a set of run-of-river plants, called the Paulo Afonso Complex, with 4424 MW of installed power, as is shown in figure 5.

As the system has no thermal generation, the benefit function associated with the use of water for hydroelectric generation was considered to be simply the reduction in the social-economic cost of load shedding.

In order to represent the benefit function associated with the use of water for irrigation the government irrigation programs were analysed. The irrigation area as a function of the amount of water used for irrigation was considered as the benefit function for irrigation. Both benefit functions were fitted by quadratic functions, one for each month of the year. The optimization was carried out for a 5-year horizon and only the first year was considered for analysis. The optimal solution obtained was used to simulate the system behaviour over the 50 years of historical data in order to evaluate the reliability of the electric energy system. The statistics obtained by these simulations were the risk (r) and the expected value (v) of load shedding, and, on the other hand, the average amount of water for irrigation (i), measured in equivalent energy. The procedure was repeated for different values of the weight ω and the results are shown in the trade-off curves of figure 6.

The results show that, although the risk of load shedding greatly increases with the increase in the irrigation use of water, the expected value of load shedding does not suffer so much. For instance,for $\omega = 5$ one have an average irrigation corresponding to 450 MW of energy implies a high risk of load shedding (24 %) but a low expected value for this load shedding (30 MW aproximately or 1.2 % of the average load demand).

CONCLUSIONS

In this paper the optimal operation of a water resources system with two conflicting objectives, the hydroelectric generation and irrigation, was considered. The system was represented by a composite reservoir of energy. The two-objective problem was approached by the weighting method. The resulting single-objective

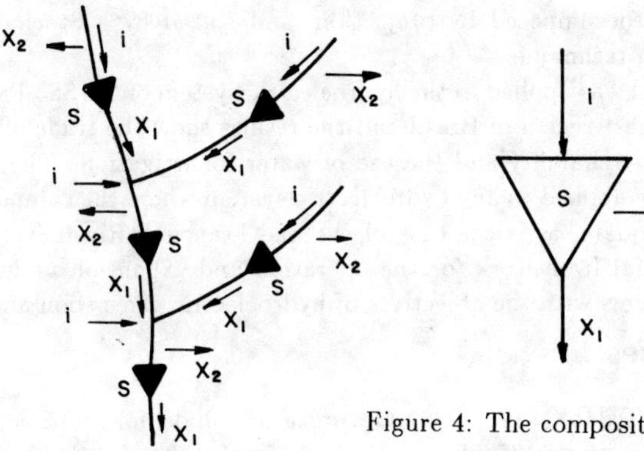

Figure 4: The composite reservoir

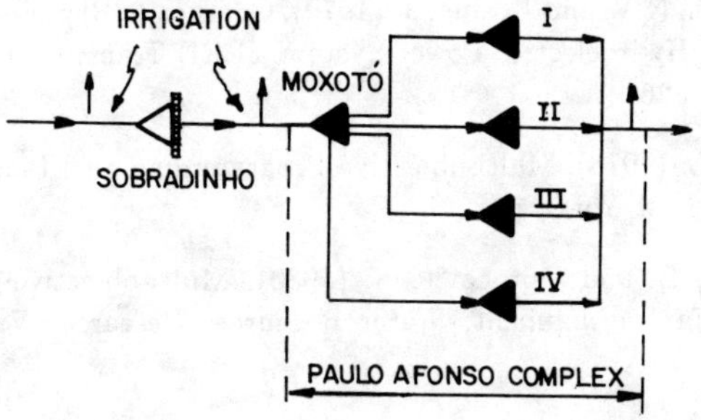

Figure 5: São Francisco River Hydroelectric System

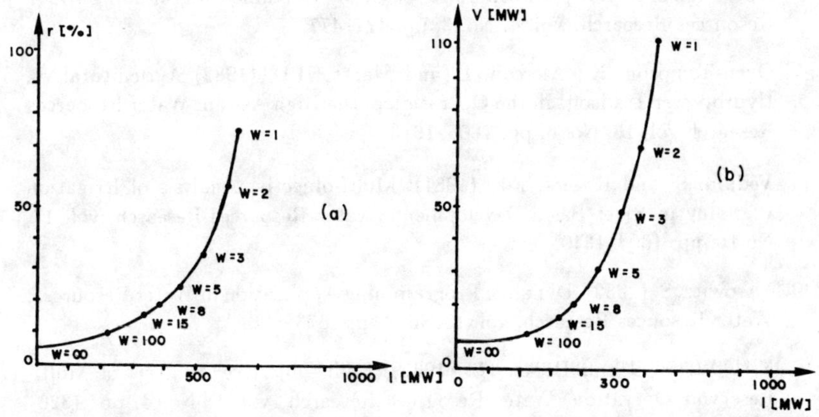

Figure 6: Trade-off curves — (a) risk of load shedding x irrigation — (b) expected value of load shedding x irrigation

problem was decomposed by projection, and solved by a Stochastic Dynamic Programming technique.

The model was applied to the hydroelectric system of the São Francisco River in the Northeast region of Brazil and the results show the trade-off between the energy supply reliability and the use of water for irrigation. The proposed approach can be applied to any hydroelectric system where the composite reservoir model is adequate, as is the case of the São Francisco River System, and represents a useful framework for the operation and expansion planning of water resource systems with the objectives of hydroelectric generation and irrigation.

References

[1] Andrade Filho, M.G. (1986), Optimization Model for Multiple Use on Water Resources of the São Francisco River , MSc.Thesis, Faculty of Electrical Engineering, State University of Campinas — UNICAMP, São Paulo, Brazil.

[2] Arvanitidis, N.V. and Rosing, J. (1970), Composite Representation of a Multireservoir Hydroelectric Power System, IEEE Trans. on PAS, vol. 89, No. 2, pp. 319–326.

[3] Cohon, J.L. (1978), Multi-objective Programming ancl Planning, Academic Press Inc. New York.

[4] Duckstein, L. and Oprocovic, S. (1980), Multi-objective Optimization in River Basin Development, Water Resources Research, Vol. 16,No. 1, pp. 14–20.

[5] Gershons, M.; Duckstein, L. and Meaniff, R. (1982), Multi-objective River Basin Planning With Qualitative Criteria, Water Resources Research, vol. 18, No. 2, pp. 193–202.

[6] Hall, W.H.; Butcher, W.S. and Esoughue, A. (1968), Optimization of the Operation of a Multiple-Purpose Reservoir by Dynamic Programming Water Resources Research, Vol. 4, No. 3, pp. 471–477.

[7] Oven-Tompson, K.; Alercon, L. and Marks, H.D. (1982) Agricultural vs. Hydropower Tradeoffs in the Operation of the High Aswan, Water Resources Research vol. 18, No. 6, pp. 1605-1613.

[8] Vedula, S. and Rogers, P.P. (1981), Multi-objective Analysis of Irrigation Planning in River Basin Development, Water Resources Research, vol. 17, No.15, pp. 1304-1310.

[9] Yakowitz, S. (1982), Dynamic Programming Application in Water Resources, Water Resouces Research, vol. 18,No. 4, pp. 673–696.

[10] Yen, W. W-G and Becker, L. (1982), Multi-objective Analysis of Multi-Reservoir Operation, Water Resources Research, vol. 18,No. 4, pp. 1326–1336.

A Flexible Polyhedron Method with Monotonicity Analysis

Shu-yu Wang and Zhang-lin Chen
Civil Engineering Department, Zhejiang University, Hangzhou, People's Republic of China

1. INTRODUCTION

In engineering optimization, objective functions and constraints are always not differentiable or even not continuous. Some of them may not be explicit functions. Therefore, the direct search methods are still widely used because they are of robustness, ease of problem preparation, no requirement of derivatives, etc.. The chief drawback of them is that they converge very slowly. This can be improved if we utilize the information fully and reasonably which has been accumulated during the iterative process.

The algorithm presented in this paper consists of following main components: (1) To pretreat the given problem by using global monotonicity analysis, form a mathematical model which is equivalent to the problem but simplifies the problem to a great extent. (2) To generate an initial polyhedron in the feasible region randomly. (3) To calculate a weighted centroid and obtain a descent direction. (4) To estimate step-length for a new better point(vertex) based on local monotonicity analysis around the centroid, and reconstruct a new polyhedron. (5) Using the gradient projection direction to further improve convergence of the algorithm at final search stage. We emphasize the discussions on (3)-(5) because of page limitations.

2. SELECTION OF SEARCH DIRECTION USING WEIGHTED CENTROID

The original Complex method only employed the coordinates of vertices to calculate a centroid vector X^c. Efficiency of the algorithm will be improved if the objective function values of these vertices can be introduced as "weight coefficients". Hence, the following relation is adopted to determine the centroid

$$X_W^c = \sum_h w_k X^k \tag{1}$$

and

$$w_k = (c-f(X^k))/\sum_h (c-f(X^k)) \tag{2}$$

where $\sum_h(\cdot)=\sum_{i=1,i\neq h}^{p}(\cdot)$, p is the number of vertices, X^k denotes the kth vertex of the current polyhedron, X^h the worst vertex (with the largest value of $f(X^k)$ for minimization problems), c is a reasonable large number, such as $\geqslant 2f(X^h)$, for minimization problems. It can be proved that the direction from X^h through X_w^c is a better descent direction than that through X^c for convex and quasi-convex functions.

Let us consider a differentiable and convex function for simplicity. The search direction (X^c-X^h) satisfies the following inequality

$$\nabla f(X^h)^T(X^c-X^h)\leqslant \frac{1}{p-1}\sum_h f(X^k)-f(X^h) \qquad (3)$$

and the search direction through the weighted centroid $(X_w^c-X^h)$ satisfies

$$\nabla f(X^h)^T(X_w^c-X^h)\leqslant \sum_h w_k f(X^k)-f(X^h) \qquad (4)$$

Obviously, the right term value of inequality (4) is less than that of inequality (3). The difference between these two terms is

$$\sum_{k=1,k\neq h}^{p-1}\ \sum_{j=k+1}^{p}\ (f(X^j)-f(X^k))^2/((p-1)\sum_h(c-f(X^k))) \qquad 5)$$

and is positive(Wang[1]). Although we are not certain whether any reflection direction from the worst or worse vertex through X_w^c is better than that through X^c, the value domain of $\nabla f(X^h)^T(X_w^c-X^h)$ is always smaller than that of $\nabla f(X^h)^T(X^c-X^h)$. So the algorithm using weighted centroid, as a whole, will enhance the rate of convergence of $f(X^k)$ in search process, especially in early stages. This effect will reduce in latter stages when the polyhedron becomes smaller.

Comparisons between the algorithms with weighted and general centroid have been made via several test functions, such as Rosenblock's functions. Results, shown in Fig. 1, clearly demonstrated that the above conclusion is correct. The number of function evaluation required by the algorithm here is only 50% of that by the original algorithm for the same terminal criteria.

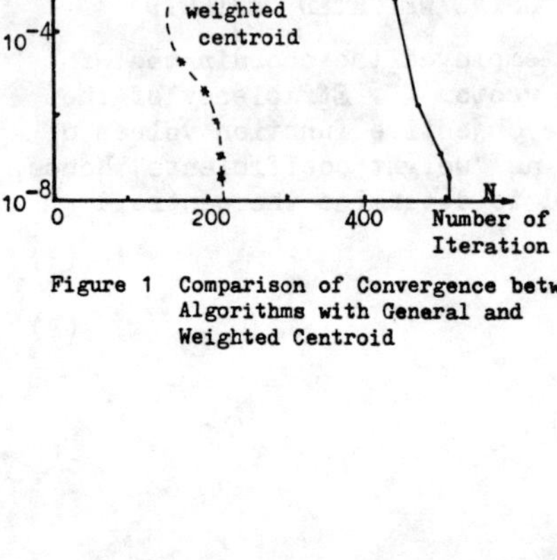

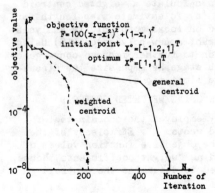

Figure 1 Comparison of Convergence between Algorithms with General and Weighted Centroid

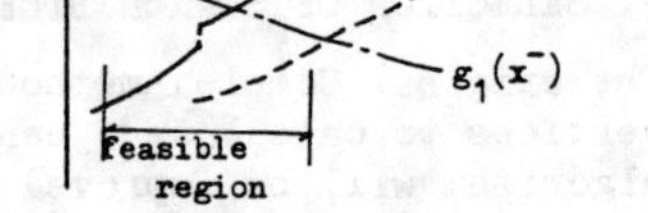

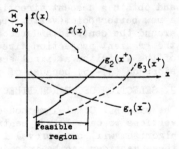

Figure 2 Illustration of Monotonicity Analysis

3. EVALUATION OF STEP LENGTH VIA MONOTONICITY ANALYSIS

The original algorithm employed four basic operations of reflection, expansion, contraction and reduction to search for a better vertex and form a new polyhedron. In essence, the search step-length was determined by trial and error. Naturally, this took much more CPU time. Monotonicity analysis can be applied to predict which constraint will be active and to determine the proper step-length for a new better vertex. Let us consider the following problem.

$$\min_{X \in S} \quad f(X) \tag{6}$$

$$S = \{X \mid X \in E^n, g_j(X) \leq 0, j=1,2,\cdots,m\}$$

If $f(X)$ is monotonic(e.g. continually decreasing) with respect to one of the design variables x_i, at least one constraint $g_j(X)$ having opposite monotonicity(continually increasing) with respect to x_i has to be active at the optimum. When more than one constraint have opposite monotonicity, the constraint providing the most severe limit on the increase of x_i will be active and is referred to as a dominant constraint(Papalambros[2]) Fig.2 illustrates the situation for a single variable optimization. The notation $f(x_i^-, x_j^+, x_k^0)$ denotes that the function f decreases with x_i, increases with x_j, and is independent on x_k. Because $g_2(x)=0$ provides the least upper bound on x and limits the decrease of $f(x)$ most severely, an optimum is expected where $g_2(x)=0$. $g_2(x)$ is known as a dominant constraint. The preceding discussion of monotonicity and dominance with respect to any design variables can be extended to a given search direction. For application to practical problems, the monotonicity of functions should be calculated in advance. For linear functions, the monotonicities can be easily determined in terms of their linear coefficients. They are constant and global. For nonlinear differentiable functions, the gradient $\nabla f(X^k)$ or $\nabla g_j(X^k)$ are used to evaluate local monotonicities at X^k. If the functions are not linear and differentiable, the finite difference is used to estimate local or regional monotonicities approximately.

After calculating the weighted centroid and determining the search direction

$$D = (X_w^c - X^h)/\|X_w^c - X^h\| \tag{7}$$

the step-length can be evaluated as follows: (1) Evaluate monotonicity of the objective function at X_w^c along the direction D. If it identifies with the monotonicity at X^h, that means $f(\alpha^-)|_{X_w^c}$ and a forward search (positive α) is needed to decrease the function value. Otherwise, it means $f(\alpha^+)|_{X_w^c}$ and a backward search (negative α) is needed. (2) Evaluate monotonicities of constraints at X_w^c along D. Those constraints having opposite monotonicity to objective function are candidate active constraints. (3) Estimate move limitations α_j to each candidate active constraint by linear approximation, and define $\alpha_m = \min\{\alpha_j\}$. The corresponding $g_m(X)$ is a candidate dominant constraint. (4) Determine a new point

$$X^{p+1} = X_W^c + (\beta\alpha_m(X_W^c - X^h)/\|X_W^c - X^h\| \qquad (8)$$

in which β is an adjustable coefficient depending on the non-linear degree of functions and the value of α_m.(5) If the sign of monotonicity of objective function at X^{p+1} along D does not alter, $g_m(X)\leqslant 0$ can be identified with a dominant constraint. Otherwise a change in the regional monotonicity has occurred, a minimum may exist between X_W^c and X^{p+1}, and it can be approached by quadratic interpolation through X^h, X_W^c and X^{p+1}.

4.FINAL SEARCH STRATEGY

Experiments in engineering application show that the flexible polyhedron method (or the Complex method) improves objective function values much slowly in latter stages as the polyhedron becomes smaller and smaller, so the gradient projection direction at X^l (the best vertex of the current polyhedron) will be employed to further search for a better vertex. That is

$$D = - Q\nabla f(X^l) \qquad (9)$$

where Q is a projection operator from the design space onto the intersection set of all active constraints at X^l. Its matric form is

$$Q = I - N(N^T N)^{-1} N^T \qquad (10)$$

in which $N=[\nabla g_j(X^l)], j\in I$, I is the index set of active constraints at the concerning point. Since the polyhedron has shrunk rather small at that time, $\nabla f(X^l)$ and $\nabla g_j(X^l)$ can be calculated approximately by solving a set of linear equations

$$f(X^k)-f(X^l) = \overline{\nabla f(X^l)}^T (X^k-X^l) \qquad k=1,2,\cdots,n \qquad (11)$$

where $\overline{\nabla f(X^l)}$ denotes the estimation of gradient of objective function at X^l. Similar relationship can be obtained for $\nabla g_j(X^l), j\in I$. The step-length is still determined by monotonicity analysis described previously.

5.EXAMPLES

Example 1. An optimization of planning for irrigation project. There are two classes of variables associated with this problem. (1) design variables: The capacity of reservoir for irrigation V_i, and irrigated area W. (2) Operational variables: The storage volume in reservoir at the end of each period V_j, $j=1,2,\cdots,7$.

The objective is to maximize the annual profits Z.

$$Z = B - Cr - C_i \qquad (12)$$

where C_i is the annual cost of irrigation system ($C_i=\gamma W$), Cr, the annual cost of reservoir $f(V_i)$, B denotes the annual revenues from agriculture

$$B = \alpha pW + \beta(1-p)W \qquad (13)$$

in which α and β are the net worths of the increase in yields per mu irrigated during guaranteed and unguaranteed year respectively, p is the percentage of guarantee. The constraints imposed are: (1) Limits of water supplies. The amount of water released for irrigation must not exceed the summation of the quantity of water flowing into reservoir I_j and the decrease in storage volume of reservoir $(V_j - V_{j-1})$ in the period j.

$$m_j W/\eta + (1+\tfrac{1}{2}\mu)V_j - (1-\tfrac{1}{2}\mu)V_{j-1} \leq I_j \qquad j=1,2,\cdots,n \qquad (14)$$

where m_j is the duty of water in the jth period, η irrigation application efficiency, n the number of time intervals for a design year, and μ the percentage of loss by the evaporation, etc.. (2) Limits of storage volume of reservoir. Such as the storage of reservoir should be within acceptable limits during flood period. (3) Bounds of design variables.

According to the practical investigation, the parameters mentioned above are adopted as follows:

γ=2.76 yuan/mu α=50 yuan/mu β=20 yuan/mu p=0.7

$$f(V_i)=0.06604V_i-1.1598*10^{-3}V_i^2+8.3462*10^{-6}V_i^3$$

$$\eta=1+0.1106W-9.4838*10^{-3}W^2+7.9965*10^{-4}W^3-2.8201*10^{-6}W^4$$

W_{max}=0.60 million mu V_{imax}=105 million cubic meter

Table 1 Data on I_j, m_j and μ_j

month	1-2	3-4	5-6	7-8	9	10	11-12
period	1	2	3	4	5	6	7
I_j million M^3	4.50	6.0	13.5	45.0	45.0	30.0	7.5
m_j M^3/mu	0	60	70	100	0	50	0
μ_j %	1	2	3	4	1	1	1.5

The mathematical model of this optimization problem is

$$\min.\quad -Z = 1.70 + f(V_i) - 38.24W$$

$$\begin{aligned} \text{s.t.}\quad & 1.005v_1-0.995v_7\le 4.5, \quad -0.99v_1+1.01v_2+60\eta W\le 6.0 \\ & -0.985v_2+1.015v_3+70\eta W\le 13.5 \\ & -0.98v_3+1.02v_4+100\eta W\le 45.0, \quad -0.995v_4+1.005v_5\le 45.0 \\ & -0.995v_5+1.005v_6+50\eta W\le 30.0, \\ & -0.9925v_6+1.0075v_7\le 7.50, \quad 0\le V_i\le 105, \quad 0\le W\le 0.60 \end{aligned}$$

After pretreating, it can be simplified to a problem with 2 design variables (V_i, W), 2 operational variables (V_2, V_6) and only 7 inequality constraints. The optimum solved by the algorithm is V_i^*=44.7851 million M^3, W^*=0.3162 million mu, Z^*=9.0114 million yuan. The number of iterations, objective function and constraint evaluation are 137, 289 and 296, respectively, against 254, 2351, and 716 required by the Complex method. The results are the same as the initial point changes.

Example 2. An optimum design of a concrete dam with broad joints. Nine design variables adopted are shown in Fig.3a. The volume of concrete is considered as the objective function. There are three kinds of constraints. The first is a geometric constraint, such as upper and lower bounds of design variables, a limitation of dam width at the top, and a condition of holding global stability of dam buttress. The second is a limit on the resistance to sliding failure. The third is a stress constraint. Principal stresses at points A, B, C and D are limited to the allowable strength of concrete (tensile 100T/M^2 and compressive 600T/M^2). Tensile stresses on the contact surface of dam with foundation are not allowed. Forces caused by uplift loads are taken into account in the analysis.

Initial point X^0=[52.0, 15.0, 0.24, 0.76, 0.48, 15.0, 9.5, 5.4, 42.5]^T, the corresponding objective value f^0=83886.2 M^3. The optimum solved by the algorithm is X^*=[53.45, 20.42, 0.298,

0.693,0.149,14.62,6.32,4.57,41.77]^T,f^*=66166.5 M^3.The optimum solved by the Complex method is X^*=[62.04,37.67,0.165,0.753, 0.422,14.94,7.83,5.42,46.53]^T, f^*=68961.39 M^3.The number of objective and constraint evaluation is about 490 and 608 against 737 and 1202 required by the Complex method.

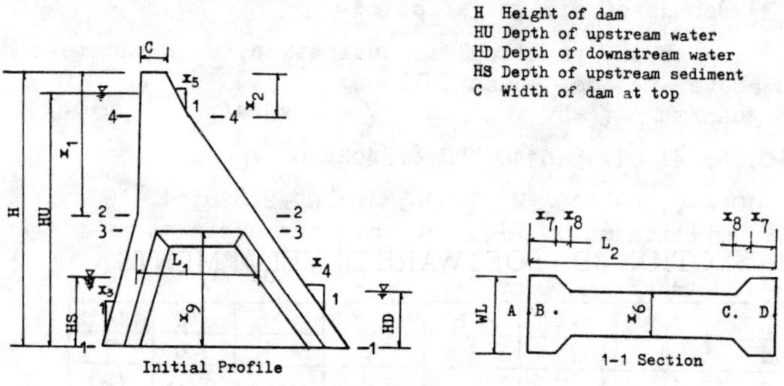

(3 - a)

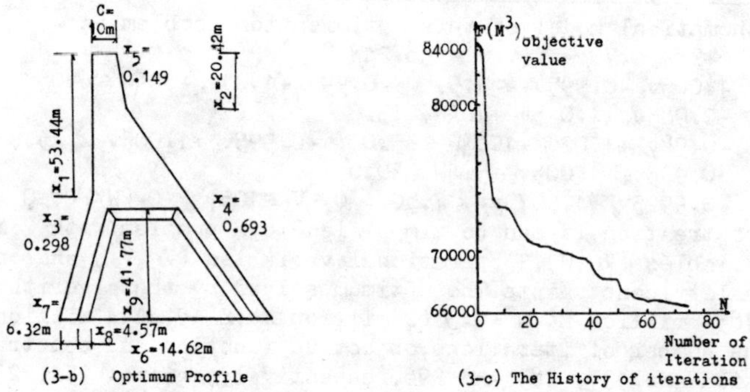

(3-b) Optimum Profile (3-c) The History of iterations

Figure 3 Optimum Design of Concrete Gravity Dam with Broad Joints

REFERENCES

1. Wang S.Y. and Li H.Y.(1980), An Improvement of the Complex Method, References of the 1st National Conference on Computational Mechanics,Hangzhou,P.R.C.
2. Papalambros P. and Li H.L.(1983), Notes on the Operational Utility of Monotonicity in Optimization,ASME,J. of Mechanisms, Transmissions and Automation in Design,Vol.105,pp.174-180

SECTION 3D - SOFTWARE DEVELOPMENTS

A Software Package for the Computer Aided Design of Sewer Systems

W. Bauwens
National Fund for Scientific Research (Belgium), Laboratory of Hydrology, Free University of Brussels

ABSTRACT

The paper describes the main characteristics of a software package for the computer aided design of open sewer networks according to the rational method, the timearea method and a rainfall runoff model. The software runs on a microcomputer, makes use of GEM and is coupled to a generalized CAD package with a relational database.

INTRODUCTION

The increasing complexity of the hydrologic and hydraulic models used for design tends to build up a discrepancy between the official institutions acting as builders and the consulting engineering enterprise on the one hand and theoreticians on the other.

This discrepancy is mainly caused by the intrinsic complexity of the new design methods, by the scepticism of the practical designer concerning the benefits of new methods and by the cumbersome additional data inputs more advanced methods require.

The availability of software packages that allow for an easy comparison between both traditional and more recent design methods and for an up to date data collection may help to overcome this gap. Moreover, such packages would be ideal educational tools for students in civil and hydraulic engineering and in hydrology.

Although several design models are readily available, only few make the most of the possibilities the micro- or minicomputers and their peripherals offer in the field of

A Software Package for the Computer Aided Design of Sewer Systems

W. Bauwens

National Fund for Scientific Research (Belgium), Laboratory of Hydrology, Free University of Brussels

ABSTRACT

The paper describes the main characteristics of a software package for the computer aided design of open sewer networks according to the rational method, the time-area method and a rainfall-runoff model. The software runs on microcomputer, makes use of GEM and is coupled to a generalized CAD package with a relational database.

INTRODUCTION

The increasing complexity of the hydrologic and hydraulic models used for design tends to build up a discrepancy between the official institutions acting as builders and the consulting engineering enterprises on the one hand and theoreticians on the other.

This discrepancy is mainly caused by the intrinsic complexity of the new design methods, by the scepticism of the practical designer concerning the benefits of new methods and by the cumbersome additional data inputs more advanced methods require.

The availability of software packages that allow for an easy comparison between both traditional and more recent design methods and for an up to date data collection may help to overcome this gap. Moreover, such packages would be ideal educational tools for students in civil and hydraulic engineering and in hydrology.

Although several design models are readily available, only few make the most of the possibilities the micro- or minicomputers and their peripherals offer in the field of data handling and processing and of computer aided design. This is clearly in contrast to the increasing use of smaller computers in civil engineering, urban planning and topographic surveying.

It is believed that the reasons for this are twofold. At first, most existing programs are merely revisions of programs that were originally developed for mainframe computers. The second reason relies in the fact that the design of large sewer systems requires extensive amounts of computer memory and calculation time, especially if advanced hydraulic calculations or design optimization are involved.

Although the latter reason still restricts the use of mini- and microcomputers for the design of large and complex sewer systems, the initial computer aided design of sewer slopes and diameters on these machines should at least enable the consulting engineer to establish a draft of the sewer system in most practical design problems.

SEWER DESIGN METHODS

With the presented package, the sewer slopes and diameters are dimensioned assuming steady uniform flow in the sewer, according to the Manning-Strickler formula. The design filling ratio can be set by the designer, up to full pipe flow. The use of the package is restricted to open networks of stormwater sewers.

Where possible, the pipes are set at a minimal depth defined by the designer and the slope is set equal to the ground slope or -if required - to a slope set by the designer. The slope is adapted in case the user defined level constraints or flow velocity limits are not met.

Both traditional and more advanced design methods are implemented :
- the rational method;
- a time-area runoff routing method;
- a rainfall-runoff model in combination with a Muskingum-Cunge pipe routing model.

The time-area method shows a great similitude to the TRRL-method (Watkins, 1962). The rainfall-runoff model is a simplified version of the model used in SWMM (Metcalf and Eddy, 1971). Manning's equation in combination with the continuity equation is used to route the excess rainfall over the surface. Excess rainfall is calculated taking infiltration characteristics and depression storage of the soil into account.

For the latter two methods, the design of pumps, overflows and detention basins may be included.

SOFTWARE AND HARDWARE CHARACTERISTICS

In concurrence with the philosophy of computer aided design, special attention is paid to the userfriendliness and the flexibility and to the use of graphical expedients for the input of data and the representation of computed results.

To achieve these goals, the hydrological and hydraulic components of the package are coupled to a generalized CAD package - ARKEY - which may be linked to a relational database.

Preprocessing and postprocessing are performed under the Graphic Environment Manager (GEM) operating environment (e.g. Balma and Fitler, 1986) allowing an easy use of drop-down menu's, query forms, information and alarm boxes, file selecting windows and graphics. The computational components are written in Pascal.

The package has originally been developed on Atari ST, and is now being adapted for use on IBM-compatible microcomputers. Plotters, graphical screens and digitizer tablets are used as peripherals in the pre- and postprocessing phase.

NUMERICAL INPUT

The preprocessor of the package is completely managed by GEM. GEM offers an ideal environment for the construction of a userfriendly and well structured preprocessor by which the user may be questioned, informed or warned, depending on the information he provides. It thus becomes possible to incorporate a certain level of intelligence into the preprocessor, helping the user to avoid incorrect or incomplete input data sets.

One out of four available generalized intensity-duration-frequency (IDF) relations may be parametrized, allowing the user to meet most of the IDF relations found in literature. Facultatively, the return period and an area reduction coefficient may be set. Storm profiles may be derived from the IDF relations or historical rainfall data may be entered.

The hydraulic parameters to be set are the type of roughness coefficient (manning or sand roughness), the upper and lower velocity limits and the design filling ratio.

A runoff coefficient may be assigned to each individual subarea in an interactive-graphical way (see below), or else, up to five runoff coefficients or -alternatively- infiltration parameter sets may be set, to be used in combination with the areas of different land use and soil type. The latter are circumscribed graphically with the CAD package (see below).

For the methods that require an inlet time, a global inlet time may be set for all upper manholes. However, the inlet time can also be set or calculated -using Kerby's formula (Kerby, 1959) - for each individual upper manhole. In this case, the appropriate information is to be provided in the graphical input phase (see below).

Finally, a set of available pipe diameters with their roughness values is entered and additional design parameters such as slope and cover limits and a non-decreasing diameter option have to be set.

GRAPHICAL INPUT

The input for the design is performed to a large extent in an interactive-graphical way, through the CAD package and its relational database. GEM is also used at this stage to provide query-forms for entering pipe-, manhole- or structure specific numerical information at the user's request. Appropriate forms are displayed by clicking the mousebutton on the desired component. Alternatively, all the information mentioned in this section may also be defined through keyboard input.

The basic input for the package consists of the ground-plan of the area under study. This plan may be entered by digitizing an existing plan or loaded from an automated topographic survey through the CAD package.

Using the same package, the layout of the sewer system is superposed to the latter ground-plan. Pump, overflow and detention basin locations complete the layout.

Intrinsic functions of the CAD package are used to circumscribe subareas of different land use and soil type, to calculate lengths and areas and to allocate these to the respective pipes or manholes.

Grounds levels at the manholes may be calculated by triangulation or entered numerically by completing a query-form. Query forms are also used to define additional pipe, manhole, structure or subarea characteristics. Examples of parameters that may be assigned at this stage of the design are the levels and diameters of existing pipes, pipe level circumscriptions, overland flow characteristics at manholes, user-defined manhole or pipe numbers, flow limits etc.

The graphical input is processed automatically and stored on file, where it is ready to be used for the design computations.

COMPUTATION OUTPUTS

The text output may be visualized on monitor and/or printer and consists of :

- a clear overview of the basic design hypotheses and parameters used for the design, providing a clear echo of the input;
- the detailed list of the sewer segments, lengths, levels, slopes, drained areas of different land use or soil type, diameters, flow and concentration times, flow velocities, discharges and Froude numbers.

The graphical output may be visualized on monitor and/or via plotter. The standard output consists of the annotated groundplan of the area and the sewer system as well as of long profiles. The used symbol library may be edited by the user, if required.

Finally, the standard output may be manipulated by using the editing functions of the CAD package. This may prove helpful e.g. to add construction details on the plan.

CONCLUSIONS

The presented package allows the comparison of sewer design as calculated by different widespread techniques for most of the practical design problems, using up to date computer hardware and software techniques.

It must be clear however that the design techniques that are used may not always yield ideal results, especially if complex hydraulic phenomena are likely to occur in the system. Under any circumstances, the designer should be cautious about the computation results and perform control simulations with hydraulically more advanced models if any doubt exists about the design. Possibly an expert system could be developed to assist the engineer for the quality assessment of the design.

REFERENCES

Balma P. and Fitler W. (1986). Programmer's guide to GEM, Sibex, Berkeley.

Kerby W.S. (1959). Time of Concentration for Overland Flow, in Civil Engineering, Vol. 29, No. 3, p. 174.

Metcalf and Eddy, Inc. (1975). Storm Water Management Model - User's Manual, Version 3, Report No. EPA-670/2-75-017, U.S. Environmental Protection Agency, Washington D.C.

Watkins L.H. (1962). The Design of Urban Sewer Systems, Road Research Technical Paper No. 55, Department of Scientific and Industrial Research, Her Majesty's Stationery Office, London.

Interactive Design of Irregular Triangular Grids

R.F. Henry

Institute of Ocean Sciences, Sidney, B.C. Canada

INTRODUCTION

This paper describes an interactive software package in use at IOS for construction of irregular triangular grids suitable for shallow water models. The grids obtained represent a compromise among the following design requirements: (a) close fitting of coastlines,(b) near-uniform element area/water depth ratio, (c) near-equilateral element shape.

The main innovations in the present work are an indirect method of creating appropriately-spaced vertices for the triangles, an interactive graphical editor which facilitates alterations to a grid on the basis of visual judgments, and the dual use of the same software for both depth and model grids. Extensive use is made of colour graphics throughout the initial grid construction phase as well as in subsequent interactive editing. The principal output file contains a record for each grid vertex, specifying the location of the vertex, the local mean water depth, a computational code indicating whether the vertex is in the model interior or requires some specific boundary computation, and a list of neighbouring vertices to which it is connected. Subsidiary outputs, such as triangle lists, can also be produced if required. An example of a grid produced with this software is shown in the paper by Foreman[1] in Volume 1.

INTERACTIVE AND AUTOMATIC METHODS

Of the many methods available for computerised irregular grid construction, (see for instance the review by Thacker[2]), almost all are essentially fully automatic, that is, after initial specification of bathymetry and coastline geometry, grid elements are computed with little or no intervention by the modeller, apart from initial setting of some controlling parameters. Many of these methods produce quite satisfactory grids, but have the drawback that any subsequent local changes required by various modelling considerations involve error-prone manual alterations of vertex positions and interconnections. Recent improvements in graphics hardware and software, and the growing availability of colour graphics, make it more practicable to develop interactive graphics editors to handle alterations to existing grids.

The original intent of this work was to develop a flexible graphical editor for making local adjustments to grids, including addition or deletion of triangles, in order to improve the overall uniformity of the element area/water depth ratio throughout a grid. However, even with the aid of a fast interactive editor, this is a tedious task for a large model unless the initial distribution of nodes is at least roughly correct. Since none of the available methods placed sufficient emphasis on this feature for the writer's purpose, a new automatic method was devised to provide a suitable initial grid for input to the interactive editor. In practice, the automatic and interactive portions of the package are complementary, though the interactive editor can still be used to edit grids produced by other automatic grid generators.

MAIN FEATURES OF THE AUTOMATIC GRID GENERATOR

The principal consideration in design of the generator was to obtain a distribution of grid nodes (triangle vertices) whose spacing is inversely related to local water depth, without resorting to time-consuming iterative addition or deletion of nodes with repeated grid relaxation. The basic idea used can be explained best by considering first a model where water depth is uniform. Imagine the model domain overlaid with a honeycomb pattern of cells of uniform area, either hexagonal or circular, but sized so that the required element (cell) area to water depth ratio is achieved. The centroids of these cells provide a basis for a regular grid of equilateral triangles, the most appropriate triangular grid required for a model with uniform water depth. The important feature is that a complete triangular 'tiling' of the model domain is obtained irrespective of the precise shape of the original cells or of small gaps between them. The extension of this idea to a model with varying depth is fairly obvious. The model domain first has to be subdivided into contiguous cells whose individual sizes (areas) are proportional to local water depth. A certain degree of irregularity in cell shape can be tolerated, as can occasional gaps between cells. After the model domain has been covered with such cells, the centroids of the cells together with suitably selected points on the model boundaries form the basis of an irregular triangular tiling, whose elements have about the same degree of uniformity of area/depth ratio as the irregular cells do.

To construct the irregular cells, the entire model domain is covered with a regular grid of square elements, whose mesh size is chosen approximately equal to the linear dimension of the smallest triangles required in the final model grid. Starting at the square with greatest water depth, a cluster of squares is built up in a spiral pattern until the area of the cluster reaches a specified ratio to the mean water depth over the cluster. Another cluster is then built up around the square with the next deepest water depth, but omitting squares already used in the previous cluster. This process is continued until insufficient free squares remain to build clusters of suitable area. Now taking the centroids of all the clusters together with a selection of suitably-spaced points from the digitized boundaries, a triangulation algorithm developed by Nelson[3] is used to form an initial model grid. The resulting triangles not only have fairly uniform area/water depth ratio but also generally do not depart too seriously from equilateral shape, provided that the clustering algorithm limits formation of clusters with straggling shape. Accurate fitting of coastlines follows automatically from inclusion of selected boundary points with the cluster centroids as nodes for the triangulation. A grid adjustment program is available for use at this stage or later to adjust vertex positions iteratively according to an arbitrarily weighted combination of area/depth and equilateralness criteria.

From the formation of the initial model grid onwards, a standard file format, which will be termed 'neighbourhood format', is used for all grid descriptions. Use of this standardized format increases the versatility of the whole software package, as will be seen later in the discussion of depth calculations.

MAIN FEATURES OF THE INTERACTIVE EDITOR

The interactive editing package consists of a main editor, and grid-splitting and joining programs. The main editor permits display of the subject grid on Tektronix 4100 series and Digital VT240 or VT340 colour graphics terminals. A conventional software windowing facility is available at all times. The simplest operation which can be carried out with the editor is soliciting information. After the required vertex has been identified (by crosshairs if the vertex is visible in the current display or by entering the index number of the vertex) the contents of the corresponding record in the neighbourhood format file are displayed in a screen window beside the displayed grid. This information includes the coordinates of the vertex, an integer computation code indicating the nature of the point (i.e. interior or boundary), the water depth at the vertex and a list of neighbouring vertices to which it is connected. The computation code, water depth or coordinates of the current vertex can be changed at will by keyboard entries. The following types of operations can also be carried out:-

- moving a vertex: after the vertex to be moved and its new position have been identified by successive crosshair entries, the display is revised to show the new configuration. At this point the user can either confirm or cancel the change. In this and all other editing operations, prompt messages lead the user through the appropriate actions and visual aids such as temporary coloured markers appear in the display to indicate the current stage reached. The digitised boundaries and depth contours can be superposed in different colours on the display to assist revision of the grid.

- addition or deletion of a vertex and its connections to the neighbouring vertices.
- deletion or addition of connections between pairs of vertices.
- merging of two vertices: if a vertex is moved to within a prespecified short distance of a neighbouring vertex, it is assumed that the two vertices are to be replaced by a single vertex connected to all the vertices to which the two merged vertices were singly or jointly connected.

All of these operations require only visual judgments and crosshair inputs from the user. All associated bookkeeping details are handled automatically by the editor program.

Colour is used in two further ways to help check and edit the grid. Markers can be placed at all vertices satisfying particular criteria. For instance, different ranges of water depth can be flagged in different colours, to facilitate detection of any anomalous depths. Other vertex properties such as number of neighbours or computational code can also be displayed. With limited additional programming effort, any other derived vertex property could also be shown in this way. Triangle properties can be monitored by colouring individual triangles according to any criterion of interest, for example, area/water depth ratio. Even where a geometric property can be judged by eye, in the case of grids containing hundreds of triangles it is generally worthwhile also to express the geometric property numerically and shade the triangles in various colours corresponding to different numerical ranges of the criterion. The ranges used and the corresponding colours chosen can be set either beforehand or in the course of an editing session. The user is given the option of placing coloured markers at the centroids of the triangles as a fast alternative to full colour shading.

The splitter program facilitates division of an existing grid into two (or more) separate self-consistent sub-grids. First a closed polygon with up to ten sides is defined over the displayed grid, using crosshair input. Wherever any side of the polygon lies across the grid (i.e. along some required open boundary) vertices lying near the polygon must then be selected and moved to coincide with the polygon, since a split can be made only along a line consisting of connected line segments in the existing grid (otherwise the sub-grids formed will contain non-triangular elements). The splitter program leads the user through all the steps required to design the splitting polygon and move vertices to the polygon. It then automatically carries out all the bookkeeping operations required to divide the neighbourhood format file for the initial grid into separate self-consistently numbered files for the portions of the grid lying inside and outside the polygon. The splitter program is useful in the following circumstances:-

- to extract part of an existing grid for use in a smaller model.
- to divide a very large grid temporarily into smaller parts for more convenient local editing (see comments below on joiner program).
- in order to position or reposition open boundaries. To obtain appropriately large triangles in deep water at open boundaries, it is necessary to let the initial model grid (and depth grid) extend one or two elements beyond the intended open boundary line. These surplus parts are removed using the splitter. Delaying the precise placement of open boundaries to this stage proves very convenient in practice, since initial tests with a model often indicate some need to reposition the open boundaries.

The sub-grids produced by the splitter program have vertices at identical points along their common boundary. In order to rejoin the temporarily separated sub-grids referred to above, a joiner program is provided, which can automatically combine two grids into a single consistently numbered grid, provided they have one or more common boundaries with identically positioned vertices. The joiner can of course be used to combine two grids of any origin, provided their vertices can be brought to match exactly on a common boundary, but the merge facility in the main editor provides a much more flexible means of doing this. Two adjacent grids to be joined are first run through an auxiliary program which renumbers the vertices of one grid to follow consecutively on those of the other and outputs a single combined file in neighbourhood format. To the main editor, the two grids then appear as a single grid, though in fact there are no connections between the two sets of vertices. Using the screen display, the user can merge adjacent vertices from both sets and so set up a properly connected single grid for the whole domain. This technique can be used even if the grids to be joined overlap.

DEPTH GRID

It is clear from the above description that construction of a model grid with element size geared to mean water depth poses the need for some means of evaluating water depth at any point in the model domain. The problem is met here by prior construction of an irregular triangular depth grid covering exactly the same domain as the model grid but otherwise quite distinct from the latter. Currently, the mean water depth at any required point is found by linear interpolation within that triangle of the depth grid containing the point. To permit construction of sufficiently large triangles at any open boundary of the model grid, the depth grid is extended beyond the intended site of the boundary in the same way as the initial model grid.

The method used to obtain nodes (vertices) for the depth grid is to take every m^{th} digitized boundary point and every n^{th} digitized depth contour point, where m and n are chosen so as to ensure a good fit to the coastlines and to give very roughly the same number of triangles in the depth grid as in the model grid. The same boundary points are used in both model and depth grids, to ensure that computed water depths are non-negative. Once vertices for the depth grid have been selected, triangulation is carried out using Nelson's algorithm. The resulting depth grid information is held in a file of exactly the same neighbourhood format used subsequently for the model grid. In consequence, a depth grid can be checked and, if necessary, edited, using the interactive editor, before it is used to produce depths for the model grid. Other benefits which follow from this identity of depth and model grid formats are:

- when existing model grids are combined, using the joiner, their depth grids can be combined using the same program to give a unified depth grid for the larger domain. The converse applies when a large grid is divided with the splitter.
- since a model grid file contains the depths at the (model) vertices, it can be used as a depth grid, if necessary. For instance, if vertices are moved or added, water depths at the new vertex locations can be evaluated using the input model grid as a depth grid, if no original depth grid is available.

TECHNICAL DETAILS

The software described is written in VAX FORTRAN and uses the ISSCO DISSPLA graphics software package. As implemented under the VMS operating system on the VAX 11/785 at IOS, the package is menu-driven, with a HELP facility.

REFERENCES

1. Foreman, M.G.G. (1988), A comparison of tidal models for the southwest coast of Vancouver Island, in Proceedings of the 7th Int. Conf. on Computational Methods in Water Resources, Cambridge, Mass., 1988, Vol.1.

2. Thacker, W.C. (1980), A brief review of techniques for generating irregular computational grids, Int. J. Num. Meth. Engng, Vol.15, pp. 1335-1341.

3. Nelson, J.M. (1978), A triangulation algorithm for arbitrary planar domains, Appl. Math. Modelling, Vol.2, pp. 151-159.

FLOSA - 3FE: Velocity Oriented Three-Dimensional Finite Element Simulator of Groundwater Flow

M. Nawalany
Warsaw Technical University, Warsaw, Poland and TNO-DGV, Institute of Applied Geoscience, Delft, The Netherlands

INTRODUCTION

In contrast to the classical approach, which uses a piezometric head as a state variable, the Transport Velocity Representation (TVR) of groundwater flow (Zijl, 1984) works on the velocity components as a primary variable. The representation, when supplied with additional boundary conditions, is equivalent to the mass balance equation and the Darcy Law. FLOSA-3FE is the first three-dimensional finite element simulator which approximates the solutions of the TVR-equations for groundwater problems defined on a regional scale. The simulator supports arbitrary boundary conditions and arbitrary spatial distribution of pumping wells. The heart of the simulator is an automatic translation of the user-specified boundary conditions (piezometric heads and/or normal fluxes) into the additional ones necessary to make the TVR and the classical theories equivalent. The resultant large and sparse algebraic equations, with the velocity components as the unknowns, are being solved by outer iterations and the ICCG solver. The special searching scheme makes the ICCG solver very fast even for the PC version of the simulator. Also a user-friendly grid generator and graphical display of groundwater flowlines have proved to be handy and versatile in making qualitative and quantitative analysis of complex groundwater systems. The results obtained have shown high accuracy of the simulator on relatively sparse finite element grid used for discretizing groundwater systems.

THEORETICAL ASPECTS

A theoretical background for the simulator consists of the Zijl equations (W. Zijl, 1984; W. Zijl et al., 1987) for the velocity components q_x, q_y, q_z and the corresponding boundary conditions which both are reiterated here for the sake of completeness. Introducing the following notations

$$\beta = \mu/k\varrho \quad \text{and} \quad \sigma = 1/\beta \,, \quad \alpha = \ln\beta \tag{1}$$

$$\underline{S} = \beta\underline{q} \tag{2}$$

where ϱ - water density, (kg/m^3)
μ - dynamic viscosity, $(Pa \cdot s)$
k - intrinsic permeability of the porous medium, (m^2)
$\underline{q} = (q_x, q_y, q_z)$

the TVR model can be written in the form of the Laplace-like equations

$$\nabla.(\sigma\nabla S_x) - \sigma S_x\frac{\partial^2\alpha}{\partial x^2} = \sigma S_y\frac{\partial^2\alpha}{\partial x\partial y} + q_z\frac{\partial^2\alpha}{\partial x\partial z}$$

$$\nabla.(\sigma\nabla S_y) - \sigma S_y\frac{\partial^2\alpha}{\partial y^2} = \sigma S_x\frac{\partial^2\alpha}{\partial x\partial y} + q_z\frac{\partial^2\alpha}{\partial y\partial z} \tag{3}$$

$$\nabla.(\beta\nabla q_z) + \beta q_z\left(\frac{\partial^2\alpha}{\partial x^2} + \frac{\partial^2\alpha}{\partial y^2}\right) = S_x\frac{\partial^2\alpha}{\partial x\partial z} + S_y\frac{\partial^2\alpha}{\partial y\partial z} \,.$$

Equations (3), however, when supplemented only with physical boundary conditions (specified pressure and/or specified normal flux) do not represent the well-posed problem. In order to make the TVR approach equivalent to the continuity equation and the Darcy Law one has to add the following auxiliary boundary conditions:

i) if the pressure p is specified on the boundary ∂D it is necessary to prescribe in addition

$$\nabla \cdot \underline{q} = 0 \quad \text{on} \quad \partial D \tag{4}$$

ii) if the normal component of the flux $q_n = \underline{n} \cdot \underline{q}$ is specified on ∂D it is necessary to prescribe also

$$\underline{n} \times (\underline{\Omega} - \nabla \times \underline{q}) = \underline{0} \quad \text{on} \quad \partial D \tag{5}$$

where $$\underline{\Omega} = (\nabla(\varrho g) \times \nabla z - \nabla\beta \times \underline{q})/\beta. \tag{6}$$

Because of the complicated nature of the auxiliary boundary conditions (4) and (5) the FLOSA-3FE simulator has to translate automatically the physical boundary conditions supplied by the user into the auxiliary ones. The translation consists the main task for the simulator. The equations (3) themselves are being solved by the finite element method in a straightforward manner. Since the TVR equations, when decoupled, are symmetric and positive definite, the ICCG can be applied resulting in fast and accurate approximations for q_x q_y and q_z. And finally, given the numerical approximation of the velocity field, the trajectories of the water particles are computed with the England method (see M. Nawalany, 1986).

COMPUTATIONAL ASPECTS

The FLOSA-3FE package is intended to be used for simulation of three-dimensional steady state groundwater flow on a regional scale. It is also a new product from the FLOSA family. Distinct features of the simulation package are its theoretical background - the Transport Velocity Representation (TVR) of groundwater flow (W. Zijl, 1984) - and the numerical method used to solve the TVR equations - the three-dimensional Galerkin finite element method (M. Nawalany, 1987). The TVR-approach allows to approximate all the three components of the velocity field very accurately, hence the application area for the package consists of those problems in which groundwater transport phenomena are important (e.g. groundwater pollution and salinisation problems). Such characteristics of the groundwater velocity field like spatial pattern of the flow subsystems, travel time distributions, recharge-discharge relationships etc. can be accurately calculated with the simulator. Since the model operates on velocity components rather than on piezometric head it must be considered as an extension to existing, piezometric-head-oriented models. Moreover, the finite element environment applied in the package makes it possible to increase the accuracy of the numerical approximation of the velocity field by refining of the discretization mesh in the vicinity of wells or abrupt changes of boundary conditions. This is done during the preprocessing stage by generation of a tetrahedral finite element grid - see Figure 1. During the simulation stage the numerical approximations of the three components of the velocity in discrete points of the finite element mesh are computed. These numbers are used as data in the postprocessing stage in which the graphical image of groundwater pathlines is presented to the user. Also the hard-copy results (groundwater velocity field in discrete points of the system) are available after completion of the simulation stage.

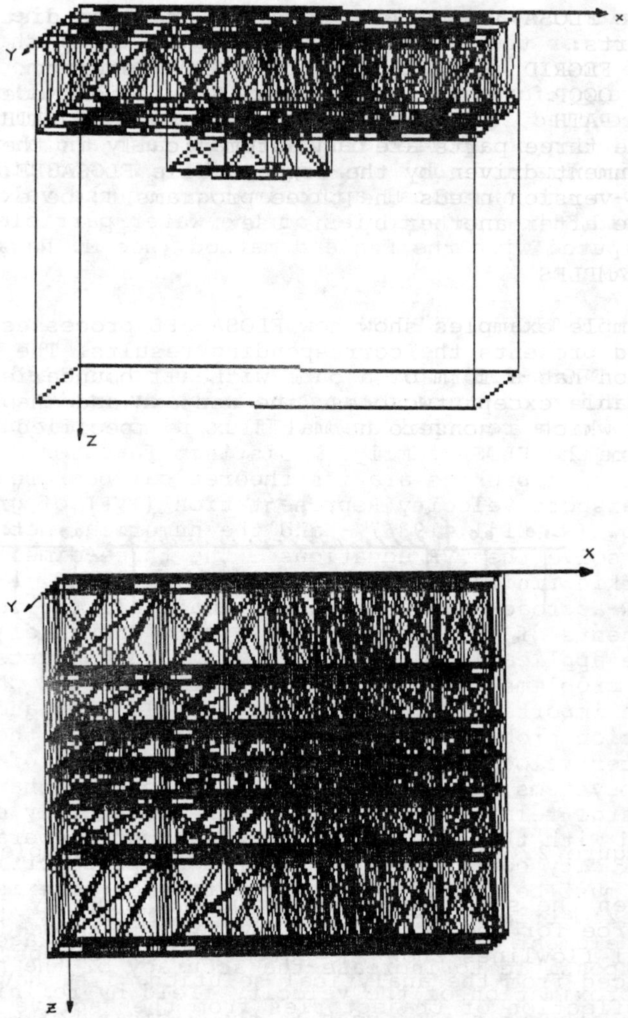

Figure 1. Finite element discretization mesh used in the FLOSA-3FE simulator.

The FLOSA-3FE package consists of three distinct parts:

1. FEGRID - Finite Element GRID Generator
2. QQQP - Finite Element Simulation Model
3. PATH - Graphical Presentation of PATHlines

The three parts are run automatically in the VAX environment driven by the command file FLOSA3FE.COM. The PC-version needs the three programs to be executed one after another by the user.

EXAMPLES

Simple examples show how FLOSA-3FE processes the data and presents the corresponding results. The flow region has a form of a cube with all boundaries impermeable except two ones (the western and the eastern) on which a nonzero normal flux is specified - see Figure 2.

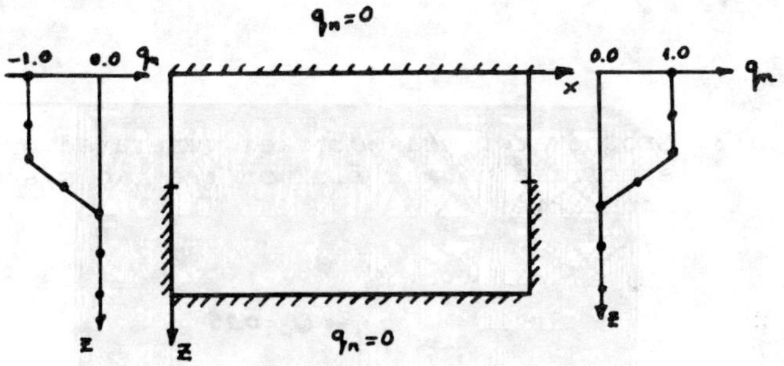

Figure 2. Boundary conditions in the example.

When the specified normal flux is the only driving force for the groundwater system the three-dimensional flowlines show the pattern which can be also deduced from the analytical solution, i.e. having a deflection of trajectories from the "active zone" towards the "dead zone" - see Figure 3.
On the other hand, when the well is installed additionally to the previous boundary conditions (Figure 4) the pattern changes considerably showing some pathlines to be attracted by pumping - see Figures 5, 6 and 7.
In fact, an introduction of a well in the TVR model can only be done by removing some internal finite elements in the vicinity of the well and specifying corresponding, physical and auxiliary boundary conditions on the surfaces of the elements removed. In the examples the third-type boundary conditions have been simplified by assuming the constant-pressure conditions at the well.

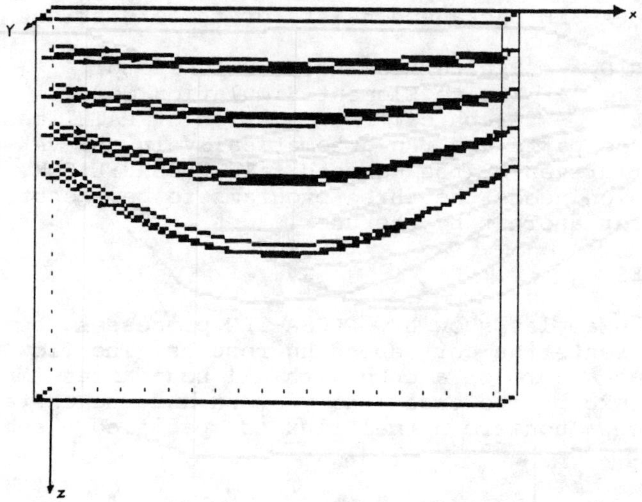

Figure 3. Groundwater trajectories generated by the specified normal flux on the two system´s boundaries.

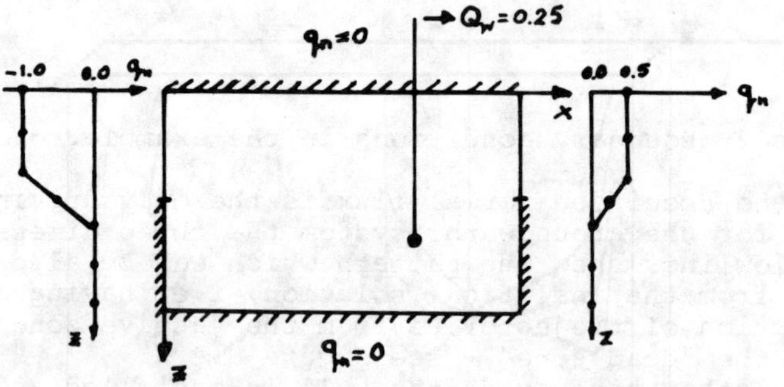

Figure 4. System driven by the boundary conditions and the well.

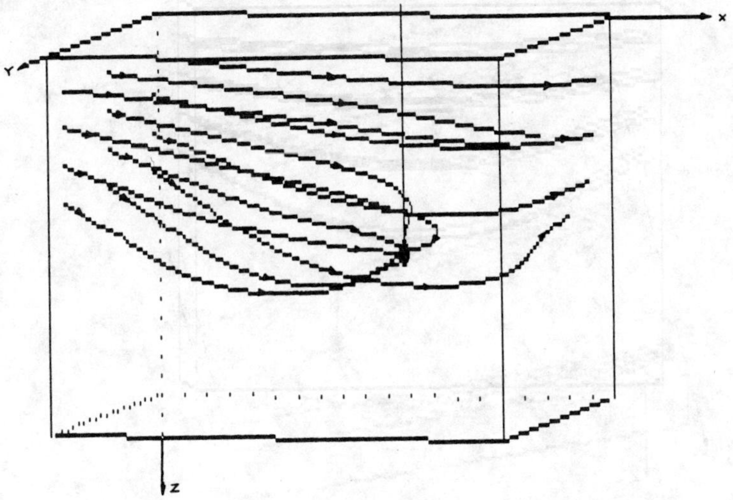

Figure 5. Groundwater trajectories generated by the specified normal flux on the two system´s boundaries and a constant pressure at the well - southern view.

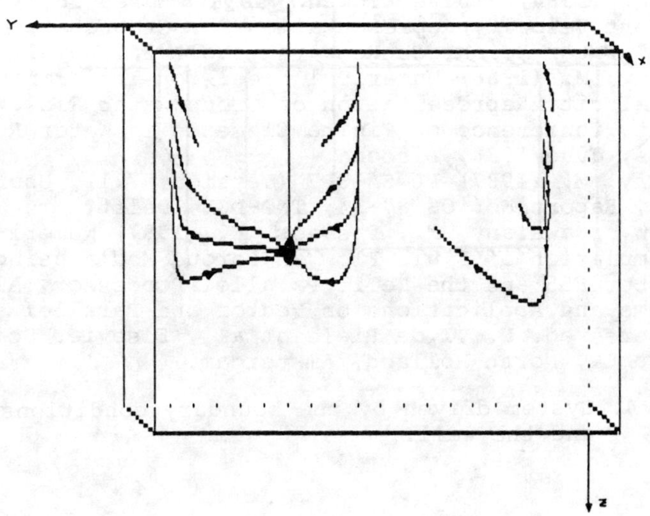

Figure 6. Groundwater trajectories - eastern view.

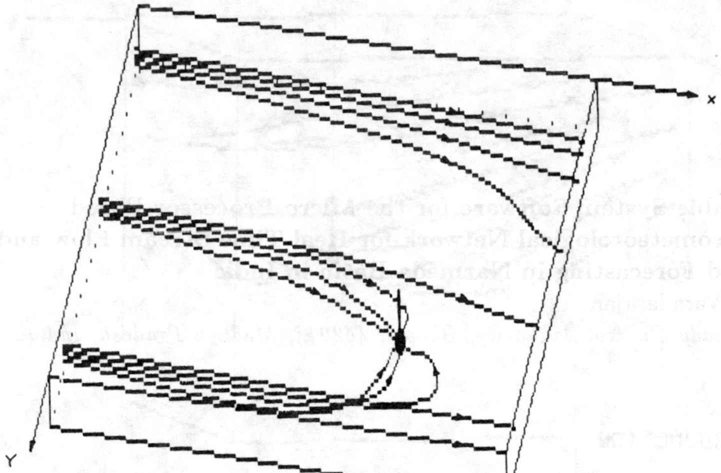

Figure 7. Groundwater trajectories - top view.

REFERENCES

Zijl, W. (1984) Finite Element Method Based in a Transport Velocity Representation for Groundwater Motion, Water Resources Research, Vol. 20, No. 1.
Nawalany, M. (1986) Numerical Model for the Transport Velocity Representation of Groundwater Flow, 6th Int. Conference on Finite Elements in Water Resources, June 1986, Lisbona.
Nawalany, M. (1987) FLOSA-3FE (Version 87.1), Users Manual, Raport No. OS 87-34, TNO-DGV, Delft.
Zijl, W., Nawalany, M., Pasveer, F. (1987) Numerical Simulation of Fluid Flow in Porous Media using the Cyber 205 and the Delft Parallel Processor, Algorithms and Applications on Vector and Parallel Computers, ed. H.I.I.de Riele et al., Elsevier Science Publ., North-Holland, Amsterdam.

Reliable System Software for the Micro-Processor Based Hydrometeorological Network for Real Time Stream Flow and Flood Forecasting in Narmada Basin in India

R.S. Varadarajan
Narmada Control Authority, Bhopal, 462014, Madhya Pradesh, India

INTRODUCTION

A real time computer system is adopted in situations where the computer system is required to be operated concurrently with a physical process such that the results of the computations are available in time to usefully influence or control the process (WAPCOS[1]). Such a system is used especially in computerised reservation systems, bank transactions, stream flow and flood forecasting in a river basin and the like, where an immediate response is required. Development of relevant software, ensuring its reliability to work as envisaged is one of the important requirements of any software industry.

This paper presents discussions on the computer facilities envisaged in the proposed net work of hydrometeorological stations, the functions to be performed by the computer network, the related system software requirements and its reliability to carry out the tasks as specified for the system.

HYDROMETEOROLOGICAL NETWORK

A micro-processor based system comprising 56 Hydrometeorological stations (HMS) including irrigation and power project sites has been proposed for Narmada river basin in India for collection, processing, storage and transmission of data to a Master Computer Control Centre (MCC) for stream flow analysis and flood forecasting in real time. The MCC is proposed to be located at Indore City, which is in the lower part of the Narmada Basin. The data communication is provided through a satellite-UHF-VHF telecommunication network configuration. A

Cluster Control Centre (CCC) is proposed which will group a number of Remote Processing Units (RPU) in the network to form a cluster. The CCC will be designed to interrogate its slave RPUs and report to MCC on interrogation from MCC. The RPUs will be scanned by the corresponding CCC in some predetermined intervals of time (PROJECT REPORT 2).

Remote Processing Unit.

General The Remote Processing Unit (RPU) is a micro-processor based data acquisition system proposed on all the 56 locations. Each RPU functions as a remote data acquisition module and as secondary data storage. The local data storage is envisaged in situations where the telecommunication links are not provided or to ensure data salvage in the case of link failure. The data storage also remainsas a parallel activity in the network.

The RPU serves as an interface between the hydrometeorological sensors and the digital data transmission system. It is a slave data acquisition system for a remote central computer system. The RPU accepts inputs from the hydrometeorological sensors, converts data into scientific units, stores them in the memory, and delivers a serial data stream for asynchronous communication.

As a backup facility a semi conductor based Insitu Data Storage Cartridge (DSC) is provided to store data locally, for audit purposes. It is detachable from the RPU so as to facilitate sending it through a courier to a suitable location where an appropriate reading unit is available for transferring the DSC contents to the computer system. During such a transfer a spare DSC is plugged into the RPU for data storage.

FUNCTIONAL REQUIREMENTS AT CLUSTER CONTROL CENTRES

The CCC are designed to perform the following functions:

CCC will acquire and validate data from its RPUs by interrogation at certain predefined intervals through VHF communication channels. CCC will transmit data to MCC on interrogation through a high speed data communication media. In the case of any communication breakdown between MCC and CCC, the latter is required to store the data collected from its RPUs and transmit them to MCC as and when the communication is restored.

The CCC will permit entry of manually measured parameters and message required for management purposes. This creates a file which can be transmitted in block mode to MCC when interrogated. The cluster facilitates hydrological analysis and preparation of rating tables, curves etc., as also its updating.

Assessment of sediment inflow from streams into reservoirs, cross check of silt surveys carried out which serves as a planning and monitoring tool.

Message transmission and receiving to and from MCC for effective flood forecast and operational needs using electronic mail.

FUNCTIONAL REQUIREMENTS OF RPU INTERROGATION ANSWER BACK

Each RPU will be interrogated by an external computer system by sending a pre-defined bit pattern. The RPUs will be interrogated on broadcast basis and will be identified by an identity code. The RPU can be programmed such that it responds only when it finds its ID matching with the ID part of the interrogation signal. In the interrogation answer back protocol, every interrogation from the computer will provoke an answer back. The answer back shall consist of a header followed by a number of message entries.

The following minimum information must be conveyed by the RPU in the header part of the answer back :

Date and time
Station identification
Control information

The RPU can be programmed for the above message header. Message entries, each of 4 bytes length, are meant to communicate actually measured values or derived values. Each message entry will be complete in itself and would provide information like type of measurement being reported, the value itself and any other information.

Since it is possible to have several measurements reported during one interrogation cycle, the RPU inserts a time tag entry for each set of measurands belonging to the same sampling instant. The interrogation of RPUs by the central computer will be every half an hour.

FUNCTIONAL REQUIREMENTS OF MCC

MCC will acquire data from all RPUs through suitably designed communication channels. The capacity to acquire data from 100 RPUs is envisaged. It will carryout data processing, both primary and secondary processing. MCC has to store data in the form of rating tables of gauge discharge-sediment measurement stations as also level versus reservoir storage. It will also perform the initial task of catchment calibration for a distributed model.

Periodic report for dissemination of forecast will be sent by MCC out of the data base. It will also transfer real time data in the form of input files for running models or for forecasting of floods. Model files shall be collected and kept in the centre for ready use and shall be accessible. MCC is designed to handle data in appropriate format for long term trend forecast for planning and operation of reservoirs. It will perform tasks relating to flow routing, back water computation and so on by transferring data and related message to remote Visual Display Units (VDUs) located at project sites.

It is designed to transfer electronic mail to remote VDUs at project stations for giving operational information to reservoirs.

It is also effectively designed to meet future needs which may relate to connecting public data networks for exchanging the hydromet data with other agencies.

The computer control centre is suitably designed to run mathematical models for hydrological simulation, fecilitate reservoir operation and the like in addition to the maintenance of a large hydrometeorological data base. Results of several calibrated mathematical models need presentation in graphical form at MCC for proper understanding and taking quick decisions. MCC is designed to retain data of previous 31 days online before it is transferred on to a tape.

SYSTEM SOFTWARE AT MCC

To perform the functions of MCC as specified in the preceding paragraphs, MCC is designed to have a sophisticated system software. The main components of system software required at MCC would be as discussed below (PROJECT REPORT 2).

Operating system and utilities

The essential features of the operating system for the computer at MCC would be :

Multi-user, multiprogramming environment with efficient process scheduling and resource allocation facilities.

User friendly interactive procedures for invoking operating system facilities.
Executive call through high level languages.

Batch and time-sharing modes of operation.

Virtual memory management.

Device spotting functions.

Software priorities.

Inter process communication facilities like event flags, mail boxes, shared memory and shared files.

On-line documentation.

Run time library : Single copy of library shall be shareable by different users.

File management system

The file management system of the operating system will provide the following facilities :

Inter-user protection

User friendly procedures for file manipulations such as transfer, conversion, protection and so on.

Sequential, direct and indexed sequential file structures.

Interface to high level languages.

Programme development tools

In order to facilitate development of several application programmes required by the users for data checking, model calibration, model running, for forecasting and taking operational decisions for Narmada basin, programme development aids are essential for developing and debugging the application programmes. Appropriate tools such as text editors, symbolic debuggers etc., are provided.

SYSTEM RELIABILITY

To ensure reliability of operation of the entire system, following safeguards have been proposed in the hardware and software of the computer and the micro-processors :

System generation and reconfiguration

This is necessary for incorporating changes due to hardware or system software modifications.

System diagnostics

This is necessary for early diagnosis of malfunction of subsystems (disk, memory or central processing unit). The system shall provide the necessary routings for automatic error logging at the request of the system manager.

Recovery

In case of unexpected failure of any of the sub-systems due to internal malfunction or due to external reasons, such as power failure, it would be possible for the system manager to salvage any erroneous programmes or files.

Backup and transfer

The system shall facilitate taking periodical backups of disk files by providing convenient procedures. Routines for conversion between standard formats such as ASCII and EBCDIC shall also be provided.

Accounting

The system manager is provided with the means of accounting the uses of the various sub-systems, CPU, disk, tape, printer, terminal etc. Monitoring and accounting as also billing for the users of these sub-systems is automatic. The system manager will have access to confidential procedures for creating new user accounts and to specify the necessary protection devices such as pass words etc.

The system software as well as the hardware of the master teleprocessor are made modular so that more elaborate functions could be carried out in future by expandability of the memory of the master. The entire slave hardware action is governed by the software resident within the slave teleprocessors. By changing the software, the slave can be made to work either with the satellite or ground link.

The software is written in such a manner that the slave station not only repeats the data but also attempts to get the data three times, before flagging any failure indication to the master.

The software and the associated standby configuration are designed in such a way as to dispense with the common switch element to ensure 100% redundency.

The software for the RPUs and the master computer are adequately designed to guard against communication over-reach problems during transmission of data through data communication channels.

The system failures are adequately described on the console provided in the system.

Remote diagnostics

Facility to conduct some limited diagnostics by sending an appropriate command to a particular RPU, recognising the diagnostic command, conducting appropriate self-test and returning a signature containing indication about its health is an essential feature and the following information must be available:

Station battery voltage
Voltage at some key points in the RPU hardware
Local real time clock reading
Sensor status
RAM/ROM status
CPU card health
Input-output card health
Key board and LCD display status.

This self diagnostics is appropriately formulated and returned as the answer back on interrogation.

Clock management

The RPU is fitted with a real time clock. The clock value should be accessible from the RPU software for time dependent functions. The clock can be reset periodically from the external computer using interrogation facility.

CONCLUSIONS

The principal task of the forecasting software system is to produce operational real time forecasting. It uses mathematical models and data from operational data bank. It is very important that the software development to run the real time system is most reliable and efforts have been made to achieve 100% reliability by adequately designing the software in the case of the proposed hydrometeorological network of Narmada basin.

REFERENCES

Proceedings of a workshop

1. Pituman resources development group : Water and Power Consultancy Services (India) Ltd; Proceedings of development and management training course on 'Real time data for water resources projects planning and operation'.

Draft project report

2. Detailed project report on hydrometeorological network of Narmada basin for real time data collection and flood forecasting.